The VEGETABLE Producer's Manual

A practical guide for cultivating vegetables profitably

The VEGETABLE Producer's Manual

A practical guide for cultivating vegetables profitably

Piet Stork

Published by Kejafa Knowledge Works
35 Piet Retief Avenue
Noordheuwel, Krugersdorp, 1739 South Africa
www.kejafa.com
014 577 8006

Copyright © 2017 Piet Stork

All rights reserved. No part of this book may be reproduced or transmitted in any form or by any means, electronic or mechanical, including photocopying, recording, or by any information storage and retrieval system, without permission in writing from the publisher.

Disclaimer: Although great care was taken to ensure that the information in this guide is as accurate as possible, the publisher gives no warranty, expressed or implied, for the information given, nor do we accept any liability for any loss, direct or consequential, that may arise from whatsoever cause.

Proofreading by Peter Boynton and Duncan Stork
Layout and cover by Christina van Straaten
Printed and bound by IngramSpark, England

First published 2017
First print 2017

ISBN: 978-0-620-72378-7

FOREWORD

I wrote The Vegetable Producer's Manual to assist vegetable producers on all levels and from any background to overcome the various problems, issues and challenges they face on a daily basis, both outside in the field and inside their businesses.

This is a user-friendly but comprehensive manual, which will educate and inspire producers to produce higher yields and better quality vegetables, naturally resulting in higher profit margins and greater business sustainability.

It took me more than 20 years to write The Vegetable Producer's Manual simply because I did not realise the amount of work that would be involved in compiling the decades of information and experience I had accumulated.

I was extremely privileged to have been employed at the South African Department of Agriculture - Plant and Quality Control Division for 20 years and at the Agricultural Research Council of South Africa - Vegetable and Ornamental Research Institute for 25 years.

My expertise ranges from land inspections for certification of vegetable seed production, to the evaluation, multiplication, production and maintenance of all types of open-field and under-cover crops and cultivars, to hydroponics research, development and production, and many other associated roles and responsibilities too numerous to mention.

I have worked with many seed and chemical companies, attended countless farmers' days and assisted many hydroponics producers. I have trained many disadvantaged, aspiring, beginner, small and large-scale producers which included government departments, colleges, schools and commercial producers. The outcomes from my hands-on training were always excellent resulting in me earning many prestigious awards.

Over the decades of my interactions with all kinds of producers, I have witnessed many mistakes being made, some small, some substantial. The causes of these mistakes usually originate from ill-informed decision making and limited knowledge especially related to aspects such as planning, irrigation, fertilization, the control of weeds, insects and diseases, the evaluation and selection of different cultivars, labour issues, handling, packaging, marketing and transport, amongst others.

This manual assists vegetable producers to recognise and avoid the many different types of mistakes and pitfalls, so that they can achieve commercialization, profitability and success sooner.

This manual includes almost 40 chapters and close to 350 photographs. It is designed to be read mostly in chronological order, where later chapters build on the knowledge of previous chapters. The reader is guided from pre-production all the way through to the collection of profits from markets.

By using this manual and following the detailed, step-by-step guidelines provided within, initially the result may be higher input capital costs and therefore higher production costs. However, producers can look forward to higher yields, better quality and more eye-catching produce which will result in more consumer interest and ultimately higher profit margins.

The time to be a vegetable producer could never be better.

Piet Stork, Author

ABOUT THE AUTHOR

Piet Stork

Piet Stork has more than 40 years' experience in agriculture and was employed by the Department of Agriculture – Division: Plant and Quality Control Division (1967-1985) and by the Agricultural Research Council – Vegetable and Ornamental Plant Research Institute (1985-2012).

OTHER EXPERIENCE
- Studied Plant Science (1967-1969).
- 16 years of field inspections across South Africa for the certification of seed production of registered producers on their seed propagation land(s). Worked in partnership with the South African National Seed Organisation (SANSOR), the seed trade as well as seed companies and corporations.
- 19 years of hydroponics theory and practical experience, including the construction and design of greenhouses. The author was involved in many research projects as well as crop and cultivar evaluations. He also training all types of producers including large organizations, extension officers from the Department of Agriculture, schools, colleges, existing producers, small-scale farmers, private persons, as well as upcoming, beginner and underprivileged farmers.
- Involved in hybrid maize management in the Bethlehem, Clarens, Lindley, Fauriesburg areas within South Africa. This included field inspections and taking seed-lot samples to determine the grading and classification of maize seed as well as implementing germination and purity tests, in order to approve certification (1970-1982).
- The inspection of vegetable seed breeding lands in order to approve the production of crops which plant breeders' present for approval to the SA Certification Scheme.
- Maintained and supplied disease-free nucleus sweet potato plant material of various sweet potato cultivars. Issued this plant material to members of the National Vine Producers Association (NPV). Involved in field inspections of NPV members' lands across South Africa to approve nucleus multiplication and nursery production lands.
- Maintained and reproduced vegetative nucleus sweet potato plant material (mother plants) which are stored in glasshouses and various sweet potato gene banks.
- Maintained a vegetable gene bank to invigorate breeder seed of many crops, cultivars and breeders' lines, as well as new and outdated cultivars which are established to assist in the reproduction and refreshment of the seed.
- The continuous production, maintenance, germination and storage of mostly vegetable crops for the refreshment of harvested seed as well as for cultivars that have special genes.
- Inspected nurseries that supplied shrubs, pot plants, trees and other plants to the public and organisations such as government departments (forestry). This was to ensure that the nurseries do not distribute harmful diseases and soil nematodes.
- Elimination of garlic viruses and maintenance and evaluation of new cultivars. Performed field inspections at producer's production and propagation lands where mostly virus and disease-free bulbs were bred on a yearly basis and issued to the South African Garlic Growers Association (S.A.G.G.A.) for propagation. These bulbs were then supplied to commercial farmers as improved, higher-yield plant material. The author received an achievement award at SAGGA's national congress for assisting them to improve and maintain the garlic industry. The Author also received 5 separate awards from the Agricultural Research Council (ARC) for his outstanding contributions to ARC.
- Propagation of disease and virus-free mini/small potato tubers from various new and standard cultivars, and distributing them to plant breeders' and evaluating them. Tubers of the new, improved cultivars are then supplied to reliable, commercial potato producers all across the country for establishment in their production lands.
- Attended farmers' days, gave presentations and lectures at congresses and trained many different levels of

producers including underprivileged and upcoming farmers, students and groups. Training dealt with how to practically cultivate vegetable crops on specially designed small plots as well as the implementation of hydroponic systems.
- Held telephonic conferences with various producers all across the country, advising them about vegetable and hydroponic production. Also visited all types of farmers to train them and learn from them.
- Designed, maintained, transported and erected small NFT hydroponic table units (systems). Erected 6 tables in a specially designed glasshouse in order to supply fresh vegetables to the Big Brother house in 1999.
- Held demonstrations at agricultural shows on the production of vegetable crops in small NFT hydroponic table units which were designed by the author and set-up with various vegetable crops.
- Created exceptionally high yield compost to produce high-value indeterminate tomato and pepper plants in Shade Net Houses.
- Arranged phytosanitary certificates (import and export permits) for the import and export of disease-free sweet potato plant material to other countries. Plant material was maintained in special glasshouses and regularly tested, especially for the presence of viruses.
- Additionally, good results and practical experience has been achieved in the following areas: The erection of greenhouses and the installation and maintenance of different types of hydroponic systems. This involved working with different cultivars, formulating fertilizer solutions, monitoring irrigation systems, using electrical conductivity and pH readings, spacing of plants, the implementation of different planting times and growth mediums as well as the trellising, pruning and pollination of crops. Other greenhouse operations performed include heating and cooling (climate control), controlling insects and diseases, harvesting, estimating costs and yields, handling, sorting, grading, packing, transporting, marketing and so forth.

INTRODUCTORY REMARKS

The author designed this manual to be as user friendly and practical as possible and to assist all kinds of vegetable producers to produce good quality vegetable crops. If this is realised, then high yields will automatically follow, provided that the correct cultivation practices are followed as set out in this manual. The amount of information contained in this manual may be an eye-opener for many types of producers, and will inform their decision making with regards to vegetable production and achieving high yields and high quality.

The author is positive that this manual will assist all types of producers, especially beginners, that have the intention to produce vegetables profitably.

The single biggest privilege that SA and its citizens have, is the relatively good quality cheap food (especially vegetables) that is produced by over 39 000 farmers. However, many more vegetable producers are required to meet the demands of a fast-growing population. We need to assist these vegetable producers to become profitable and avoid failure due to a lack of knowledge on their part.

There are many vegetable production experts that are specialists in their fields and who are willing to assist producers with valuable information about their specific areas of expertise.

At times it will be necessary to spend large amounts of money, for instance on labour, expensive tractors and equipment - knowing that these investments can only be afforded if high yields are achieved. Therefore, the aim of this manual is to inspire and encourage all kinds of producers to do their best at all times, so that they are able to make maximum profits, while at the same time looking after their land. After all, the main aim of commercial vegetable production is to make a profit in order to earn a good living and remain sustainable.

HOW TO USE THIS MANUAL

This manual consists of two parts – Part 1 and Part 2.

Part 1 provides comprehensive guidelines which deal with all the aspects related to profitable vegetable production. Each aspect is dealt with one at a time, in its own chapter. For example, soil preparation, equipment and implements, irrigation, fertilization and so on, all have their own chapters.

It is highly advisable that vegetable producers study Part 1 carefully, as it contains important information which producers should know, before they begin to produce vegetable crops.

Part 2 provides crop-specific vegetable production guidelines. For instance, a tomato producer would refer to the chapter on tomato production, to learn about all the specific requirements to produce this crop successfully.

The production of non-vegetable crops such as garlic, sweet corn and green mealies are also included in this manual. Green mealies and sweet corn are classified as grain crops while garlic is classified as a herb.

Estimated costs and yields of all vegetable crops are specified in a table format for each crop in Part 2. The cost estimations will assist producers to design a budget, cash flow and business plan.

This manual was also designed for education purposes, allowing educators to present chapters in a sequential order from the first to the last chapter. It is recommended that the manual is provided to students beforehand, in order to allow them the opportunity to inform themselves and ask meaningful questions during instruction.

CONTENTS

FOREWORD i

ABOUT THE AUTHOR..ii

HOW TO USE THIS MANUAL iv

PART 1
A COMPLETE GUIDE TO VEGETABLE PRODUCTION...1

CHAPTER 1
VEGETABLE PRODUCTION IN OPEN FIELDS, SHADE NET HOUSES AND TUNNELS 2
1.1 GENERAL ASPECTS TO BE AWARE OF BEFORE ESTABLISHING CROPS.............2
1.2 REQUIREMENTS FOR COMMERCIALISATION 6
1.2.1 HOUSING 6
1.2.2 LABOUR 6
1.2.3 COMMUNICATION 6
1.2.4 SOIL...6
1.2.5 VEHICLES AND WORKSHOP 6
1.2.6 IMPLEMENTS..6
1.2.7 SEED AND SEEDLINGS...7
1.2.8 SEED STORAGE...7
1.2.9 WATER AND IRRIGATION 7
1.2.10 FERTILIZER...7
1.2.11 INSECT, DISEASE AND WEED CONTROL 7
1.2.12 EQUIPMENT FOR INSECT AND DISEASE CONTROL 7
1.2.13 TRELLISING PLANTS...7
1.2.14 HARVESTING AND DRYING CROPS...7
1.2.15 VEGETABLES UNDER PROTECTION..7
1.3 MONOCROPPING 8
1.3.1 ADVANTAGES OF MONOCROPPING 8
1.3.2 DISADVANTAGES OF MONOCROPPING 8
1.4 RISK FACTORS 8
1.5 ESTIMATED PRODUCTION COSTS, YIELDS AND PROFITS 9
1.6 VEGETABLES AND HEALTH 9

CHAPTER 2
PLANNING AND RECORD KEEPING...11
2.1 PRIORITISE PLANNING 11
2.2 ESSENTIAL ASPECTS FOR PROPER PLANNING..12
2.2.1 MORE ADVICE FOR BETTER PRODUCTION PLANNING..................................12
2.2.2 TALK WITH MARKET AGENTS WELL AHEAD OF TIME 13
2.3 LABOUR REQUIREMENTS..14
2.3.1 FARM SAFETY..14

CHAPTER 3
CLIMATE FOR VEGETABLE CROPS 15
3.1 CLIMATIC EFFECTS AND ZONES..15
3.1.1 CLIMATE ZONES..15
3.2 COOL AND WARM SEASONAL CROPS 15

3.3	TEMPERATURE AND TEMPERATURE RANGES	16
3.3.1	SOIL TEMPERATURE	16
3.3.2	AIR TEMPERATURE	16
3.4	HUMIDITY	17
3.5	LIGHT AND DAY LENGTH	18
3.5.1	LIGHT REQUIREMENTS FOR CROPS	18
3.6	FROST AND FROST HARDY VEGETABLES	18
3.7	WIND DAMAGE	19
3.8	RAINFALL AND WATER REQUIREMENTS FOR IRRIGATION	19
3.9	CULTIVAR CHOICES IN PRODUCTION AREAS	19
3.10	BOLTING OR PREMATURE FLOWERING	20

CHAPTER 4
SEED, SEEDLINGS AND NURSERIES 23

4.1	THE TREATMENT OF VEGETABLE SEED	23
4.1.1	SEED TREATMENT	23
4.2	USE CERTIFIED TESTED SEED	24
4.2.1	PALETTED SEED	24
4.3	PRODUCING YOUR OWN SEED	25
4.3.1	PRODUCING TOMATO SEED	26
4.3.2	PRODUCING CUCURBIT SEED	26
4.3.3	PRODUCING CARROT, BEETROOT, CABBAGE AND ONION SEED	26
4.4	SEED STORAGE AND LIFESPAN	27
4.5	PERFORMING GERMINATION TESTS	27
4.5.1	GERMINATION TEST FOR SMALL SEEDED CROPS	28
4.5.2	COUNTING GERMINATED SEEDS	28
4.5.3	GERMINATION TEST FOR LARGE SEEDED CROPS	29
4.5.4	GERMINATING SEED IN THE PRODUCTION LAND	29
4.6	SEEDS AND SEEDLINGS FROM NURSERIES	29
4.6.1	SEED	29
4.6.2	COSTS OF SEEDLINGS	29
4.6.3	SANITATION	30
4.6.4	REPUTATION OF THE NURSERY	30
4.6.5	CULTIVAR CHOICES	30
4.6.6	VISITING THE NURSERY	30
4.7	OPEN POLLINATED AND HYBRID SEED	32
4.8	PRODUCING SEEDLINGS	33
4.8.1	PRODUCING SEEDLINGS IN SEEDLING TRAYS	33
4.8.2	ADVANTAGES OF USING SEEDLINGS	34
4.8.3	SEED TRAY SIZES	34
4.9	FLOOR SPACE REQUIREMENTS FOR SEEDLING TRAYS	35
4.10	STERILIZATION AND REUSE OF SEEDLING TRAYS	35
4.11	ESTABLISHING AND CARING FOR SEED IN SEEDLING TRAYS	36
4.11.1	ENVIRONMENTAL FACTORS WHICH AFFECT SEED GERMINATION	36
4.11.2	ESTABLISHING AND CARING FOR SEED IN SEEDLING TRAYS	36
4.12	BUILDING A SEEDLING SHADE NET HOUSE FOR SEEDLING PRODUCTION	38
4.13	USING A GERMINATION CHAMBER	39
4.14	TEMPERATURES DURING SEED PRODUCTION	39
4.14.1	WARM WEATHER CROPS	39
4.14.2	COOL WEATHER CROPS	39
4.14.3	STEPS TO REDUCE SEEDLING DISEASES	40
4.15	TIMING FOR TRANSPLANTING SEEDLINGS FROM SEED TRAYS	40

4.16	METHODS TO HARDEN SEEDLINGS	41
4.16.1	HARDENING VIA FERTILIZER AND WATER SUPPLY REDUCTION	41
4.16.2	WITHHOLDING NITROGEN N FERTILIZER	41
4.16.3	MOVING SEEDLINGS TO HARDEN THEM	41
4.17	SEEDLING DISEASES	41
4.17.1	STAGES OF SUSCEPTIBILITY TO DISEASE	41
4.17.2	PATHOGENS WHICH MAY BE RESPONSIBLE FOR DISEASE	42
4.17.3	CONTROL MEASURES	42

CHAPTER 5
SOIL REQUIREMENTS FOR VEGETABLE CROPS 43

5.1	INTRODUCTION	43
5.1.1	FACTORS THAT INFLUENCE PLANT NUTRITION	44
5.1.2	FACTORS WHICH CAUSE TOXICITY AND HAMPER PLANT GROWTH IN SOIL	44
5.2	HEALTHY, LIVING SOIL AND SOIL ORGANISMS	45
5.3	SOIL REQUIREMENTS FOR VEGETABLE CROPS	46
5.3.1	SOIL THAT LACKS ORGANIC MATTER	47
5.3.2	COMPACTED SOIL	47
5.3.3	MANAGING SOIL PROPERLY	47
5.3.4	SOIL TYPES	47
5.3.5	SOIL PARTICLE SIZE NOT SUITABLE FOR VEGETABLE PRODUCTION	48
5.3.6	SOIL ACIDICY AND PH	48
5.3.7	LEAF AND SAP ANALYSIS	49

CHAPTER 6
LABORATORY SOIL ANALYSIS TESTING 51

6.1	TAKING A SOIL SAMPLE FOR A NUTRIENT ANALYSIS TEST	51
6.2	LABORATORY SOIL ANALYSIS TEST	52
6.2.1	SOIL SAMPLING	52
6.3	HOW TO TAKE A SOIL SAMPLE	52
6.3.1	INFORMATION REQUIRED FOR LABORATORY SOIL ANALYSIS TESTS	53
6.4	ADDITIONAL INFORMATION REQUIRED FOR A SOIL SAMPLE	53

CHAPTER 7
IMPLEMENTS AND EQUIPMENT 54

7.1	INTRODUCTION	54
7.2	KNOW YOUR IMPLEMENTS AND EQUIPMENT	55
7.1.1	TRACTORS	55
7.2.2	TRACTORS WITH PLOUGHS AND MOULDBOARD PLOUGHS	55
7.2.3	DISC PLOUGHS	55
7.2.4	CHISEL PLOUGHS	55
7.2.5	ROTOVATOR (ROTARY TILLER)	55
7.2.6	TINED CULTIVATORS	56
7.2.7	HARROWS	56
7.2.8	RIPPERS	56
7.3	OTHER EQUIPMENT AND IMPLEMENTS	56

CHAPTER 8
SOIL PREPARATION, SEEDLING PRODUCTION AND TRANSPLANTATION 58

8.1	INTRODUCTION	58
8.1.1	KINDS OF SEEDBEDS IN SOIL	58
8.2	STEPS TO PREPARE SOIL BEFORE ESTABLISHING CROPS	59

8.2.1	LAND PREPERATION	59
8.2.2	SOIL PREPARATION	59
8.2.3	RIPPING (CHISELLING)	60
8.2.4	PLOUGHING, DISCING, TILLING AND ROTOVATING	60
8.2.5	DRAINAGE OF SOIL	60
8.3	SIMPLE SOIL DRAINAGE ABILITY TESTS	60
8.4	PRODUCING SEEDLINGS IN SOIL	61
8.5	TRANSPLANTING SEEDLINGS FROM SOIL SEEDBEDS TO THE PREPARED LAND	63
8.6	ESTABLISHING SEED DIRECTLY IN WELL-PREPARED LAND	64
8.6.1	SOWING SEED DIRECTLY INTO SMALL PORTIONS OF LAND	65
8.7	MULCHING OR COVERING OF CROPS	66
8.7.1	THE PURPOSE OF MULCHING	66
8.7.2	THE PROBLEM WITH MULCHING	66
8.7.3	MULCHING WITH PLASTIC COVERS	66
8.7.4	THREE IMPORTANT FACTORS FOR MULCHING	66
8.7.5	PUROSE OF REFLECTIVE SURFACE PLASTICS	67
8.7.6	OTHER MULCHING MATERIAL FOR SOWING SEED IN SEEDBEDS	67
8.8	TRANSPLANTING SEEDLINGS FROM SEED TRAYS	67
8.9	FILLING GAPS DUE TO A POOR STAND	68

CHAPTER 9
ORGANIC MATTER, GREEN MANURE AND COMPOST PRODUCTION ... 69

9.1	ANIMAL MANURE AND THE VALUE OF ORGANIC MATTER	69
9.1.1	THE ADVANTAGES OF KRAAL MANURE OVER FRESH MANURE	69
9.2	USING GREEN MANURE TO RESTORE CROPS	69
9.3	COMPOST AND COMPOST PRODUCTION	71
9.3.1	HOW GOOD COMPOST IMPROVES SOIL	71
9.3.2	THE HOT AND COLD METHOD OF COMPOST PRODUCTION	72
9.3.3	THE QUANTITY OF COMPOST REQUIRED FOR PRODUCTION	72
9.3.4	AN EXAMPLE OF COMMERCIAL COMPOST PRODUCTION	73
9.4	COMMERCIAL COMPOST PRODUCTION	73
9.4.1	PHASE 1	73
9.4.2	PHASE 2	73
9.4.3	PHASE 3	74
9.4.4	IMPORTANT ASPECTS TO LOOK OUT FOR WHEN PRODUCING COMPOST	74

CHAPTER 10
ORGANIC FARMING ... 75

10.1	INTRODUCTION	75
10.2	SHORT OVERVIEW OF ORGANIC FARMING	76

CHAPTER 11
SPACING OF CROPS ... 79

11.1	INTRODUCTION	79
11.1.1	THE IMPACT OF OVERLY DENSE PLANT SPACING	79
11.2	CALCULATIONS FOR PLANT SPACING AND PLANT POPULATION	80
11.3	SPACING CROPS IN GREEN HOUSES	81

CHAPTER 12
TRELLISING (TRAINING) AND RIDGING ... 83

12.1	INTRODUCTION	83
12.2	ADVANTAGES AND DISADVANTAGES OF TRELLISING	84

12.2.1	ADVANTAGES OF TRELLISING	84
12.2.2	DISADVANTAGES OF TRELLISING	84
12.3	RIDGING OF CROPS AND RAISED SEED BEDS	84
12.3.1	RIDGING	84
12.3.2	RAISED BEDS	84

CHAPTER 13
WATER 85

13.1	INTRODUCTION	85
13.2	WATER SOURCES	85
13.3	APPLYING THE CORRECT QUANTITY OF IRRIGATION WATER	86
13.3.1	THE EFFECTS OF OVER-IRRIGATION	86
13.3.2	THE EFFECTS OF UNDER-IRRIGATION	87
13.4	RAINFALL AND MEASURING METERS	87
13.4.1	USING A SPADE TO ASSESS THE WATER CONTENT IN THE SOIL	88
13.4.2	**TENSION WATER MEASURING METERS**	**88**
13.4.3	MOISTURE DATA LOGGERS AND SMARTPHONES	88

CHAPTER 14
IRRIGATION 91

14.1	INTRODUCTION	91
14.2	POINTERS FOR IRRIGATING CROPS	92
14.3	GUIDELINES FOR IRRIGATION	93
14.4	IRRIGATION AND ROOT DEPTH	93
14.5	IRRIGATION SYSTEMS	94
14.5.1	IRRIGATION SYSTEMS AVAILABLE IN SOUTH AFRICA	94
14.6	REASONS FOR A POORLY DESIGNED IRRIGATION SYSTEM	96

CHAPTER 15
FERTILIZER.. 101

15.1	INTRODUCTION	101
15.2	INCORRECT FERTILIZING	104
15.3	PLANT NUTRIENTS AND REQUIREMENTS	105
15.4	ORGANIC AND INORGANIC FERTILIZER	106
15.4.1	STRAIGHT FERTILIZER	106
15.4.2	MIXED OR COMPOUND FERTILIZER	107
15.4.3	NITROGEN FERTILIZER CARRIERS	107
15.4.4	POTASSIUM FERTILIZER CARRIERS	107
15.4.5	MICRO OR TRACE ELEMENTS	107
15.4.6	APPLYING MICRO OR TRACE ELEMENTS	107
15.5	APPLYING THE CORRECT TYPE OF LIME	108
15.5.1	THE FOUR TYPES OF LIME	108
15.6	TOP OR SIDEDRESSING	109
15.7	AN EXAMPLE OF AN AVERAGE TOMATO FERTILIZER PROGRAM	110
15.8	FERTILIZER FOR INDETERMINATE OPEN-FIELD TOMATO CULTIVARS	111
15.9	APPLYING FERTILIZER OVER LARGE AREAS	111
15.10	APPLYING FERTILIZER OVER SMALLER AREAS	112
15.10.1	MEASURING FERTILIZER BY HAND	112
15.10.2	APPLYING THE FERTILIZER	112
15.11	FOLIAR SPRAY FERTILIZER	113
15.11.1	REASONS FOR MICROELEMENT SHORTAGES	113
15.11.2	THE CONDITIONS UNDER WHICH FOLIAR SPRAYS MAY BE APPLIED	113

15.11.3	FERTILIZER THROUGH A DRIPPER SYSTEM IN SOIL	114
15.12	CHEMICAL SPRAYING THROUGH IRRIGATION WATER	115
15.12.1	CHEMICAL SPRAY CLASSIFICATIONS FOR WATER-BASED APPLICATIONS	115
15.12.2	ADVANTAGES OF CHEMICAL SPRAY	115
15.12.3	DISADVANTAGES OF CHEMICAL SPRAY	115

CHAPTER 16
SOWING SEED AND THINNING SEEDLINGS ... 119

16.1	SOWING SEED	119
16.2	THINNING SEEDLING PLANTS IN THE PRODUCTION LAND	119

CHAPTER 17
WEED CONTROL ... 121

17.1	INTRODUCTION	121
17.1.1	WEED REPRODUCTION	121
17.2	WEED CONTROL WHEN USING HERBICIDES	122
17.2.1	TYPES OF HERBICIDES	123
17.2.2	HERBICIDE ACTION	124

CHAPTER 18
CROP ROTATION AND SOIL IMPROVEMENT ... 125

18.1	INTRODUCTION	125
18.1.1	THE FOUR GROUPS OF VEGETABLES CROPS	125
18.2	CROP ROTATION AND CROP NUTRIENT REQUIREMENTS	126
18.2.1	LIGHT NUTRIENT USERS - ROOTED CROPS	126
18.2.2	MILD NUTRIENT USERS - BEAN FAMILY	126
18.2.3	HEAVY NUTRIENT USERS - LEAF AND FRUIT BEARING CROPS	126
18.3	REASONS FOR CROP ROTATION	127
18.3.1	INSECT AND DISEASE CONTROL	127
18.3.2	INCREASED SOIL FERTILITY	127
18.3.3	IMPROVED SOIL STRUCTURE	127
18.3.4	PREVENTING WATER AND SOIL LOSSES	127
18.3.5	WHEN CROP ROTATION BECOMES UNNECESSARY	127
18.4	SOIL IMPROVEMENT	128
18.4.1	GREEN MANURE	128
18.4.2	LEGUMINOUS CROPS THAT FIX NITROGEN INTO THE SOIL	129
18.4.3	BIO-FUMIGATION CROPS MAY BE INCORPORATED INTO THE SOIL	129

CHAPTER 19
SANITATION AND SCANNING FOR INSECTS AND DISEASES ... 130

19.1	INTRODUCTION	130
19.1.1	HIDING PLACES OF INSECTS AND WHERE TO FIND THEM	131
19.1.2	SCANNING	131

CHAPTER 20
OVERVIEW OF INSECTS, DISEASES, VIRUSES AND PHYSICAL DISORDERS ... 133

20.1	INTRODUCTON	133
20.1.1	OUTDATED CHEMICALS	134
20.1.2	VARIOUS WAYS THAT DISEASES SPREAD	134
20.1.3	THE CAUSES OF INSECT OUTBREAKS	134
20.1.4	HOW INSECT AND DISEASE INFECTIONS MAY CAUSE HEAVY CROP LOSSES	134
20.2	THE MAIN PARTS OF PLANTS THAT INSECTS/DISEASES ATTACK	135

20.2.1	LEAVES	135
20.2.2	STEMS	135
20.2.3	ROOTS	135
20.2.4	FRUIT AND FLOWERS	135
20.2.5	A SELECTION OF MAJOR DISEASES TO TAKE NOTE OF	135
20.3	GMO CROPS	135
20.4	GUIDES FOR CONTROLLING INSECTS AND DISEASES	136
20.4.1	PLANT RESISTANCE AGAINST DISEASES AND INSECTS	137
20.5	INSECTS	137
20.5.1	THE MOST COMMON TYPES OF INSECTS	137
20.6	DISEASES	142
20.6.1	THE 3 PRIMARY DISEASE CLASSIFICATIONS	142
20.6.2	CHEMICAL REGULATIONS	143
20.6.3	DISEASE DESCRIPTIONS	143
20.7	VIRUSES	149
20.7.1	FACTORS TO CONSIDER WHEN CONTROLLING VIRUSES	150
20.7.2	PLANT VIRUS DESCRIPTIONS	150
20.8	PHYSICAL DISORDERS	150
20.8.1	PHYSICAL DISORDER DESCRIPTIONS	151

CHAPTER 21
INTRODUCTION TO INSECT AND DISEASE CONTROL 154

21.1	INTRODUCTION	154
21.2	CHEMICAL GUIDELINES FOR CONTROLLING INSECTS AND DISEASES	154
21.3	NATURAL METHODS FOR CONTROLLING INSECTS AND DISEASES	155

CHAPTER 22
GUIDELINES FOR INSECT AND DISEASE CONTROL 157

22.1	INTRODUCTION	157
22.2	COMMON REASONS FOR FAILURE TO CONTROL INSECTS AND DISEASES	157
22.3	ADVICE FOR EFFECTIVE INSECT, DISEASE AND VIRUS CONTROL	158
22.4	SPRAYING EQUIPMENT USED FOR INSECT AND DISEASE CONTROL	158
22.5	SAFEGUARDING AGAINST POISONING	159
22.5.1	THE BASIC RULES WHEN USING CHEMICALS	159
22.5.2	THE CORRECT PROTECTIVE CLOTHING WHEN SPRAYING CROPS	159
22.6	TYPES OF SPRAYING EQUIPMENT FOR INSECT AND DISEASE CONTROL	159
22.7	CHEMICAL SPRAYING VOLUMES AND DROPLET SIZES	160
22.7.1	INSECTICIDE TIMING	160
22.7.2	SPRAYING VOLUMES	160
22.7.3	DROPLET SIZES	160
22.8	PH OF CHEMICAL WATER	161
22.9	RESISTANCE OF INSECTS AGAINST CHEMICALS	161
22.10	BIOLOGICAL CONTROL OF INSECTS AND DISEASES	162
22.11	SPRAYING PROGRAM GUIDELINES	163
22.12	BUILDING A CHEMICAL STORE	163
22.12.1	WARNING SIGNS	163
22.12.2	EQUIPMENT INSIDE THE STORE	163

CHAPTER 23
PRODUCING VEGETABLES UNDER PROTECTION 165

| 23.1 | PRODUCING VEGETABLE CROPS UNDER PROTECTION IN SOIL | 165 |
| 23.1.1 | ADVANTAGES AND DISADVANTAGES OF SHADE NET HOUSES | 166 |

23.2	CONSTRUCTING A SHADE NET HOUSE	167
23.2.1	POLES FOR A 3.7M HIGH SHADE NET HOUSE	167
23.2.2	CABLES AND WIRES FOR SHADE NET HOUSES	167
23.2.3	FITTING THE SHADE NET	167
23.2.4	ANCHORS	168
23.3	PRODUCTION OF HIGH-VALUE TOMATO OR PEPPERS IN A SHADE NET HOUSE DIRECTLY IN SOIL	170
23.3.1	SOIL PREPARATION IN SHADE NET HOUSES	170
23.3.2	SHADE NET HOUSE INSTALLATION	171
23.3.3	ESTABLISHING SEEDLINGS IN A SHADE NET HOUSE	172
23.3.4	IRRIGATION IN GREEN HOUSES	172
23.4	PRODUCING CROPS IN TUNNELS AND IN SOIL	172
23.4.1	POINTERS FOR PRODUCING CROPS IN TUNNELS IN SOIL	172
23.4.2	TRELLISING AND PRUNING OF CROPS IN TUNNELS AND SHADE NET HOUSES	174
23.4.3	INDETERMINATE TOMATO PRODUCTION IN GREENHOUSES	174
23.5	PEPPER PRODUCTION IN GREENHOUSES	175
23.6	CUCUMBER PRODUCTION IN GREENHOUSES	175
23.6.1	ISRAELI CUCUMBERS AND GHERKINS	176
23.7	TEMPERATURE CONTROL AND VENTILATION UNITS FOR TUNNELS	176
23.7.1	TEMPERATURE CONTROL IN TUNNELS	176
23.7.2	TUNNEL VENTILATION	177
23.8	NUTRIENT FLOW TECHNIQUE (NFT) HYDROPONIC SYSTEMS IN GREENHOUSES	178
23.9	MULCHING IN GREENHOUSES	178
23.10	ESTIMATED COST OF ERECTING A 1HA SHADE NET HOUSE	179
23.11	ESTIMATED COSTS, YIELDS AND PROFITS WHEN PRODUCING HIGH-VALUE CROPS IN TUNNELS DIRECTLY IN SOIL (2014)	182

CHAPTER 24
HARVESTING, HANDLING AND PACKAGING 186

24.1	WHEN TO HARVEST	186
24.2	POST HARVESTING	186
24.2.1	CAUSES OF POST HARVESTING DAMAGE	186
24.3	GRADING	187
24.4	PACKAGING	187
24.5	COOLING AND DETERIORATION	189
24.6	STORAGE AND THE PACK HOUSE	189
24.7	PLANNING THE PACK HOUSE	189

CHAPTER 25
MANAGEMENT AND MARKETING 191

25.1	MANAGEMENT	191
25.1.1	THE TWO MAIN FACTORS FOR GOOD MANAGEMENT	191
25.2	MARKETING	191

PART 2
GUIDELINES FOR CROP-SPECIFIC VEGETABLE PRODUCTION 195

CHAPTER 26
PRODUCTION GUIDELINES FOR TOMATOES 196

26.1	INTRODUCTION	198
26.2	DIFFERENT TYPES OF GROWING BEHAVIOURS	199
26.2.1	*DETERMINATE GROWERS*	*199*

26.2.2	INDETERMINATE GROWERS	**199**
26.2.3	BUSH OR ROMA TYPES	199
26.3	CULTIVAR SELECTION AND GRAFTING OF SEEDLINGS	199
26.3.1	GRAFTED SEEDLINGS	200
26.4	ESTIMATED PRODUCTION INPUT COSTS AND YIELDS FOR DETERMINATE CULTIVARS	200
26.4.1	EXPECTED YIELDS FOR DIFFERENT SYSTEMS OF TOMATO PRODUCTION	202
26.5	QUICK STEP-BY-STEP PRODUCTION PLANNING GUIDE FOR TOMATOES AND OTHER CROPS	203
26.6	TOMATO PRODUCTION GUIDELINES AND REQUIREMENTS	205
26.6.1	SOIL REQUIREMENTS	205
26.6.2	CLIMATE REQUIREMENTS AND OPTIMAL PLANTING AND SOWING PERIODS	205
26.6.3	TOMATO CULTIVARS	206
26.6.4	SEEDLING PRODUCTION IN SEED TRAYS	207
26.6.5	SOIL PREPARATION	208
26.6.6	TRANSPLANTING SEEDLINGS	208
26.6.7	WEED CONTROL	208
26.6.8	FERTILIZATION	208
26.6.9	IRRIGATION	209
26.6.10	TRELLISING AND SPACING METHODS	210
26.6.11	CROP ROTATION	218
26.6.12	MULCHING	218
26.6.13	HARVESTING AND PICKING	219
26.6.14	SORTING, CLASSING AND MARKETING	222
26.6.15	INSECTS	222
24.6.16	DISEASES	222
26.6.17	VIRAL DISEASES	223
26.6.18	PHYSIOLOGICAL DISORDERS	223

CHAPTER 27
PRODUCTION GUIDELINES FOR CUCURBITS: PUMPKIN, SQUASH, WATERMELON AND MUSK MELON 225

27.1	INTRODUCTION	225
27.1.1	PROTECTION AGAINST SUNBURN	227
27.2	CROP PRODUCTION	227
27.3	ESTIMATED PRODUCTION COSTS FOR CURCURBITS	227
27.4	PRODUCTION GUIDELINES FOR CURCURBITS	229
27.4.1	SOIL REQUIREMENTS	229
27.4.2	PH OF THE SOIL	229
27.4.3	CLIMATE REQUIREMENTS	229
27.4.4	CULTIVARS	231
27.4.5	SOIL PREPARATION	233
27.4.6	PLANTING TIMES	233
27.4.7	SEED AND THE QUANTITY OF SEEDS PER HA	234
27.4.8	PLANTING METHODS	234
27.4.9	SPACING	236
27.4.10	DURATION BETWEEN TRANSPLANTATION AND MATURATION	238
27.4.11	FERTILIZING	238
27.4.12	IRRIGATION	240
27.4.13	CROP ROTATION	241
27.4.14	WEED CONTROL	241
27.4.15	WIND DAMAGE	241

27.4.16	POLLINATION AND ABORTION OF FRUIT	241
27.4.17	YIELDS	242
27.4.18	HARVESTING AND STORAGE	243
27.4.19	INSECTS	244
27.4.20	DISEASES	245
27.4.21	VIRUSES	246
27.4.22	PHYSICAL DISORDERS	247

CHAPTER 28
PRODUCTION GUIDELINES FOR CABBAGE, CAULIFLOWER, BROCCOLI, BRUSSELS SPROUTS, TURNIPS AND CHINESE CABBAGE ... 248

28.1	INTRODUCTION	248
28.2	ESTIMATED PRODUCTION COSTS AND PROFITS FOR COLE CROPS (CABBAGE)	249
28.3	INTRODUCTION TO COLE CROP PRODUCTION	251
28.3.1	CABBAGE	251
28.3.2	CAULIFLOWER	251
28.3.3	BROCCOLI	251
28.3.4	BRUSSELS SPROUTS	251
28.4	PRODUCTION GUIDELINES FOR COLE CROPS	253
28.4.1	SOIL REQUIREMENTS	253
28.4.2	PH OF THE SOIL	253
28.4.3	CLIMATIC REQUIREMENTS AND OPTIMUM PLANTING TIMES	254
28.4.4	CULTIVARS AND CULTIVAR CHOICES	255
28.4.5	SEEDLING PRODUCTION IN SEEDBEDS AND IN THE LAND	258
28.4.6	SOIL PREPARATION AND TRANSPLANTING OF SEEDLINGS	259
28.4.7	AMOUNT OF SEED P/HA	259
28.4.8	SPACING	259
28.4.9	IRRIGATION	260
28.4.10	FERTILIZER	261
28.4.11	THE IMPORTANCE OF TRACE ELEMENTS	262
28.4.12	CROP ROTATION	262
28.4.13	WEED CONTROL	262
28.4.14	YIELDS AND DAYS UNTIL HARVESTING	262
28.4.16	HARVESTING, PACKAGING AND MARKETING	262
28.4.17	STORAGE	264
28.4.18	INSECTS	265
28.4.19	DISEASES	265
28.4.20	PHYSIOLOGICAL DISORDERS	265

CHAPTER 29
PRODUCTION GUIDELINES FOR ONIONS ... 267

29.1	INTRODUCTION	267
29.2	BOLTING AND SPLITTING OF ONION BULBS	268
29.3	ONION SET PRODUCTION	268
29.3.1	SEED PRODUCTION	269
29.3.2	MEDICAL USES	269
29.4	ESTIMATED PRODUCTION COSTS AND PROFITS	269
29.5	PRODUCTION GUIDELINES FOR ONIONS	271
29.5.1	SOIL REQUIREMENTS	271
29.5.2	CLIMATE REQUIREMENTS AND PLANTING TIMES	271
29.5.3	PH OF THE SOIL	272
29.5.4	CULTIVARS	272

29.5.5	SOWING SEED AND TRANSPLANTATION TIMES OF SEEDLINGS	272
29.5.6	SEED P/HA	273
29.5.7	SEEDLING PRODUCTION	273
29.5.8	SPACING AND PLANT DENSITY IN THE LAND	274
29.5.9	TRANSPLANTING SEEDLINGS IN THE LAND	275
29.5.10	FERTILIZATION	275
29.5.11	IRRIGATION	276
29.5.12	CROP ROTATION	276
29.5.13	WEED CONTROL	276
29.5.14	DURATION BETWEEN TRANSPLANTATION AND HARVESTING	277
29.5.15	HARVESTING AND DRYING	277
29.5.16	GRADING, PACKAGING AND STORAGE	278
29.5.17	YIELDS	279
29.5.18	MARKETING	279
29.5.19	INSECTS	279
29.5.20	DISEASES	279

CHAPTER 30
PRODUCTION GUIDELINES FOR SWEET POTATOES 281

30.1	INTRODUCTION	281
30.2	ESTIMATED PRODUCTION COSTS AND PROFITS (2014)	283
30.3	PRODUCTION GUIDELINES FOR SWEET POTATOES	284
30.3.1	CLIMATE REQUIREMENTS AND PLANTING TIMES	284
30.3.2	CULTIVARS	285
30.3.3	PROPAGATION OF PLANT MATERIAL	285
30.3.4	HANDLING OF VINES OR CUTTINGS	286
30.3.5	SOIL PH	287
30.3.6	SOIL REQUIREMENTS	287
30.3.7	SOIL PREPARATION	287
30.3.8	PLANTING METHOD AND SPACING	287
30.3.9	PRODUCTION OF PLANT MATERIAL AND VINES	288
30.3.10	FERTILIZER	289
30.3.11	IRRIGATION	289
30.3.12	GROWING PERIOD	290
30.3.13	WEED CONTROL	290
30.3.14	CROP ROTATION	290
30.3.15	HARVESTING	291
30.3.16	STORAGE, PACKAGING, SORTING AND MARKETING	292
30.3.17	YIELDS	292
30.3.18	INSECTS	292
30.3.19	DISEASES	293
30.3.20	PHYSIOLOGICAL DISORDERS	293

CHAPTER 31
PRODUCTION GUIDELINES FOR PEPPERS, PAPRIKA AND CHILLIES 294

31.1	INTRODUCTION	294
31.2	SUNBURN AND PEPPER PRODUCTION	295
31.3	PRODUCTION OF PEPPERS UNDER PROTECTION	296
31.4	PRODUCTION COSTS FOR PEPPERS	297
31.5	PRODUCTION GUIDELINES AND REQUIREMENTS	298
31.5.1	CLIMATE REQUIREMENTS, MAIN AREAS OF PRODUCTION AND PLANTING TIMES	298
31.5.2	CULTIVARS	300

31.5.3	SEED P/HA	301
31.5.4	YIELDS	301
31.5.5	SEEDLING PRODUCTION AND TRANSPLANTING OF SEEDLINGS	301
31.5.6	MULCHING	302
31.5.7	SPACING, DENSITY AND TRELLISING OF PLANTS	302
31.5.8	FERTILIZATION	303
31.5.9	IRRIGATION	305
31.5.10	WEED CONTROL	305
31.5.11	CROP ROTATION	305
31.5.12	HARVESTING, COOLING AND STORAGE	305
31.5.13	MARKETING AND PACKAGING	307
31.5.14	INSECTS	307
31.5.15	DISEASES	307
31.5.16	VIRAL DISEASES	308
31.5.17	FRUIT DISORDER	308

CHAPTER 32
PRODUCTION GUIDELINES FOR SPINACH 310

32.1	INTRODUCTION	310
32.2	PRODUCTION COSTS AND POSSIBLE YIELDS	311
32.3	PRODUCTION GUIDELINES AND REQUIREMENTS FOR SPINACH	312
32.3.1	SOIL REQUIREMENTS	312
32.3.2	PH OF THE SOIL	312
32.3.3	CLIMATE REQUIREMENTS AND OPTIMUM PLANTING TIMES	313
32.3.4	CULTIVARS	313
32.3.5	SOIL PREPARATION	314
32.3.6	PLANTING METHODS AND SPACING	314
32.3.7	THINNING OF PLANTS	316
32.3.8	SEED P/HA	316
32.3.9	FERTIGATION	316
32.3.10	IRRIGATION	317
32.3.11	WEED CONTROL	317
32.3.12	CROP ROTATION	317
32.3.13	YIELDS	317
32.3.14	HARVESTING AND MARKETING	317
32.3.15	STORAGE	318
32.3.16	INSECTS AND DISEASES	318

CHAPTER 33
PRODUCTION GUIDELINES FOR BEETROOT 319

33.1	INTRODUCTION	319
33.1.1	HEALTH AND FOOD VALUE	319
33.2	ESTIMATED PRODUCTION COSTS, YIELDS AND INCOME	320
33.3	PRODUCTION GUIDELINES AND REQUIREMENTS FOR BEETROOT	322
33.3.1	CLIMATE REQUIREMENTS AND PLANTING TIMES	322
33.3.2	TYPES AND CULTIVARS	323
33.3.3	SOIL REQUIREMENTS AND PREPARATION	323
33.3.4	SEED P/HA	324
33.3.5	SEED SOWING METHODS AND SPACING	324
33.3.6	TRANSPLANTING SEEDLINGS	326
33.3.7	DURATION UNTIL MATURITY	326
33.3.8	MULCHING	326

33.3.9	FERTILIZER	327
33.3.10	IRRIGATION	328
33.3.11	PH OF THE SOIL	329
33.3.12	CROP ROTATION AND SOIL IMPROVEMENT	329
33.3.13	THINNING OF SEEDLINGS	329
33.3.14	WEED CONTROL	329
33.3.15	YIELDS	330
33.3.16	HARVESTING, HANDLING, MARKETING, STORAGE AND PACKAGING	330
33.3.17	INSECTS	331
33.3.18	DISEASES	331

CHAPTER 34
PRODUCTION GUIDELINES FOR CARROTS 332

34.1	INTRODUCTION	332
34.1.1	SUGAR (SWEETNESS) IN THE CARROT ROOT	333
34.2	ESTIMATED PRODUCTION COST FOR CARROTS	334
34.3	GUIDELINES FOR CARROT PRODUCTION	335
34.3.1	CLIMATE REQUIREMENTS AND OPTIMUM PLANTING TIMES	335
34.3.2	TYPES AND CULTIVARS	336
34.3.3	SOIL REQUIREMENTS AND PREPARATION	337
34.3.4	PH OF THE SOIL	338
34.3.5	SEED SOWING METHODS, PLANT POPULATION AND SPACING	338
34.3.6	MULCHING	340
34.3.7	CROP ROTATION	341
34.3.8	THINNING OF PLANTS	341
34.3.9	SEED PER HA	341
34.3.10	FERTILIZER	341
34.3.11	IRRIGATION AND IRRIGATION SYSTEMS	342
34.3.12	WEED CONTROL	343
34.3.13	BOLTING	343
34.3.14	YIELDS	344
34.3.15	DURATION BETWEEN PLANTING AND HARVESTING	344
34.3.16	HARVESTING AND MARKETING	344
34.3.17	STORAGE	345
34.3.18	INSECTS	345
34.3.19	DISEASES	345

CHAPTER 35
PRODUCTION GUIDELINES FOR BEANS 346

35.1	INTRODUCTION	346
35.2	ESTIMATED PRODUCTION COSTS AND PROFITS	347
35.3	PRODUCTION GUIDELINES FOR GREEN BEANS	349
35.3.1	SOIL REQUIREMENTS	349
35.3.2	PH OF THE SOIL	349
35.3.3	SOIL PREPARATION	349
35.3.4	CLIMATE REQUIREMENTS AND PLANTING TIMES	349
35.3.5	CULTIVARS	350
35.3.6	PLANTING METHOD, SPACING AND DENSITY	350
35.3.7	AMOUNT OF SEED P/HA	352
35.3.8	DAYS TO HARVEST	352
35.3.9	FERTILIZATION	352
35.3.10	IRRIGATION	353

35.3.11	CROP ROTATION	353
35.3.12	WEED CONTROL	353
35.3.13	YIELDS	353
35.3.14	HARVESTING, STORAGE, COOLING AND PACKAGING	354
35.3.15	INSECTS	355
35.3.16	DISEASES	355

CHAPTER 36
PRODUCTION GUIDELINES FOR LETTUCE 357

36.1	INTRODUCTION	357
36.2	ESTIMATED PRODUCTION COSTS AND PROFITS	358
36.3	PRODUCTION GUIDELINES FOR LETTUCE	360
36.3.1	SOIL REQUIREMENTS	360
36.3.2	SOIL PREPARATION	360
36.3.3	SOIL PH	360
36.3.4	CLIMATE REQUIREMENTS AND PLANTING TIMES	360
36.3.5	GROWTH PERIOD	361
36.3.6	IRRIGATION	361
36.3.7	FERTILIZER	362
36.3.8	TYPES OF LETTUCES AND CULTIVARS	362
36.3.9	SEED P/HA	363
36.3.10	SOWING METHODS, SPACING AND PLANTING	363
36.3.11	CROP ROTATION	364
36.3.12	WEED CONTROL AND THINNING OF PLANTS	364
36.3.13	YIELDS	365
36.3.14	BOLTING OR PREMATURE SEEDING	365
36.3.15	HARVESTING, PACKAGING AND MARKETING	365
36.3.16	INSECTS	365
36.3.17	DISEASES	365
36.3.18	PHYSICAL DISORDERS	366

CHAPTER 37
PRODUCTION GUIDELINES FOR SWEET CORN AND GREEN MEALIES 367

37.1	INTRODUCTION	367
37.1.1	CLASSIFICATION OF SWEET CORN SWEETNESS	368
37.1.2	GREEN MEALIES	368
37.2	ESTIMATED PRODUCTION COSTS AND PROFITS (2014)	368
37.3	PRODUCTION GUIDELINES FOR SWEET CORN AND GREEN MEALIES	370
37.3.1	SOIL REQUIREMENTS	370
37.3.2	SOIL PH	370
37.3.3	CLIMATE REQUIREMENTS AND PLANTING TIMES	370
37.3.4	SOIL PREPARATION AND PLANTING	371
37.3.5	CULTIVARS	371
37.3.6	IRRIGATION	371
37.3.7	FERTILIZER	372
37.3.8	GROWING PERIOD	372
37.3.9	PLANT POPULATION AND SPACING	373
37.3.10	POLLINATION	373
37.3.11	PLANTING DEPTH	373
37.3.12	SEED P/HA	373
37.3.13	WEED CONTROL	374
37.3.14	YIELDS	374

37.3.15	HARVESTING AND GRADING	374
37.3.16	STORAGE	375
37.3.17	MARKETING	375
37.3.18	INSECTS	375
37.3.19	DISEASES	375
37.3.20	VIRUSES	376

CHAPTER 38
PRODUCTION GUIDELINES FOR GARLIC ... 377

38.1	INTRODUCTION	377
38.1.1	MEDICAL USES	378
38.2	PLANNING FOR GARLIC PRODUCTION	378
38.3	ESTIMATED PRODUCTION COSTS AND PROFITS	381
38.4	PRODUCTION GUIDELINES FOR GARLIC	383
38.4.1	SOIL REQUIREMENTS	383
38.4.2	SOIL PREPARATION	383
38.4.3	PH OF THE SOIL	383
38.4.4	CLIMATE REQUREMENTS	383
38.4.5	AMOUNT OF SEED PER HA	384
38.4.6	CULTIVARS	384
38.4.7	PLANTING TIMES	384
38.4.8	PREPARATION OF BULBS AND PLANTING METHODS	385
38.4.9	SPACING	386
38.4.10	CROP ROTATION	387
38.4.11	WEED CONTROL	387
38.4.12	YIELDS	387
38.4.13	IRRIGATION	387
38.4.14	FERTILIZATION	388
38.4.15	HARVESTING THE PLANTS	388
38.4.16	DRYING OF PLANTS	389
38.4.17	CLEANING, SORTING, GRADING AND PACKAGING	389
38.4.18	STORAGE	390
38.4.19	MARKETING	391
38.4.20	QUALITY STANDARDS	391
38.4.21	PLANT MATERIAL (SEED) SORTING AND STORAGE	391
38.4.22	INSECTS	392
38.4.23	DISEASES	392

REFERENCES ... 395

INDEX ... 396

SUPPLIERS LIST ... 404

LIST OF TABLES

TABLE 1:	IDEAL GROWING TEMPERATURES FOR VEGETABLE CROPS	17
TABLE 2:	THE EXPECTED LIFE SPAN OF VEGETABLE SEED WHEN IN STORAGE	27
TABLE 3:	ROOT PENETRATION DEPTH FOR DIFFERENT VEGETABLE CROPS	93
TABLE 4:	THE NUTRIENT REQUIREMENTS FOR CROPS	105
TABLE 5:	MACRO AND MICRO NUTRIENT REQUIREMENTS OF PLANTS	105
TABLE 6:	RELATIVE TOLERANCE OF VEGETABLE CROPS TO THE ELEMENT BORON	106
TABLE 7:	DESCRIPTION OF STRAIGHT NUTRIENT FORMULA FERTILIZER	107
TABLE 8:	AN AVERAGE STRAIGHT FERTILIZER PROGRAM FOR TOMATOES	110
TABLE 9:	SYMPTOMS OF ELEMENT DEFICIENCIES	116
TABLE 10:	AN EXAMPLE OF A 4-CROP ROTATIONAL SYSTEM THAT IS BASED ON A 3 YEAR ROTATIONAL PERIOD (PLOT A: CARROTS, PLOT B: BEETROOT, PLOT C: CABBAGE AND PLOT D: BEANS)	127
TABLE 11:	ESTIMATED COST OF ERECTING A 1HA SHADE NET HOUSE (2014)	179
TABLE 12:	PROJECTED PRODUCTION COST TO PRODUCE INDETERMINATE TOMATOES OR PEPPERS IN A 1HA SHADE NET HOUSE IN SOIL (26 000 PLANTS P/HA) (2014)	181
TABLE 13:	ESTIMATED YIELDS AND PROFITS FOR TOMATOES AND PEPPERS IN SOIL IN A 1HA SHADE NET HOUSE	182
TABLE 14:	ESTIMATED COSTS AND YIELDS IN TUNNELS FOR PRODUCTION IN SOIL OVER 1HA (2014)	183
TABLE 15:	POSSIBLE YIELDS AND PROFITS IN TUNNELS FOR PRODUCTION IN SOIL P/HA	184
TABLE 16:	PROBLEMS RELATED TO TOMATO PRODUCTION IN GREENHOUSES	185
TABLE 17:	ESTIMATED PRODUCTION COSTS AND YIELDS FOR DETERMINATE TOMATO GROWERS IN THE OPEN-FIELD FOR 18 500 PLANTS P/HA (2014)	200
TABLE 18:	ESTIMATE YIELDS AND PROFITS FOR DETERMINATE GROWERS	202
TABLE 19:	QUICK STEP-BY-STEP PRODUCTION PLANNING GUIDE FOR TOMATOES AND OTHER CROPS	203
TABLE 20:	ESTIMATED PRODUCTION COSTS FOR PUMPKINS (FLAT WHITE BOER AND GREEN HUBBARDS) (2014)	228
TABLE 21:	YIELD AND POSSIBLE PROFIT FOR PUMPKINS	229
TABLE 22:	ESTIMATED COSTS AND PROFITS FOR CABBAGE PRODUCTION (2014)	250
TABLE 23:	POSSIBLE YIELD AND PROFITS FOR CABBAGE	251
TABLE 24:	ESTIMATED PRODUCTION COSTS FOR ONIONS (1HA) (2014)	270
TABLE 25:	POSSIBLE YIELDS AND PROFITS FOR ONIONS	271
TABLE 26:	PLANT DENSITIES AND BULB SIZES	275
TABLE 27:	ESTIMATED PRODUCTION COSTS AND PROFITS (2014)	283
TABLE 28:	ESTIMATE YIELDS AND PROFITS P/HA	284
TABLE 29:	PRODUCTION COSTS FOR PEPPERS ON 1HA (2013)	297
TABLE 30:	POSSIBLE YIELDS AND PROFITS	298
TABLE 31:	EXAMPLE APLICATION RATES FOR N, P AND K (KG/HA)	303
TABLE 32:	ESTIMATED PRODUCTION COSTS FOR SPINACH (2014)	311
TABLE 33:	ESTIMATED YIELDS AND PROFITS FOR SPINACH	312
TABLE 34:	ESTIMATED PRODUCTION COST FOR BEETROOT (2012)	321
TABLE 35:	ESTIMATED YIELDS AND PROFITS FOR BEETROOT	322
TABLE 36:	ESTIMATED PRODUCTION COSTS AND PROFITS FOR CARROTS (2014)	334
TABLE 37:	ESTIMATED YIELDS AND PROFITS FOR CARROTS	335
TABLE 38:	A FERTILIZER APPLICATION SUGGESTION FOR A YIELD OF 40 TON P/HA	342
TABLE 39:	ESTIMATED PRODUCTION COSTS AND PROFITS FOR BEANS (2013)	348
TABLE 40:	ESTIMATED YIELDS AND PROFITS FOR BEANS	348
TABLE 41:	ESTIMATED PRODUCTION COSTS AND PROFITS FOR LETTUCE (2014)	359

TABLE 42:	ESTIMATE YIELDS AND PROFITS FOR LETTUCE	360
TABLE 43:	ESTIMATED PRODUCTION COSTS AND PROFITS FOR SWEET CORN AND GREEN MEALIES (2013)	369
TABLE 44:	YIELDS AND POSSIBLE INCOME FOR SWEET CORN AND GREEN MEALIES	370
TABLE 45:	PLANNING FOR GARLIC PRODUCTION	379
TABLE 46:	ESTIMATED GARLIC PRODUCTION COSTS PER HA (2006 VS 2014)	381
TABLE 47:	ESTIMATED YIELDS AND INCOME FOR GARLIC (2014)	382

Various Vegetable Production Systems

1 Various open-field vegetable crops produced on a small plot. Top-left: green mealies, front-left: beetroot and carrots, front-right: spinach.

2 An indeterminate long-shelf tomato cultivar produced in soil using the one-stem pruning method and trellised up to the 2.5m horizontal wire in a shade net house.

3 Tomato production in a 10 x 30m hydroponics tunnel using the Open Bag System (OBS). Plants are to be pruned to the one-stem method and trellised up to the 2.5 meter wire.

4 Hydroponic indeterminate tomato production in a shade net house. Plants are trellised up to the 2.5m horizontal supporting wire and pruned to the one-stem method.

5 Indeterminate long-shelf tomato production in soil trellised up to the 2m horizontal supporting wire and boxed with baler twine. Special wide spaced cultivars are established with this method.

6 Lettuce production in a Gravel Flow Technique NFT hydroponics system. The plastic hydro lines are 16m long, 1m wide and 4.5cm deep and are filled with a thin layer of gravel. This system was designed by the author and built by labourers of ARC Roodeplaat. **See Chapter 23.8.**

7 Various crops in a good condition on a small plot. From left to right: tomatoes, spinach, carrots and butternuts.

8 Various colours of new and standard pepper cultivars displayed at a farmers' day. Green, orange, red, yellow and purple cultivars are displayed above.

9 A small NFT hydroponics table unit displayed at the Agricultural Show, The Dome, Randburg in 2001. The unit is designed and built by the author. When it is dismantled it fits on a 1 ton pickup.

10 Hydroponics cucumber production inside a 10 x 30m tunnel. Crops are established in 10 litre plastic bags and filled with sawdust. A dripper system is inserted into each 10 litre bag.

11 Various pepper/chilli cultivars are boxed using stakes (poles) in 3 rows using cheap baler twine with 0.5m foot paths.

12 Practical training offered by the Author to assist producers to establish crops in small seedbeds. In front is a raised seedbed covered with mulch into which carrot seed was sowed. Next to the carrot seedbed is cabbage seedlings which have been transplanted to a well-prepared seedbed. Next to the single poles are tomato seedlings which are to be trellised by using the single-pole method.

Kejafa Knowledge Works has been involved with the agricultural industry for the past 15 years. Our hands-on publications has helped numerous farmers reap the rewards of successful farming over the years. Visit us online for more available publications and take farming to the next level.

Tel: 014 577 8006 | Email: accounts@kejafa.co.za | www.kejafa.com

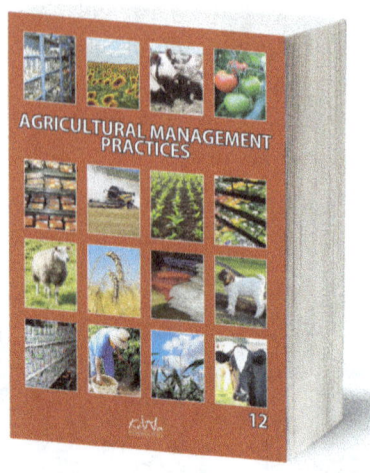
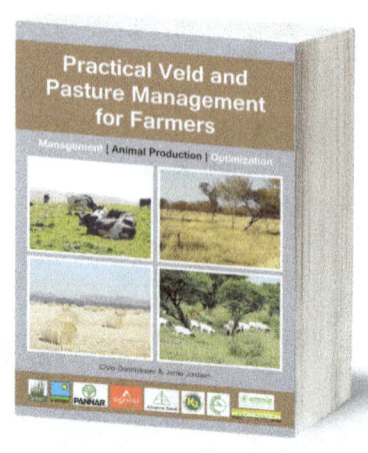
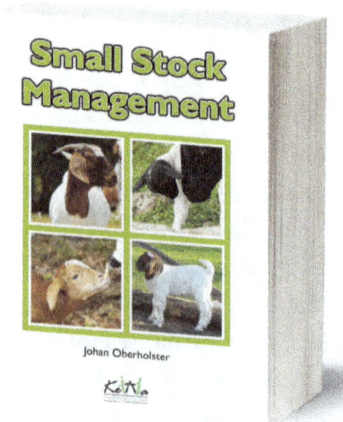

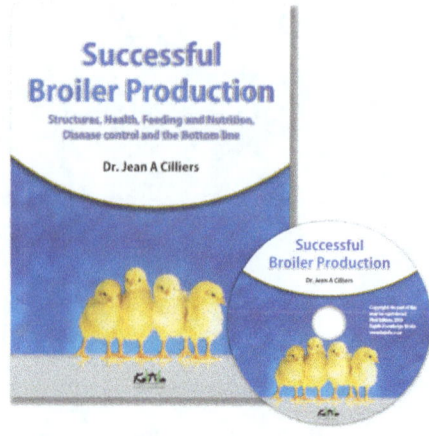
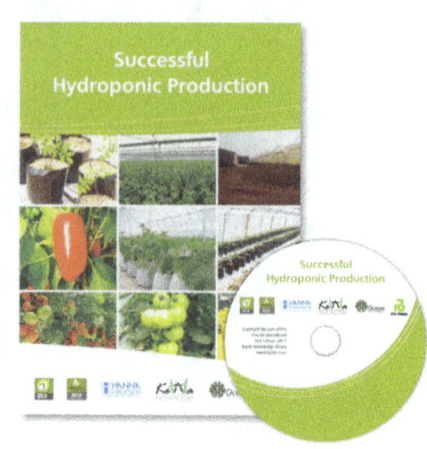

PART 1

A COMPLETE GUIDE TO VEGETABLE PRODUCTION

PART 1 CONSISTS OF CHAPTERS 1-24 AND INCLUDES THE FOLLOWING:

- Planning, record keeping, market research and labour requirements.
- Climate requirements, cool and warm seasonal crops, humidity, bolting and premature flowering of plants.
- Seed, F1 hybrid seed, nurseries and seedling production, establishing seed in seed trays, transplanting seedlings, temperatures for seedlings, hardening of seedlings and seedling diseases.
- Soil types, soil analysis tests and interpretation, soil preparation, soil acidity and pH.
- Farming equipment and implements.
- Organic matter, compost, animal manure and green manure.
- Organic farming.
- Training, trellising and ridging of plants.
- Spacing of plants and spacing of crops in greenhouses.
- Water, water sources, water quantities, rainfall and other measuring instruments.
- Irrigation, irrigation guidelines, irrigation systems and dripper systems.
- Fertilization, incorrect fertilizer, application of fertilizer, foliar sprays, microelements, nutritional requirements, organic and inorganic fertilizer, using the correct type of lime and using top or side dressings.
- Sowing seed and thinning of seedling plants.
- Weed control, the types of herbicides, their applications and actions.
- Crop rotation, soil improvement, leguminous crops that correct N and biofumigation.
- Scanning or scouting of insects and diseases.
- Vegetable production under protection: shade net houses (construction and cost thereof, the establishment of high value crops, soil preparation, estimated costs, yields and profits), green houses (irrigation, mulching, estimated costs, yields and profits, issues with tomato production), tunnels (production in soil, temperature control, estimated cost, yields and profits) and hydroponic systems.
- Descriptions of insects, diseases, viruses and physical disorders, GMO crops, guidelines for insect and disease control, resistance of diseases and insects.
- Harvesting, handling, grading, sorting, classing, packaging, cooling and deterioration of produce.
- Marketing of produce.

CHAPTER 1

VEGETABLE PRODUCTION IN OPEN FIELDS, SHADE NET HOUSES AND TUNNELS

1.1 GENERAL ASPECTS TO BE AWARE OF BEFORE ESTABLISHING CROPS

When producing vegetable crops it is necessary to keep in mind that production begins at the market (the markets need to be identified) and also ends at the markets (supplying produce to these markets).

When a producer has not yet identified a market for their crops, it is not advisable to embark on production in the hope of finding a market later. This is a sure recipe for financial failure and may put a producer permanently out of business.

Remember it is expensive starting from scratch and becoming a fully commercialized vegetable producer. Even producing crops on a small-scale and making a reasonable profit may be challenging.

A land area of 0.25 Ha may for instance yield an average of 8 000 cabbage heads that may weigh as much as 15 tonnes when they are harvested.

Many producers may find it unaffordable becoming fully mechanised. A tractor may be more expensive than a Mercedes motor car.

To learn about becoming a fully commercialized vegetable producer *see Chapter 1.2*. This list indicates the type of infrastructure, machinery and implements that may be necessary to become fully commercialized. It may not be possible to be successful starting with large production lands due to the prohibitively high costs.

Many beginners and small-scale producers throw in the towel before their production even reaches halfway due to the following factors:
- Weeds are allowed to seed and take over or when weed control becomes difficult, for example, in narrowly spaced rows or when small seed is sown by hand. When the common nutty weeds are present this could become a huge problem.
- Expensive irrigation, scarcity of water, high labour costs or when crop development is poor due to various reasons, resulting in producers losing confidence, giving up and losing lots of money.
- Some producers make use of shade-net houses and tunnels but do not realise the expense of production, or that certain crops are not economically viable to produce in tunnels. *See Chapter 23.4.*
- Many producers, especially beginners, do not understand what the cost of vegetable production is and the expected yields of crops. Because of this many producers fail when they begin producing crops on a scale too large or a land area that they cannot manage because of a lack of irrigation, insect/disease/weed control and labour costs and as a result put themselves into debt.
- The unavailability and high prices of equipment and implements.

When homeowners produce for instance tomatoes and proudly recall how they achieved excellent yields. However, generally they do not even harvest more than 1.2kg or 10 medium size tomato fruit per plant. Unfortunately, if a commercial producer had to achieve similar outputs per plant, they would soon be out of business.

Members of the public think that vegetable production a simple task and an easy way to make money, especially when they have a piece of land somewhere. However, many producers fail when they start producing vegetable crops.

Many farmers run their business inefficiently and money is wasted in many ways. For instance, if water is inefficiently used and unevenly applied, chemical spraying of crops is incorrect or unnecessarily expensive chemicals are used. Also, sorting, classification and packaging of produce is sometimes chaotic.

Producers should understand and know their crops well, before selecting the best crops and cultivars for their region. Most of all it is important to know what the producer and markets prefer, before decisions can be made to produce the best cultivars.

Vegetable production is a business and has become a highly scientific operation. The aim of producers should be to make maximum use of their land resources by keeping up to date with technology and other new developments.

It is a fact that vegetable production may be the most unstable and difficult of all farming industries. Some producers are not conversant with proper cultivation techniques.

It is important to keep up to date with F1 hybrid cultivars as some seed companies, that are also affiliated to overseas companies or organizations, invest considerable amounts of money and resources into research as well as farmers days that are open once a year to all producers and the public. Farmer's days are important for the commercial vegetable producers, because new improved and standard cultivars are displayed in demonstration plots that have been established by leading seed companies. In addition, new technologies about all aspects of vegetables are dealt with, especially newly bred cultivars in comparison with previous generations. This is because cultivar evaluation and selection is important for producers. Farmer's days provide an important opportunity to ask questions and initiate discussions with experts in their own field.

Some commercial producers invest significantly and even make use of helicopters to visit some of these farmers' days as they understand the importance of it.

Market trends should be taken into consideration to assist producers visualise and understand the supply and demand of vegetable crops. This allows producers to avoid the price crashes related to the oversupplying of the markets.

Knowledge of many aspects of production are required including general cultivation, planning, production, record keeping, transplanting seedlings, soil, water, irrigation, sowing seed, soil preparation, spacing of crops, plant population, training/trellising of plants, fertilizer and chemicals. Also important is controlling insects/diseases/weeds, crop rotation, organic matter, the correct stage to establish crops, harvesting, sorting, grading, packaging and marketing.

Producers should be able to determine the best time to produce and achieve the best yields and prices on the markets, so that producers meet the demands of the markets.

There is an increasing demand for healthy vegetable products by the public and consumers. Unfortunately, it is expensive to begin commercial vegetable production. *For a better understanding of the costs involved, see production costs and yields for each crop in the relevant chapters.*

It is often the case that many beginner producers do not plan correctly and do not know what the cost to produce their crops are. For instance, cabbage may cost more than R30 000 p/Ha.

To produce vegetables, especially small-scale farmers who cannot afford mechanical implements (mechanization), care should be taken to not over-produce on huge land areas which cannot be managed or cultivated properly.

To produce vegetables is hard work, time consuming and labour intensive. Crops such as spinach are favourable to produce, as it is an easy fast growing cash crop, but the market is limited and the crop has a limited shelf life.

Southern Africa has a harsh climate and can be classified as semi-arid to arid. There is limited good quality land and water for vegetable production available. Average rainfall is low (in the region of 450mm per year) compared to the rest of the world that has an average of 850mm per year.

In South Africa 60% of the country receives only 500mm rainfall per year and is characterized by periodical droughts, floods, winds and sometimes heavy hail storms.

Earth warming and the El Nino factor (2015) was the worst ever that influenced this country and still continues. It was predicted by Wiggs and Holmes (2011) that earth warming, stronger winds, extremely high temperatures and heavy thunderstorms were going to continue. As yet, there is no long-term solution even though many world organisations have come together to attempt to find solutions. The reality is that the climate situation is likely to remain challenging.

The Western Province, Eastern parts of the Free State and many other parts of South Africa, have been declared as drought disaster areas by the Government, similar to 1982 and 1985. However, certain vegetable crops that require small areas of land, may overcome this problem provided that ground or dam water is available. Water pollution has also become a serious problem. It is known that climate change is a fact and according to scientists the globe has already become warmer and if nothing is done, this could result in the planet earth becoming 2-3°C warmer as soon as 2019.

Soil, climate and water conditions should be monitored before any decisions are made. Labour, equipment and access to credit facilities must also be considered.

A cash flow plan should be designed long before planting time. In this manual cash flow plans have been designed for many crops. See the relevant chapters in this regard.

It is not always possible to obtain credit, but communities may organise themselves into co-operatives to overcome the problem of obtaining funds and securing loans.

It is well-known that many South African farmers are among the best producers of agricultural products.

Mining activities and mining water acids contaminate water supplies. Also, drought, wind, hail, storms, labour problems, theft, insects/diseases infections etc. may influence yields negatively and sometimes drastically. Therefore, to overcome these problems, vegetables can be produced under protection in greenhouses, shade net houses or in tunnels. Successful vegetable production requires fairly large amounts of water and as a result there are many areas in South Africa where vegetable crops cannot be grown successfully.

When producing crops on a small-scale (land area) or retailing vegetable products locally, this may not be profitable and may be financially fatal if not managed properly. Therefore, the producer should consider what crop, how big the area for a specific crop should be and most of all the consumer demand and retail prices at the markets. The producer should be aware of the production cost of individual crops and prices that can change daily at the markets, especially when some crops are over supplied.

Plants do not have an immune system like humans and therefore plants cannot build immunity as humans and animals do. Some diseases/insect infections cannot be healed (especially virus and bacterial infections) because once the plant is infected, it does not have the capacity to resist or remove it. In such cases it is best to remove the infected plant from the land and destroy it thereby preventing further infections from occurring.

Sometimes plants do not show any signs of diseases or infections. Sometimes there may be symptoms, other times not. It is important that producers should carefully inspect the production land on a regular basis for insects/diseases and weeds.

It should be remembered that soil improvement is necessary to achieve high yields. It is therefore important to continuously implement programs to improve the soil. It is best to make use of well-prepared compost, but this can be expensive especially for larger production lands.

A simple method of soil improvement is by means of adding good compost, animal manure, green manure (or crops) and also non-seeding weeds. These can be worked into the soil via crop rotation or other methods. The above mentioned methods are necessary for the improvement of soil in the long-term and is a must for all organic producers, because of the strict rules and regulations that organic farming requires.

Research over the world is mainly based on breeding programs to cultivate new and improved cultivars. These cultivars should enhance good quality, higher yields and better adaptation to deal with fluctuating climate conditions, especially droughts. There are partly tolerant cultivars for extreme environmental conditions which are also resistant or partially resistant against insects/diseases/viruses and so on.

The plant breeders' in South Africa and over the world share their knowledge with each other. Their focus is mostly on newly advanced methods as well as genetically modified DNA and GMO crops. This is to meet the increasing demand for food by improving crop production.

Note that it is undesirable to continue applying granular fertilizer year after year on the same soil. The average producer is not aware of this or is not willing to improve their soil due to the mindset that if one piece of land is spoiled there is always enough other land to replace it.

It should be taken into consideration that fertilizer is made out of salts and may spoil your soil when used repeatedly. Salts builds up in the soil and after a couple of years becomes contaminated (salt infected) and plant growth is therefore negatively affected.

When soil improvement is practiced correctly, the yield and quality should increase season after season. This is why organic farmers produce excellent quality and high yields, because they are obliged by the rules and regulations to improve the soil at all times, using organic soil improvement practices. *See Chapter 10 Organic Farming.*

For success, commercial farmers should plan their production properly to improve crop production year-after-year. They need to know their crops before establishing them, where and what the size of the market is, what the best cultivar will be to produce in their region, where and how production is going to be managed and so on. Consumers may for instance prefer a smaller tomato fruit rather than a larger one or a green pumpkin instead of a white one.

Most commercial vegetable growers rip certain soil, that are known to form a hard crust, quite deeply every 3 to 4 years. Breaking the hard pan or layer below the top soil level improves drainage, penetration of roots and aeration, all of which are important for vegetable crops.

It is important for small-scale and commercial farmers to plan production properly. The kind of vegetables that are going to be produced and how many production cycles per season are possible. For instance, beans, a summer orientated crop, can be produced in the Gauteng area every 3 months when planting is started between September and February. Similarly, spinach can also be established all year round in warmer areas.

A problem producing summer orientated crops, when temperatures become favourable after cold winters, is when many producers establish these crops all at the same time of the year. (Usually starting mid-September in the Gauteng areas.) The harvesting of produce such as tomatoes, peppers, cucurbits and beans subsequently all takes place at the end of November. This may result in overproduction at the markets and cause a rapid price decrease to below the production cost threshold.

It is better practice to start production with 1 or 2 crops, instead of producing too many that may confuse the labourers, especially when the next season starts.

1.2 REQUIREMENTS FOR COMMERCIALISATION

Not many vegetable producers are able to afford to become fully mechanized (commercialized), due to the expenses involved, especially tractors, equipment and labour.

The following list is an indication what kind of infrastructure, housing, communication, seed, irrigation, fertilizer, transport, machinery, implements and so on, may be necessary to become fully commercialized. Note that the list below only provides an overview of the items that could be used. Most items may not necessary for the beginner producers or small-scale farmers.

1.2.1 HOUSING
- A house for the farmer and his family.
- Very important is security to protect the producer, family and labourers.
- An office in the house or not too far from the house.
- Housing for permanent labourers.
- Toilets and a washing area.

1.2.2 LABOUR
- Affordable labour should be in place and available at peak times or on holidays if necessary.
- Usually 3 labourers p/Ha and 10 or more temporary labourers at peak times for some crops.

1.2.3 COMMUNICATION
- Telephone.
- A computer, fax, e-mail, Internet and very important, a smart phone/cellphone.

1.2.4 SOIL
- When a potential producer wants to hire or buy a farm, it is important to look out for good potential soil and availability of water sources.
- Good drainage ability and ideally fairly level soil.
- Good texture. Avoid too low or a too high pH soil content or too sandy, clayish and stony soil.

1.2.5 VEHICLES AND WORKSHOP
- Workshop and garage large enough for tractors, transport vehicles, implements, tools and machinery.
- At least 1 tractor to start with and an additional one to be acquired at a later date.
- A pickup, motor bike or bicycles (cheap transport).
- A lorry to dispatch produce to the markets.
- A flatbed trailer to transport crops from the land to the packing store and for general use.

1.2.6 IMPLEMENTS
- Seed and seedling planter.
- Small diesel hand driven tractor with plough, trailer and rotovator.
- Slasher to cut or slash weeds, grass and young bush in and around the land.
- Ripper.
- Plough.
- Disk.
- Rotovator or harrow.
- Grub.
- Bed, ridge maker.
- Under cut blade, or modified potato lifter for harvesting rooted crops.

- Back actor.
- Trailer.
- General tools and hand implements.

1.2.7 SEED AND SEEDLINGS
- Preferably certified and approved seed from reliable Seed Companies and the best, most adaptable cultivars.
- Seedlings from nurseries for some crops or vegetative plant material (sweet potato runners or garlic bulbs).

1.2.8 SEED STORAGE
Seed storage that is dark, cool and controlled.

1.2.9 WATER AND IRRIGATION
- Good quality and available water. Take note that to irrigate 1mm water p/Ha, 100,000 litres is necessary.
- Water or borehole pumps.
- Concrete, high flex, ground dams or plastic water tanks (to assist in capturing water from the roofs).
- The most economical irrigation system installed correctly should be a priority.
- Electrical or diesel water pump motors to supply irrigation water from the river or borehole to the holding tank or dam as well as motors that supply irrigation water from the dams or tanks to the production land.
- At least 6 rainmeters p/Ha.
- Tension meters, data loggers, slim cell phones or other measuring apparatus to measure the soil-water capacity.

1.2.10 FERTILIZER
- A fertilizer store room.
- Sort and order fertilizer long before planting time.
- Soil analysis and a recommended fertilizer expert to identify the correct kinds of fertilizer needed.
- Foliar sprays for specialized producers, buffers, stickers and activators.
- Good compost if available, to improve soil fertility.

1.2.11 INSECT, DISEASE AND WEED CONTROL
- Chemicals to control insects/diseases and herbicides to control weeds. Store in chemical room.
- Weed control implements.

1.2.12 EQUIPMENT FOR INSECT AND DISEASE CONTROL
- Tractor drawn chemical spray power pump.
- At least 2 hand lever spray pumps. One for insecticides and the other one for weed control.
- A mist blower sprayer is a good option for all farmers and especially handy for Green houses.
- Protective clothing. Overalls, hand gloves, boots, safety glasses, chemical protection masks and filters.
- A chemical room designed according to the safety regulations.

1.2.13 TRELLISING PLANTS
- Poles and a wire puller to tension wire, needed to trellis tomatoes/peppers in the production land.
- Galvanised non-stretch wire or baler twine that is used for certain trellising methods.

1.2.14 HARVESTING AND DRYING CROPS
- Pack store.
- Plastic crates.
- A trailer to carry loads from the production land to the pack store.
- A store to dry crops such as onions and garlic even if they are mostly dried outside in the land.
- Cold room and cold room trucks. Usually used by commercial farmers for certain crops.

1.2.15 VEGETABLES UNDER PROTECTION
- Shade net house or tunnels and plastic UV treated strips on ridges.

- Trellising material for high value crops such as tomatoes, peppers and cucumbers. Tunnels and shade net houses already have structures therefore only 2.5m horizontal wires in tunnels and extra carrying poles for shade net houses are necessary.
- Water tanks at least 3 x 5 000 litres cement or soil dams and high flex plastic dams.
- Soluble and granular fertilizer.
- Irrigation/fertilization. Electrical motor (diesel or electric) and water pump, from the river or dam.
- Other equipment: PVC pipes (main and sister lines), spaghetti tubes, arrow drippers, electrical boxes, timers and suitable electrical cables (especially to the main supply), fittings, timers, solenoid valves etc.

1.3 MONOCROPPING

Monocropping is when only one crop is produced.

1.3.1 ADVANTAGES OF MONOCROPPING
- Farmers become skilled when producing only one crop.
- Labourers become skilled, because they become well acquainted with the crop and its requirements when repeatedly growing the same crop year after year.
- Inexpensive mechanical equipment/implements are ideal when cultivating a single crop.
- It is easier to identify insects and diseases, as some insects are crop specific.

Cultivar choices for vegetables are always important. Certain cultivars can be established throughout the season (in certain areas) and other cultivars can be established early in the season allowing producers to reach the markets before the seasonal oversupply.

1.3.2 DISADVANTAGES OF MONOCROPPING
- Pests and diseases can build up, especially in the soil.
- If a disaster such as hail or drought occurs the producers may lose the entire crop.
- For a specific crop, many farm workers are needed during peak times. For example, when transplanting seedlings, harvesting, packaging, marketing and so on.
- More fertilizer is necessary because the plants use the same type of nutrients year after year.

Monocropping is sometimes not possible because market requirements change and over production may occur. Some producers make use of a high risk crop and then establish a low risk one, such as sweet potatoes or carrots, to spread the risk.

1.4 RISK FACTORS

Many risk factors are involved when producing vegetables, however the most important ones are mentioned below.
- South Africa has a variable climate and this is a disadvantage when producing vegetable crops,
- Without sufficient management inputs, producers make mistakes and which may result in them not being profitable. Examples of this include inefficient water and fertilization applications, inefficient weed/insect/disease control and so on.
- Vegetable production is risky and there are many factors that may influence vegetable yields and quality. Examples include hail, strong winds, floods, drought, using the incorrect fertilizer, incorrect soil choices, theft, inefficient insect, disease and weed control and so on.
- Labour is expensive and availability may fluctuate depending on public holidays, after-hours or overtime. Labour may not be available when they are needed at critical times. Examples include harvesting perishable crops at prime time or when some crops that have a short maturing or marketing time. Also, when crops need to be harvested regularly and at intervals.
- It seems that weather conditions also are subject to chance. Producers should be prepared and adapt to this.
- Producers should be prepared and adaptable to constantly changing weather conditions.

- When a market has not yet been established for a crop, it is not advisable to produce it in the hope of establishing a market. This may be a recipe for financial failure.
- The producer is largely responsible for the appearance of his product on the market.
- Sometimes large farming businesses have contracts with Supermarkets and other large distributors. They may then control the markets and create a monopoly, because their produce receives first opportunity and the ordinary producer then comes second and may receive poor prices.
- When inheriting a farm or joint partnership with the family, this may be helpful but not always.
- The recommended way for the ordinary man that has a piece of land and has ambitions to farm, should be to start small, learn and expand over time.
- It should be taken into consideration that vegetable production is labour intensive and hard work. It requires effective and responsible management, up to date record-keeping and good planning and organisation, especially when larger land areas are involved.
- Mechanical powered machinery is generally out of reach or unprofitable for most small-scale, upcoming and poor farmers. Therefore, production is limited to smaller areas within the capacity of these farmers.
- Yield, prices, correct packaging (classification and sorting) and the quality of crops are the main factors to consider when marketing crops to maximise the chances of making a good profit.
- Be aware that vegetable production is an expensive enterprise. For instance, to produce 1 Ha cabbage heads may cost approximately R36 000 p/Ha (2016). The average weight for a medium size head is about 2.5kg and when retailed for R1 each, the producer may break even but only on input costs. On the other hand, a good cabbage cultivar could retail for R10 and producers usually establish 30 000 medium size heads p/Ha.
- Certain cabbage producers establish smaller pieces of land, for instance every 3 weeks planting early, medium and late cabbage cultivars. This may avoid the oversupply and the risk of very low market prices, because the harvesting time is spread over a much longer period.
- Cabbage heads can also be delayed and left on the land for a few weeks without loss of quality so as to extend the harvesting and marketing time.
- Many producers establish crops and cultivars that are not suitable for their region. They should make sure to select the best cultivar and planting time for their region and conditions.

Always consult market agents, local farmers and seed companies. They will assist all producers. Be aware that there may be seed merchants that will promote their own products or cultivars. If a popular cultivar is out of stock or sold out, then they will likely promote another one or the second best cultivar. In these cases make sure to approach other merchants to obtain the best cultivar that is suitable for your region.

1.5 ESTIMATED PRODUCTION COSTS, YIELDS AND PROFITS

The first thing that most businesses, especially beginner producers, want to know is the costs and profits involved when producing vegetable crops. Assessing potential yields and related profits will assist with these calculations. Generally calculations have been based on producing crops on a 1Ha portion of land.

Estimated costs, yields and profits have been calculated for each individual vegetable crop in this manual.

1.6 VEGETABLES AND HEALTH

Vegetables play an important role all over the globe as they are included in the diet of many people and their intake together with fruit is increasing all the time. This is mainly due to ever-increasing populations and the number of health conscious people who prefer organic fruit and vegetables.

Millions of underfed children across South Africa and Africa suffer from vitamin A and other nutritional deficiencies (30% are between the ages of 1-10 years). Consequently their development is stunted and are often underweight.

In fact, less than half of the recommended vegetable intake is consumed by these children. According to the general household security survey published in April 2015, one in five children in this country goes to school hungry, 34% die from malnutrition, 5.9% of the population experiences severe food shortages and about 12 million people live in extreme poverty and go to bed hungry.

Vegetables are important for people of all ages. Vitamin A, B6, riboflavin, niacin, folic acid, iron, magnesium, potassium and zinc are all found in vegetable crops.

Health Organisations and specialized doctors recommend vegetables and fruit in a healthy diet program, especially if the vegetables are eaten raw. Many people become vegetarians for health reasons.

More than 3 million children in the developing countries have a serious vitamin A deficiency (2012) which greatly increase their risk of dying from infections such as flu, measles, tuberculosis, immunity against common diseases, diarrhoea etc.

According to Dr Smith from Grain SA, production should be promoted via self-gardening in communities to produce their own vegetable and other crops. However, there is limited knowledge on the subject, even worldwide.

In order to address this knowledge gap, The Agricultural Research Council (ARC) at Roodeplaat provides assistance in many forms such as excellent research programmes, practical production techniques and many services including training courses. They also assist and train communities in their own areas and even have their own field and seed gene banks for medicinal plants as well as training for traditional healers.

Many people still believe in 'muti', or traditional remedies including the author. In South Africa we have our own herb queen called Margaret Roberts who produces many herbal health products that are popular.

It is interesting to note that the Koisan community in southern Africa, were among the first on record to use traditional medicinal plants. However, there is concern that these traditions are busy being lost due to the newer, more westernised medical practices.

CHAPTER 2

PLANNING AND RECORD KEEPING

2.1 PRIORITISE PLANNING

A producer without a budget and cash flow plan is a producer that may fail to produce a crop by the end of the production period. Without a plan a financial crisis may happen to any producer at any time. This becomes evident when producers experience difficulty paying back loans, due to the lack of profits. Planning should begin at least 3 months before establishing any crops.

Even small-scale farmers need to manage vegetable production as a business. Therefore, careful planning and especially record keeping should be a high priority. Planning begins by identifying the correct markets and ends when the produce arrives at the market. Determine which markets are to be served. Examples include local, national, supermarkets, hotels, restaurants, communities, green grocers, street sellers as well as contracts and any other potential buyers.

Not having a plan is a recipe for failure and is a bad business option. Producers should aim to improve their vegetable production year after year and keep up to date with the latest technology. They should also continuously investigate other aspects such as resistant cultivars, how to obtain higher yields, regularly attend farmers days and learn about chemicals and insect/disease/weed control measures. Producers should also continue to learn about fertilization, irrigation, labour productivity, harvesting methods, packing material, transport, better cultivating practices (by comparing cultivars against each other). It is advisable to ask questions and find the best options for your particular vegetable production.

It is advisable for small-scale and commercial farmers to make use of a simple record system that is easy to refer to at any time.

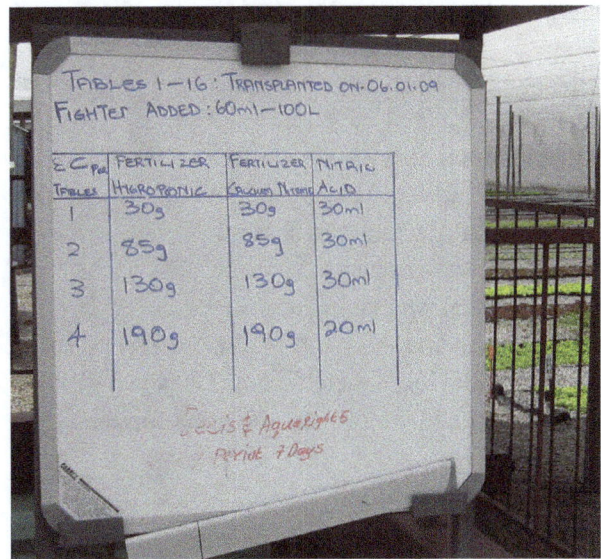

13 A simple way of performing day-to-day planning is to make use of a chalk/white board. It can be used to provide instructions to labourers or to get feedback from them, if for example, they wish to report any problems. The above board could also be used to indicate how much fertilizer has been placed into the tanks, thereby making it easier for managers to do quality checks at any time.

A well-known commercial producer once said to me that when he started vegetable production he often ran out of cash, but never gave up. He began planning and implementing more economical ways to become successful. He decided to start small and maximise his learning and today he is one of the largest tomato producers in the country.

It is advisable to continue seeking information, attending farmer's days and agricultural shows in order to expand your knowledge. More often than not, there are many other producers who want to extend their knowledge and via this network you may be surprised at how much information is freely available as all good farmers are always looking for new methods to do things better. For example, some will modify implements, make use of different chemical control measures, evaluate new cultivars, improve his/her soil and train labourers. It is well-known that farmers are an innovative group and produce "farmer patents" that could save thousands and assist with the reduction of labour costs. The internet is also a valuable source of information.

It is recommended to make use of a laptop or smart phone that is portable and which can be used to easily refer to your records anytime and anywhere.

2.2 ESSENTIAL ASPECTS FOR PROPER PLANNING

- Planning begins at the market, by finding a market for your produce and ends at the market, when you deliver your product to the market.
- Fully understand the requirements to become a fully commercialized producer. *See Chapter 1.2.*
- Determine the distance to market and how large the market is for your specific crops that you intend to produce.
- Determine what the best crops and cultivars are for your region as well as which cultivars the house wives and the markets prefer.
- Take note that the climate, water and soil should be suitable and the optimal planting time for each crop the producer intends to produce. Many crops could be established out of the optimal planting times, however care should be taken to select the best cultivar. The best F1 hybrid cultivars are adapted to this process.
- When producers determine what kinds of crops they want to produce, they plan production carefully long before planting time and determine the production costs, possible yields and profits for each type of crop. They then design a cash flow and if necessary a business plan.
- Implement the most optimum spacing for your crops and cultivars. See the chapter on spacing for crops as well as under cover production.
- Be careful not to over produce crops when establishing too large lands. Sometimes it may be financially fatal when over supplying crops as it may cause prices to drop rapidly.
- Consider competition from other producers in your area and do not underestimate them.
- Determine when your crop would more or less be ready to be harvested.
- Determine current and past market prices and carefully study the market trends when supplying crops to the National markets.
- If necessary make use of different planting times or divide the planting times by planting seedlings or sowing seed every 3-4 weeks to extend the harvesting period of some crops.
- Be aware of the quality standards that are approved by the national markets.
- Producers should always follow the market requirements carefully. Protocol inspectors that control batches of vegetable crops may degrade or reject your produce for one reason or another and this may be at great expense to the producer.
- Take into consideration transportation, the distance to the markets, the time to transport your crops to the markets as well as labour costs for the driver and assistant.
- Transporting produce is expensive and time consuming. Sometimes heavy loads are to be transported, for example 1 Ha cabbage heads may weigh approximately 60 tons, tomatoes 40 tons, onions 40 tons, sweet corn 25 tons, sweet potato 45 tons, beans 10 tons, garlic 6 tons, pumpkins 50 tons and so on.
- Plan for temporary labourers especially when peak times and holidays occur.
- If it is necessary to produce crops out of season then prices are usually much higher.
- Make use of the internet, a laptop and smart phone to investigate and talk with market agents, potential buyers, neighbours, local dealers, seed and chemical companies, 'The Vegetable and Ornamental Plant Institute' as well as to agricultural departments for advice, the best cultivar choices and other topics.

2.2.1 MORE ADVICE FOR BETTER PRODUCTION PLANNING
- How large do you plan on making your production area? Do this kind of planning carefully.
- What time of the year do you intend to establish your crops? **Refer to the optimal planting times in the relevant chapter.** Investigate and make sure to select the best cultivar for your region.
- Do not produce too large areas without mechanisation (tractors, implements, irrigation systems etc).
- Obtain the correct amount of seed or seedlings at least 8 or more weeks before sowing or transplanting your crops.
- When producing certain crops out of season, this may result in favourable prices at the markets. However, make sure to investigate all possible consequences before undertaking such an initiative.
- Are you going to produce your own seedlings, produce seedlings directly in the soil or obtain seedlings from nurseries?

- Identify when the best times are for sowing seeds or transplanting seedlings and what the price of seed and seedlings would be. This is important for your budget and cash flow plan.
- It is important to place your order for seeds and seedlings long before planting time. This is because some cultivars are popular and the seed companies may run out of stock due to high demand during peak planting times. In these cases the second best cultivar may be suggested.
- It is imperative to confirm with the nurseries the best cultivar and exact time when your seedlings would be ready for collection.
- It is important to make use of a soil analysis.
- Is there enough quality water to irrigate the intended production area?
- Is your irrigation system ready and suitable? (Test it). What type of irrigation system are you going to install? Make sure to select the best system possible. Consult S.A.B.I. (South Africa Irrigation Institute) for the best and most economical irrigation systems.
- Most vegetables crops use about 350mm (or 350 000 litre water p/Ha) throughout the entire growing season. If rainfall is low in your area then water could be costly. Take note that 1mm water per Ha equals 10 000 litres of water.
- Keep yourself up to date with the main insect pests and diseases and confirm with the chemical companies the most appropriate chemicals to use to control them. Most established producers more or less know what kind of insects and diseases are to be expected and also know how to control them.
- Are your tractors, implements and equipment in good working order?
- Correct crop spacing of certain crops is a high priority and should be adapted to the implements and tractors that are going to be used.
- Make sure that your production land is ready when receiving your seedlings and do not wait too long before transplanting them.
- Can you extend or reduce the planting time to avoid the seasonal over supply that sometimes occurs at the markets?
- Certain areas in SA are suitable to produce summer orientated crops throughout the year, because the temperature permits it.
- Do you want to produce consistently, supplying volumes throughout the season or only produce one product?
- Estimate how much the production cost would be and the related affordability for the full growing production period. Design a cash flow plan and a business plan when required. *See Part 2.* All the info for production costs and estimated profits are included in the relevant chapter.
- Would labour be available during peak times and holidays periods and what type of salary are you willing to pay. Make sure to meet at least the minimum wages.
- Is your packaging store in good order? Make sure to clean and sterilized.
- If you have a tractor/lorry driver, they may be able to train other labourers through their experience.
- Do you have transport from the land to the pack store as well as to the markets? Be aware when harvesting for instance an average of 60 tons of cabbage heads, then a 5 ton truck may travel 12 times to the markets. Fuel is expensive. Also, try to avoid the National market when the distance is very far away. However, note that ZZ2 for example, markets their quality tomatoes and onions all over the country.

2.2.2 TALK WITH MARKET AGENTS WELL AHEAD OF TIME
See Chapter 25 Management and Marketing.
- Seek the best and most suitable cultivars for your region. This is what the Markets and consumers prefer.
- If possible gather info and do research at the time when your crops are going to be delivered to the markets. Estimate how much of your crops volumes are going to be delivered and at what specific time.
- Producers should perform production planning, such as which crops to establish and what the best cultivars for your region are.
- Establish the ideal time to establish your crops and in what sequences.
- All these considerations should be linked to the financial requirements to ensure income, cash flow and hopefully a good profit.
- Near harvesting time, the producer should move from production to marketing. Everything should be aimed to produce quality vegetables that are good looking, correctly sorted and graded, clean, nicely packaged in

the correct prescribed containers and most of all, have a striking logo and good branding as prescribed by a class 1 product.
- Begin doing business with market agents a week or two before the first volumes of your crops are ready for delivery and advise the agents that you will soon be sending your produce to market. The agents should then keep you regularly updated.
- Make use of the internet or smart phone where necessary. It is important to understand the Market requirements which mostly operates on pricing, demand and supply.
- Aim to send only the best quality produce (class 1 or grade A) to the markets and make use of a strikingly visible logo that is easy to remember by the consumers.
- The key of producing and supplying to the Market is to be early and consistent.
- Buyers will remember your produce if you consistently supply good quality vegetable products. They would buy your produce even without seeing or opening the containers once they have confidence in your brand.

2.3 LABOUR REQUIREMENTS

It is not always possible to forecast how much labour is necessary to produce vegetable crops on large areas. Vegetable production is the most labour intensive enterprise when compared with other farming industries. It is well-known that the production of vegetable crops may differ considerably from each other.

One cannot compare tomato production with other crops such as for instance cole crops, carrots, onions, spinach, beans, carrots or beetroot. This is because the production requirements are totally different for each crop and the amount of labour also differs for each crop, especially when producing more than one crop.

Commercial farmers tend to practice mechanization due to the expensive labour costs and prohibitive labour laws. When machinery and implements are not in place, much more labour is necessary. Permanent labourers are paid more and also have much more legal benefits than temporary labourers. Permanent labourers should for instance be tractor and truck drivers who should assist in the training of fellow labourers such as how to calibrate implements and spraying equipment as well as the preparation and installation of irrigation systems. Permanent labourers should also preferably be skilled and have knowledge of all farming activities. They can also be in charge when the manager is not available for long periods and should also be able to train and assist labourers, especially temporary labourers.

Most vegetable producers agree that generally 2 to 3 permanent labourers are necessary to produce 1-3Ha for most vegetable crops. During the shorter peak periods, for instance when controlling weeds, establishing and harvesting crops, then about 10 temporary labourers may be necessary.

Be aware that the labour law or act is not negotiable and the regulations should be studied and followed carefully, (such as registration and a service contract, minimum wage, working hours, leave and sick leave and overtime payment). They should also be allowed to be members of the Unions etc.

2.3.1 FARM SAFETY

Be aware that farm safety is the responsibility of the farmer. Safety measures are not a warranty against the farmer and his family becoming victims of violent crime. However, there should be a plan in place in emergency situations. Safety measures will not be an absolute safeguard against the farmer and his family experiencing incidents of crime, but it may restrict criminals, actions which could allow the farmer to counteract them.

There are many security measures to be familiar with and this should be rehearsed by the farmer, his farm workers and their families. They should be familiar with the danger of intruders and criminals. It is important to know your neighbours and have immediate communication with them in case of emergencies.

Be aware that organised agriculture as well as the local police station have an adequate safety plan in their area. Contact them for more information. This should be taken seriously.

CHAPTER 3

CLIMATE FOR VEGETABLE CROPS

3.1 CLIMATIC EFFECTS AND ZONES

Cool and warm seasonal crops refer to the kinds of temperatures that crops may be able to withstand as well as ideal climate conditions under which they would be able to thrive to produce high and quality yields.

Some vegetable crops are hardy under high and low temperatures, some are semi-hardy and others are very sensitive. Hardy crops can tolerate ordinary frost without being damaged but sensitive crops would be damaged or killed by cold and freezing conditions.

The severity of frost damage is shown in the differences between hardy and sensitive crops. On the other hand, certain hardy crops do not grow well in hot dry conditions.

It may be advisable to establish certain summer orientated crops early in spring or later in the summer in order to improve your chances of experiencing the ideal temperature conditions.

Seedlings are mostly used in the summer orientated areas as early as the beginning of September. Seedlings are produced in hot houses to be ready for the early establishment of seedlings in the summer orientated areas.

Some crops will withstand frost and also do well during hot weather conditions.

3.1.1 CLIMATE ZONES
The weather or climate plays an important role when producing agricultural crops, as the better the climate conditions are for a specific crop, the better the results.

The South African Weather Bureau divides the country into 12 different climatic zones regarding rainfall, temperatures and incidence of frost etc. Info is available from the Bureau.

3.2 COOL AND WARM SEASONAL CROPS

Cool seasonal crops are those where the edible part is a root, stem, leaf or an immature flowering crop such as broccoli, cauliflower as well as seed stalk flowering crops such as carrots, beetroot and onion. Some garlic plants go over to flowering but produce vegetative seed that cannot reproduce.

Warm seasonal crops are those that prefer warm days and nights where the edible part is mostly a fruit such as tomatoes, cucurbits pumpkins, squash, water/muck melons, peppers, sweet corn, beans and especially sweet potatoes. There are 2 exceptions: peas and broad beans which are cool seasonal crops.

Each crop has its own ideal temperature conditions. Some cool seasonal crops are relatively small in size and shallow rooted but a few are moderately deep rooted.

Shallow rooted crops need much more **careful and frequent irrigations than deep rooted crops.**

When fertilizer is applied in cool soil, most of the plants normally respond slowly, as they have a slow transpiration uptake of water and elements.

Breeding programs for most vegetable crops exist mainly to improve and create new cultivars that include features such as good quality, higher yields and better adaptation to climate conditions. This is especially true for drought fluctuations, exposure of tolerant or part tolerant cultivars to extreme environmental cold and warm conditions.

3.3 TEMPERATURE AND TEMPERATURE RANGES

Many factors that affect plant growth are important, but none is more important than temperature.

Temperature affects plant development in some of the following ways: transpiration, nutrient uptake, quality, yield, root development, pollination, fruit set, harvesting time, length of the harvesting period, shelf life, shape as well as inner or outer colour of fruit, bulbs, globes, roots and so on.

Factors which relate to climatic requirements are soil, air temperatures and humidity.

Some vegetable crops may be damaged by early frost or killed by moderate frost conditions or when very high temperatures occur.

Temperature, light, wind and rainfall are all factors that may have an influence and should be given special attention when producing certain crops.

Vegetable crops should be established when the soil and air temperatures are ideal for the specific crop/s and plant growth. However, some producers find ways to establish certain crops out of season such as crops and cultivars that are more tolerant to temperature and tolerant or resistant to diseases, insects and viruses.

3.3.1 SOIL TEMPERATURE
Soil temperature has a huge influence on germination of seed and also for root growth.

The optimum lower soil temperatures for summer orientated crops are 16-18°C. At this temperature, the seed of most crops emerge in 7-10 days. Seed of summer orientated crops germinate poorly below 15°C and requires 14-20 days to emerge. None or very little germination takes place below 10°C.

The optimum soil temperature for summer orientated plant root growth is 16-20°C while temperatures below 14°C may have a negative effect on root growth.

3.3.2 AIR TEMPERATURE
Vegetative growth, flowering and fruit-set is mainly effected by air temperatures. Summer orientated crops are sensitive to frost and may be badly damaged at temperatures below 0°C. Also, no growth would take place below 10°C while for temperatures above 18-28°C plants would grow rapidly. Plants grow more vigorously when higher temperatures and humidity occurs.

Plant breeders' all over the world and also certain local breeders' have breeding programs for crops that are more tolerant to warm or cold conditions.

TABLE 1: IDEAL GROWING TEMPERATURES FOR VEGETABLE CROPS

CROP	15°C	20°C	25°C	30°C
Tomato and pepper		XXX	XXX	
Sweet melon		XXX	XXX	XX
Pumpkin, squash, sweet corn and green mealie		XXX	XXX	X
Eggplant, beans and cucumber		XXX	XXX	
Carrot, cole crops, garlic, beetroot, spinach, lettuce, watermelon, potato and onion	XXX	XXX	X	
Sweet potato			XXX	XXX

3.4 HUMIDITY

Summer orientated crops that are produced in areas that have a high humidity are more prone to be infected with fungus diseases, especially when moist conditions occur at the end of the growing period.

Producers should make use of an effective disease spraying control program when producing vegetable crops in humid areas or conditions near the coast.

Long rainy periods may cause the fruit of pumpkins, watermelons and musk melons to rot. This is because the bottom of the fruit that are in contact with the moist soil (especially in clayish types of soil), may rot. To avoid this, it is better to plan production in such a way that the maturity of the fruit should not take place during periods when cool, cloudy or moist weather is prevalent.

14 Different determinate tomato cultivars exhibited at a farmers day using a drip system and mulch in black plastic. Note the low growing cultivars on the right. At farmer's days the seed companies often compare standard cultivars against the newly bred or improved ones. Producers evaluate the new improved cultivars and then switch systematically to the new ones. Most large farming enterprises have a special evaluation program in place.

15 Pepper cultivars are neatly boxed on a small plot. This is to prevent them from falling over especially in windy areas. When the plants grow vigorously, then less poles will be required as shown in the above photo. Even crooked poles or iron rods may be used. Also, remember to save materials for the next production cycle or for other trellising projects to save money.

Bees do not work in moist and rainy conditions, especially when these conditions persist for long periods. Poor pollination, blossom drop and fruit-set could be the result.

3.5 LIGHT AND DAY LENGTH

Light quality and the duration of light influences growth and production. Too little sunlight results in reduced transpiration, water uptake and plant growth and may cause plants to grow long and thin.

Light intensity effects transpiration and reduces production also when clouds, dust and smoke occur.

3.5.1 LIGHT REQUIREMENTS FOR CROPS
- **Crops with high light requirements:**
  Maize, pumpkin/squash, cucumber, tomato, pepper, chillies, beans, maize, sweet potato and brinjal (egg fruit).
- **Crops with medium light requirements:**
  Onion, garlic, carrot, beetroot, lettuce, cabbage and spinach.
- **Crops with low light requirements:**
  Ginger and mushrooms.

The length or hours of sunlight per day have a huge influence on crops such as onions and garlic. Onions are classified into short and long day cultivars.

3.6 FROST AND FROST HARDY VEGETABLES

Cold damage is a difficult problem to manage. Even light frost may cause considerable damage when winds occur (wind-chill).

When light frost is a possibility, then plants could be irrigated early in the morning. This causes a slight rise in temperature and may prevent injury to some plants.

When heavy frost occurs then irrigation water may freeze on the surface of plants, causing plants to be frozen for a much longer time.

16 Spinach crops where cold conditions spoiled the leaf quality, however most of the leaves including the younger ones will recover when warmer conditions occur.

17 A new cauliflower F1 hybrid cultivar was tested and evaluated at Roodeplaat, Pretoria (2009). This producer achieved exceptionally good summer production. The cultivar has been bred for this purpose and good quality and high yields were achieved.

Drought related stress is more likely when cold conditions occur. It is important to avoid water on plants, especially when the soil is already wet. Winds may evaporate the moisture in the soil and cause a temperature drop. Frost covers are an option. Crops such as lettuce and spinach are moderately frost tolerant and covers help protect the crop against sudden intense temperature drops. Vegetable plant covers also provide protection to susceptible crops where light frost occurs. Plastic tunnels only provide 2-3°C differences in temperatures, but covers inside tunnels provide considerably more protection.

3.7 WIND DAMAGE

Physical damage of certain crops may sometimes be considerable. When heavy winds occur for example, it may affect seedlings, soft and high growing plants, flowers, pollination of certain crops and even crops that are spaced too far from each other, such as beans and peas. Crops may even collapse or break clean off at the soil level.

Producers should be aware that winds may also effect overhead irrigation tremendously. Even when slight winds occur, water loss can take place due to the evaporation of the water molecules. Evaporation may be as high as 40% when moderate winds occur.

Shelter may be provided by a wooden fence or establishing the well-known perennial Napier grass around the production land. However, high growing plants that are trellised may still collapse when heavy storms occur.

3.8 RAINFALL AND WATER REQUIREMENTS FOR IRRIGATION

Rainfall is limited in most parts of South Africa. Dry land production for vegetables is not possible due to the critical irrigation times most crops require. Certain areas in South Africa may have nearly enough rain water to support their crops throughout the entire growing season, but this is a gamble especially in these changing, drought plagued weather conditions that occurs in most parts of this country.

Plant growth should not be stressed due to a lack of water. Any stress should be avoided at all times.

The soil water content should always be satisfactory, because water is one of the main factors for high yields and quality production.

Many vegetable producers over irrigate their crops and because of this fertilizer could be washed downwards out of the reach of the roots. Also as a result, oxygen would be replaced by the water in the soil root zone and certain fruited crops could burst and rot. In these cases Blossom End Rot (BER) could also be a problem.

Be aware that vegetable crops need plenty of water. For instance, when irrigating 1mm water p/Ha then 10 000 litres of water is necessary. Most crops need about 350-400mm water p/Ha during their entire growing cycle. That is 3.5 to 4 million litres of rain and irrigation water.

Take into consideration that if rainfall is low in your area, then much more irrigation water is needed. Even when water is to be pumped from a dam or borehole it is still expensive.

3.9 CULTIVAR CHOICES IN PRODUCTION AREAS

It should be a high priority to select the best cultivar of crops that are suitable for your area. There are many cultivars available of most types of crops. When a producer learns that a cultivar does well in his area then they should not switch to another one. It is however wise to evaluate other cultivars as well. Plant one or two rows together in the production land and when it is found that in evaluating a new cultivar it is in some way better, then change systematically to the better one.

A producer from Gauteng consulted me and wanted to know why her onion production that covered 1Ha did not produce any developed onion bulbs. After a long discussion I determined that she established a long-day length onion cultivar in a short-day area and the result was zero bulb development. This was a serious mistake and because she selected the incorrect cultivar for her area, a substantial amount of money was lost.

Temperatures, insects and diseases differ all over the country. Make sure that you are at least aware of the main insects and diseases that may attack the crops that you intend to establish. It is sometimes difficult, especially for the beginner producers, to control especially diseases.

Cultivars of certain crops and many F1 hybrids are partly adapted or fully resistant to high or low temperatures. Therefore, plant breeders' are continuously busy breeding and improving crops and cultivars all over the world. This results in better quality and higher yields, better uniformity, better inner and outer coloured fruit, resistance or tolerance to insects/diseases, early or later harvesting, short or high plant growth, adaptation of crops to short or long day lengths, thicker fruit walls and long shelf life, among other things.

18 The spinach plant (left) has gone over to bolting. The production was established in February and the bolted lettuce has gone over to flowering only 9 months later.

19 Leaf lettuce which grow quickly and easily go-over to bolting (flowering), especially in warm summer conditions. Front-left is a butter lettuce and in the back there are leaf lettuces. They were all established at the same time at a farmers' day near Pretoria, South Africa in December 2010. This photo was taken 3 months later in March 2011.

3.10 BOLTING OR PREMATURE FLOWERING

Bolting or premature flowering is the elongation of the seed stem of some crops, because of the reproductive (flowering) process. The plant tries to reproduce fast and this produces early seed.

There may be several reasons for bolting:
- Wrong or poorly adapted cultivar.
- Establishing the crop out of season.
- Too high or too low temperatures especially when combined with water stress.
- Water shortages or water stress especially later in the plant development stage.
- Soil should be kept moist until harvesting, especially crops such as broccoli to keep the curds crispy.

Some crops produce seed stalks that emerge from within the crop and grow through the heads such as cabbage and lettuce.

Spinach and beetroot produce seed stalks and start to flower bolt. *See photo 18.*

The leaf lettuce types bolt very easily and quickly, especially in warm conditions. *See photo 19.*

Some chinese cabbage cultivars start to flower and produce short stems that are used in crispy salads. Cauliflower and broccoli curds begin to rise and become loose and later becomes brownish especially when the producer waits too long before harvesting. Following this, the curds go over to flowering, which happens faster when warmer conditions occur.

Onion bulbs produce a thick hollow stalk and when established for seed production, the bulb breaks up and produces 2 to 6 hollow seed stalks. *See photo 40.* Carrots produce seed stalks that are covered with many flowers. *See photo 20.*

The F1 hybrid crops are more resistant to bolting and climate conditions. In areas where crops are established out of their normal temperature ranges or out of season, the producers should be aware that early bolting of certain crops and cultivars may occur.

20 A summer production carrot trial. The labourers count the number of bolting or premature plants in the different cultivar plots. It is mainly the F1 Hybrid cultivars which are more resistant to bolting in comparison with the open pollinated cultivars. In front next to the lady is a very early bolting plant that is already in full flower.

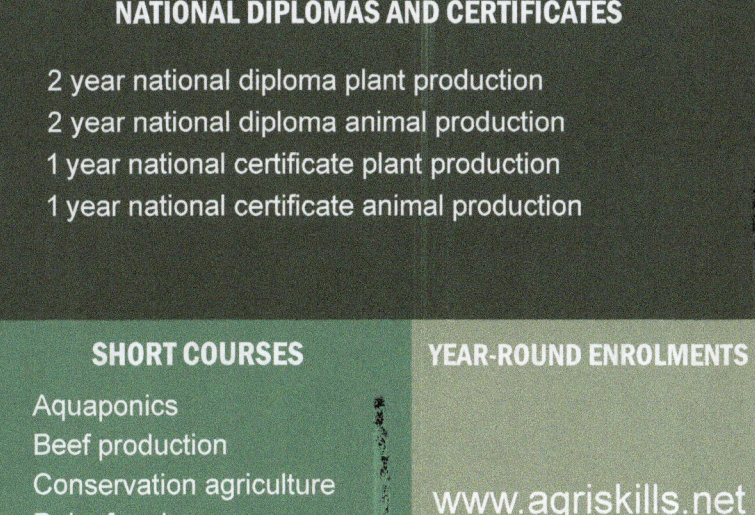

NATIONAL DIPLOMAS AND CERTIFICATES

- 2 year national diploma plant production
- 2 year national diploma animal production
- 1 year national certificate plant production
- 1 year national certificate animal production

SHORT COURSES

Aquaponics
Beef production
Conservation agriculture
Dairy farming
Empowerment projects
Farm management
Grazing for profit
Hydroponics
Organic vegetable production
Poultry farming
Small stock production
Tilapia farming
Tunnel farming
Vegetable production

YEAR-ROUND ENROLMENTS

www.agriskills.net
info@agriskills.net
0860 10 36 35

learn. grow. prosper

DISTANCE LEARNING

AGRISKILLS TRANSFER OFFERS A UNIQUE DISTANCE LEARNING EXPERIENCE. OUTCOMES-BASED, EXPERIENTIAL LEARNING FOR ALL WALKS OF LIFE.

GET A QUALIFICATION IN AGRICULTURE WHILE YOU FARM

SETA ACCREDITED

Sakata
Quality Vegetable Seed

Optima

Pluto

Robin*

Falcon

*Experimental: This variety does not appear on the current South African Variety list, but has been submitted for registration.

SAKATA®
PASSION in Seed

Tel: 011 548 2800
www.sakata.co.za
e-mail: info.saf@sakata.eu

QUALITY — **RELIABILITY** — **SERVICE**

CHAPTER 4

SEED, SEEDLINGS AND NURSERIES

4.1 THE TREATMENT OF VEGETABLE SEED

Seed, especially certified seed and certain F1 hybrid cultivars, are expensive and should be handled with care.

Never expose seed to full sunlight or leave it in a warm place such as inside a motor vehicle, as high temperatures may damage the germination of the seed. This will not occur in cool or cold conditions.

4.1.1 SEED TREATMENT

Make sure that seed is treated with a fungicide chemical such as Thiram or Captab. This is because soil fungi such as phytium, fusarium and phytophthora may affect the germination of seedlings or affect them when they are in the process of emerging.

Many pathogens may be introduced to healthy production areas through infected seed. Sometimes diseased organisms (pathogens) infect the seed which then infects the seedlings when they emerge.

To control diseases such as the mosaic virus that may reside inside the seed itself, a method that is called Hot Water treatment may be used. This is a simple process, but should be done very carefully. This method is used when especially tomato seeds are submerged in warm water. It is very important that the temperature should be kept at a constant 50°C for 30 minutes to eliminate the virus inside the seeds. Cauliflower seeds should for example, only be submerged for 8 minutes.

Take note that only fresh seeds should be used in the above treatment, because seeds older than 2 years may be at risk and their germination could be affected negatively.

21 An instructor from a seed company explains the importance of good quality seed and what the best cultivar would be for production in certain areas.

22 Seedlings may be pulled out of seed trays and packed in plastic bags so as to make it possible to transport a reasonable amount of seedlings in a small pickup fitted with a canopy. These young seedlings should be transplanted as soon as possible in well-prepared production land.

23 A small hand operated vegetable seed planter that can be pushed. Only one labourer is required to directly plant the seeds in a well-prepared smooth production land.

4.2 USE CERTIFIED TESTED SEED

It is always a good decision to use certified seed if it is available, due to some of the following factors:
- Producers are required to be registered to produce certified seed. These producers are committed to several land inspections by trained inspectors who will investigate and determine if the crop complies with the minimum certification requirements. The purity and germination are tested in laboratories. Seed samples of the harvested seed are also sent to be tested by Plant and Quality Control in Roodeplaat to determine if the cultivar is true-to-type, stable and homogeneous.
- Land inspections are carried out by the Division of Plant and Quality Control together with the South Africa National Seed Organization (SANSOR). Generally 2 land inspections take place. The first one when the plants are flowering and the second one just before harvesting.
- Certified seed must comply with the minimum requirements. The crop and cultivar seed are normally graded. It is well-known that sometimes the larger graded seeds of some crops germinate better and stronger seedling plants.
- If seedborne diseases are found at the time of land inspections, the land is rejected by the seed inspectors. Therefore, this seed cannot be used for the purpose of seed certification until corrective action has been taken.
- Certified seed is treated with chemicals that destroy various fungal diseases on the seed coat.

Always purchase seed with a good record, especially seed that has a good germination count and the best genetic cultivars that are true to type. This is a good production practice which will bring good returns in the longer run.

Be aware that certified seeds are issued with a certificate that is characterised with a blue label and seal. This provides evidence that the seeds have complied with the South African certified scheme regulations.

4.2.1 PALETTED SEED
Small seeds, such as carrots and lettuce, sometimes tend to stick together and may germinate in two's. One way to overcome this problem is to make use of certified graded seed or paletted coated seed.

24 This small seedling production shade net house is mainly used for leafy crops. However, when seedlings are required earlier in the season, they are produced in a heat controlled tunnel. They should then be transplanted when the outside temperature is warm enough. This is for the summer orientated crops.

25 An ideal seedling plug. The roots of the seedling should be white and not root bound. The seedlings should be pulled out of the seedling trays smoothly. If the trays are bumped to loosen the seedling plugs from their cavities, then be aware that if the trays are turned upside down, the seedlings may fall out of the trays.

26 These leaf lettuce plants are only 3 weeks old and very healthy. Large, healthy, strong seedlings were transplanted in a NFT system. Take note to not underestimate the roots of plants as they may penetrate deeply into the soil, especially if the soil is properly prepared.

27 Seed in packets, tins and other containers. It is always advisable to use certified tested seed.

28 Top-left is a bag of soluble multifeed that includes 13 nutrient elements. This is ideal for seedling production. The sizes of the seed tray cavities are, from left to right, 128 to 200 and on the right 300, which are the standard seed tray cavity sizes.

29 Seed trays established with cucumber seed. A little indentation can be made in the soil of each tray cavity with a finger or by using a roller. Afterwards the indentations are usually covered with vermiculite.

Many kinds of seed, especially the smaller varieties are covered or coated with a substance that makes the seed round and colourful like a small ball-bearing. A variety of bright colours are used to help differentiate different crop seeds from one another.

Only the best quality seeds, which are coated, are used. The seed coating contains a fungicide and some nutrient substance that dissolves and is available when the seed starts to germinate.

Seeds are also conditioned. This is to allow them to germinate rapidly even under high temperature conditions. It is because of this that paletted seed can generally be directly established in well-prepared land.

The main advantage of paletted seed is that it is ideal when used with hand pushed or mechanical planters. This will result in an excellent stand and plant population.

Paletted seed is expensive but in the long run it is worthwhile as it results in accurate spacing, good germination and emergence of seeds.

4.3 PRODUCING YOUR OWN SEED

South Africa has a long history of traditional small holders. Some farmers save a portion of seed from their harvested crops to be used for the following planting season. This has been done for many decades. Some producer's, especially home owners and small-scale farmers, produce their own seed which is a good and cheap option.

Traditional leaders often genetically preserve many veld plants and herbs that would otherwise be lost. It is recommended to not reproduce F1 hybrid cultivars. This is because originally, two or more parents were cross-bred with each other to make up a Hybrid cultivar. When reproducing a F1 hybrid cultivar seed, often the result may be that the plants, bulbs, roots or fruit are uneven, poor yielding and may for instance lose their resistance against diseases and insects as well as their weather adaptability. However, this is not always the case.

Only the open pollinated crops and cultivars should be reproduced. Open pollinated cultivars are those that are open for anyone who wants to reproduce seed. This is because there are no breeders' rights involved. These open cultivars were usually selected to be true-to-type over a time span of many years by some private farmers.

Leading seed companies still maintain, reproduce and supply the seed of well-known cultivars. For example, for Tomatoes there is Rodade and Floradade, for beans there is Wintergreen and Rolito, for runner beans there is Lazy Housewife and Witsa, for carrots there is Cape Market and Chantenay Karroo, for beetroot there is Detriod Dark Red and Early wonder, for onions there is Bon Accord and Texas Grano, for pumpkins and squash there is Flat White Boer, Green Hubbard, Long White Bush and Gems Rolet and so on.

There are two pollination actions for crops, self pollination and cross pollination.

Self pollination means that the plant pollinates itself. When the flower opens, it is already being pollinated by itself. Cross pollination happens when the flowers open and needs to be pollinated from other plants or by itself by means of insects. This occurs mainly with the assistance of bees and also in some cases by the wind (such as for Maize). Self-pollinated crops include tomatoes, capsicums (peppers and chillies), beans and peas. Cross pollinated crops include pumpkins, squash, water and muck melons, carrots, beetroot, lettuce, onions, maize, cabbage, cauliflower broccoli and brussels sprouts.

4.3.1 PRODUCING TOMATO SEED
- Pick the tomato fruit when fully ripe and squash the pulp from the fruit into a plastic drum or other container. Leave the pulp to stand for 2 to 3 days at room temperature (do not add any water).
- The pulp will ferment in the container, however be aware that the seed may begin to germinate as soon as the second day under warm conditions. This should be avoided.
- Stir the pulp twice a day to release the seeds that are still captured in the pulp.
- After the second or third day, stir the pulp thoroughly and fill the container with water. Allow the pulp to settle. The seed will sink to the bottom of the container and the pulp will float to the top of the container.
- Continue to stir the pulp thoroughly before tilting the water-filled container, as some seeds could still be captured in the pulp.
- Wait approximately 1 to 2 minutes to allow the seed to sink to the bottom of the container.
- Tilt the container sideways to allow the pulp and water to pour out slowly. Do this repeatedly until only clean seed is left at the bottom of the container.
- Pour the clean seed and a little water out from the container onto some gauze. Let the water drain through the gauze. Squeeze or press the water out of the seed with your hands. Then spread a thin layer of seed evenly on the gauze or on a plastic sheet.
- Dry the seed as quickly as possible using a fan or under the sun, but not for long periods. Stir the seed regularly on the gauze until it is nearly fully dried.
- When the seed is fully dried they stick together and should be rubbed to separate the seeds from one another.
- Dry the seed at least a week or two before storing it to allow the seed to become fully dried.

This method could also be used for small quantities by squeezing the pulp of 1 or 2 tomato fruit in a glass or any other small container and then simply drying the pulp on cloth or paper. Finally, when the seed is dry, it may be rubbed from the pulp.

4.3.2 PRODUCING CUCURBIT SEED
The process for cucurbits is more or less the same as for tomatoes, but the seed is scratched out of the fruit and placed in a container that is **half filled with clean water**. Take note that the fermentation process takes about 2 to 3 days before the seeds may be separated from the pulp using water. If the fermentation process goes on for too long then the seeds will turn a brownish colour. The pulp in the containers should be stirred every day to release some seed from the pulp. For small-scale production, the seed could also be scratched out of the fruit and dried in a store until completely dry. The seeds are then rubbed to release the chaff from the seed. Finally use a fan to release and blow the chaff away from the seeds.

4.3.3 PRODUCING CARROT, BEETROOT, CABBAGE AND ONION SEED
Harvest the seed from the land by stripping it when the plants are fully dry. Take care when the seeds becomes too dry as they may fall to the ground especially when windy. Spread the seed and some stalks in a store on

a clean floor or on a canvas and then proceed with drying them completely. The seed should now be released from the stalks by rubbing them. One method is to fill a bag with some of the harvested seeds and stalks and then to jump on the bags repeatedly. Alternatively, to hit the bags or harvested material with iron rods or wood stakes to release the seed. Next, pour the seed in front of a fan. The wind from the fan will blow the chaff and dull seed away.

Collect the seed near the fan and repeat the blowing process until the seed is separated from the chaff. A suitable sieve should be used to sieve the broken seeds and larger stalks. Large seed companies use a Keep Kelly to separate seed from the dull seeds, dust and chaff. This also improves seed germination, because the lighter seeds are separated from the heavier seeds.

4.4 SEED STORAGE AND LIFESPAN

Always store seeds in cool dark places. The germination time of the seeds should be good before storing them. When seeds are stored in optimal storage conditions (more or less at 5ºC in a cold room) then the above mentioned germination times will become much longer. This is provided that the seed is very dry and sealed and that the temperature and humidity levels are ideal and kept constant inside the cold room.

TABLE 2: THE EXPECTED LIFE SPAN OF VEGETABLE SEED WHEN IN STORAGE

SEED WITH A SHORT LIFE SPAN UP TO 3 YEARS	SEED WITH A MODERATE LONG LIFE SPAN UP TO 4/5 YEARS	SEED WITH A VERY LONG LIFE SPAN UP TO 5/6 YEARS
Onions, leeks	Spinach, cole crops	Tomato, egg fruit
Lettuce (only 2 years)	Beans, watermelons, musk melons	Beetroot
Peppers	Carrots, sweet corn, green mealies	
	Peppers chillies/paprika	

Ideal storage for seed is in a dark, cold room and sealed within containers with a humidity of 45. The seed contents inside the container should be dry enough to avoid fungal development. It is a good practice to vacuum pack the seed. The seed germination period would then be extended. Storing seeds in high humidity conditions should be totally avoided (vacuum packed seed is less affected). When storing seed in small containers use containers that have a sealed lid. Confirm that the seeds are dry enough before inserting them in the sealed containers or before storing them in a cold room.

Technology to freeze seed is actually quite simple and is being practiced by gene banks especially in overseas countries. To be able to freeze them in these conditions requires that the seed embryo be dried to contain almost a zero moisture. The seed is then vacuum packed and placed in a freezer which would allow the germination period of the seeds to be much longer.

4.5 PERFORMING GERMINATION TESTS

When seed becomes old, a certain percentage or even all of them may not germinate. If there is any doubt about the germination of seed the following simple germination test can be performed:

30 A small seedling unit. Note the sprinklers in front which are 1.5 to 2m apart from each other. The sprinkler nozzles could be fitted to cover half (180°), quarter (45°) and full (360°) circles to avoid wasting irrigation water.

31 A cheap way to build tables. Note the buckles to tension the wires. Do not use wooden poles to build the seedling house as they would rot within a few years due to the high humidity inside the unit. Take note that seed trays should be level and that all planting surfaces should be irrigated evenly. The buckles can be used to continuously tension the wires.

32 Healthy leaf lettuce seedlings ready to be transplanted into a Nutrient Film Technique (NFT) hydroponic system. The seedlings above are only 4 weeks old and have been established in a 128 cavity tray. The larger cavities allow the seedlings to become larger so that when they are transplanted they would mature quicker. The above seedlings will be ready to be harvested only 3.5 weeks after they are transplanted.

4.5.1 GERMINATION TEST FOR SMALL SEEDED CROPS
Carrots, spinach, beetroot, lettuce, brassicas, capsicums, tomatoes and onions
- Use a small Tupperware container such as a lunch box with a sealed lid or any other flat type of plastic container.
- Line the container on the bottom with cloth or paper towels.
- Moisten the cloth or paper towel properly and drain excessive water from the container. The paper towel should be damp but not wet. When pressing your finger on the paper, only a thin film of water should be visible on your finger.
- Place for instance 20, or preferably 100 seeds apart from one another on the damp paper inside the container. It is better to use more than one container for each cultivar.
- Place the lids onto the containers to prevent humidity exchange. Label the containers and also the date of establishing the seed and place it in a dark place. The ideal temperature is about 20°C at night and 30°C during the day.
- Remember to handle the containers with care and to keep them constantly level at all times to avoid, especially the round seeds, from shifting in their containers.
- The paper towel or cloth should never be allowed to dry out. If this happens moisten it immediately. It is handy to use something like an eye dropper to do this, as light seedlings may drift or move when watering them.
- The seeds begin germinating quickly. Most seeds provide a good germination count after about 5 days and the final count can be obtained just after 7 to 10 days.

4.5.2 COUNTING GERMINATED SEEDS
- Divide the germinated seedlings and remove them where necessary, as their roots may tangle.
- Inspect the roots carefully.
- If the roots are to any extent under-developed or brown coloured then take this as a good indication that they will not develop properly out in the land.
- Some seeds will also not likely germinate if they are hard scaled, muffed or rotten.
- If 20 out of 25 seeds germinate (sprout) then the germination is still good.
- If, for instance, only 15 out of 25 seeds germinate, plant 25% more seeds in the production land.

4.5.3 GERMINATION TEST FOR LARGE SEEDED CROPS
Cucurbits, maize, beans and peas.
- Larger seeds could be rolled or lined onto a damp paper towel and placed in a black plastic bag to avoid moisture loss and exposure to light. Clean sterilized sand that is usually used in children's play boxes is also a good option.
- Use large, flat, sealed Tupperware or other flat plastic containers.
- Establish the seed 1-1.5cm deep in the clean, moist sand.
- Store the containers and inspect the germination as described above.

4.5.4 GERMINATING SEED IN THE PRODUCTION LAND
To achieve a good stand or plant population, when planting seed in soil in the production land, be aware of the following:
- Preferably seed should be fresh to maximise the germination count.
- It is important that the seed is treated with a recommended anti-fungal chemical.
- The soil should be well-prepared, smooth and clot-free in order to ease germination.
- It is fatal for seeds that are in their germination phase in the land to experience any moisture-related stress. Seeds should never dry-out especially during hot conditions. The top soil layer should be carefully monitored and irrigation scheduled to maintain acceptable moisture levels.
- The planting depth of seeds is very important. They should not be planted or sowed to shallow or to deep. The depth of planting varies according to the seed size, soil type, moisture content of the specific soil and particularly soil temperature. When hot conditions occur plant seeds deeper and when cooler plant seeds shallower.
- Generally most vegetable seeds are planted more or less 3 times deeper than the length of the seed.

4.6 SEEDS AND SEEDLINGS FROM NURSERIES

When large farming enterprises or commercial producers order seedlings from nurseries then the following aspects should be taken into consideration:

4.6.1 SEED
All seeds should be treated with a fungicide before establishing them. Some vegetable crops cannot be established in seed trays, such as carrots, beans, maize and to a lesser degree. Cucurbits (pumpkins and squash), onions and beetroot.

4.6.2 COSTS OF SEEDLINGS
It is expensive when a large amount of seedlings are necessary to produce 1Ha. Producers should take note that some crops require a very high plant population.

Be aware that when establishing 1Ha seedlings then the seedling transplanting rate is as follows: spinach is 280 000 to 300 000 seedlings, lettuce is 60 000 to 90 000 and onions it is 600 000 to 900 000 seedlings.

Onion seedlings do not develop very well when producing them in seed trays, as often 3 to 5 seeds are sown in one small cavity to save seed trays, growth mediums and room space.

Many producers sow onions and other vegetable seed in a well-prepared, small seedbed in soil and then produce their own seedlings. This to save money. ***See Chapter 8.4 to produce seedlings in soil.***

Costs to produce seedlings differ between cultivars and nurseries all over the country. In the Western Cape, South Africa, certain large organizations supply especially onion seedlings to producers.

An open pollinated cabbage seedling for instance may cost R0.25 (2013). When ordering smaller amounts of seedlings then the price would be much higher.

Pepper open pollinated seedlings, may cost approximately R0.35 per seedling (or R350 per 1 000 seedlings) at most reliable nurseries.

Open pollinated determinate tomato seedlings are approximately R350 per 1 000 seedlings (2013).

When using F1 hybrid cultivars then 25 000 to 26 000 tomato seedlings p/Ha are required. For production in shade net houses and tunnels, the high value F1 hybrid indeterminate long-shelf-life tomato seedlings would cost R1.80 to R2.90 per seedling and higher. Seedling prices even differ between cultivars. Therefore, when 25 000 indeterminate tomato seedlings are required then they may cost 25 000 x R2.90 = R72 500 p/Ha.

High-value grafted indeterminate F1 hybrid **tomato** seedlings may cost up to R3 per seedling or more.

4.6.3 SANITATION
It is advisable that labourers wash their hands when they are involved with seedling production.
When seed trays are going to be reused, then the nursery should thoroughly sterilize them.
Seedling mediums should be stored in a clean room underneath a roof.

4.6.4 REPUTATION OF THE NURSERY
A nursery should be reliable and should always supply disease/insect free quality seedlings.

Take note that it is sometimes necessary for nurseries to harden seedlings, especially when warm conditions are expected during transplanting time in your area.

The appearance of summer-orientated seedlings that have been hardened are usually short, leathery, sometimes with a pale green colour and also fairly stunted. Be aware that it may be fatal when transplanting long-legged, brittle, pale coloured seedlings in very hot summer conditions and losses of 25% or more may be expected. *See Chapter 4.14 for how to harden seedlings.*

4.6.5 CULTIVAR CHOICES
Long before planting time, it should be decided together with the nursery manager which are the best cultivar/s for your region. Then provide the best choice cultivar seed to the nursery manager.

Sometimes nurseries supply seedlings of open standard cultivars which are still popular. This is for small-scale home owners and upcoming farmers. However, when large amounts of seedlings are required then producers should place their order long before planting time to avoid being disappointed.

Be aware cultivar choice plays an important role. Be informed and don't be fooled by misleading information.

4.6.6 VISITING THE NURSERY
Visit the nursery when your seedlings are still in the developing stage.
Inspect your plants and discuss with the manager when any problems or aspects need attention.

4.6.6.1 SEEDLINGS AND WHAT TO LOOK OUT FOR
- Seedlings should have straight relatively thick stems.
- Well developed leaves should not be too narrow or folded. This is a sign of under watering.
- The roots should be white, thick and must fill the seed tray cavity plug from the top to the bottom. Discoloured roots that do not reach the bottom of the cavity are a sign that the plants have undergone a water stressed period and this will delay root development when transplanting them.
- When transplanting seedlings that are too old or root bound for too long, usually results in a reduction in yield. This is because the roots take a while to recover and thereby support the plant. They will also absorb nutrients more slowly at the start after they are transplanted in the production land.
- When a purple colouring appears on the seedlings leaves, especially on cole crops, then this is a a sign of

phosphorus P deficiency and may delay early growth of seedlings when transplanting them.
- Once it has been determined when the seedlings will be ready to be transplanted in the prepared production land, the producer should organize labour as soon as possible to transplant the seedlings.
- Nurseries supply seedlings mostly in plastic bags which should be handled and transported carefully. *See photo 42.*

4.6.6.2 GRAFTED SEEDLINGS

33 Seedling mediums that are used to fill the cavities of seed trays. In front is a bag of multi feed, on the left is a bag of seedling medium to fill the cavities of the trays and on the right a bag of vermiculite.

34 The use of a roller to make indentations in the seed trays. The seeds are then placed in the cavities and covered with vermiculite.

35 Younger to older pepper seedlings that are in a good condition. Older pepper seedlings can be transplanted deeper into the land because the seedling stems are able to produce roots when planted deeply.

36 Tomato seedlings in a 128 cavity seedling tray. The seedlings are a little too old and long-legged. They were not hardened, are soft and brittle and should not be transplanted when very hot conditions prevail in the production land.

37 A small half-moon tunnel for summer-orientated seedlings which are needed early in the season. Seedlings should be produced in a small temperature controlled unit or in warmer areas. Note that it is very expensive to heat even small half-moon tunnels. In front is an electric heater which can result in high energy consumption.

38 This is one of the largest organic nurseries in Pretoria, South Africa. Organic producers only use organic produced seedlings. This unit is very isolated from any other crops and is situated high up on a hill. This makes it ideal for this type of production.

When seedlings are grafted the upper stem of a chosen cultivar is grafted on to a root stock (under-stem) that has a strong root system. Such a cultivar may be, for instance, resistant to nematodes or soil diseases. This is a considerable advantage for producers that produce tomatoes and peppers under protection.

The under-stem is usually a strong rooted cultivar that will transfer water and nutrients more effectively via the strong root system forcing the plant to grow better and to transpirate more water. This plays an important role in producing higher quality and better yields.

When grafting tomatoes, peppers and other crops, the choice of the rootstock could also be from a cultivar that is chosen for the duration or length of the growing season. Grafted seedlings are mainly used for hydroponic production or vegetables under protection. Histill is one of the leading grafting seedling companies.

4.7 OPEN POLLINATED AND HYBRID SEED

Some of the cultivars mentioned in this manual are so called 'open cultivars' that are older but still in the trade. Because they are open cultivars and were produced successfully for many years, they have no breeder's rights and are open for any person to produce seed for their own use. However, when they trade with this seed then it must comply with the standard seed requirements.

Open cultivates are especially popular amongst small-scale farmers, home gardeners, underprivileged and upcoming producers. This is due to the cultivars uneven maturing, longer harvesting period of certain crops and also much lower cost of seed and seedlings.

F1 hybrid vegetable crops and certain cultivars are created when cross-breeding two or more breeding or sister lines of parent plants.

Fertile pollen is captured from the male plants in order to fertilize the female flowers. Certain parent plants from the same genetic family have special cross-pollinating requirements that differ from each other. When cross-breeding the results are usually but not necessarily higher yields, stronger root systems, better leave canopies, better inner or outside fruit colour, better shaped smaller or taller fruit, uniformity, stability, tolerance to heat or cold, drought tolerance, disease/insect tolerance or resistance and so forth.

39 A F1 hybrid green mealies crossing. Female plants are detasseled and the males provide the pollen. This is one of the first green mealie hybrid crossings.

40 A hybrid onion seed multiplication crossing. The female plants are sterile while the males provide the pollen which are mainly propagated by bees in a closed, isolated shade net house.

Some crops are self-pollinated where the pollen is captured and applied by hand. This is a time consuming and expensive operation. For certain crops, female flowers are infertile and the male plants are the pollinators, while for other crops, the female parts have to be detasseled, as the males are then the pollinators.

Open pollinated cultivars are similar to their parents meaning they could be reproduced. The seed could be multiplied by the producer, then be used for the following season or stored for a few years. *See Chapter 4.3 Produce your own seed.*

Breeding programs of certain crops and cultivars is a complex business and usually requires several years to produce a new cultivar that has certain special characteristics. Newer, improved F1 cultivars usually have other special characteristics such as heat/cold resistance, better yields and quality, disease/insect tolerance or resistance, a strong root system and so forth. This makes the hybrid cultivars better than the standard cultivars.

Thousands of crossings of various crops are made. Afterwards they become established and are evaluated against each other and against the standard cultivars.

It has been reported that when an American breeder develops a new improved cultivar within his life-time (for instance a tomato cultivar), then all his expenses would be recovered and a good profit would normally be the result.

When F1 hybrid cultivars are bred and improved genetically by plant breeders', then some of the inbred characteristics may be difficult to identify with the naked eye. These include attributes such as heat, cold, full/partial tolerance and disease, insect and virus resistance. This is because these characteristics are genetically captured within the genetic structure of the seed.

4.8 PRODUCING SEEDLINGS

Small-scale farmers and home owners could use egg trays, car tires, wood or bricks boxes or pots to produce seedlings for their own purpose.
- It is important to use a balanced combination of medium mixtures, such as good compost, peat/bark, vermiculite, lime and small amounts of fertilizer.
- It is better to make use of ready mixed mediums from suppliers. Note that the medium should be slightly coarse, loose and crumbly in structure. Always read the instructions when ready mixed mediums are used.
- Be aware that certain F1 hybrid crops do not produce seed such as seedless cucumbers, water melons and musk melons, due to the modified genetic structure of these crops by human intervention.

Even small tomato plants that are established in seed trays, may reproduce themselves by producing small tomato fruit about 2cm in diameter. This occurs when they are left too long in the seedling trays. It has been noted in a small experiment that after about 2.5 months the small fruit could be harvested, have their seeds extracted and the seeds established in seedling trays. The seedlings developed into healthy normal seedlings.

4.8.1 PRODUCING SEEDLINGS IN SEEDLING TRAYS
It is advisable that labourers should wash their hands before handling seedlings.

Production of seedlings using seedling trays is a popular and effective way to produce them. Some crops need more than 800 000 seedlings to produce 1Ha. This could make it too expensive to produce seedlings. The medium in seed trays should be well aerated, have a good water holding capacity and preferably have a pH of around 6.5. Peat, bark or vermiculite are usually used in the mixture. This to assist in the retention of water.

Medium mixtures can be pre-enriched during preparation. Alternatively, fertilizer can be applied through the irrigation system.

Be aware that certain mediums include a fine seedling medium that has a low porosity and which may result in a poor drainage ability. In such cases it is likely that a green mould or algae may build up on top of the medium. This often causes a damping-off disease to establish in the trays and when transplanting the seedlings, the plants usually fall over and do not recover. When this happens spindly seedlings could also be the result. Also, often the pH is not correctly adjusted which causes the microelements to not be equally available to the plant.

Tannin in bark mediums may cause problems such as the plants not developing and changing colour. Bark is processed to eliminate tannins, but small amounts can remain and cause problems, especially when high temperatures occur.

When germination problems occur, producers can send samples of the medium to be analysed. Mediums could be enriched using selected soluble fertilizer such as Multifeed which could be supplied via the water tank to the sprinklers.

4.8.2 ADVANTAGES OF USING SEEDLINGS
- Light weight seed trays, polystyrene, plastic or Styrofoam trays are available.
- Seedlings from nurseries are easily removed from the trays and are usually packed in plastic bags. Large amounts of seedlings can then be transported, for example in a pickup truck.
- Seedling plants provide the producer with a head start. Crops may be harvested a month or more ahead when compared with sowing seed directly in the field.
- Summer orientated seedlings could be produced earlier in Hothouses and be transplanted into the production land when the soil and air temperatures are favourable.
- If nematodes are presented in the soil then seedling plants also have an advantage. Nematodes start to attack the first root after the seed germinates. However, because seedlings have a head-start they develop a mass of roots that continue to develop quickly even after the seedlings are transplanted. Transplantation usually occurs about 4 weeks after planting the seed in the trays.
- Seedling production reduces labour and weed control costs considerably.
- When the plug of the seedling is properly moistened, it helps the seedlings survive for longer periods after transplanting them in the production land, especially during hot days. This is due to the moisture that is present in the plug. Just make sure that the seedling trays are irrigated properly before transplanting them.
- Losses of seedlings after transplantation from seedling trays are very rare. Therefore, the spacing of plants in the field can be done well, allowing more efficient land use.
- Seedlings can be hardened-off. This is a good option because the seedlings are then better adapted to outside warmer or colder conditions and should therefore survive better when transplanting them in hot or cool climate conditions. *See Chapter 4.16 Hardening Seedlings.*

4.8.3 SEED TRAY SIZES
Commonly used seed trays have 128, 200 or 300 cavities.

- 128 cavity seed trays are used for plants that have a strong roots system such as cucurbits and high value cucumbers plants. Some producers make use of larger cavities to keep the seedlings in the seedling trays for longer. This gives the plants a better head-start, such as for the cucurbits families.
- Seed trays with 200 cavities are normally used for tomatoes, spinach (small-scale), lettuce, capsicums and cole crop families.
- Seed trays with 300 cavities are used for spinach, onions, spring onions and leeks. Accordingly, 3-5 seeds per cavity are established. Sometimes cole crops are treated in this way, due to the high number of seedlings required. Be aware that the smaller the seed tray cavities are, the sooner the seedlings will need to be transplanted. This is because root bounding quickly takes place in smaller cavities which causes the seedlings to develop poorly.
- When using seed trays with larger cavities, much more seed trays, floor space and growth medium is necessary.
- It is true that better quality seedlings, yields, plant growth and plant development occurs when seedlings are established in larger cavities for crops that require a smaller number of plants. However, it is also more expensive.
- Producers and nurseries use seed trays with smaller cavities to save floor space, seed trays and growth medium. This is primarily done to save money.
- Research has shown that when 128 cavity seed trays are used then larger, healthier and better quality seedlings are the result. With certain crops the seedlings could also remain longer in the larger seed tray

cavities and therefore spend a shorter amount of time in the field. The larger seed tray cavities also allow the seedlings to be transplanted much later into the production land.
- A good practice is to dip the seedlings trays in water mixed with a fungicide chemical and soluble nutrients, before transplanting them. This is to destroy fungal diseases.

4.9 FLOOR SPACE REQUIREMENTS FOR SEEDLING TRAYS

The size of the most commonly used seed trays are 40cm wide, 70cm long and 6cm deep. The number of seed trays with different cavities which are required to produce 30 000 seedlings are as follows:
- 150 seed trays with 200 cavities cover a floor space of 8.5 x 8.5m without walking spaces.
- 100 seed trays with 300 cavities cover a floor space of 5.5 x 5.5m without walking spaces.
- Add 20% for walking spaces between the seed trays.

4.10 STERILIZATION AND REUSE OF SEEDLING TRAYS

When seed trays are to be reused, then the nursery or producer should sterilize them thoroughly to destroy all fungi and bacteria.

Seedling trays could be reused for many years if properly cared for. Labourers should be trained to clean and sterilize seedling trays correctly when reusing them. All old soil and roots should be removed from the seed trays before sterilizing begins. Some nurseries use ordinary Jik (Sodium Hypochlorite) or a chemical such as Spore Kill to destroy fungi and bacteria. Seed trays should be dipped into these solutions for at least 25 minutes. Some producers use high pressure spray-pumps and high concentrations of chemicals to spray the seedling trays. The trays should be washed properly afterwards using clean water.

41 A seedling chamber stacked with seedling trays. This chamber is used for fast and uniform seed germination which is possible due to the constant temperature and humidity in the chamber. This is an enormous advantage as all fertile seeds (even the hardened ones) are then triggered to germinate at the same time.

42 The removal of seedlings from seed trays in order for them to be packed into special plastic bags for transportation in large quantities. A pickup truck fitted with a canopy or other similar transport may be used.

Seed trays could be used repeatedly, however older seed trays cavities may become porous. When this happens the roots of seedlings tend to grow into the rough, porous cavities, making it difficult to remove them. A remedy called Styroseal is available that has a double action of destroying fungi/bacteria and sealing the seed trays at the same time. This makes the cavities smooth again. A cheaper way to seal reused seedling trays is to use a good PVA paint mixed with water. Dip the trays for at least 15 minutes in this solution and then dry them.

4.11 ESTABLISHING AND CARING FOR SEED IN SEEDLING TRAYS

4.11.1 ENVIRONMENTAL FACTORS WHICH AFFECT SEED GERMINATION

4.11.1.1 Water

The first step for ideal seed germination is the absorption of water. This is because seeds have the ability to absorb surprisingly large quantities of water once the germination process begins. Take care to avoid water stressing seeds once they emerge, as this can be fatal if the embryo inside the seed dies.

It is very important that the seedling medium in the seed trays should never dry out, especially when seedlings begin to germinate. The timing of watering the seed trays is critical. When they are irrigated automatically, the timing is usually 10am and 3pm for 10-15 minutes in shade net houses. When using sprinklers to irrigate the seedlings, care should be taken to turn off the irrigation once the water begins to drip at the bottom of the trays. Be aware that too much watering may introduce algae that builds up and which may become a problem.

It is important to apply soluble fertilizer such as Multifeed by means of a watering can with a fine nozzle. When using a Dozetron or a separate fertigation tank system, the seedlings should be fertilized about 2 weeks after the seed has emerged, because some mediums have no nutrients.

4.11.1.2 Oxygen

When germination begins then oxygen is very important to the development of the root system. Therefore, the seedling medium should be loose and not compacted after watering.

4.11.1.3 Heat-Temperature

The importance of maintaining optimal temperatures to achieve maximum germination cannot be over emphasized.

Temperatures not only affect the germination percentage but also the rate of the germination of seeds. Temperatures for crops sometimes differ dramatically and producers sometimes fail to achieve and maintain the ideal temperatures, for instance when heating tunnels.

It is a good option for commercial producers to produce their own seedlings, especially when a producer requires considerable quantities of seedlings in seed trays. **See Chapter 4.8 for how to produce your own seedlings.**

Seedlings could also be kept longer in seed trays when using larger seed tray cavities. This would result in the seedlings being in the field for a shorter period of time. This has been proven to be generally advantageous for the producer when transplanting seedlings in this way. Growth mediums for seed trays should not be too coarse or too fine. If it is too coarse then irrigation water will drain too quickly through the medium. If the medium is too fine, then water drains too slowly and the seed trays may develop a green algae on the surface of the trays and various fungi may cause the development of diseases such as the well-known damping-off disease.

Seed should always be treated with a fungal chemical such as Thiram or Captab. This prevents fungal diseases that are carried on the surface of the seeds and also prevents a damping off disease.

If algae occurs on top of the seed trays, especially when the characteristically green algae is spotted, then act immediately. This should be controlled using a chemical that is applied in the irrigation water, or a medium called Hygro-mix that includes trichoderma.

Trichoderma contains a living organism that protects the seedlings from fungal infections. Producers may also use a chemical called Spore Kill which is effective in sterilizing seed trays.

4.11.2 ESTABLISHING AND CARING FOR SEED IN SEEDLING TRAYS

- Be aware that the pH of the medium is very important and should be between 5.6 and 6.6.
- Growth mediums from suppliers are usually too dry and therefore should be moistened. Mix thoroughly

- with water and test to determine if enough water is present. Test by squeezing the medium in your hand- only a few drops should leach out.
- Always use fresh growth medium once the bags have been left open for long periods. The medium dries out and bacteria in the medium may break down.
- Start filling the seed tray cavities with the moist, sterilized growth medium.
- Press or rub the medium slightly with the palm of your hand to remove any air that may be present. However, be aware not to compact the moist medium at all.
- One method to determine if the medium is too dry is to make an indentation with your finger in the cavity. If the medium pops up then the medium is still too dry. On the other hand, it may float when being watered, especially when using a watering can with a coarse nozzle.
- Sprinkle irrigation is used to irrigate seedlings. **See photo 30 and photo 98.**
- Press or punch holes into the medium by means of a suitable object, such as a match stick, or a 3-4mm pencil when establishing small seeds. Be careful when using a match stick, so as to not press the holes too deeply (use a stopper).
- Another common method is to use your finger to make indentations into the cavities of the seed trays. This is commonly used by small-scale farmers.
- Press-plates, vacuum seed planters and rollers are also available to make indentations, but are also expensive. **See photo 34.**
- A general rule for the planting depth of seeds is normally 3-4 times the size of the seed when measuring the length of the seed.
- It is recommended to plant only one seed of the smaller seeds in the seed tray indentations, especially when using expensive F1 hybrid seeds.
- Use a thin layer of fresh vermiculite to cover the indentations, after sowing the seed in the cavities.
- Rub the vermiculite gently to level the seed trays. Fresh vermiculite absorbs a lot of water and stays soft and non-sticky which is ideal for seeds to emerge.
- Other seedling mediums (to cover the seed tray cavities) may also be used as long as the medium is soft, can hold water and does not stick together to form a hard crust. Clean washed sand is also a good option.
- Irrigate the seedling trays thoroughly with a watering can or a hose with a very fine nozzle. Be careful not to wash the medium and seeds out of the seed tray cavities.
- Some large nurseries usually pack the trays on top of each other in a temperature controlled growing chamber for a few days. This is to ensure easy, even emergence of seedlings, resulting in faster germination. **See Chapter 4.13 for how to build a growth chamber and photo 41.**
- Shift the seedlings from a chamber to a seedling production unit, or heat controlled tunnel.
- Some small-scale producers use layers of newspapers and spread them all over the seed trays. Next the trays are moistened until emergence takes place. Newspapers keep the medium humid and soft and the result is normally a better, faster and better uniform germination rate.
- The medium should never dry out in the seed trays. If this happens it may hamper germination and the tender plants will start wilting. The shock may be considerable and the seedlings may only start recovering after 24 hours.
- Too much water may lead to algae build up and also nutrients could leach. Once algae starts to develop it may become a problem. The green algae sometimes emerges unbelievably quickly, especially in hot, sunny conditions. The green layer of algae may cover the seed trays completely in a few days. Be aware that algae extracts a lot of nutrients from the medium.
- It is important to feed the young seedlings normally about 2-3 weeks after emergence. Be aware that most mediums do not contain enough nutrients and may last only until the first true leaf stage which is more or less 8 days after emergence. Thereafter nutrients should be provided to the young seedlings.
- Various soluble fertilizer is available to feed the young seedlings such as Multi-feed and Hyper feed. Seedling hydroponics is also suitable for a basic nutritional mixture. Read the instructions of the soluble fertilizer carefully before using them.
- Remove double seedlings as soon as they appear by removing the weakest double seedling as early as possible after emergence. Use a sharp knife or scissors to cut the weakest ones.

- Be aware F1 hybrid doubled seedlings could be carefully divided and be transplanted in the production land. They should be watered immediately after transplanting them to ensure a full stand.
- Soluble fertilizer could also be mixed in a separate plastic water tank and applied automatically through the irrigation water. A suitable irrigation pump, solenoid and a time switch is used to set the timing to activate the irrigation periods.
- The amount of soluble fertilizer needed in the water is supplied by fertilizer dealers, seed or chemical companies. The amount of fertilizer required is usually indicated on the fertilizer bags.
- Follow the instructions of the manufacturers carefully when applying soluble fertilizer. An option is to use a watering can and then mixing approximately 25g/10 litres of water or 5-6 teaspoons in 10 litres of water. Do not apply too much nutrients because the tender roots of the seedlings may burn.
- The application of fertilizer through the water should start about 10 days after emergence.
- The ideal applications of nutrients through the water should be 6 times a week and then 1 day clear water. Always look at the colour and health of the seedlings.
- A power failure will stop the automatic water irrigation system therefore a manual valve should be in place to open the water by hand. Usually the automatic solenoid valve has a lever to open and close the valve by hand. Water from a high stationed standby tank could supply the required pressured water to the seedling unit.
- Be aware that the ideal air temperature for the summer orientated crop seedlings until emergence is 24-32°C. Deviations may result in losses.
- Summer orientated seedlings cannot be produced when temperatures fall below 15°C in a shade net house. Producers should wait until the temperatures become favourable. This is usually mid-September in the Gauteng (South Africa) region.

4.12 BUILDING A SEEDLING SHADE NET HOUSE FOR SEEDLING PRODUCTION

- Design and plan the size of the seedling unit to be large enough for your purposes.
- Construct a concrete floor with a slight slope to allow excess water to drain away.
- Use 2.5m iron poles and plant them 0.5m deep 3 x 6.5m from each other. Do not use wooden poles as they may not last long due to the moisture inside the house.
- Seed trays are 70cm long, therefore 2 tables of 1.4m wide and 1 of 0.7m can be fitted between the 6.5m poles, leaving 1m walking spaces between the tables.
- The seed trays should not make contact with the floor to reduce risks related to diseases, infections, insufficient draining and poor aeration.
- Tensioned wires could be used as they are a cheaper option. They may be spaced close to each other about 15cm above soil level. However, because the wires will begin to stretch it will cause the seed trays to not be level on the seedling tables. A better option is to use tables that are covered with diamond mesh.
- Benches or tables could also be built lower and closer to the soil surface. However, it is advisable to construct them at least 1m high, because it is more comfortable for the labourers to work on.
- It is important to install the seedling tables as level as possible to allow water to disperse and spread evenly inside the unit.
- Use small overhead sprinklers inside the unit and space them 1.5-2m square apart from each other. Install them on the structure or on the tables. The main water lines are then placed on the floor underneath the tables so that they are out of the way.
- The sprinklers should rise at least 30cm above the seedling trays.
- It is important to use the correct type and size water pump (for the correct water pressure) and sprinklers. The water pressures for the sprinklers are indicated by the manufacturers, but usually not more than 2 bars is required. The water pump should deliver enough pressure **according to the number of sprayers as well as the quantity of irrigation water required by the area.**
- Be aware that certain sprinklers only irrigate corners, others half a circle and others all round in a full circle, therefore make sure to install them correctly.
- Set the timer correctly to control the irrigation water. The time switch should only open a solenoid valve, usually at 10am and 2pm for a few minutes, until water starts to drip out from underneath the seed trays, at which time the irrigation should be stopped.

- Determine your own water applications and do not allow that the seedling medium to dry out.
- Nutrients to feed the seedlings could be provided automatically through the irrigation water using an injector apparatus such as a Dozetron, a solenoid valve and a timer switch.
- Seedlings should only be given soluble nutrients through the water after about 7-10 days after emergence.
- The correct amount of soluble fertilizer are mixed in the water tank. The water pump then supplies the fertilizer to the sprinklers to irrigate the nutrient water to the seedling trays.
- For small-scale farmers, soluble nutrients could be given by means of a watering can with a fine nozzle.

4.13 USING A GERMINATION CHAMBER

A commercial seedling producer may consider it worthwhile to own a growth chamber. The benefit of this chamber is that the temperature is kept at approximately 24°C day and night while the humidity is kept at approximately 90%. The seed then germinates evenly and all at once, especially when the temperatures are controlled correctly. The humidity inside the chamber should be kept high, as this allows the seedlings to germinate fully before any irrigation water is necessary. The duration time for crops to be in the chamber are generally very short and differs. For instance, tomato seed duration time is only 24 hours, cucurbits and lettuce 48 hours and capsicums 4-5 days.

The chamber should be airtight and isolated, so that the temperatures remain constant inside the room. Heating in the chamber should be installed by a qualified electrician due to the high humidity inside the room. Usually one wall is fitted with electrical heating equipment. Seed trays should be irrigated thoroughly before they are stacked on top of each other inside the chamber. *See photo 41.*

4.14 TEMPERATURES DURING SEED PRODUCTION

Be aware that it will take longer than the normal 4 weeks when producing seedlings in cool weather conditions. Also, take note that older seedlings are more vulnerable to diseases.

Low humidity during hot days may affect the development of seedlings in the seed trays. In these conditions plant growth is usually stunted and the colour may become pale. If this happens the seedlings could be moved to a temperature controlled tunnel that has a higher inside humidity in order to correct the situation.

To produce seedlings for summer orientated areas, the producer should be aware that seedlings are usually required around mid-September. To achieve this, seed should be established in seed trays by mid-July, usually when it is still freezing cold. Therefore, the seedlings should be in a small heat controlled tunnel where the ideal temperature in the tunnel or glass house should be 20°C at night and 30°C during the day. This process will assist in producing seedlings for early transplanting in the prepared production land.

Note that heating is expensive therefore nurseries sometimes use small half-moon tunnels to save energy and money.

4.14.1 WARM WEATHER CROPS
Ideal temperatures for most warm weather or summer orientated crops are 20°C at night and 30°C during the day. When night temperatures fall below 8°C then plants such as tomatoes, cucumbers, peppers, chillies, paprika, beans, maize, sweet potatoes, brinjal, muskmelon, watermelon and pumpkin/squash will stop growing. Temperatures below 0°C could freeze the plant cells and plants will be partly or completely killed. Certain producers who aim to be early to market prepare seed trays with larger cavities within heat-controlled tunnels. When soil temperatures are favourable they transplant these seedlings in the land.

4.14.2 COOL WEATHER CROPS
Most cool weather crops could be produced all year round, except when it is very cold or very hot for long periods.

Producers mostly prefer to establish crops at optimal planting times, however when planting at optimal times, the market may be overstocked and prices may then fall rapidly, which is not desirable.

Certain producers establish crops out of the optimal planting times in certain regions to avoid the market over supply, hoping for better prices and they are often successful.

Some producers take a risk and plant or transplant certain summer orientated crops very early in the season when it is still cold. They transplant the seedlings into specially constructed north-facing ridges on the land. This is generally not recommended, because the seedlings develop slowly. However, this technique has sometimes been shown to be successful.

It should be noted that in the cooler winter areas when temperatures fall below 4°C, that even the colder-orientated crops normally grow very slowly in the production land.

4.14.3 STEPS TO REDUCE SEEDLING DISEASES
- Use fungicide treated seed to reduce seedborne diseases.
- Use commercial growing mediums and sterile seed trays.
- It is a good option to drench the seedling medium by using a fungicide immediately after the seed has been established in the seedling trays.
- Avoid over watering, because this may increase humid conditions, creating algae and promoting particularly damping-off diseases.
- Maintain good air movement throughout the seedling unit. This is especially important when the seedlings become larger.

4.15 TIMING FOR TRANSPLANTING SEEDLINGS FROM SEED TRAYS

- Make sure to transplant seedlings at the correct stage of development.
- The time for planting seed in seed trays and when they should be transplanted differs from crop to crop. The ideal temperatures are approximately 18°C at night and 29°C during the day for the summer orientated crops. Under these conditions for instance tomato seedlings are normally ready to be transplanted after 5-6 weeks, cucumbers 3-4 weeks, peppers 6-7 weeks and lettuce 4-5 weeks.
- It is safe to keep the seedlings in the seed trays for a longer period, but growth will slow down when the roots run out of room to grow in the tray cavities. In this situation when these seedlings are then transplanted, their roots are compacted and the seedlings will recover more slowly which is a disadvantage.
- Commercial producers and nurseries normally use seed trays with smaller cavities to save floor space, growth medium and money.
- Plants that are established in seed trays with smaller cavities must be transplanted earlier than those produced in larger seed cavities.
- It is sometimes a problem to transplant soft, young, vigorous seedlings to the production land as the conditions in the land may be very hot or cold. The soft, tender seedlings may then be damaged.

The first priority is to achieve a full-stand in the production land, when transplanting seedlings. Hardening of seedlings is also a good option for seedlings to survive the heat or cold as well as the transplanting shock. **However, take note that even if you pamper a soft seedling in severe weather conditions, it may still not guarantee your success.**

Healthy seedlings should develop quite quickly after transplanting them. This is in contrast with root-bound seedlings which recover slowly because they must still produce new roots. This is not always possible and may be a disadvantage. When hardening seedlings be aware that they should survive. Also, be careful when extreme warm conditions and a power failure occurs. A backup of irrigation water should always be on hand.

4.16 METHODS TO HARDEN SEEDLINGS

The reason for hardening seedlings is to make them tough and resilient during extreme hot or cold weather conditions.

Hardened seedlings may produce crops that have an outer skin that is thicker and tougher. The plants may appear more leathery, stunted and less succulent. However, they are then much more resistant to transplanting shock. Be aware do not over harden seedlings, as they may recover slowly when they are transplanted in the production land.

At no stage should the seedling medium become too dry and the seedling plants become wilted, as this will restrict the development of the seedlings. Not all seedling plants react the same when hardening them. For instance, a tomato plant may still grow well while a pepper plant may develop very poorly after transplanting. When hardening seedlings they are usually darker, pale green of colour and may have a leathery appearance. The leaves of some crops such as the cole crops may develop a reddish or purple colour.

4.16.1 HARDENING VIA FERTILIZER AND WATER SUPPLY REDUCTION
- Producers should do their own research to harden seedlings. It may differ from area to area and it also depends on temperatures and the time when they are transplanted.
- It is sometimes only necessary to harden seedlings slightly. This is mostly to prevent them becoming spindly, long-legged or soft and brittle.
- A practical way to harden seedlings is to decrease the amount of water and fertilizer before the seedlings are going to be transplanted.
- When for instance seedlings have been irrigated twice a day for the first 2 weeks, then irrigate once a day up to the 4th week before transplanting them.
- It may be necessary to restrict fertilizer and remove any kind of shading from the seedling unit.
- It is not always easy, because although the irrigation system should be very uniform, some sections inside the unit may be irrigated unevenly, especially at the corners of the unit.
- It is also necessary to protect the seedlings from rain when hardening them.

When doing this correctly the hardening process should be effective. Be careful, hardening should not be carried out for a too long period and the plants should not be over-hardened, because plants need to grow rapidly when transplanted. The seedlings should not take too long to start developing new roots when they transplanted in the production land. When the seedlings are too old after hardening them then this may affect the yield and quality. It is very important that seedlings should not become root bound in the seed trays cavities when transplanting them as this will reduce the yield.

4.16.2 WITHHOLDING NITROGEN N FERTILIZER
Another method to harden seedlings is to withhold nitrogen N, causing the plant to grow slower. Withhold nitrogen N and also irrigate less than usual starting 2 weeks before transplanting the seedlings. A disadvantage of this method is that the plant has no spare nutrients when transplanting them, **therefore a Nitrogen N application boost should be given** shortly before transplanting the seedlings.

4.16.3 MOVING SEEDLINGS TO HARDEN THEM
Seedling plants could also be moved outside in the sun to hardened them for several hours each day. Lengthen the intervals systematically for longer periods approximately 14-16 days before transplanting them.

4.17 SEEDLING DISEASES

Seedlings could be attacked by diseases in the production land even when the plants are still very young.

4.17.1 STAGES OF SUSCEPTIBILITY TO DISEASE
Seedlings may be infected with diseases during one of two stages.

- **Before pre-emergence:** Directly after emergence and before the seedling stem appears above the soil. This is connected with bacterial and fungal rotting that is present in the soil. Even seed could be infected with fungi spores.
- **Post-emergence:** Pathogens may penetrate into the stem bases and causes the seedlings to fall over. This may occur, even in the later stages after transplanting the seedlings.

4.17.2 PATHOGENS WHICH MAY BE RESPONSIBLE FOR DISEASE
- Phythium species: This fungi is mainly responsible for problems that occur before emergence. The cause is mostly due to over-irrigation. Later, when the plants mature, only a bruise on the stem may cause seedlings to die.
- Rhizoctonia Solani: Leads to plant death before and after emergence.
- Fusarium and Veticillium Wilt: Plants are small and underdeveloped and the leaves start to become yellowish.

4.17.3 CONTROL MEASURES
- Avoid hot or warm top soil and too much water.
- Water quality should be good and a pH between 6.0 and 6.5 is best.
- River or dam water should be properly filtered by using the correct and most effectively filters for your specific requirements. Be aware that some river water is sometimes badly contaminated and may not be suitable for many types of vegetable production.
- Fertilizer such as the Nitrogen N should only be given after the second week of the emergence of the seedlings and also usually 3 weeks after transplanting the seedlings.
- If necessary, sterilize soil to eliminate nematodes.
- Consider using hardened seedlings to help the plant withstand the planting shock and be much more adaptable to hot or cold conditions.

43 A seedling planter mounted behind a tractor. The planter can transplant more than 3 000 seedlings per hour. It can assist in feeding the seedlings, making holes at the required spacing, planting the seedling and then compacting them with the the 2 iron wheels at the back.

44 An automatic irrigation/fertigation system to irrigate seedlings in a tunnel or shade net house. When seedlings are required early before the summer planting season the tunnels are heated so that the seeds may be established at the end of July or mid-August when the outside temperatures are still cold. The ideal temperatures for seedling production are 30°C in the day and 20°C at night. The system operates with an electric motor solenoid valve and a timer to open and close the valves automatically at pre-set times.

CHAPTER 5

SOIL REQUIREMENTS FOR VEGETABLE CROPS

5.1 INTRODUCTION

Soil is definitely one of the most precious resources that has been bestowed upon mankind. However, many farmers do not appreciate its richness and character nor do they take the responsibility for caring and nurturing it for themselves and future generations.

Soil is not a dead, inert thing. It is the living home of all plant roots and its condition and treatment depends largely on our farmers. It is important that farmers realise that the way the soil is developed or damaged will have a serious impact on the future of humankinds food production capacity and overall survival.

Soil is the storage place of minerals, nutrients, microbes, earth worms, nematodes and billions upon billions of bacteria and all this can be found in only 1 meter of good, fertile top soil.

Farmers need to know and understand their soil so that they may manage and make use of it without damaging it. However, some producers do not even realize how they are damaging and degrading their soil tremendously, year after year.

South African soil are steadily deteriorating and dying due to the lack of understanding and awareness regarding the biological aspects of it. The main reasons for this continued soil degeneration is that soil is overworked year-after-year and repeatedly applied with fertilizer to achieve higher yields. In this process, soil fertility and improvement is neglected. These aspects should be a high priority for all farmers.

Soil fertility is the ability of the soil to supply nutrients to plants as well as the quantity and quality of the macro and microelements in the soil, all of which determines plant development.

The nutrients that are necessary for plant growth are to be identified by a soil analysis. When repeatedly using inorganic granular fertilizer year-after-year, salts will build-up in the soil. Sometimes to such an extent that the soil eventually becomes toxic to plant growth. Also, when using heavy machinery to cultivate soil, especially in smaller areas, you should be careful as this may compact your soil heavily. Refrain from cultivating your soil unnecessarily.

It is important that producers should continuously improve their soil. After about 6 seasons, crops may not survive when fertilized repeatedly using granular fertilizer. Soil may become toxic to varying degrees, as fertilizer are essentially made out of salts.

In many overseas countries, it is against the law to drain the soluble fertilizer into the soil, especially soluble fertilizer from hydroponic systems. As a result, producer's collect their soluble fertilizer and either reuse them or make compost with them.

Fertile soil are well aerated and have good drainage ability which allows the plant roots to penetrate the soil comfortably. Most vegetable crop roots penetrate the soil deeply. For instance, in ideal soil conditions, carrot roots can easily penetrate 60cm deep into the soil. Take note that when the carrot root is pulled out of the soil, all its hairy roots remain behind.

The soil condition may be altered by drainage, tillage, aeration, incorporating organic matter especially good compost, soil improvement, cover crops, lime and fertilizer.

If the soil is not in a good condition, high yields and good quality produce should not be expected. Even if the producer followed good fertilizer practices, selected quality seed from the best, most adaptable cultivars in their region and followed the best cultivation practices, all of this would not ensure success, unless the soil is of a good fertile texture and is well-prepared.

Above all else, soil should be continuously improved. *See Chapter 18.4 for soil improvement.*

According to the World Conservation Agricultural Association, when agricultural production systems do not take soil health seriously they are unlikely to be profitable. The soil should be cared for. For example, too much tillage is definitely harmful for certain soil, even if this is not always possible when producing certain vegetables crops.

5.1.1 FACTORS THAT INFLUENCE PLANT NUTRITION
- Water quality.
- Acid or alkaline soil.
- Heavy clay content (affects certain crops).
- Organic material (microorganisms).
- Soil moisture, soil temperature and soil aeration.

When all the above factors have been addressed, vegetable crops should then be produced successfully and profitably.

5.1.2 FACTORS WHICH CAUSE TOXICITY AND HAMPER PLANT GROWTH IN SOIL
- Aluminium, chlorine and sodium.
- Too much manganese and boron.
- Too acidic or too alkaline soil conditions.
- Too high salt concentrations and brackish soil (except for beetroot).
- Mining activity causing acid soil, air pollution and a lack of trees or bushes.
- Changing weather conditions, as can be seen all over the world such as the unprecedented drought conditions which have occurred in South Africa in 2015/2016.
- Wrong balance of nutrients.

When beginner producers start to produce vegetable crops in virgin soil, they may find that the soil is quite soft. However, after approximately 6 or less seasons of production, the soil may begin to change. Incrementally, the soil may become compacted, forming a hard crust or a hard top layer and/or a hard layer or pan underneath the soil itself. When repeatedly using fertilizer of the same formula, salts may build up in the soil and contribute to these conditions.

Poor irrigation leads to the lack of water in soil, which may cause dry vertical spots to occur in the soil, due to the poor lateral movement of water in the soil.

A hard pan is a layer of dense soil below the soil level. A hard pan forms when the soil particles are compacted together in a layer that stays hard even when the soil is moist. This causes a water table to form, resulting in poor drainage and the drowning of plants, especially in high rainfall areas. The spoiling of soil happens to many producers and also to some commercial producers. Some producers dismiss this problem as they may have enough other unspoilt soil available.

Increasing fertilizer applications to achieve the same yield as the previous year, is not a sustainable or successful tactic in the longer term. Applying only inorganic fertilizer year-after-year on the same soil, may permanently harm the soil. This is a sure recipe for failure and the producers will eventually be out of business.

It is even more important for intensive vegetable producers (especially when limited land is available), to practise crop rotation, soil improvement by applying compost, working cover crops into the soil as well as resting the soil.

Most producers give the majority of their attention to the growth and appearance of the plant above the ground. However, it is also important for producers to realise that the part of the plant that is in the soil also needs and deserves attention.

The soil anchors the plant, provides water, nutrients and oxygen for favourable growth. Producers need to examine the soil to ensure favourable water content, which is the main factor for healthy plant growth.

It is important for plant roots to comfortably penetrate the soil to the required depth. Adequate water should always be available, because if the soil is not moist or irrigated properly, the result may be poor root penetration. When a crop extracts water from a depth of 60cm in the soil, then it will use much more water than a crop growing in the same soil with a shallower root system.

Tomatoes for instance, is a deep rooted crop and uses more water when compared with sweet corn, beans or lettuce which all have a shallower root system.

Individual farmers who cultivate their own soil, should understand the efficient use of fertilizer. Soil and the processes that occur within the soil are not always easy to understand, but an experienced farmer knows that every soil tells its own story.

Soil exhibit various characteristics, such as soil colour. For example, in large prepared lands where the land is against a steep slope, the soil colour may be different from the highest to the lowest point. This formation is usually reddish at the highest point and yellow and black at the lowest point.

The soil colour is directly related to the quality of the water supply, drainage, sunlight, direction of the slope and the amount of organic material in the soil. The best way to restore the soil balance would be to improve the soil structure. You do this by balancing the soil ingredients, improving the soil micro and macro elements and making use of a proper soil analysis.

5.1.3 CHARACTERISTICS OF THE IDEAL FERTILE SOIL
- Soft to the touch, crumbles easily and drains well. It also warms up quickly in spring.
- Stores moisture and soaks up heavy rains resulting in little water washing away.
- Minimum erosion and nutrient losses.
- Few clots and no hard pans below the soil level.
- High populations of soil organisms and a rich earthy smell.
- Requires much less inputs to achieve healthy, high yielding, quality and good looking crops.

5.2 HEALTHY, LIVING SOIL AND SOIL ORGANISMS

One Ha of good living fertile organic top soil, can unbelievably contain more than 1 000kg of earthworms, 2 400kg of fungi, 1 700kg of good bacteria, 108kg protozoa and 100kg of algae per square cm. Organic soil matter contains millions of bacteria, dead organisms, plant matter, other organic matter, nutrients and earth worms in various phases of decomposition. This is an ideal environment to get your soil into a lovely fertile condition.

Time and experience is required to develop fertile soil conditions that produce excellent quality and yields of crops. Note that too much fertilizer may kill earthworms.

Organic producers are not allowed to produce organic vegetables until they have improved their soil for at least 2-3 seasons. Only then would they be able to qualify or be accepted as a certified organic producer (2012). Soil improvement is a necessity and is always practiced by organic producers, because they are bound by the regulations of the Organic Association.

Soil texture affects the drainage and cultivation of soil. Sandy and loamy soil are ideal for most vegetable crops. Drainage in clay and silty soil tends to be poor.

Heavy, clay soil is very sticky when wet. When ploughing wet, clayish soil they have a shine to them, which is an indication to wait a few days for the soil to dry out adequately. On the other hand, when the shiny soil is left to dry for too long, then it can become stone hard which can make it very difficult or near impossible to obtain a good, smooth seedbed without creating unwanted clots.

The ideal soil pH for most vegetable crops is usually 5.8 to 6.8. When the pH is too low or too high, it will have an impact on the development of plant growth which could influence yields and quality, sometimes quite drastically.

The texture of the soil may largely determine or sometimes influence the choice of the crop to be established, the type of equipment used, the conditions under which the soil should be prepared and the time it should be left to rest.

45 An ideal fertile loamy soil that has good drainage ability. This soil has been improved before and should produce quality crops and high yields.

46 A red, sandy, loamy soil that produces high yields when cultivated properly. These sweet potato storage roots have produced a yield of 70 ton p/Ha.

5.3 SOIL REQUIREMENTS FOR VEGETABLE CROPS

Soil that are not properly prepared prior to planting can create difficulties at a later stage. Fertilizing is often difficult to correct after establishing the crop. As a result, producers will likely have problems for the entire growing season.

Good drainage of soil is essential for most crops. Early development of crops is impossible when producing them in wet clayish soil, as the wet soil takes longer to heat up, which influences aeration which is important for plant development. Good drainage is also beneficial for organisms which assist in making certain nutrients available to the plants. Usually open ditches are dug at the lower end of the land to drain any excess water away from the land. It is advisable to rather take an extra day or two in order to start correctly.

Some crops and early cultivars are known to do better in lightly textured soil. These early cultivars grow better and mature faster in light sandy soil, such as watermelons that could be harvested much earlier when establishing them in sandy soil, mainly because these soil warm-up quickly after the winter.

Some crops that require longer growing periods or cultivars that take longer to mature when established in heavier, clayish soil will grow slower and mature later. In these cases much higher yields and good quality produce may be expected, especially for the F1 hybrid cultivars. Good quality, large size, heavy curds of cauliflower may then be possible. *See photo 254.*

5.3.1 SOIL THAT LACKS ORGANIC MATTER

Soil functions at its best when it contains soil-born microorganisms that are found in organic matter. Organic matter also keeps the soil loose and aerated.

The best possible soil that a producer could hope for, which organic producers always strive for, is a good, loose structure that can supply enough oxygen. Preferably a loam or sandy loam type of soil that is free from nematodes. *See photo 47.*

5.3.2 COMPACTED SOIL

Compacted soil sometimes reduces water drainage and therefore may hamper root penetration. Incorrect plough methods, for instance, when tractors wheels are not adapted correctly to the type of soil, can cause compacting.

The top few centimetres of the soil may look good, however producers are sometimes totally unaware that the soil below is compacted. Activities in the land such as labourers walking across planting beds is completely unacceptable. They should make use of the walking spaces that are clearly marked in the production land.

Compacted soil contains much less oxygen (air), which is essential for root growth and microbial activities. Good productive soil is loose and crumbles easy. Also, do not prepare heavy loam or clay soil when it is too wet. Be aware that heavy tractors and tillage equipment damages huge amounts of soil worldwide. This causes the soil to become very hard and compacted which is very undesirable for plant growth.

5.3.3 MANAGING SOIL PROPERLY

Good soil management programs should be implemented such as proper fertilization, liming, tillage practices, crop rotations, regular applications of organic matter, weed control and proper irrigation management. The use of winter cover crops (rotation) and resting of the land are essential in preventing the deterioration of a soil structure that is necessary for high levels of production and yields.

Unproductive soil that have been spoiled by humans can sometimes be corrected, but it is very costly. Soil that are left unmanaged, full of weeds and rubbish are sometimes common on large farms.

Lands full of weeds provide millions of unwanted weed seeds which may emerge in series as some weed seeds lie latent in the soil. These seeds may not germinate the first season but only when ploughing the land the following season. The latent weed seeds should be removed when germination takes place or when conditions become favourable. Be aware that some weed seeds can germinate after many years. Keep the weeds out of your lands and do not allow them to produce seed.

It is not advisable to plough everything into the land and leave the soil exposed for long periods to the environment and the baking sun. Rather improve the soil by establishing rotational crops.

5.3.4 SOIL TYPES

All classes of soil could be used to produce vegetable crops, to some extent. Each class has its own advantage and disadvantages. One class of soil may be desirable for crops that are grown under certain conditions, but undesirable for others.

Soil texture is a term that is also used to indicate coarseness or fineness of the soil and is tested in laboratories.

5.3.4.1 SOIL IS SEPARATED INTO 3 GROUPS
- **Sandy soil:** Includes gravel, coarse, medium, fine and very fine sand.

- **Loam soil:** Includes medium, fine and very fine sandy loam soil (nearly clayish).
- **Clay soil:** Includes gravel and soil with a heavy clay content.

Inspect your soil structure before producing anything.

Sandy soil use much more water and fertilizer, because of the fast drainage abilities. Sandy loam or slightly clayish soil use much less water because of the slower drainage of the water and therefore also hold the water and nutrients at the root zone much longer.

Drainage of water through coarser, sandy soil is fast and carries, especially N fertilizer, downwards quite quickly. Due to their fast drainage abilities, the fertilizer may become out of the reach for the plant roots. If sandy soil is too course or alternatively, when the clay content of soil is very high, it is recommended to avoid these soil for most crops.

5.3.5 SOIL PARTICLE SIZE NOT SUITABLE FOR VEGETABLE PRODUCTION
When sand particles of a soil are 0.2-2.0mm in diameter it is not suitable for vegetable production. This is because water and fertilizer drains too quickly and soluble salts such as Na, Ca, Mg and K may remain behind in quantities that are harmful for plants. Na (Sodium) is usually the main problem.

5.3.6 SOIL ACIDICY AND PH
The pH is simply a method to advise the producer how alkaline or acidic the soil-water content is. The pH level of soil ranges from 0 which is strongly acid to 14 that is very alkaline and then pH 7 which is neutral. When soil contains too much acid, the plant roots develop poorly and may remain small and weak. It is important to know what the pH level in your soil is. A proper pH level for each crop in the soil is directly related to the healthy development of the plants. Plants cannot properly absorb nutrients when the pH is too high or too low.

All soil contains salts. Some are the same salts that we eat and some are different. When soil contain higher levels of hydrogen H it is classified as acidic and when it contains too much sodium N the soil-water content would be considered saline. Usually the pH of sandy soil is low, because the salts drain fast through the coarse sand particles, while clayish types of soil capture the nutrients and salts which usually result in higher pH levels.

Certain crops are tolerant to high levels of salt content in the soil, such as the brassica or cole crop families. They can tolerate a pH level of 5 and have no reduction on yields. On the other hand, red cabbage is sensitive to high salt levels and may cause poor yields and quality.

The higher the salt content in the soil, the more susceptible the plants may be to certain diseases. Soil acidity is by no means unique and it is recorded that 30% of the land available in South Africa is acidic. Acid soil restrict plant growth and are widely spread in the eastern parts of South Africa. In the higher rainfall areas, the soil is often naturally acidic. Human intervention accelerates acidification significantly.

Many producers find it difficult to understand the various soil parameters that are listed in the soil analysis report. They may then be further confused by conflicting advice regarding the use of products such as lime and gypsum. The hydrogen ion concentration of pure water is pH 7.0, however be aware that with each unit change (higher or lower) the pH changes with a factor of ten. For example, a pH of 5.0 is 10 times more acidic than a pH of 6.

The pH of the soil-water content should always be considered as a factor, especially when producing vegetable crops on a larger scale.

The correct pH is important for plant growth and affects solubility (uptake of water and nutrients). The up-take of nutrients/elements are necessary for all plant growth and when the pH is ideal, this process happens naturally.

The ideal pH range for plant growth is normally between 5.8 to 6.6 for most vegetable crops. However, be aware that there are exceptions. When the pH is higher or lower than the normal range, this could restrict the uptake of certain elements, for certain crops, which could result in poor plant development.

To correct the soil acidity make use of a soil analysis to determine the recommended fertilizer and the correct kind of lime to be used, **See Chapter 15.5 for liming correctly.** When the pH is too high then ammonium sulphate or N fertilizer could help bring it down slightly. Urea or LAN/KAN as a topdressing is also then recommended.

When the soil is too alkaline and too much sodium is present then add sulphur, aluminium sulphate, sulphuric acid or gypsum. This is however very expensive and not practical to use on large areas. Gypsum may only be effective after the 2nd year (or even later) and should be applied in high quantities, especially on heavy, clayish soil in order to improve the pH level.

For instance, two tons of limestone could change a pH from 5.0 to 6.0 for a sandy loam soil, however for sandy types of soil it is normally double that amount to achieve the required pH level. When the pH is low then the correct type of lime should be worked into the soil at least 1 month before establishing the crop.

When reducing the acidity of the soil it should be done gradually, as excessive applications may result in a soil reaction that binds nutrients and makes them partly unavailable to the plants. Be aware that when improving the soil fertility, much less fertilizer input is required. Consult fertilizer companies in this regard.

5.3.6.1 RANGE OF PH LEVELS FOR GROUPS OF VEGETABLE CROPS
Group 1 pH 5.5-6.8 Beetroot, cabbage, pumpkins, tomatoes, beans, carrots and chillies.
Group 2 pH 6.0-6.8 Lettuce, onion, maize, cauliflower, peas and spinach.
Group 3 pH 5.5-6.5 Watermelon.
Group 4 pH 5.0-6.5 Potatoes.

The ARC Institute for Soil, Climate and Water and many other similar organisations will interpret your soil analysis, including the pH status of your soil. A fertilizer expert should interpret the samples and issue the producer with a proper certificate indicating the amount of fertilizer and liming to be applied on the production land for the specific crop.

5.3.7 LEAF AND SAP ANALYSIS
There are many different types of soil analysis and each analysis provides its own kind of information. The producer should ask according to his/her needs. Normally, a soil analysis will provide the producer with information relating to the nutrient status for best plant growth.

A leaf sap analysis is not always an accurate indication of what kinds of nutrients the plant has actually taken up. For this reason, plant analysis tests have been developed that takes a closer look at the plant itself.

A leaf sap analysis indicates the nutrients that the plant has already used as well as any shortages of elements that may need to be addressed to correct the situation. The necessary elements should be applied immediately, as the plants develop fast and it is important to correct shortfalls as early as possible, especially when the plants are still young. The results of a leaf analysis should be made available to the producer as quickly as possible.

Contact Omnia suppliers and other leading fertilizing companies in this regard.

CHAPTER 6

LABORATORY SOIL ANALYSIS TESTING

47 An example of good quality topsoil. This topsoil is a loose, sandy loam that is fertile, smells of earth and has previously been improved to get it to this stage.

6.1 TAKING A SOIL SAMPLE FOR A NUTRIENT ANALYSIS TEST

The purpose of a soil analysis is to determine the kinds of elements/nutrients that are present in the soil and which are available for the plants. A sufficient quantity of balanced elements should be available for the most optimum plant growth and development.

Performing a soil analysis is recommended and would keep the producer up to date on his soil status. From the analysis it is possible to determine the optimum fertilizer application levels to ensure a high yield and good quality produce for all types of crops. It could also reduce excessive applications of expensive fertilizer.

Soil analysis is therefore an aid to both the farmer and the agricultural expert.

Most commercial farmers make use of soil analysis tests every season which they then get interpreted to understand their soil status, what kind of nutrients are to be applied or withheld and whether cover crops are necessary.

The pH of soil indicates the acidity or alkalinity of the soil water content. Lime is required when the pH is too low.

The magnesium content indicates whether the soil should be applied with a dolomite or calcitic lime application.

The percentage of clay in soil is important, especially when applying herbicides. It also regulates the timing of irrigation for nitrogen N containing fertilizer. Be aware, excessive levels of one element may result in poor uptake of another.

When taking soil samples or performing soil analysis, it should be at least 7 weeks before establishing any crops.

The reason for this is that when there is for instance a shortage of lime, it is then necessary to apply it 5-6 weeks before establishing the crop. ***See Chapter 15.3 for plant nutrients.***

Without a soil analysis test, it is like driving a car in dense fog, you are never sure where you are going.

When producers do not know what the present fertility level of their soil is, then the rate of application of lime and/or fertilizer could be too high or too low. Either situation is undesirable and expensive.

The most efficient production application rates of lime and fertilizer comes from knowing the current or existing soil fertility levels.

6.2 LABORATORY SOIL ANALYSIS TEST

Plants require nutrients from the soil to develop properly and to reproduce in order to survive. Nutrients consist of different types of elements and are divided into Macro and Microelements. The nutrients that have been consumed by plants during their growing period should be replaced in the soil after the crops have been harvested. This is to supply the correct nutrients for the next crops. Producers should realise that a couple of N, P and K bags of fertilizer are not enough to produce crops year-after-year on the same production land or soil.

A laboratory soil analysis test includes the Ph, N, P, K, Mg and Ca levels in the soil and when required, the water, sand, silt and clay content of soil.

Microelements are not routinely tested but the producer may want to include them in the tests, especially for certain crops such as cole crops, which may require small amounts of boron B, molybdenum Mo and magnesium Mg.

When the results of a soil sample have been analyzed, it is important that the results be interpreted by a fertilizer expert. He/she can then advise the producer on the correct amount of fertilizer to use according to the soil analysis test. This is to ensure that the producer applies and incorporates the correct amounts of the various types of fertilizer into the production land.

6.2.1 SOIL SAMPLING

To ensure that the soil analysis is as accurate as possible, special care should be taken when a soil sample is taken.

The accuracy of a soil sample analysis is as good or as bad as the soil sample that has been taken.

It is very important to have representative soil samples, because recommendations are based on analytical data.

When the soil structure and type in the land appears to differ from one another, use your own discretion and consider taking more than one soil sample so that they may be analysed separately. Be aware that each soil sample analysis test is charged for separately.

6.3 HOW TO TAKE A SOIL SAMPLE

- Remove the top 1-2cm of the soil surface when taking a sample, as this soil may contain plant residues.
- Next, a top soil sample should be taken 30cm deep, using a spade or auger. An auger works like a cork screw, rotating into the soil 30cm deep or even deeper.
- All the layers from the top soil samples must be placed in a dry container and should be spread and mixed thoroughly on a clean floor or table.
- Only 1kg of the representative soil sample is required for a laboratory test.

- Use strong plastic, paper or other types of bags and mark them clearly as 'top soil'.
- Under certain conditions a sub-underground soil sample may be taken. This is usually performed by large commercial producers who would like additional information. Certain vegetable roots may penetrate quite deeply, for example, a carrot's side roots may penetrate to a 70cm depth in favourable soil conditions. When pulling the carrot root most of these side roots remain behind.
- Often, more than 80% of roots are captured in the 30cm top soil.
- Usually sub-soil samples which are taken up to 60cm deep, are mostly intended for the deep rooted and permanently established crops, such as grapes, peaches, lemons and other similar trees.
- If the preference is to go 60cm deep, follow the same procedure as described above for sub-soil samples and clearly mark sub-soil on the containers/bags.

6.3.1 INFORMATION REQUIRED FOR LABORATORY SOIL ANALYSIS TESTS
- Indicate the crop that you intend to produce.
- Indicate the kind of yield that you are aiming for. For instance, if you are aiming for a tomato production yield of 40t/Ha then the analysis test is specific for the production of 40t/Ha. On the other hand, if you are expecting 50t/Ha then it is obvious that the analysis tests will need to be adapted and more fertilizer will likely be necessary.
- Apply for a nutrient element results test, especially when establishing cole crops (cabbage, cauliflower/broccoli), beetroot and carrots. This is because a boron B, magnesium Mg and calcium Ca deficiency is often observed.

6.4 ADDITIONAL INFORMATION REQUIRED FOR A SOIL SAMPLE

- Name and postal address,
- Land size and number,
- Date of sampling,
- If possible, the application of lime and the pH readings from the previous production lands.
- The type of crop that is intended to be produced.
- The yield that is expected for the production crop.
- The amount of fertilizer that have been applied previously and the yields that were achieved.
- The pH and salt index of the soil, if known.
- The texture of the soil and % clay content, if known.

Soil samples may be sent to various institutions associated with the Agricultural Research Council (ARC), as well as the fertilizer companies themselves.

It is suggested that a soil sample be sent by a courier or Postnet to the chosen laboratories.

CHAPTER 7

IMPLEMENTS AND EQUIPMENT

48 Tools that may be used for vegetable production. From left to right: a rake, grub, hay fork and watering can with a fine nozzle. In front is a crate with T-markers, twine and pegs.

49 A tractor and rotovator preparing a smooth seedbed. A rotovator is normally the best implement to prepare a smooth seedbed without any clots. However, be aware that the water content in the soil should be just right.

7.1 INTRODUCTION

There is always new or modified equipment, tools, machinery and implements available on the markets.

Due to labour being expensive in certain overseas countries, they have become quite advanced in the use of implements and equipment. Producers try to automate as much as possible to lower these labour costs.

The list of tractors, implements and mechanically driven motors mentioned below are only a basic list of equipment that could be used on a farm.

Farmers use many other kinds of implements and also modifications of implements. Make sure that you are aware of the type of tractor and implements suitable for your specific operation or conditions.

It is of little use to buy a small tractor to plough large lands or to buy a large tractor to pull a 2 scar plough. Contact your local dealer before buying expensive equipment or look for second-hand equipment.

It is necessary to train tractor operators, not only to drive the tractor, but also the practical side of using the implements. This is especially the case for calibrating implements, chemical spraying equipment and general maintenance. Producers should contact the relevant training centres and equipment/implement manufacturers to assist them in training their operators.

To cultivate lands is an expensive operation, due to the high diesel/petrol prices. It is therefore important for the commercial producer to know how much fuel a tractor and implement may use, what the ultimate cost factor is, so that they may budget for these costs.

The Directorate Agricultural Economics have a guide by Koos le Roux called 'Guide to Machinery' which may be ordered from the Directorate Agricultural Information. This guide includes all the costs related to tractors and agricultural machinery.

The Agricultural Research Counsel's Engineering Department situated in Pretoria will also assist producers regarding general information about irrigation, tractors and implements as well as the best, most economical methods for specific tasks.

7.2 KNOW YOUR IMPLEMENTS AND EQUIPMENT

Secondary tillage implements are used to control weeds and prepare seedbeds.
Primary implements are the basic implements that are used to loosen and improved the soil structure. They include mould board, ploughs, discs, chisels and rippers.

New implements and equipment are expensive and prices are rising all the time.

7.1.1 TRACTORS
One cannot use a small tractor to handle a huge plough. When a farmer is looking to buy a tractor then he/she should buy to achieve value for money. Look for optimal performance in relation of the cost effectiveness of the particular farming operation. Make sure you know which tractor, implements and equipment are needed for your purposes.

Tractors may be classified into three groups:
- Small: 10-49Kw
- Medium: 50-60Kw
- Large: 70Kw and above

Tractors may vary from basic to highly sophisticated.

Speak with the surrounding farmers and dealers to determine which is the best tractor and implements for your specific purposes and operations.

7.2.2 TRACTORS WITH PLOUGHS AND MOULDBOARD PLOUGHS
This plough is used to turn soil up to 30cm deep and is particularly useful for heavier, well-structured soil. Mouldboard ploughs are however not recommended for sandy soil, as poorly structured soil units that exist, could be destroyed and wind erosion may be promoted. Under too dry soil conditions, large clots may form that could hamper seedbed preparation. When soil conditions are too wet smearing may result followed by a hardening of the soil during dry conditions. *See photo 56.*

7.2.3 DISC PLOUGHS
This plough has a cutting/slicing action. Its main advantage is that it works well in dry, hard conditions. Fertilizer could also be worked into the soil, if the soil structure allows it. *See photo 54.*

7.2.4 CHISEL PLOUGHS
Chisel ploughs are used mainly to loosen the soil up to a depth of about 25cm. The best results are obtained when the soil is relatively dry, as the chisel breaks the soil and creates structural units. If the conditions are too dry, big clots are usually formed.

7.2.5 ROTOVATOR (ROTARY TILLER)
Ideally used on sandy loam and clay soil. This useful implement can prepare the seedbed in one operation as well as cut fertilizer into the soil after ploughing. However, if not careful, the cutting action could destroy the soil structure in a short period due to the rotating action of the blades. *See photo 49, photo 55 and photo 58.*

7.2.6 TINED CULTIVATORS
Tined cultivators include a variety of hoeing implements that are used mainly for controlling young weeds and for breaking the soil surface, especially for hard crusts. It is also often used for seedbed preparation.

7.2.7 HARROWS
Harrows include a variety of implements. The tined harrow is primarily used to level seedbeds once the seedbeds are nicely established. The primary objective of the harrow is to break surface crusts, but it can also be used to break clots and produce fine seedbeds.

7.2.8 RIPPERS
Rippers are used to break hard pans below the soil surface. When soil is tilled annually to the same depth, a plough sole or a hard pan may develop. This plough sole or hard pan prevents water drainage and also proper root penetration. When this layer appears, it should be broken regularly. The main disadvantage of the ripper is that it compacts the soil laterally and inwards which can limit lateral root development, especially when wet clay soil are to be ripped. *See photo 53.*

7.3 OTHER EQUIPMENT AND IMPLEMENTS

SMALL GARDEN SPADE/TROWEL: Used to create holes when transplanting seedlings in the land.
HAND HOES: To cover ridges and furrows especially when creating potato ridges.
HAND GROB OR 3 TOOTH RAKE: To break hard topsoil layers, allow better water penetration and remove smaller weeds. When applying top dressings the fertilizer is grubbed slightly into the soil.
DUTCH HOE: A small hoe which is very useful for controlling small weeds in the rows. Labourers would stand upright when controlling weeds with this implement.
WHEEL BARROW: Necessary for general use.
WATERING CAN WITH FINE NOZZLE: To water seedlings in seedling trays as well as seedling plants after transplantation.
CRATES, T-MARKERS, PEGS, TWINE and MEASURING TAPE: Crates to harvest and transport crops. Twine or builders fish line and pegs to measure beds and foot paths. This is mostly for small-scale farmers.
HAND PUSHED HOES WITH WHEELS TO CONTROL WEEDS: To control weeds between rows.
MACHANICAL PLANTERS: Cuts the soil open, places the seed in the opening, covers the seed and compacts the soil slightly. Various types of seed planters are available, some more complex than others in terms of how they measure seeds.
HAND PUSH PLANTERS: Planters which can sow almost any kind of vegetable seeds. The planters make use of an interchangeable, circular, sowing plate which allows it to be configured for different size holes for each crop. The planter has 2 wheels and when it is pushed the circular plate rotates at a set circulation ratio. The planter is very handy for small-scale farmers and will save a considerable amount of time, labour and money, due to the fact that it controls seedlings from thinning. *See photo 23.*
SEEDLING PLANTER: Transplants various seedlings in the prepared land. *See photo 43.*
TRACTOR WITH SLASHER: Used to slash grass, weeds and bushes in and around the land. *See photo 52.*
TRACTOR WITH RIPPER: Breaks the hard layer or hard pan below soil level for better drainage of soil. *See photo 53.*
TRACTOR WITH DISK: Breaks coarse clots in order to make the land smoother. *See photo 54.*
TRACTOR WITH TINED IMPLEMENTS: Usually used for hoeing young weeds.
TRACTOR WITH A HURROW: Level lands before land preparation and maintains or assists in creating new roads.
TRACTOR WITH A LAND LEVELER: Level lands before preparation and maintains or makes new roads.
TRACTOR WITH A BED MAKER: Creates raised seedbeds for certain crops. *See photo 63.*
TRACTOR WITH A CUT BLADE: Cuts and loosens soil beneath roots, bulbs and tuber crops.
TRACTOR WITH A RIDGE MAKER: Makes ridges for sweet potatoes, tomatoes and other crops.
SMALL SELF-DRIVEN ROTOVATOR: Small hand driven diesel rotovator. Ideal to cultivate soil for small-scale farmers. *See photo 58.*
TRACTOR WITH A GRUB: Used to break hard layers of soil between rows. Also used for general loosening

of soil for effective water penetration and to control certain types of weeds. *See photo 50.*
FERTILIZER SPREADER: *See photo 103 and photo 105.*
TRACTOR WITH A POTATO LIFTER: Used to lift potatoes out of the soil. It could be modified for harvesting.
TRACTOR WITH A SPRAY PUMP: *See photo 159.* Used to spray chemicals on a large scale.
KNAPSACK HAND LEVER SPRAY PUMP: *See photo 156.* Used to control pests/diseases and weeds using herbicides on a small-scale.
POWERED DRIVEN MISTBLOWER: *See photo 158.*
POWERED DRIVEN MISTBLOWER MOUNTED ON A TRACTOR: To control insects/diseases.
BACK ACTOR: Very useful for turning-over compost, making deep furrows to drain water away from lands and in general use, but it is a very expensive piece of equipment. *See photo 69.*

50 A tractor with a grub used to control weeds and to break the hard, crusty soil surfaces to allow water to penetrate into the soil and to improve aeration of soil, especially clayish types of soil.

51 A medium size tractor and 3-scar plough. A medium sized tractor is adequate to handle the 3-scar plough. This prevents unnecessary compacting which a larger tractor would otherwise cause.

CHAPTER 8

SOIL PREPARATION, SEEDLING PRODUCTION AND TRANSPLANTATION

8.1 INTRODUCTION

It is important to clear the surface of the production land of weeds, grass, shrubs, clots, debris, trash and any other unwanted material to provide a smooth seedbed when establishing crops.

Soil preparation, especially on new virgin soil, is very important for commercial farmers. Soil preparation involves various activities such as slashing, levelling, ripping, draining, ploughing, disking, rotovating, tilling/harrowing, dragging and so on. Most vegetables require good soil drainage and soil preparation can be very expensive. However, do not take shortcuts, because it is not worthwhile in the longer run.

8.1.1 KINDS OF SEEDBEDS IN SOIL
- Seedbeds to produce **seedlings** for transplantation into production land are usually small areas of land. These areas should be well-prepared to be able to produce healthy, strong and quality seedlings for transplanting into the production land.
- **A production land** is usually a large piece of land that should be well-prepared so that seedlings or seed may be transplanted to it. Seedlings may be transplanted from seed trays or from the soil while seed may be planted with a seed planter or sown by hand in order to establish crops directly into the soil such as vines (sweet potato), cloves (garlic) and tubers (potatoes).
- **A raised seedbed** is created by hand or with a mechanical raised bed maker. The result is a 1m or 1.2m smooth, flat, raised bed.
- **Tractor wheels may also be used to make seed beds.** The heavy wheels are used to cut into the soft soil. However, the heavy wheels may also damage the soil where the tractor wheels cut into the soil. When using this method be aware that a deep, slightly loose sandy soil is required that has a good drainage ability.
- Seedbeds are also created on **flat, well,prepared soil.** Producers usually make use of foot paths for the narrower spaced crops, while for wider-spaced crops, foot paths are normally not required.

52 A slasher to slash (cut) weeds, grass, the remains of previous crops and small bushes in and outside the production land.

53 A two-tooth ripper to break-up hard, compacted soil layers in order to promote aeration and root penetration.

54 After ploughing, a disk is used to further break-up rough clots of soil. Next a rotovator or harrow is used to provide a smooth seedbed.

Difficulty is sometimes experienced in providing smooth seedbeds when sowing seed in small furrows or when using a seed planter. This is mostly due to the stickiness of certain types of clayish soil. For these types of soil, the soil-water content should be in the correct range (not too wet) before the soil preparation takes place.

Be aware that when ploughing crops, cover crops or weeds into the soil, the producer should wait at least 30 to 40 days before establishing the next cycle of crops. This is because the plant material needs to adequately decompose in the soil. During this stage, the soil should at all times be kept moist to allow the bacteria to do their job.

8.2 STEPS TO PREPARE SOIL BEFORE ESTABLISHING CROPS

8.2.1 LAND PREPERATION
Level the land if necessary and incorporate all the residues.

Well before planting time, follow the practice of cleaning the land inside and around the edges. This will assist in preventing insects and diseases from establishing and weeds from multiplying. During this process most insects still remaining after the winter, will be eliminated.

Plough the soil and let the topsoil settle at the bottom.

8.2.2 SOIL PREPARATION
Before beginning soil preparation, the soil should be well irrigated to about 30mm deep, to leave it reasonably moist. You can also wait for rain before beginning soil preparation.

To maximise the effectiveness of the seedbeds, the soil particles should be fine and moist enough to properly surround the seeds.

Soil should be well-prepared to provide the best possible conditions for seed or seedlings to be established. Sub or top soil should never remain long without aeration. Oxygen should be available to the roots for as long as possible. Do not allow the roots to become waterlogged for very long periods, especially in clayish soil. Prepare and maintain the soil at a good tilt, ploughing in the correct direction to control erosion. The prepared land should be fumigated when problems with nematodes (e.g. eelworm) are expected, especially when nematodes were previously a problem.

The most difficult challenge facing a vegetable grower is to maintain the organic matter content in the soil. The type of crop that is grown is important to know, since crops differ in the amount, rate and loss of organic matter they utilise.

For instance, cabbage uses more nutrients from organic matter than sweet corn. Therefore, most of the organic matter produced by cabbage is solid and will be harvested. When harvesting the solid heads, not much solid material remains to plough into the soil. In contrast, sweet corn leaves more solid material behind to work back into the soil.

The layout and size of the land should be planned carefully to consider aspects such as spacing, plough direction, turning spaces for a tractor, walking spaces, distance between the rows, harvesting as well as transport in the land and to the markets. Most importantly, there should be enough money and cash flow to carry the costs according to the size of the production land to be established. Note that labourers should be able to move around the production land without damaging the plants.

The irrigation system should be planned and laid-out correctly. The use of machinery and turning spaces for the tractor on the ends of the land should be taken into consideration.

The availability of labour and cost should be measured against the rental of a tractor and implements. During cooler months, it is sometimes better that the plant rows should be north-east facing, especially for high growing plants. In the summer months, north-south is recommended in order to make best-use of sunlight, however this is not critical, especially in very warm areas.

Together with raised seed beds, this provides an optimum environment for germination and plant growth. Raised beds can be created with a bed maker pulled behind a tractor to improve drainage and aeration. However, be aware that a bed-maker often accelerates the drying of the top soil layer.

8.2.3 RIPPING (CHISELLING)
Break the hard, compacted layer or pan below the level of the soil in order to promote soil drainage and aeration.

It is worthwhile to rip (chisel) virgin soil, as sometimes a hard layer or pan may be present deep underneath the soil surface. This is important, especially when the soil is prepared year after year. Certain more advanced commercial producers rip their soil every third year.

Hard layers of soil mostly occur in sandy loam or in slightly clayish soil. It is advisable to rip the heavier soil with a tractor and ripper every 3-4 years, especially for soil that are known to produce hard layers or pans.

8.2.4 PLOUGHING, DISCING, TILLING AND ROTOVATING
Do not prepare soil (especially with a rotovator) when it is too dry, as the soil particles may be damaged if they are broken-up to a fine powder. This may change a soil structure over a period of time. Higher tractor wear and tear as well as high diesel costs occur when working with too dry or too wet soil. The danger when ploughing the soil in wet conditions cannot be over emphasized. Be aware when ploughing, especially clayish soil, when the turned-over soil is shiny, this is an indication that the soil is too wet. On the other hand, when this soil dries, it becomes extremely hard which makes it near-impossible to prepare a smooth seedbed without clots.

Only light sandy soil can be prepared soon after rain or irrigation. Commercial farmers usually irrigate the soil when necessary and plough it 7-8 weeks before the crop is to be established. They also get a soil sample to be analyzed before establishing the crop. If lime is necessary it should be applied 4-5 weeks before establishing any crops. Make sure to apply the correct type of lime. **See Chapter 15.5 to apply the correct lime.** The seedbed should be prepared just before establishing the crop.

Irrigate the land and if necessary, plough it. Thereafter, disk the soil to break-down the rough clots. Next, apply fertilizer, rotovate, disk, till or harrow the fertilizer into the soil to provide a fine, smooth seedbed. When preparing sandy soil, ploughing is usually not necessary. For this type of soil, usually only a disc needs to be used while the fertilizer can be applied before using the disc.

A rotovator usually provides a smooth seedbed on certain soil. **Fertilizer can be worked into the soil ideally by using the rotovator** which is a popular implement to use. Do not plough, disc, or rotovate your soil unnecessarily.

8.2.5 DRAINAGE OF SOIL
Sometimes when heavy rains occur, it may cause the water to drain slowly. Plants may then drown, due to the lack of oxygen in the root zone.

The symptoms of drowning usually begin after a few days and plants become pale yellow and may wilt. When plants begin to drown, the stomata of the leaves close. Most vegetable crop plants then struggle to take-up water, oxygen and nutrients.

Heavier, clayish types of soil are more prone to drowning and water may dam at the lower-end of the land. The plants in that area may drown or partly drown and show discolouration symptoms. This problem can be corrected by installing drainage pipes, or digging furrows, in order to drain the water away from the land.

8.3 SIMPLE SOIL DRAINAGE ABILITY TESTS

Dig a square hole in the production land about 30cm deep.
Fill the hole with water and take note of the following:

- If the water drains in **3 minutes** or less, then the soil has excellent drainage ability.
- If the soil drains in **5 minutes** or less, then it has good drainage ability.
- If the soil drains in **10 minutes** or less, then it has fair drainage ability.
- If it takes longer than **20 minutes** for the soil to drain, then it has poor drainage ability.

55 A rotovator producing a smooth seedbed. Fertilizer have been applied before rotovation. For sandy soils, only a disc is required.

56 Ploughing the soil with a 3-scar plough. This soil is a perfect example of a sandy, loam soil which is excellent for most vegetable crops.

57 After ploughing, when the soil is loose, the soil may be harrowed. Usually fertilizer is worked into the soil before harrowing.

58 Rotovating the land with a small rotovator diesel tractor to produce a smooth seedbed. Such a tractor is useful for small-scale farmers. Attachments such as ploughs and trailers are available.

8.4 PRODUCING SEEDLINGS IN SOIL

A soil seedbed is a small area only used to produce seedlings that are going to be transplanted into prepared production land. Producing seedlings in soil seedbeds are somewhat outdated. Some producers however still make use of soil seedbeds, especially when producing onion seedlings, due to the fact that a large amount (650 000 to 900 000) seedlings are required to establish 1Ha.

Small-scale producers usually sow small seeds directly in the land by hand. This is for crops such as carrots, beetroot, spinach and to a certain degree head lettuce. Some producers use expensive paletted seed for this purpose.

Most commercial producers plant seeded crops by using a mechanical or a precision seed planter. When producing your own seedlings in soil seedbeds, it is a relatively specialized job. However, when managed well, it is easy and much cheaper to produce seedlings in soil seedbeds.

It is important to note that F1 hybrid seed and seedlings are very expensive and therefore not suited to production in soil seedbeds.

Organizations, especially in the Western Cape, produce onion seedlings which are ordered well ahead of time from specialist onion seedlings producers.

- A well drained sandy loam soil, especially virgin soil, is ideal for seedling production.
- The soil should be fumigated to control nematodes (eelworm), fungi and certain weeds especially when soil is reused. It is important to follow the instruction of the manufacturers carefully when fumigating the soil to control nematodes.
- If good cattle manure or compost is available, it should be worked into the soil at the rate of at least 8kg per m² (80 ton p/Ha.) 1-2 weeks before sowing the seed.
- When producing seedlings in soil seedbeds, correct application of fertilizer is often neglected which then produces poor results.
- When a soil analysis is not available, then the following estimate may be used: 60g Super phosphate, 30g Potassium Sulphate and 90g Dolomitic lime per m². If the soil is very acid, use more lime. If kraal manure is not available use 120g 2.3.2 (22) Zn per m².
- The soil should be moist but not too wet before producing raised seedbeds.
- Create the raised seedbeds 1m wide and not longer than 10m. The paths between the beds should be 50-60cm apart from each other.
- Use the surrounding soil from the foot paths to create the raised beds, but not more than 6cm high.
- Sometimes seedlings in soil take longer than 8 weeks before they are ready to be transplanted, due to the cooler soil temperatures and the general soil conditions. Higher fertilizer N requirements are therefore necessary, due to the longer periods that the seedlings are left in the soil. However, be careful not to apply too much N or irrigation water to reduce the chances of long, brittle, soft and 'leggy' seedlings.
- The best way to irrigate the seedbed is by using sprinklers that are stationed to spread the water evenly over the entire seedbed area. This is to ensure the same size and good quality seedlings.
- It is important when warm dry conditions occur to make use of mulching to cover the seedbeds. Thatch grass is a good option to keep the soil damp, cool and soft. This assists in the emergence of the seedlings. However, the mulch should be removed when the seedlings begin to emerge. ***See photo 12.***
- When sowing seed by hand, the density in the soil seedbeds should be monitored. It is of utmost importance to train only one labourer who is responsible for this job. This is because when sowing seed too densely, it becomes an uncomfortable and time consuming task to thin the young seedlings out. When making use of more than one labourer to sow seed, when something goes wrong, they may not take individual responsibility.
- When seed is sown too densely, it becomes important to thin them out approximately 2 weeks after emergence. Do not wait too long, as weeds may overgrow the seedlings and use most of the nutrients from the soil, especially N.
- The density of the seedlings in the seedbed is very important. Seedlings that are too densely planted lead too long legged plants that are undesirable. When transplanting these soft, long-legged seedlings in warm conditions, expect that more than 30% of the seedlings may be lost.
- The seedbeds should be worked to a fine tilt, be smooth in appearance and as as level and even as possible.

- Measure the beds with twine or a building line. Create the raised seedbed 1-1.2m wide and raise them not higher than 6cm. Use the soil between the beds which eventually becomes the foot paths. The foot paths should be 0.5m from each other.
- Level the soil on top of the raised beds with a rake. It is important to make the seedbeds level to prevent any hollow areas.
- Some producers create the beds to be slightly elevated on one side to assist with draining excess water. However, be careful of this, as heavy rains may cause the sides to get eroded and curved at the lower end of the beds.
- Commercial producers use a bed maker to make the seedbeds. The bed maker usually covers a 1.2m wide seedbed and can even be adjusted to make wide ridges. *See photo 63.*
- When preparing the soil for large areas, the following implements may be used: plough, disc, seedbed maker, tiller, harrow or rotovator and so on. When the area is small then a fork, spade and rake may be used for this purpose.
- Create furrows across the seed beds that are 15cm apart and 1 to 2cm deep. Sow, for example, cole crops and tomato seeds more or less 3cm apart from each other.
- When sowing onion seed, try to sow the seed 1cm from each other in the furrow. About 70g seed is necessary to cover an area of 10m².
- Be aware that weeds may overgrow tender seedlings which may result in long-legged seedlings.
- Every attempt should be made with seedlings of tomatoes, peppers, cole crops and onions, to achieve a stem diameter of 2.5mm or thicker and a height of about 10-12cm.
- When the stand (spacing) of the seedlings are too dense, the plants should be thinned, about a week or two after emergence. At this stage, seedlings of most crops should be ready to be transplanted after approximately 7-8 week.
- Irrigate the soil seedbed after sowing the seed and be aware that the soil should not be too wet or too dry. Irrigate slightly one to two times a day, for the first 2 weeks to keep the top soil damp and soft. This is to allow easy emergence of the seedlings.
- Some producers irrigate the soil to full capacity, sow the seed and cover/mulch the seedbed. They do not irrigate again until the seed has emerged.
- To harden seedlings in a soil seedling bed in order to diminish the effect of transplanting shock, reduce the irrigation slightly for the first 2 weeks to twice a day until emergence. Then in week 3 only once a week and week 4 only once in 10 days. This also depends on the soil structure and water capacity of the soil.
- Be aware seedlings should not be allowed to wilt at any stage. Use reliable water measurement apparatus such as tension meters to monitor the soil water capacity.

8.5 TRANSPLANTING SEEDLINGS FROM SOIL SEEDBEDS TO THE PREPARED LAND

- Prepare and irrigate the soil in the production land properly. *See soil preparation above.*
- Insert pegs at the top and bottom ends of the land. Use twine or building line between the top and bottom pegs where the furrows are going to be made. Stretch and fasten the lines.
- When transplanting seedlings such as onions, make furrows at the required spacing next to the planting line. *See photo 12.*
- When transplanting seedlings from seed trays, make holes in the prepared moistened production land at the required spacing in the rows, next to the twine. Establish the seedlings in the holes.
- Transplant the seedlings just slightly deeper than they stood in the seedling bed.
- Press the soil slightly around each plant stem to reduce aeration. Not too close to the stems however, as the seedlings are tender and may be damaged.
- Transplant only healthy seedlings, discard the weaker, unhealthy and small ones.
- It is good practice to give approximately a quarter litre of water per plant directly after transplanting the seedlings. Use a watering can or hose when warm, dry conditions occur. This is however not critical, especially when transplanting onion seedlings.

- Irrigate the land as soon as possible after transplanting, preferably during warm days. This is to give the seedlings a good start.
- It is better to transplant the summer-orientated seedlings in soil when it is either cloudy, early in the morning or late in the afternoon. This is because the plants roots that have been produced in soil seedbeds are bare and exposed to light and the general outside conditions. The seedlings should recover during the night from the transplanting shock.
- If the soil seedlings are not going to be transplanted soon after removal, put them in a wet hessian bag or place them in a bucket with water that covers the bare roots. They should never be allowed to dry out.
- It is interesting to note that sweet potato cuttings can produce roots in a bucket of water after only 2 days.

59 Transplanting a cabbage seedling from a seed bed to the land. Note the hole should be deep enough to prevent the roots from getting bent or twisted.

60 Good quality onions seedlings that have been produced in a soil seedbed. They are ready to be transplanted to the production land. Behind the seedbed is a small cabbage plot that was established by trainees who were trained by the author.

61 Mulching seedbeds using thatch grass. When sowing seed and when warm conditions occur then the seedbeds should be covered using mulch to keep the soil moist and cool. This is to facilitate the emergence of the seedlings. Note that it is not necessary to cover the rows with thick layers of mulch as their only purpose is to keep the soil moist and to provide cover from the direct sunlight.

8.6 ESTABLISHING SEED DIRECTLY IN WELL-PREPARED LAND

Commercial farmers use mechanical planters to plant seed accurately and directly in the soil. Various planters are available for this purpose.

When using planters it is important that the seed should be graded to ensure even sizes of the seed lot. Producers should make sure that the germination of the seed is good when using planters. Sowing density should be adapted according to the germination of the seed.

Pallet coated seed are an option, but are very expensive. When using paletted seed the results should be successful. A near perfect stand could be the outcome, provided that all the correct cultivation requirements has been followed.

Onion and other small seeds could also be planted directly in the production land. This method requires precision plantings practices by professional producers.

Precision seed plantings is risky and expensive and inexperienced producers should seek reliable advice.

Do your own evaluations, research and experiments with regards to sowing seed in soil seedbeds. For example, normally 550 000 to 900 000 onion plants are needed to establish 1 Ha. Small-scale farmers usually establish seed directly in the soil by making furrows. Crops such as spinach, beetroot, carrots, beans, peas, maize, field cucumbers, watermelons, musk melons, squash and pumpkins could be established directly in soil.

Some old fashioned farmers successfully produce good quality seedlings by sowing seed in soil seedbeds such as cole crops, tomatoes, brinjals, onions, leeks, peppers, chillies, paprika and so on.

Some commercial producers use a tractor with a 3 or 4 tooth grub and set it at the correct spacing (distance) behind the tractor to make furrows. They then sow larger seeds such as beans, maize, pumpkin/squash and garlic cloves in the furrows.

When a planter is not available, small seeds are sown 2cm deep in furrows and for larger seeds 3-4cm. Be careful and do not sow small seeds too deeply or too shallow during very hot conditions.

8.6.1 SOWING SEED DIRECTLY INTO SMALL PORTIONS OF LAND

- Measure your prepared land. Use twine or building line to make straight lines.
- Measure the rows to match the required spacing between the rows and insert pegs on both ends of the land. The twine must be fastened and stretched between the top and bottom end of the pegs where the furrows are going to be made.
- Make furrows next to the twine or building line with a spade or rake about 2-2.5cm deep. This is to sow spinach, beetroot, carrots and lettuce.
- Sow the seed at the required depth and space them apart from each other in the furrows. Next, smoothly cover the seed with soil using a spade or rake.
- Remove any soil clots that may be on top of the covered furrows as they may prevent emergence of the tender seedlings.
- Irrigate and cover the seed in the furrows as soon as possible when hot conditions occur.
- When making use of the full capacity irrigation method, wait until the seed has emerged before irrigating again.
- Make sure to use cutworm bait or other chemicals to prevent cutworm plant losses. This may be critical, especially when this was a problem in the past.
- Apply N usually 3 weeks after emergence and if necessary again after 5 weeks.

62 Sowing seed in a raised seedbed to produce seedlings in soil. Seedlings are usually transplanted when they reach 6-7 weeks old.

63 Making a smooth, raised bed with a modern bed maker. The bed maker can be adjusted for example to make narrow ridges.

64 A mulch liner. This machine could also punch holes into the liners. Holes could also be punched by using a sharpened PVC pipe to transplant the seedlings into the holes.

8.7 MULCHING OR COVERING OF CROPS

Mulching is the layer of material used to cover a soil surface. When temperatures are high, tender seedlings in seedbeds may become damaged by the baking sun. It is therefore sometimes necessary to cover or mulch the seedbeds, or at least keep the top soil moist and cool until the crop has emerged.

When producing carrots, spinach, lettuce or beetroot in the production land, only the rows that have already been sown should be covered.

Materials used for mulching includes plastic layers, grass (thatch grass), grain straw, corn-cobs, sawdust, sunflower kernels, peanut shells, grass clippings, newspapers etc.

Black mulch is ideal to prevent weed seeds from emerging.

8.7.1 THE PURPOSE OF MULCHING
- Increase the capacity of soil to absorb water.
- Maintain a uniform soil temperature.
- Reduce evaporation by as much as 60%.
- Increase yields.
- Assist in controlling weeds and certain diseases.
- Conserves moisture and prevents nutrients leaching.
- Provides protection from mice and other pests.

8.7.2 THE PROBLEM WITH MULCHING
- It is labour intensive and costly, especially when using UV protected plastics.
- Materials are not readily available.
- It could be a fire hazard.
- Mulch may cause a temporary N deficiency.
- Plastic mulch can sometimes be extremely warm and clear plastic has the highest soil warming capability. *See photo 61 and photo 64.*

8.7.3 MULCHING WITH PLASTIC COVERS
An increasing number of vegetable growers use black plastic mulching liners. Drip irrigation is then usually laid underneath the plastic liners. For effective use of plastic mulching, the soil should be even and smooth with no sharp objects, stones, large hard clods of soil and wooden branches that could punch unwanted holes through the liners. Plastic liners may also provide a protective barrier between fruit and the soil, such as for pumpkins, squash and melons. Mulch may also prevent some soil-borne diseases.

8.7.4 THREE IMPORTANT FACTORS FOR MULCHING
- The soil has a good water holding capacity, because of minimum evaporation.
- Before laying plastic strip mulching down or applying fertilizer through a drip irrigation program, granular fertilizer should be applied according to a soil analysis.
- The soil should be brought up to its proper water carrying capacity before installing plastic mulching and the crop should be established immediately after laying the plastic strips.

Take note that it is not the air temperature, but the soil temperature that controls seed germination. Apply plastic mulch after the land has been levelled and smoothed. Plastic liners are usually established on ridges. In the case of black plastic mulch, good uniform soil contact is essential, because the soil is warmed through the plastic, from outside heat sources such as the sun.

Commercially the simplest way to apply a mulch film is with a mechanical mulch layer. Hand application is also an option, however when applying more than 0.5 Ha this can be a difficult and time consuming process.

A mechanical liner is then a better option. ***See photo 64 for a mulch liner.*** The mulch liner is not too expensive and should be considered when large lands are going to be mulched. This saves labour costs. The use of plastic mulches and drip irrigation also reduces water consumption considerably.

Some producers believe that black plastic liners do not warm the soil effectively enough. This is because even though the air under the plastic liners is warm, it penetrates slowly and tends to cool-off during the day. This could be a problem when early plantings are established in cool soil. Poor root development and nutrient uptake may then become a problem.

Producers should perform their own research and keep themselves to date regarding other colours and types of plastics mulches that may be available for their specific crop within their area.

Create holes in the plastic strips, usually with a sharpened PVC pipe, at the required spacing distances to allow the seedlings to be transplanted through the holes.

Brown plastic sheeting that allows heat to pass through is effective and results in warm soil. However, this plastic is more expensive than black plastic but it is more effective when used in cold soil. Another type of plastic liner that is also available, is a reflective liner that is either black or brown on one side and white or silver on the other side.

8.7.5 PUROSE OF REFLECTIVE SURFACE PLASTICS
- They reflect light, especially in greenhouses. White plastic also confuses aphids.
- It cools the soil beneath the plastic by reflecting the heat and sun.

The surface of black plastic can get very hot and may burn seedlings as hot air emerges from the planting holes.

8.7.6 OTHER MULCHING MATERIAL FOR SOWING SEED IN SEEDBEDS
It is worthwhile to wait until the soil temperature reaches the desired degree, before transplanting the summer-orientated crops. This is especially true when the soil temperature is still cold.

When sowing seed in seedbeds during very hot days, small-scale farmers usually use a layer of dry grass (thatch grass) or other similar types of mulches. This is to keep the soil cool and prevent the soil from drying out or warming up too quickly. Remember the main purpose of mulches is to keep the soil cool, soft and damp to allow the seedlings to emerge comfortably and evenly.

Once the seed begins to emerge it is important to remove the grass or other mulches, usually 5-6 days after having sowed the seed. Never allow the seedlings to emerge and become long-legged and bend because of the weight of the mulch.

When waiting too long before the mulch is removed, the tender soft seedlings may be damaged by sunburn and hot soil surface conditions.

Mulch for covering seed beds for seedling production should be economical, readily available and easy to handle. It should also be stable and not wash or blow away.

During the summer months the mulch has a cooling effect on the root system of plants and during winter it prevents frost from penetrating into the soil.

8.8 TRANSPLANTING SEEDLINGS FROM SEED TRAYS

- It is important that seed trays should be **soaked or watered thoroughly** before removing the seedlings from the trays. It is then easier and a smoother process to pull the seedlings out of their seed trays.

- It is advisable to dip the seedlings in water that are treated with a chemical compound to prevent fungal diseases.
- The moisture that is captured in the seedling plug will last at least 5 hours, until the seedlings are to be transplanted, especially on hot or warm days.
- If seedling plants do not smoothly pull-out from seed trays, it is usually due to old reused seedling trays that have porous cavities. The roots of the seedlings partly grow into the porous sides of the seedling tray cavities and make them difficult to remove.
- To ease the release of seedlings from the seed tray cavities, the seed trays could be tapped or bumped on a solid surface. Do this while the roots fill the cavities but before they get root bound. Now when turning the trays upside down, the seedlings should release easier. When using this method, be aware that the seedling roots should just about have become fully root bound in the cavities.
- When the seedling roots fill the cavities of the trays and penetrate through the cavity holes at the bottom of the seed trays, this is acceptable and are called 'air roots'. This is why some seedling producers raise the seed trays in the seedling unit so that the roots may hang in mid-air. This is desirable when the seedlings are slightly past their transplanting time. However, note that some crop seedlings quickly penetrate the bottom of the trays but do not fill the cavities fully, as in the case of the leaf lettuce which should be transplanted after only 4 weeks from sowing the seeds.
- It is acceptable when seedlings are kept a week or two longer in the larger seed tray cavities, as it becomes economical to sometimes do it this way.
- Seedlings could also be pressed or pushed out of the seed trays, by using a stick that is the thickness of a pencil. Press them out from the bottom of the trays.
- Avoid transplanting too young seedlings as it may not be worthwhile. Instead wait another week or two.
- The planting depth of the seedlings is very important. Plant the seedlings plugs level with the soil surface, otherwise when they are planted too deeply, the stem of the seedling may burn, due to the hot soil surface. Infections may then also take place. On the other hand, when the seedlings are planted too shallow and part of the plug extends above the soil, it is then exposed to sunlight and wind. The plug may then dry-out and the seedlings may experience moisture stress.
- Small producers usually use a broom stick, a flat iron bar or the edge of a spade to create the holes for the seedlings at the required spacing. However, be aware that the holes should be deep and wide enough to accommodate the seeding plug comfortably. Also, the soil in which the plug is established should be moist. **See photo 59.**
- Transplant only healthy seedlings and discard the weak, unhealthy and underdeveloped ones.
- Place the plug in the hole, without bending or twisting the roots. Compact the soil slightly around the roots so that the seedlings can grow immediately.
- It is important to irrigate the prepared land before sowing seed or transplanting seedlings.
- Seedling planters are available that can transplant 3 000 or more seedlings per hour. **See photo 43.**

8.9 FILLING GAPS DUE TO A POOR STAND

Seedlings that have been removed from the soil seedbeds are fragile, as they do not have a plug or soil around the roots. Therefore, seedlings may easily be damaged when transplanting them in the production land.

To fill the gaps when seedling plants die or are damaged by cutworms after transplantation, the producer could plant bubble plants every 15th row, especially for the capsicum families, cole crops, tomatoes etc. This is especially applicable when transplanting seedlings directly from soil seedbeds.

Before the seedlings are to be removed in order to fill the gaps, the land should be irrigated. Use a spade when lifting the seedlings up and carry them with some attached soil so that the gaps may be filled adequately. This practice should ensure a uniform stand.

CHAPTER 9

ORGANIC MATTER, GREEN MANURE AND COMPOST PRODUCTION

9.1 ANIMAL MANURE AND THE VALUE OF ORGANIC MATTER

Organic matter increases the porosity of soil, which in turn increases the water absorption of plants. The increased porosity causes aeration which is beneficial for the right types of bacteria. If animal manure can be secured cheaply, it is a good material to use for maintaining the organic content in soil.

It is particularly valuable for small-scale farmers, due to the fact that it may sometimes be quite scarce. Animal manure contains a fairly high amount of nitrogen N but is low in phosphorus P. Fresh animal manure releases CO^2 gasses which interferes with the oxygen in the soil. Therefore, do not work fresh animal manure too deeply into the soil and wait at least 10 days before establishing any crops. Manure should be protected by covering it with sheets or storing it under a roof. This is done to prevent nutrients from leaching or draining away when rain occurs. Be aware that kraal manure should be kept moist all of the time and not be allowed to dry-out.

9.1.1 THE ADVANTAGES OF KRAAL MANURE OVER FRESH MANURE

Fewer nutrients are lost through decomposition. More bacteria is available and more energy is provided for bacteria development. Buffer effects are greater and there is no burning effect on plants. Decomposed kraal manure contains urine that is valuable as it is high in nutrients and phosphorus P, provided that no antibiotic treatment was used to treat the animals. Chicken manure contains much more N. However, be aware that high concentrations may burn plants. Manure is excellent for making compost, especially when crops are widely spaced, such as the cucurbits families. This is because much less compost is then necessary, as the compost is only placed near the roots in circles or in bands.

65 A sun hemp legume crop which is to be incorporated into the soil. It is an excellent choice for incorporating into the soil and improving soil fertility. However, the crop is very sensitive to nematodes (Eelworm).

66 Incorporating green grass into the soil to improve soil fertility.

9.2 USING GREEN MANURE TO RESTORE CROPS

One of the primary uses of good quality green manure is for increasing soil fertility. This is one of the cheapest ways to improve soil fertility. The most common and best plants to use are leguminous as they are high in nitrogen N. They bind N from the atmosphere and capture it in their roots in order to make the N available for the next crop. Legume seeds should be inoculated before planting by using inoculating organisms. This gives plants a better chance to bind Rhizobium from the air in order to secure optimum N binding in the roots. Green vegetation decomposes much faster than dry material. It takes only 4 weeks in summer and 6 weeks in cooler conditions to decompose.

Green manure crops are grown for a specific time period to reach full maturity and are then ploughed into the soil to improve the soil fertility.

Sun hemp is an example of a good legume crop. **See photo 65**. Be aware this crop is very sensitive to Fusarium wild soil diseases and soil nematodes. (eelworms).

Velvet beans is also a good legume crop but grows slowly. Babala develops very fast and provides a considerable amount of green material. To a large degree it is also drought resistant, requires limited cultivation and is non leguminous.

67 Loading green material onto a pickup in order to make a compost heap. This crop is Cow Candy, a sorghum family that is slash-cut but will grow and reproduce again.

68 Packing layers of organic material to build a compost heap.

69 Turning-over compost heaps using a back-actor. The heap at the back has been turned 3 times and is ready to be used. The temperature inside the heap was measured at 60ºC.

70 Adding compost to the land at a rate of 80 ton p/Ha.

71 Compost added in bands on ridges at a rate of at least 80 ton p/Ha. A soil analysis is a must. Granular fertilizer was incorporated before making the ridges.

72 A compost heap which is only 2.5 months old and is ready to be used.

Eragrostis Curvula is also a good option if longer growing periods are required. It would also control nematodes such as Eelworm in the long run.

9.3 COMPOST AND COMPOST PRODUCTION

Compost is seldom used by large commercial producers due to the large quantities required, the labour input and expenses involved. However, it could be used for growing young plants intended for transplantation as well as by small-scale farmers, greenhouses and home owners.

According to experts, at least 80 ton p/Ha of compost should be applied to lands in warm summer areas and less during cooler times of the year. Compost should be applied every year and slowly reduced as the soil becomes more and more fertile.

In order to produce one ton of good compost it may cost the producer approximately R100 (2014). In warm, summer areas the compost should be applied repeatedly, as the organic matter breaks-down much quicker in warmer conditions.

When compost is applied to production land, the soil should never be allowed to dry-out, even after harvesting. This is in order to retain and cultivate the soil organisms that improve the soil structure so that they may remain active and alive. Certain overseas countries have much lower temperatures and usually apply less than 40 tonnes p/Ha of compost.

9.3.1 HOW GOOD COMPOST IMPROVES SOIL
- Improves aeration in soil.
- Improves the structure of soil and its water holding capacity.
- Improves the drainage of salts.
- Improves the condition of brackish soil.
- Improves the condition of acid soil.
- Cultivates a microbe population which assists in breaking down complex structures in the soil, providing the environment for the plants to take-up elements more comfortably.

Compost is not a fertilizer, but it contains nutrients that improve plant growth, improves the soil and reduces the use of fertilizer and water. Compost materials are usually low in phosphorus P and potassium K. To improve soil sufficiently is a long-term process and is not likely to be achieved in one season. Soil should be continuously improved. Compost can also not replace fertilizer, due to its lack of macro-elements. 40 tons of compost only contains approximately 37kg of N. Compost may also be used as a soil cover or mulch, as any further decomposition will occur above the roots where the plants will still benefit from it, as long as the compost is kept moist.

Composting is a rapid, self-healing process, in which organic material is decomposed. It loosens the soil structure, captures water and air and increases the soil-water holding capacity. Sir Albert Howard, the father of compost production, lived in England and found that when layering different organic materials on top of each other and mixing them, decomposition would take place more quickly and effectively. He produced his compost by layering 10 -15cm green matter, a 5cm layer of manure (blood meal, sewage sludge or other high proteins), a thin layer of rich earth and a small amount of Agricultural lime. This simple formula, when repeated layer upon layer, produces a crumbly compost that is of high value. A compost heap of 1.5m high and 3m wide is ideal. The length is optional. Sir Howard also found that decomposition was facilitated by factors such as aeration and temperature. While building the heap, he placed vertical stakes in the heap and removed them once the heap was completed to allow air to move up and down these corridors.

When the inside of the pile becomes too dry or when temperatures rise above 60ºC, water is then added. During dry periods, water is added to the pile and when the rainy season occurred he made the heap round on top. Most areas in South Africa have a warmer climate than England, were he lived.

9.3.2 THE HOT AND COLD METHOD OF COMPOST PRODUCTION

9.3.2.1 THE HOT METHOD
A **hot or active** compost heap requires careful management.

It may seem simple to produce compost, however even experienced organic producers take up to 2.5 months to do so and for this to happen, the compost should be managed well.

In order to produce good compost, the right balance of materials is essential and it should be carbon and nitrogen N rich. Piles should be turned often to provide the organisms with enough oxygen to do their work. The hot method kills weeds and pathogens quickly, because the pile reaches temperatures up to 60°C and more. However, temperatures higher than 65°C may damage the approximately 100 thousand trillion and more good bacteria.

9.3.2.2 THE COLD METHOD
If you should decide not to bother with hot piles, you could instead produce second-grade compost at home. The cold or passive method is low in maintenance, but not as good quality as any type of material is added onto the pile.

The pile should regularly be turned before the materials compact too much. Turning the heap is a labour intensive job especially when a back-actor is not available. Compacted piles should be avoided as drainage is poor. Also, too much rain restricts decomposition and some nutrients may become lost unless the pile is protected from the rain.

When using the cold method it is not possible to prepare a good quality compost from poor quality raw material. However, good quality raw material is not always available. Older, dried plant material usually does not contain all the required nutrient elements. This is poor quality material for compost.

9.3.2.3 GENERAL
Both the cold and hot methods depend on the material that is used for the compost heap. Therefore, garbage in can also mean garbage out. The output of both compost methods should be a dark brown colour, a soft crumbly texture and it should smell earthy. Once incorporated into the soil, the decomposition process of compost in most areas within South Africa is much faster due to the high temperatures.

A simple way to test the moisture content in the pile is to grab a handful of compost from deep within the pile, squeeze it and only a few drops of water should trickle from the compost. To speed up the bacterial activities it is advisable to add bacterial activators available from chemical companies or corporations. Some producers use fertilizer, especially nitrogen N-LAN/KAN or Urea to boost or activate the bacteria. Small amounts of fertilizer should be applied to the heap while the compost pile layers are made. It is interesting to note that one handful of good compost contains millions of living organisms. The ideal temperature inside the heap should be 60°C, for a few weeks. About 4-5 cubic meters of materials are required to make 1 cubic meter of compost. Compost should not be made where it may block drains. The compost should be kept moist but not too wet. When it is too wet, it may cool down too much and rot instead of break-down. The ideal moisture content in a heap should be between 55-70% to keep the organisms happy.

9.3.3 THE QUANTITY OF COMPOST REQUIRED FOR PRODUCTION
In the warm, summer areas in South Africa much more compost is required for production lands. This is because compost decomposes quickly when temperatures are high. In summer, at least 80 cubic meters, or 80 ton p/Ha or 8 large, full spades per square meter, should be applied. In England only 30 ton p/Ha is the norm, due to the cooler climate conditions.

Large amounts of material are required to produce compost for a 1 Ha area. Usually material such as sawdust, tree bark, peat, moss, bales, sunflower and peanut shells are commercially used to produce compost. Excess sawdust that is residue from hydroponic systems is a very good source material to produce compost, as it contains many nutrients and decomposes well.

9.3.4 AN EXAMPLE OF COMMERCIAL COMPOST PRODUCTION

The tomato king, Berty van Zyl from ZZ2 imported a Lindana TP 200 wood chipper (2007) from Sweden. This is capable of shredding huge black wattle trees in order to produce compost from the resultant wooden chips. This is a valuable mulch which is applied in and between rows of crops and after harvesting the chips are also incorporated into the soil. This helps to improve the soil fertility. For instance, there is a commercial farmer that produces compost by using 25% chips, 25% sawdust, 30% green plant material, 15% kraal manure and 5% potash.

9.4 COMMERCIAL COMPOST PRODUCTION

Commercial compost production can be achieved as follows:

9.4.1 PHASE 1
- Select a site near irrigation water and loosen the soil where you are going to build the pile.
- Begin by building a series of layers and adding a little old compost to introduce some bacteria into the pile.
- The pile may be built 2 to 3m wide and whatever length is convenient. Eventually the pile or heap would be about 1.3m high but will decline as soon as the bacteria starts to become active and decomposes the material.
- Make the first layer 25cm high by using dry materials such as saw dust, peat, sunflowers, peanut shells, dry straw bales, dry grass etc.
- It is important to use fairly fresh green material to make the second layer such as green leaves, green leafy weeds, green grass, or pastures such as cow candy, black mustard, maize etc.
- Water each layer slightly when building the heap.
- Some producers add N fertilizer such as LAN/KAN to the heap to make it systematically warm.
- N fertilizer works as an activator and should be washed slightly into the heap so that it may dissolve quickly into the heap. The temperature will rise because of this.
- The third layer should be an activator composed of good quality cattle, horse or sheep manure about 15cm deep. Alternatively, about 10cm chicken manure may be used. This layer should be slightly mixed on top of the second layer.
- Add a little water to each layer, but not too much.
- Repeat this procedure three times, each time building on the previous three layers until the heap is more or less 1.3m high.
- To complete the pile cover it with old sacks or a thin layer of soil.
- Within a day the heap should build-up heat and reach a temperature of between 50°C and 70°C.
- The pile should not be allowed to become to dry which can easily happen due to the high temperatures inside the heap.
- Irrigate the pile when the temperature rises too high inside the pile. Producers normally use sprinklers on top of the pile to decrease the temperatures.

9.4.2 PHASE 2
- The pile should be turned around for the first time after approximately 18 days. The temperature will still be high but aeration is necessary.
- As the pile is turned make sure that the outer materials are worked into the inside of the pile. This allows the outer materials to also decompose.
- The pile should be turned over the second time after about 34 days. Make the pile fairly high up to about 1.5m.
- Irrigate the pile to prevent the heap from becoming too dry.
- The pile will again heat-up quickly to approximately 50-60°C and should be turned around for the third time after 60 days.
- By turning compost in intervals it will be ready much sooner.

9.4.3 PHASE 3
- During this phase the compost begins to mature, stabilise and it takes longer to reach high temperatures.
- The duration of this phase is about 3-4 weeks.
- When the heap stabilizes the temperature will be approximately 20°C.
- In total the duration of the entire compost making process should be approximately 3 to 4 months or 8 to 9 months when preparing compost using the cold method.
- The turning of huge compost heaps is very labour intensive and difficult, therefore commercial farmers use machinery such as a back-actor. *See photo 69.*
- When the compost is ready to be used, it should be dark, pleasant smelling and crumbly.

9.4.4 IMPORTANT ASPECTS TO LOOK OUT FOR WHEN PRODUCING COMPOST

Acidic soil promotes the decomposition of compost in the soil at a much faster rate which may cause the soil humus to be destroyed faster, especially if too much aeration is present in the soil. The pH of soil should be 5.5 to 6.0.

When making compost it is necessary to keep the balance of the pH correct by using agricultural lime.

It is a regulation that organic producers should continually improve their soil.

9.4.4.1 PROBLEMS CAUSED BY INSUFFICIENTLY DECOMPOSED ORGANIC MATTER

When using organic matter during planting time that is still too fresh and warm and has not decomposed sufficiently the following may be the result:
- Burning of plants roots. Compost that has not decomposed fully may release CO^2 gasses that are harmful to the plants.
- When fresh compost is above 20°C wait 7 days before sowing seed or transplanting seedlings with the compost.
- Difficulty to plough as well as the formation of toxic organic compounds in certain types of clay soil.

There are other ways to reduce the production time of compost. For example, planting thick poles into the centre of the heap, then pulling them out and adding water in the holes. This also improves aeration.

CHAPTER 10

ORGANIC FARMING

73 Organic seedling production. Only organic fertilizer and pesticides are used. Seedlings should be ordered long before establishing the crop in the production land.

74 Hydroponic production is not classified as organic food because crops are not produced in the soil.

10.1 INTRODUCTION

Organic farming is quite a large, specialised field of agriculture. Entire manuals and books have been dedicated to this subject. It is for this reason, that only a short overview of organic farming is provided in this manual, in order to create a basic level of awareness.

The main aim when producing organic vegetables is to improve the soil fertility and provide a medium that provides a good balance of microbes as well as healthy soil and organic matter. This includes the correct pH and water content for the soil, optimal temperatures, correct nutrients and water applications.

An organic producer is only allowed to use organic products, such as organic compost, fertilizer and organic insect/disease control measures.

Organic producers are required to follow strict rules and regulations that are set out by the Organic Association to ensure that the consumer receives truly organic, healthy and quality produce.

It is an expensive operation to improve soil organically. The Organic Association requires that beginner producers improve their soil organically for at least 2 to 3 years before they are accepted as a truly organic producer (2012). Be aware that the Association and supermarkets do sometimes change or update their regulations on a regular basis.

Organic producers should make use of a good fertility soil program that includes compost, crop rotation, animal manure, green manure and legume crops. All of these materials should be incorporated into the soil continuously. The manure of cattle which has been treated with antibiotics is strictly forbidden.

The main idea behind organic methods is to feed the soil and the microbial life in the soil which thereby benefits the plants.

A challenge is that insect/disease control can only be managed by means of natural and often expensive products. Biological control, soil improvement, the application of organic fertilizers and chemicals as well as weed, insect and disease control are all factors that the organic producer should be able to deal with in the prescribed manner.

Soil should be constantly supplied with organic material to feed the earth worms and soil bacteria. Earthworm castings are naturally pH balanced and experts claim, that the quality and soil fertility benefits of these castings cannot be improved upon. However, the worms should always have enough plant material in the soil to consume.

Producers should be aware of the rules and regulations in order to operate as an organic producer. All aspects, such as pack stores, production land, shade net houses and tunnels are inspected by the Organic Association and the supermarkets.

10.2 SHORT OVERVIEW OF ORGANIC FARMING

Producers should educate themselves about organic farming before they make a decision to begin this type of production.

Life on earth depends on healthy, productive soil as well as a multitude of plants, all of which are needed for survival on earth. Organic production is especially focussed on promoting practises which are sustainable and in harmony with nature and all its ecosystems.

Organic production is described as a food production system in which healthy soil is the basis of production. If the soil is well-treated, healthy, strong growing plants will develop which have thick cell walls, allowing them to more easily repel attempts by insects and pathogens to penetrate their outer layers.

The quality and nutritional value between organic and conventional vegetable production differs. This is mainly due to the fact that, in general, organic produce is grown in soils are much richer in natural nutrients. The organic farmer is required to improve his/her soil structure on a continuous basis, sometimes for at least 3 seasons before organic production may take place. They do so primarily with good, self-made compost and the application of new, scientific methods of production.

In order to become an organic producer, it is required that all farming activities are inspected and certified on a regular basis. Without this certification, producers are not supplied with organic stickers. Registration and inspection fees must be paid for crops to be certified. The cost of certification and registration required to produce organic produce may be costly and is determined based on the size of the production area.

The Organic Association can guide beginners in all these rules and regulations. When a producer has a contract with a supermarket, the supermarkets will also visit the producers production lands. Inspections by supermarkets and the Association are strict and penalties may occur if producers do not comply. Many aspects may be inspected including yields, quality, weed control, packaging, hygiene in stores, cold rooms, trucks, insect and disease control, handling, labelling of produce and marketing.

The benefit of being an organic farmer is that the demand for organic produce in South Africa is increasing steadily. However, the producer is required to develop specific expertise related to such aspects as organic soil improvement and organic chemicals for the control of insects and diseases.

The labelling and packaging of organic food is strictly controlled and is more expensive than conventional food enterprises. This is because the packaging has to be of an organic standard and the labelling has to identify the grower as well as the company that has certified the food as organic.

CHAPTER 10: ORGANIC FARMING

Small-scale organic producers are rare. They try their best to control insects and diseases by using natural spraying remedies that are produced from plants, insects and other natural, biological origins. Once the recipe is properly understood, some remedies can be produced on the farm.

Organic producers make use of natural spraying remedies, such as garlic, chillies, khaki plants, herbs, liquid sun-light soap, sugar, insect remedies and other biological control measures. Insect eating predators, disease pathogens, oils that smother insects and certain herbs that act as natural repellents are all used to protect crops.

Organic producers claim that insects and fungi do not infect their plants that easily, due to the fact that their crops have healthy, thick outer cell walls that are not easily penetrated. However, it should be emphasised and producers should understand that natural remedies are not always effective in controlling certain types of insects or diseases. Conventional, more gentler kinds of chemicals are also not always effective. However, ongoing and long-term research continues to be performed on these types of chemicals so that their effectiveness may be improved.

Organically produced crops should be traceable from the farm to the table. This is of prime importance in order to promote consumer trust.

CHAPTER 11

SPACING OF CROPS

11.1 INTRODUCTION

Achieving spacing for the optimal plant population of vegetable crops is very important, however, it is not always as simple as it may sound. Many factors are involved to achieve the optimum stand and maximum yields. Temperatures also play a role in spacing, especially in warmer and hot conditions.

The spacing of certain crops is closer in warmer conditions and further apart in cooler conditions. Each crop has its own optimal spacing requirements which can differ between cultivars and species as in, for example, cole and capsicum crops. Cabbage cultivar spacing is determined by the head size of the cultivar, that is, large, medium or small head size.

The plant population for the medium large cabbage head cultivars are approximately 30 000 plants p/Ha and the small baby heads (i.e. tennis ball size) are 80 000-90 000 plants p/Ha.

Producers should make themselves familiar with the best plant spacing in order to achieve the optimal plant population in their area.

Many producers establish their crops too closely to one another between the rows, which results in a dense population of crops. This is especially true for small-scale farmers and home owners, who might have a limited planting area or sometimes a lack of water. They therefore make use of a higher plant population or narrower spacing in the hopes of achieving higher yields. It is logical to make use of higher plant populations when limited soil is available. However, in this case, more attention may be given to plants that produce runners or ranking crops such as pumpkins, green hubbard, butternuts, lemon squash, watermelons, sweet potatoes etc. It is worth experimenting with these crops within a row outside the production area, or trellising some of them on fences or trees. High-trellising indeterminate crops such as tomatoes, cucumbers, peppers and runner beans may be trellised to heights of 2m or higher. However, higher growing plants should not overshadow lower growing plants is this may impact their growth.

When the plant population is too high, the numbers of fruit (tomatoes), tubers (potatoes), bulbs (onions), roots (carrots, sweet potatoes and beetroot) may decline, as the individual plants compete for light, water, nutrients and oxygen.

Super markets and consumers sometimes prefer **smaller fruit sizes,** such as tomatoes, peppers, tubers (potato), bulbs (onions), globes (beetroot), edible roots (sweet potato and finger carrots), cole crops (baby cabbage and baby cauliflower) and even baby lettuce heads. Therefore, the plant spacing should be adapted to produce these smaller sized crops. The normal spacing for medium sized cabbage heads is 30 000-40 000 plants p/Ha and for baby cabbage, 80 000-90 000 plants p/Ha.

Certain crops should be spaced closer together in summer conditions, due to the bright, hot sunlight, while in winter, the spacing should be further apart. For instance, the population of determinate open-field tomatoes is usually 18 000 plants p/Ha while in cooler conditions it is 16 000 plants p/Ha.

11.1.1 THE IMPACT OF OVERLY DENSE PLANT SPACING
- When plants are spaced too densely, they compete against each other for essential nutrients, sunlight, water, oxygen and carbon dioxide. The result is usually long, leggy, pale green and brittle plants, especially when cooler, shadier conditions occur.

- Sunlight can be severe in warmer areas and the day-length also plays a very important role.
- Make sure to maintain the optimum spacing and plant population. If the spacing between and within rows is too wide, the yield may decline, due to the wasted space that is not being optimally used. There is an ideal minimum and maximum plant spacing requirement for each crop.
- It is more difficult to control diseases and insects when the plant population is too dense, as it creates favourable conditions for them.
- Some crops such as sweet potatoes may be used to control weeds during the later stages of development, as the plant vines overgrow and smother the young and low-growing weeds. However, be aware that some cultivars produce short runners and others much longer runners.
- Plant density influences the time of maturity. For instance, when onions, egg fruit and melons are spaced close together, the maturing and harvesting time is earlier. Cabbage heads mature later.
- Close spacing usually makes weed control much more difficult, especially when mechanical implements are used. The plant spacing between the rows should be adapted according to the distance of the tractor wheels or implements before the crop is established.
- The ideal crop populations should be known ahead of time to achieve high yields, good fruit size, firmness, large crispy leaves as well as ideal sized globes (beetroot and onions), roots (carrots and sweet potatoes), pods (beans) and so on.
- Producers are advised to do their own research and to use their common sense when determining optimum plant spacing in their particular area and production land.

Sometimes healthy, dense plant spacing tends to increase the leaf canopy of certain crops. This is sometimes beneficial, as the vigorous plant growth will protect certain fruit from sunburn. Examples of this include tomatoes, peppers and pumpkins, especially the green skinned types of pumpkins that are more vulnerable to sunburn during hot, sunny days.

11.2 CALCULATIONS FOR PLANT SPACING AND PLANT POPULATION

1 (one) Ha = 100 x 100m = 10 000 m².

When crops are spaced, for example, 1m apart from one another, between and within the rows, this results in 100 rows with 100 plants in each row = 100 x 100 = 10 000 plants p/Ha or 1 plant per m².

75 Tomato plants which are established using an arrow drip system in soil in a shade net house. Plants are trellised and pruned to the one-stem method up to the 2.5m horizontal wire.

76 Be careful with close spacing of plants. This would result in a heavy top growth and very possibly, long, thin carrot roots.

When the spacing between the rows is 50cm x 50cm and each row has 200 plants and there is a spacing of 50cm between the rows, then 200 rows need to be established (1Ha = 200 x 200 = 40 000 plants p/Ha or 4 plants per m²).

If spacing for carrot plants are too dense, the plants will compete for light and nutrients. **See photo 76.** In this case, thinning was neglected and therefore, the carrot tops became too thin. Plant growth was be higher, but the roots were thin, long and bleached. In this case, only 30% of the carrot roots were marketable.

11.3 SPACING CROPS IN GREEN HOUSES

The ideal plant population for high value crops, such as for tomatoes and peppers in tunnels (double rows on ridges), is to achieve a plant population of 26 000 plants p/Ha or 2.6 plants per m³. However, according to some experts, the plant population should not exceed 2.5 plants per m³ in summer and 2.3 plants per m³ in winter.

The spacing of plants in tunnels is closer, because when using the correct plastic covering for vegetables, the covering disperses the sunlight and creates a better spread of light throughout the entire tunnel.

Plant population of high-value crops such as tomatoes, peppers and cucumbers should not exceed 750 plants in a 10 x 30m tunnel.

Yields of high-value indeterminate tomato cultivars may easily reach 6kg per plant or 150 ton p/Ha. Remember, technology and production methods are evolving all of the time. The result is better yields, better adapted cultivars and cultivars with shorter internodes (more trusses up to the 2.5m horizontal wire). In addition, disease resistant or partly-resistant cultivars are available or becoming available.

Plant breeders' are always developing and breeding new, improved cultivars. The continuing research will eventually benefit all producers.

Producers are advised to do their own evaluations regarding spacing. From time to time, plant breeders' may release new cultivars which may require different spacing to achieve maximum yields.

77 Foot paths should be established between crops. The cauliflower (left) and the green beans (right) should have been spaced with foot paths to ease weed/chemical control and to provide better light and aeration. Surprisingly, the beans have produced high quality bean pods due to the bright sunlight during Jan/Feb.

78 Green manure or cover crops should be established as densely as possible to produce maximum foliage for later incorporation. Above is a winter crop almost ready to be incorporated. The production land should be fertilized as normal, to achieve as much plant material as possible.

CHAPTER 12

TRELLISING (TRAINING) AND RIDGING

12.1 INTRODUCTION

Fresh market tomatoes, greenhouse cucumbers, peppers and runner beans need to be trellised. There are two main types of growers: determinate or short growers and indeterminate or high growers. It is primarily the indeterminate growers that need to be trellised.

Peppers do not normally grow higher than 1.8m, depending on the cultivar. They are usually boxed to prevent the plants from falling over. When peppers are boxed, they are not pruned. *See photo 11 and photo 80.*

Green house pepper plants are trellised and pruned by using 2 or 3 stems. They are trellised up to a 2.5m horizontal wire. When production takes place in tunnels, shade net houses and in hydroponic systems, these structures are designed to allow the trellising of high growing plants that are to be pruned, such as indeterminate tomatoes, peppers and cucumbers.

Fresh market determinate tomato production in open-fields, is steadily declining (2011) due to the preference for producing indeterminate, long shelf life, high-value tomato, pepper and cucumber cultivars. The reasons for this include extended shelf live, easier insect/disease control measures, comfortable and visible harvesting of the fruit, higher first-class fruit, better yields, better quality and less spoiled and sunburned fruit.

The bushy roma and blocky tomato types, do not require to be trellised. This is due their determinate, bushy, small, low growing characteristics. The roma type of indeterminate saladette tomato plants, are however trellised and have recently become quite popular (2010). This is due to the fact that indeterminate saladette yields are much better, they have good internal colour and can be used for retail marketing as well as for processing.

79 New, improved tomato cultivars which have been trellised and established on well-prepared ridges. A drip system and a black plastic mulch covering was used.

80 A bullhorn pepper cultivar trellised and boxed in a seedbed using a drip system and black plastic mulching. This is a good producer and the F1 hybrid cultivar plants have reached a height of 2m.

12.2 ADVANTAGES AND DISADVANTAGES OF TRELLISING

12.2.1 ADVANTAGES OF TRELLISING
- Higher yields and better quality fruit.
- It is easy to harvest indeterminate tomato, pepper and cucumbers cultivars. The fruit is also clean and visible when harvested.
- The fruit of indeterminate crops tomatoes, peppers and greenhouse cucumbers are harvested over a period of approximately 2 months, due to the steady and systematic ripening of the fruit. Harvesting begins from the bottom, up to the 2.5 meter high horizontal wire, which generally produces 9-10 trusses.
- Spraying of plants to control insects and diseases is more effective and comfortable, due to the higher level of the plants.
- Insects such as aphids, red spider mites and whiteflies always hide underneath the plants leaves. When plants are trellised, it is easier and more effective to control them underneath the leaves.
- Damage of plants and fruit as well as the risk of spreading diseases is reduced to a minimum, due to the wider spacing.

12.2.2 DISADVANTAGES OF TRELLISING
- Trellising costs are high.
- Leaves take much longer to dry after irrigation, while excessive rainwater could also cause an increase in certain diseases.
- Labourers need to be trained to trellis plants correctly. Be aware, that pepper plants are very brittle and the stems break-off easily, especially when training the plants and harvesting its fruit.

12.3 RIDGING OF CROPS AND RAISED SEED BEDS

12.3.1 RIDGING
Making well-prepared ridges are beneficial for certain crops.

Ridging encourage aeration and improves the drainage ability of soil, especially when producing crops in clayish soil types (however, not in sandy soils).

When there is a chance of fungi in the soil, rather implement ridging, as it reduces the danger of diseases such as root rot (phytophthora). Because this disease attacks the root system of most plants and cannot be controlled once established, the disease should be preventatively controlled.

Crops that are normally ridged include tomatoes, greenhouse cucumbers, peppers, chillies, paprika, sweet potatoes, water/musk melons as well as potatoes after emergence and just before flowering.

It is not necessary to ridge plants in sandy types of soil, due to the already good drainage and aeration.

12.3.2 RAISED BEDS
Crops that favour being produced in raised seed beds include carrots, beetroot, lettuce, onions and to a lesser degree spinach. The beds are usually made 1 to 1.5 metres wide.

CHAPTER 13

WATER

13.1 INTRODUCTION

SA is classified as the 30th driest country in the world. Average rainfall is approximately 300mm per year compared to the rest of the world that receives more than double this amount. The agricultural sector and the mining industry are the main users of water.

Experts have indicated that global warming may reduce rainfall. Rivers, boreholes and dam water will be negatively influenced and water table levels will become lower.

Drought stricken areas are declared as disaster areas (June 2015/16) and Government is required to financially support many farmers. Unfortunately, this assistance is never quite enough, but it does assist small-scale and upcoming farmers.

Vegetable production consumes a considerable amount of water. Most crops require 350mm of water during their entire growing season. That is approximately 3.5 million litres of water p/Ha.

Ordinary household water usage has also become very expensive. Many times householders are requested not too waste water. Sometimes they are penalised with higher prices and if necessary, water restrictions. This system has already been applied in some areas (August 2015).

If you are unsure about your water quality, it should be tested before vegetable production is initiated. Water is the main factor when producing vegetable crops. It has the greatest influence on the entire farming industry. Water quality is of utmost importance and when it contains unwanted types of salts and other contaminants, it may negatively impact the physical conditions of the soil which may affect plant growth. Water that has been impacted by mining activities contain acids and other harmful elements. Plants are very sensitive to these kinds of acids.

In many areas, thousands of litres of wastewater from factories and from humans are poisoning our rivers and dams. It has become a huge problem when irrigation water is used from polluted rivers or boreholes. The presence of different types of salts in the irrigation water will not only impact the ability of plants to utilise nutrients, but may also influence the nutrient status of the soil itself. In the case of boreholes, if there is not enough water or there is an unsatisfactory delivery, only a small portion of the land may be irrigated successfully. To overcome these problems, one solution could be to build a storage dam. Irrigation systems and water pumps should be chosen correctly to suit the producer's needs and should be adapted to their water supply circumstances.

The availability and quality of soil water is an important factor and a determinant in the yield and quality of crops. Inadequate water during critical growth stages of plants, may increase the occurrence of certain diseases, cause malformation of fruit and restrict growth. On the other hand, if excessive rain or irrigation water is applied to sandy soil, it may cause leaching of mineral nutrients, especially nitrogen N fertilizer, which may be washed away from the root zone of many shallow rooted crops.

13.2 WATER SOURCES

Water sources may include the following: boreholes/wells (underground water), rivers, lakes/dams (surface water) and captured rain water in tanks or dams. Municipal water is an option for home owners, but nowadays this is very expensive and not that worth-while.

13.3 APPLYING THE CORRECT QUANTITY OF IRRIGATION WATER

Roots require large amounts of air to operate effectively. As water sinks into the soil, it draws air into the soil behind it. When poor drainage conditions occur, the water moves slowly through the soil and may even stagnate before reaching the lower levels of the plant root systems. Healthy soil is made up of around 45% minerals, 25% air, 25% water and 5% humus. Many farmers mismanage irrigation and supply either too much or too little water. It is generally known that many producers over irrigate their lands.

Plants take-up water according to two limits. The lower limit is called the wilting point. This is the point where the plant is not able to extract water from the soil, as the soil moisture level is too low and therefore wilting takes place. The upper limit is called field capacity or fully-saturated soil. This occurs when the soil is not able to retain any more water and it simply floods away.

Over irrigation displaces the air, and therefore the oxygen in the soil and causes water-logging, especially in clay types of soil. If, for example, a hard pan is present below the level of the soil, drainage will stop partially or completely, which may result in the plants drowning. In sandy soil, water and fertilizer may be wasted if it drains away too quickly from the root systems of the plants. Excessively wet soil conditions and substantial fluctuations in soil moisture levels, are unhealthy for plants and for soil organisms. Soil should not be too wet or too dry. Doing so, will keep microorganisms and plant roots alive, healthy and happy.

At all times, sufficient water and nutrients should be available to the plant roots. Experience teaches us that vegetable crops require at least 30mm water per week.

In the case of sandy soil, it is important that smaller quantities of irrigation water is applied, but on a more regularly basis. This prevents the water, fertilizer (especially the N fertilizer) and nutrients, from draining-away beyond the reach of the plant roots.

13.3.1 THE EFFECTS OF OVER-IRRIGATION
- The leaching and wasting-away of water and fertilizer, especially nitrogen (N), which drains-away too quickly through the soil.
- The replacement of air with water, resulting in the drowning of certain sensitive crops, which will show related symptoms relatively quickly.
- The reduction of root development, due to the water stress which may have taken place during germination, resulting in poor stands.
- Increases of root diseases, especially root rot.
- Retardation of the germination of seed, especially bean seed, due to the rapid way in which the seed coats crack.

Enough water should be supplied, at the correct time. During the early stages of the plants growing development, it is sometimes a good practice to withhold water to a certain extent. This will encourage growth of the deep root system for certain crops. It may be worthwhile experimenting with this.

It is important to withhold irrigation water for 1-2 weeks before harvesting certain types of crops such as garlic, onions, sweet potatoes and potatoes. This gives bulbs (onion and garlic) and roots (sweet potatoes and potatoes) below the surface of the soil, a chance to dry-off, become mature and harden their skin. This is especially applicable for crops such as sweet potatoes, potatoes and garlic. It also prevents fungi that prefer moist conditions, from developing. However, fungi are sometimes not active in the land, and may only do their damage after harvesting or in storage.

Successful crop production depends on adequate water supply throughout the growing season. Water comprises about 90% of plant tissue and therefore is one of the most important production factors. It plays an important role in physiological processes and also serves as a source of carbon, hydrogen and oxygen to the plant. An active growing plant requires large amounts of water and may replace its water content up to four times per day under optimal conditions and average transpiration rates. More than 95% of the water that is taken-up by the roots of plants, is lost through transpiration. Only a small fraction goes to plants growth. Therefore, water-stress related to too much or too little water, may have serious effects on plants.

Water stress related to under-irrigation, is usually the most serious problem, even if it occurs over a short duration. Water stress usually slows the growth rate of plants and causes the development of a smaller leaf canopy. This in-turn, may cause early maturing of plants and lower yields, especially when higher temperatures prevail.

13.3.2 THE EFFECTS OF UNDER-IRRIGATION
- The reduction of vegetative growth resulting in smaller plants, fruit, roots, bulbs and/or tubers.
- Bolting of certain crops. Plants become stressed and then go-over to flowering.
- Poor nutrient uptake, which may for instance cause blossom end rot (BER) and/or blossom drop on tomatoes, beans, cucurbits and peppers, as well as cracking of tomato fruit, potato/sweet potato tubers and carrot roots.
- A decline in the quality of fruit, leaves, bulbs, roots, globes and tubers.
- Early maturing due to the fact that the plant reproduces seed at an earlier stage, in an attempt to survive.
- Fruit may become prone or sensitive to sunburn.
- The plants become more vulnerable to insects and diseases, due to the thinner outer skin which develops. However, withholding water from certain crops may stimulate the plants to mature earlier due to the slight stress. Some producers believe that harvesting should be earlier, however, this could also backfire. Planting by the moon, is also a factor to consider.

81 Cover crops used to improve soil are spaced very densely to produce maximum top-growth. At the back is a Lucerne land which has been cut, irrigated and fertilized in order to be grown again and ploughed back into the soil for soil improvement. Also, the rhizobium bacteria which develops on the roots makes N available for the follow-up crop.

13.4 RAINFALL AND MEASURING METERS

Rainwater meters are the cheapest way to measure overhead irrigation and rainfall water. At least 3 or 4 rainmeters are recommended in a reasonably small production land. The rainmeters must be placed correctly and evenly in the land. They should be adjusted to be positioned slightly higher than the plants.

Do not install rainmeters near the edge of the land or near an overhead sprinkler. Most commercial farmers make use of overhead sprinklers which are placed 12 meters squared from one another. An easy and effective way to measure the soil water content, is by using tension meters and data logging measurement meters. Also, some of the newest smartphones may be used to assist the producer in determining the soil moisture content.

For instance, if 10mm water is to be applied on the production land, the producer should know how long the irrigation system will take to apply this amount of water across the entire land. For instance, the 12 x 12m overhead sprinklers normally apply 25mm in 4.5 hours, provided the wind conditions are still, the correct nozzles are used and the water pressure is correct.

The amount of water that is captured in the rainmeter should be checked regularly, in order to achieve the required amount of irrigation water. *See photo 82.*

However, keep in mind that the correct water applications are different for each crop. Due to these specific water requirements, incorrect irrigation could mean the difference between success or declining quality and yields.

13.4.1 USING A SPADE TO ASSESS THE WATER CONTENT IN THE SOIL
Dig a 30cm deep square hole into the production land with a spade (note the length of a standard spade is 30cm). Inspect the profile of the hole and see if the bottom is moist, semi-moist or dry. This method is difficult to employ for heavy, clay type soil and should rather be avoided.

Adjust your irrigation according to what you observe in the hole. Generally, experienced producers already know how much water to irrigate, however, upcoming farmers may make use of this method to determine water irrigation requirements. Most vegetable crop roots penetrate the soil to a depth of about 30cm. However, if the soil is well-prepared and soft, certain plant roots may penetrate the soil to a depth of 1m or more. For instance, the side roots of carrots may penetrate the soil up to a depth of 60cm, in ideal soil conditions. This is why soil improvement is so important. However, heavy clay, dead, toxic and sandy soil are excluded from this method.

13.4.2 TENSION WATER MEASURING METERS *See photo 83.*
A very useful instrument which may be used to measure the water content in the soil, is a **tension meter.** It determines when and how much water to irrigate. It is always difficult to know exactly how much water is necessary to produce crops in soil. Producers in South Africa usually over-irrigate crops and therefore waste water. On the other hand, during/after rainy days, producers tend to under-irrigate crops. The amount of soil moisture may then be misjudged, and plant stress may occur. When fully-grown and on hot summer days, plants such as tomatoes may extract more than 1 litre of water per day. Therefore, a population of 18 000 tomato plants p/Ha could extract or transpirate more than 18 000 litre water p/Ha per day.

For certain soils, too wet and too dry conditions are harmful for plants. The stomata of plants close, after which the plants may take a long time to completely recover. As a result, yield reduction and sometimes physiological problems such as fruit cracking and blossom end rot (BER) might occur. The measurement of soil moisture provides the irrigator with the basis for making intelligent decisions regarding the times and quantities that water should be irrigate. It is important to keep the soil moisture at the optimum field capacity level, which should be less than the highest soil moisture tension.

Because the most active roots of crops are not deeper than 30cm, a tension meter should be installed at the same depth. An additional tension meter may be installed at a depth of 60cm, to indicate when the soil is over irrigated. Instructions for the preparation and installation of tension meters are available from suppliers, which should be studied carefully to fully understand how to use these devices.

Tension meters measure the soil moisture water tension. This tension, may be described as the force that the soil exerts to hold water and is known as cent bars (cbs). The drier the soil becomes, the higher the cbs reading will be. For instance, for tomatoes the tension would be maintained between 10 and 20cbs. When the soil moisture tension exceeds 20cbs, irrigation should be applied.

The drier the soil, the larger this force becomes and the harder the plants need to work in order to draw the moisture is needs to overcome this force.

13.4.3 MOISTURE DATA LOGGERS AND SMARTPHONES
Computerized moisture data loggers are also available. Together with a laptop, these devices may be used in the field to measure the moisture content in the soil. This is an effective way to measure the soil water content for proper irrigation in order to save money, labour and water. Smartphone-based solutions are also available.

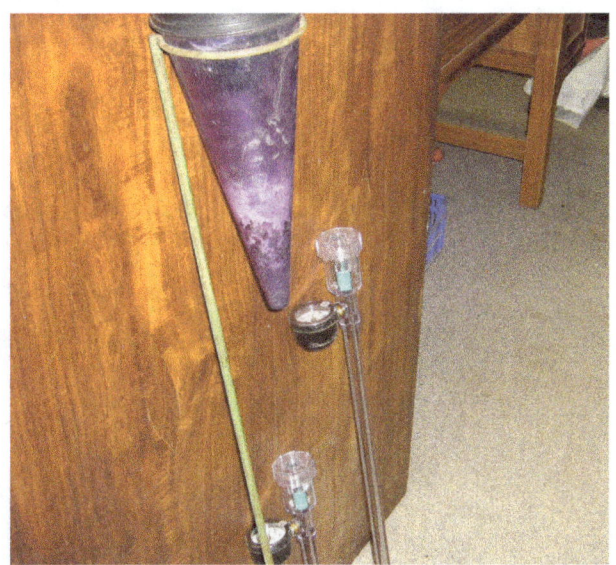 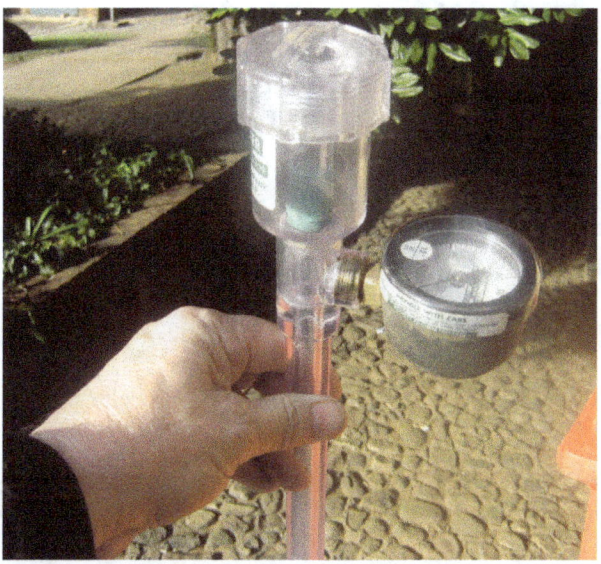

82 A rainmeter fitted on an iron rod to measure rain and irrigation water. Also, two tension meters that should be installed at different depths in order to accurately measure the water content in the soil.

83 A tension meter uses suction to measure the tension of the water in the soil. It is quite simple to read and similar to a speedometer meter. It is particularly useful for fruit trees as one meter may be established at a depth of 60cm into the soil.

25 YEARS OF GROWING MORE WITH LESS

SMART IRRIGATION SOLUTIONS

HELPING THE WORLD GROW MORE WITH LESS

With a rising world population and food demand on the one hand, and dwindling natural resources on the other, the need to grow more crops and produce more food with less resources is greater than ever. As the world's leading irrigation company, Netafim is driving mass adoption of smart irrigation solutions to fight scarcity of food, water and land. Providing farmers with the best agronomic and technical support to ensure outstanding results and peace of mind, Netafim is helping the world grow more with less.

Contact our irrigation and agronomy experts to learn more about how we can help you grow more with less.

Tel: +27 21 987 0477 • infoza@netafim.com • www.netafim.co.za

CHAPTER 14

IRRIGATION

84 An overhead sprinkler irrigation system. Water pressure should be adapted to meet the irrigation requirements. The sprinklers and pipelines are 12m apart from one another.

85 A pivot irrigation system with six towers. This system saves labour and money in the long run, but it is initially very expensive. It is also difficult to synchronise the towers.

86 A drip irrigation system used for tomato production. In front is determinate growers and at the back indeterminate growers.

14.1 INTRODUCTION

Irrigation can only produce optimum growth results from plants when the correct water (amount and quality) is used and at the correct scheduling is followed. This may seem simple, but in practice, irrigation may vary, even between cultivars of the same family. Soil structure and temperature also play a considerable role in irrigation.

The basic principles of irrigation do not change much in relation to the crop that is produced. These principles are classified into the **stages of plant development,** as follows:
Stage 1: Crop establishment stage.
Stage 2: Vegetative plant stage.
Stage 3: Fruit set and further development stage.
Stage 4: Ripening and harvesting stage.

The different stages mentioned above require different irrigation schedules to achieve optimum yields. There is a misconception by some producers that by simply applying water once or twice a week, healthy, vigorous plants will be the result. However, in many cases, in order for plants to be able to withstand adverse conditions, irrigation actually needs to be decreased during the plants initial development stages.

Getting irrigation right should be of a very high priority. This means the quantity of water, when it is applied, and how it enters and exits the soil should be carefully considered. Irrigation represents maintaining a balance between water and soil. It is the main management factor for vegetable production.

Water has become one of the most important and scarce natural resources and its prices are continuing to rise. South African agricultural farmers are the single biggest users of water. It is therefore of the utmost importance that water resources are managed as efficiently and effectively as possible. Research into water conservation and usage is therefore very important in this regard.

Drip or trickle irrigation is one of the most effective ways to supply and conserve water. However, it cannot be used for all crops. To a large extent, drip irrigation effectively reduces evaporation. The schedule of when to start and stop irrigation is also a very important aspect of good water irrigation management.

The amount of irrigation water to be applied depends on various factors such as the water retaining ability of different crops, weather conditions, soil conditions, the growing stages of plants and the depth of the root system. The water holding capacity of the soil determines the irrigation frequency, root depth and the amount of irrigation to be applied. In this regard, it is important to make use of at least 2-3 rainmeters (on a fairly small portion of land), which should be located on opposite sides of the land. This will provide a relatively accurate measurement of the amount of irrigation water and rainwater that has being received across the entire land.

It is required that producers combine the measurements of both rain and irrigation water to get an accurate result. For instance, in high rainfall areas it is hardly necessary to apply irrigation water to meet the ideal soil-water content. However, be aware that sometimes rain water does not remain in the soil for too long, because the plants do sometimes extract considerable amounts of water out of the soil. For example, when establishing 18 000 fully-grown tomato plants p/Ha, in hot conditions these plants could extract more than 1.5 litres of water per plant or 27 000 litres of water p/Ha per day. Soil moisture levels in the root zones of the plants determine the starting and stopping time of irrigation.

Soil may be compared with a sponge. When a sponge is dipped in water, it absorbs water until it is saturated. Similarly, when water is applied to the soil, it absorbs water until the full capacity of the soil is achieved. When water is squeezed out of a sponge, it dries out. The same happens when soil dries out. The plant roots are required to work harder to withdraw water and nutrients out of the soil, which leads to poor water absorption and transpiration. The result is that too little nutrients are then available to feed the plant and build/maintain its cells.

When water stress such as this occurs, the plants stomata closes, and thereafter it may take 24 hours or more to recover. This leads to poor plant development, yield losses and poor quality. Large plants withdraw water more frequently and therefore irrigation should be managed accordingly. Enough water should be applied in a scheduled manner and it should be measured as accurately as possible.

14.2 POINTERS FOR IRRIGATING CROPS

- Do not irrigate when the wind speed is above 6km per hour. Rather wait until the wind has died down, which is usually early in the morning.
- Some irrigation systems may take 3-4 hours to evenly irrigate 25mm of water.
- It is important to measure irrigation by using rainmeters which are placed at the correct locations in the land. **This measures the amount of water and monitors if the distribution is even.**
- Overhead sprinklers should be revolving at walking speed.
- Nozzles should all be the same size. Always check them, as they may be worn out.
- The basic layout of the irrigation system should be installed correctly and the distances between the sprinklers should be maintained accordingly.
- Water pressure must be controlled, because when the pressure is too low then the water will spread unevenly over the land. On the other hand, when the pressure is too high, water irrigating patterns may overlap and water would be wasted.
- Water pressure for sprinklers systems should be monitored carefully. The water pump should be chosen according to the number of sprayers and the size of the area to be irrigated.
- The main and sister lines should always be regularly checked for water leaks.
- During the hottest periods of the year, irrigation is very important as the plants have a strong need for water. Be aware that irrigation water cools the plants off.
- Producers should supply enough water to their crops throughout the day. The total volume of every irrigation should be based on the following principals:

1. Plant age
2. Water soil holding capacity
3. Runoff water

14.3 GUIDELINES FOR IRRIGATION

- Keep the top-layer of the soil moist until the seeds have emerged, especially for hot, sandy types of soil or soil that tends to form a hard crust on its surface after irrigation.
- The most critical stage for irrigating *rooted crops,* is at the early stages of their development. This is the case for crops such as tubers (potatoes), bulbs (onions, garlic), rooted crops (carrots, sweet potatoes) and globes (beetroot, turnips).
- Most vegetable crop roots are situated in the top 30cm of soil. Make sure that moisture is available at this depth, at all times.
- Evaporation and transpiration is low during the winter, but conditions may vary and producers should be adaptable.
- As a general rule, irrigate vegetable crops at least once or twice a week. However, this also depend on the soil structure, types of plants and temperatures. For example, two light irrigations would be better than one on sandy type soils.
- If possible, irrigate early in the morning to allow plants to dry. This is to avoid diseases and to allow bees and other insects to pollinate the plants effectively.
- Always take rainwater into account, as it reduces the irrigation cycles. However, after rainy days, producers sometime wait too long to irrigate, as they often misjudge the amount of rain that has fallen. Sometimes in the high rainfall areas, production can go through the entire season with little or no irrigation being required. However, this seldom occurs.
- The amount of water that is given to a crop throughout its various stages of growth development should be adapted. Water applications should be increased when some plants begin to flower and when fruit, tubers, bulbs, globes and roots begin to develop.
- Stop irrigating a week or two before **harvesting** crops such as tubers, roots and bulbs in order to give them a chance too dry-off, mature and harden their outer skins.
- It should be noted that female pumpkin flowers should receive approximately 20 visits from bees before pollination is completed and full fruit set and fruit development can take place.

14.4 IRRIGATION AND ROOT DEPTH

Vegetable crops mostly all have different root depths. The correct amount of irrigation water will moisten the soil to the depth that the roots grow. When less water is given, the moisture may not reach the roots.

The root depths that are provided below are for different crops. It is often times not necessary to irrigate much deeper than the root development system of the crops, as most vegetable crops extract around 70% of their water requirements from the top 30cm of soil.

TABLE 3: ROOT PENETRATION DEPTH FOR DIFFERENT VEGETABLE CROPS

SHALLOW ROOTED CROPS		MEDUIM ROOTED CROPS		DEEP ROOTED CROPS	
Onion	40cm	Cabbage	50-60cm	Pumpkin/Squash	70cm
Lettuce	45cm	Pepper	60cm	Watermelon	80-90cm
Spinach/Cucumber	45cm	Tomato	60cm	Green mealies	80cm
Carrot	40cm	Potato	55cm	Sweet corn	70-80cm
Beetroot	40cm	Beans	54cm	Musk melon	90cm

87 Dripper lines established on ridges. The dripper lines have been placed underneath the white plastic mulch for the production of baby marrows.

88 A dripper system for cabbage production. Note that the solenoid valves may be set manually or automatically to open and close the water flow to the different plots of land.

89 Overhead quick coupling sprinklers commonly used by most producers

14.5 IRRIGATION SYSTEMS

Producers should always remember that water is a precious resource for the benefit of all living creatures on earth. It is the biggest requirement for human and plant survival on this planet. Therefore, it is recommended to invest in a good irrigation system which applies water effectively, economically and evenly across the entire production area.

Irrigation water and irrigation systems are quite expensive. Once an irrigation system is installed, it should be effective, economical and should use the minimum amount of water possible. It is important that the layout of the irrigation system be planned in a way that is well thought-out. Water that is wasted outside the production area, due to overlapping overhead systems, should be minimised as much as possible. Producers should investigate the best irrigation system to suit their specific requirements. An irrigation system should be installed after the land has been prepared. If there is an inadequate number of pipelines, especially for larger areas, it will be necessary to continuously move the pipelines to different parts of the land. The land should be irrigated as soon as possible after crops have been established, especially if hot conditions are prevalent.

14.5.1 IRRIGATION SYSTEMS AVAILABLE IN SOUTH AFRICA
Several different irrigation system designs exist. The systems which are available in South Africa are as follows:

14.5.1.1 THE PERMANENT SYSTEM
This system is buried below the soil, but its cost is high and it is seldom used.

14.5.1.2 THE HAND IRRIGATION LINE
This system is particularly useful for the germination of small seeds, especially when the land has a difficult topography. The disadvantage of this system, is that labourers are required to move the lateral lines to the next part of the land which requires irrigation.

14.5.1.3 CENTRE PIVOT *See photo 85.*
This is one of the least expensive, but most effective systems that can be used in the longer run. It requires very little labour to operate. It is energy efficient, but very expensive.

The **disadvantage** of this system, is that the circular irrigation patterns do not reach the corners of the fields. The system can however be adapted by adding sprinklers that fold-out when they reach the corners of the land.

It is not well adapted for smaller areas and irregularly shaped lands. it is a permanent system that can only be installed in one production area at a time. The sequence of the different towers of the system are also quite difficult to calibrate.

14.5.1.4 OVERHEAD QUICK COUPLING SPRINKLERS *See photo 91.*
The most commonly used system. It is very popular for large production lands. It is easy to shift it to the next land, due to the fact that the sister irrigation pipe lines are 12m apart from one another. However, the lightweight aluminium pipes and brass sprinklers are popular amongst metal thieves.

14.5.1.5 SPRINKLERS *See photo 98.*
Some sprinklers are very light weight and can be installed and removed quickly. Sprinklers range from small lightweight plastics to large micro sprayers. There is also a wide range of static, gear-driven sprayers as well as rotating and impact sprinklers. They range from smaller, low-pressure sprinklers to large high-pressure cannons.

The pressure of the water determines the distance that the sprinkler reaches, therefore it is important to configure the system with the correct pressure. When large areas are irrigated, solenoid valves should be used to separately irrigate the various pieces of lands.

The smaller types of sprinklers are mostly effective for small-scale, upcoming and developing farmers. These sprinklers require fewer skills to operate and the layout is not too difficult to configure. They are generally used to irrigate seed beds in the land and seed trays in nurseries. Some sprinkler heads irrigate half a circle and others a quarter of a circle. This assists in irrigating the corners, prevents overlapping so that better germination can occur and also helps prevent light frost from forming early in the morning. Sprinklers are also effective when the soil topography is uneven, such as when the land is on a slope.

14.5.1.6 THE WHEEL LINE SYSTEM
This system is designed to reduce labour across the production land, as the water pipeline is driven by an electrical motor. The pipeline could also be driven with water pressure from the main irrigation line. The lateral lines that carry the water are mounted on wheels and spaced 6m apart from one another.

14.5.1.7 FURROW IRRIGATION
Furrow (or flood) irrigation is irrigation water that is confined to furrows rather than wide beds. Water wastage is much higher when flood irrigation is practiced due to the waste drainage of water in the furrows, especially in slightly sandy soil. Therefore, the furrows and their layout should be planned properly to suit the crop that is to be established.

Be aware that the longer the furrow is, the more water will be wasted. This is due to the fact that at the beginning and end of the furrows, the water usually dams, especially when the labourer waits too long to open the next furrow or to close the existing one. Labourers should be well-trained so that they may open and close the furrows at the right time, in order to prevent the water, which usually originates from a stream, from overflowing at the end of the beds.

With this system, water is used more efficiently when narrow furrows are implemented, instead of wide beds. This is because wide beds generally do not get soaked in an even manner, and evaporation is also high.

Furrow irrigation uses much more water at the beginning of the furrows, while wide beds use even more. When the gradient of the land is steep, excessive soil erosion can take place. The reason for this is that the smaller soil particles are carried from the top of the furrows to the bottom end of the beds. This occurs especially when the water flows quickly due to steep slopes. Furrow flooding cannot be practiced when the soil is too sandy.

14.5.1.8 DRIP IRRIGATION SYSTEMS *See Chapter 14.7 for dripper systems.*

90 Cabbage is an ideal crop for this dripper system as the dripper openings are spaced 30cm from one another which is an ideal spacing for medium-sized cabbage heads.

91 Many commercial producers use overhead quick coupling sprinklers. The pipelines are 12m apart from one another and 100mm in diameter. However, when large lands are to be irrigated then high water pressure is required to deliver the water. Otherwise, the pipelines are shifted to the next portion of land. A problem with this system is that the outer pipelines waste water as they do not cross the inner lines. Longer, more upright pipes could be installed to irrigate high-growing crops such as maize and tomato.

14.6 REASONS FOR A POORLY DESIGNED IRRIGATION SYSTEM

- A common mistake that producers tend to make when designing an irrigation system, is to make the main pipelines too narrow. The result is that not enough water can be delivered to the production land.
- The **water pressure** should be adjusted to the area that is to be irrigated. This should not be too high nor too low. When too high, the pipes may burst and when too low, poor, uneven distribution of water will result.
- Small irrigation systems are often purchased off the shelf with little or no advice being given, as to the most economical pipe sizes and nozzles.
- Pipelines are often underestimated and the result is that the sprinkler nozzle size is often smaller than what is required. When the nozzle size is larger, the producer should allow for shorter irrigation intervals. This is due to the larger delivery of water through the nozzles. The water pressure should be adapted if it is too low.
- Most irrigation systems that are less than 5 Ha, are operated using large diesel driven pumps. However, the cost of pumping water by using petrol or diesel has recently doubled (2014).
- Small-scale farmers are seldom trained to correctly schedule irrigation systems. This is because they do not realize the importance of the nozzle diameters and the effect that nozzle wear has on the irrigation system. Some careless producers sometimes install as much as 3 different types of nozzles in a production land. A mix of sprinklers or nozzles can have a disastrous effect on efficient distribution of water over the land, especially when soluble fertilizer is applied through the irrigation water.
- Some dealers provide poor and inaccurate advice to producers with regards to the installation of irrigation systems due to their lack of knowledge or expertise.

92 The problem with drip irrigation is illustrated above, where the dripper liners need to be removed from the land after harvesting, in order to prepare for the next crop.

93 Do not underestimate the length and density of plant roots. Above is a lettuce plant that has been pulled-out of a simple type of hydroponics system.

94 Above is an example of the roots of a single celery plant in a NFT hydroponics gravel system. The gravel is only 4.5cm deep in the plastic troughs. **See Chapter 23.8 for NFT systems.**

95 Production where overhead irrigated and fertilized was applied according to a soil analysis. Two top-dressings were applied and the soil was also previously improved.

96 The primary dripper line which only irrigates 0.4 Ha of the production land.

97 Unbelievably, this baby marrow production is only 5 weeks old from the time the seedlings were transplanted, and it is ready to be harvested for the first time.

14.7 DRIPPER SYSTEMS

- Be aware, drip irrigation systems cannot be used for all crops.
- Water pressure should be adjusted according to the dripper opening sizes and the size of the land. Solenoid valves should be installed when large lands are to be irrigated
- When using a drip system, the soil should be properly prepared and preferably be level.
- A properly designed drip system involves extensive pre-planning.
- Once a soil analysis has been conducted, fertilizer may be applied the same way as for open-field production.
- Broadcast the granular fertilizer on the planting area and work it into the soil to a depth of approximately 20cm. *See Chapter 15 Fertilizer.*
- Proper design and good management of the drip system in order to achieve an efficient operation, cannot be over-emphasized. Note that lengthy dripper lines expand during the heat of the day and shrink at night. Therefore, the **dripper lines may tend to shift from their original positions.** In order to keep them in position, they may be kept slightly tensioned by using strips of inner tubes or wire hooks. The longer the dripper lines, the more the lines will move from their original positions.
- Soluble nitrogen (N) used for topdressings is applied through the dripper system.
- Note that when preparing the soil and applying granular fertilizer (according to a soil analysis), it is done

mechanically. The soil should be moist before installing the dripper system.
- Granular fertilizer such as LAN/KAN or other mixes of fertilizer cannot be used, as they may block the openings of the dripper system.
- Only topdressings should be supplied to the plants when the correct stages of plant development have been reached. Soluble ammonium sulphate, calcium sulphate or calcium nitrate is mixed in the water tank at the required amounts.
- It is important to identify and correct any plant deficiencies during the early stages of plant development. Use M.A.P when phosphate deficiencies occur.
- Potassium K is often applied at the later stages of plant development, as it assists in building plant cells and maintaining, for instance, firm tomato fruit. However, if too much K is applied, the element becomes out of balance with the other cat-ions such as calcium (Ca) and magnesium (Mg). Note that the plants take-up elements in the proportions that they are available in the soil. However it is unable to absorb one element and leave others.
- All too often, when too much K is applied, a Mg deficiency may develop. This results in larger, thicker, lower leaves and the plants show a yellow moulting. However, a slight deficiency is not serious and could be corrected when magnesium sulphate is applied.
- A concentration of Ca in the irrigation water, especially in borehole water, may cause the pH to rise. Often this results in the development of an iron Fe deficiency in the later stages of plant development.
- When establishing wide rows, one can consider doing soil preparation with a tined implement within the planting rows, and then leaving the rest of the soil undisturbed.
- Tomato producers are particularly aware of the benefits of a drip system.
- A well-fertilized land, as confirmed by a soil analysis, will require that only N top or sidedressings require to be applied. This can be done by mixing the N in the water tank and irrigated through the dripper system.
- Monitor plants for a dark green colour, as this is a clear indication of too much N. In addition, when the tips of plants, especially for tomatoes, begin to become yellowish, N should be applied immediately.
- Some expertise and knowledge is required to manage the application of fertilizer through a dripper system.

Complex weekly applications of fertilizer through a dripper system by means of a dozetron system is expensive and should be done by using a properly designed fertigation system. Some representatives may attempt to influence you to use a weekly fertilization program through the dripper system, but this is not always necessary. It is expensive and the producer needs professional expertise when working with such a fertigation program. Approach the fertilizing experts and they will supply producers with a fertilization program.

Producers make use of a dozetron injector or a venture system and small tanks to dissolve the fertilizer. They then automatically inject the correct measured amount of fertilizer into the main water stream which flows into the drippers. 2 or 4 litre per hour drippers could be used, however this depends on the water pressure and the size of the production land.

CHAPTER 14: IRRIGATION

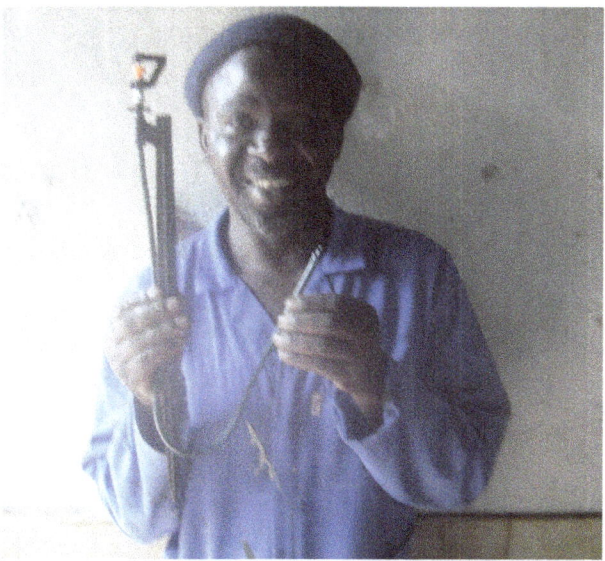

98 A small-scale gardener who is quite satisfied with his sprinkler system. Be aware that these types of systems block easily, especially the rotating types. Acceptable filters should be installed. These sprinklers are also popular with nurseries and small-scale farmers, however it is important that the water pressure balance is correct according to the number of sprinklers that have been installed.

99 Soluble fertilizers are mixed in the water tank and pumped to the sprinklers inside the seedling unit. This system could also be used in shade net houses and tunnels to irrigate high-value indeterminate tomatoes, peppers and cucumbers which are produced in soil. **See Chapter 14.5 for irrigation systems.**

Perfekte oplossings in besproeiingsbestuur

Doen navraag by jou naaste besproeiings handelaar oor Agriplas Produkte

www.agriplas.co.za

agriplas
PERFECT WATER MANAGEMENT SOLUTIONS

KAAPSTAD - Hoofkantoor
Posbus 696, Brackenfell 7561
Tel: +27 21 917 7177
Faks: +27 21 917 7200

GAUTENG
Posbus 11052, Randhart 1457
Tel: +27 11 908 2204
Faks: +27 11 908 5312

MPUMALANGA
Suite 63, Postnet X 11326, Nelspruit 1200
Tel: +27 13 755 3510
Faks: +27 13 755 3505

«Change your green into gold»

with GreenGold® 30

GreenGold® 30 is a dry fertilizer applied as top dressing after planting and post emergence with the three-in-one winning combination of nitrogen, calcium and boron.

GreenGold® 30:

- Comprises a combination of nitrogen, water-soluble calcium and water-soluble boron for rapid uptake and quick plant reaction – all in one application.
- Nitrogen is water soluble and partially quickly available and partially gives you a longer reaction.
- Enhanced efficiency can result in higher yield and profitability.
- Is highly suitable for vegetables and fruit but also suitable for grain crops, sunflower and pastures.

GreenGold® 30 is a must for vegetable growers.

Kynoch – enhanced efficiency through innovation.

011 317 2000 | info@kynoch.co.za | www.kynoch.co.za
Not trading in the Western Cape.
GreenGold® 30 K9750, Act 36 of 1947.

Farmisco (Pty) Ltd t/a Kynoch Fertilizer
Reg no. 2009/0092541/07

CHAPTER 15

FERTILIZER

100 The maize plants on the left were not limed, while the maize plants on the right were limed continuously for 3 seasons. Both plots received the same N, P and K levels. Photo: Grain SA.

101 The plants on the right are nutrient deficient. Plants with extremely slow, stunted growth usually represent a potassium or phosphorus shortage, while spindly stalks signify a nitrogen deficiency. In order to identify a particular deficiency and the best course of action, make use of plant analysis/soil testing. Photo: International Plant Nutrition Institute.

15.1 INTRODUCTION

In South Africa, vegetables are produced across a large variation of climatic conditions and soil types. This has an impact on the fertilizer and nutrient requirements for production, not just by region, but by farm, soil type and even between cultivars of the same species.

Producers find it difficult to obtain the required information from reliable sources. It is therefore becoming important for producers to know and understand their crop fertilization requirements. Understanding a crop and its requirements is not always easy. It involves understanding the growth stages of the crop, the requirements of each stage and the different types of fertilizers that are available.

Due to the high cost of fertilizer, which is approximately 25-28% of the total input cost for most crops, it becomes clear that fertilization should be well managed. More producers are changing their fertilizer programs to suit their specific soil types, as the soil type determines the cultivation and fertilizer program. Also choosing the correct nitrogen N source when applying N as a top dressing is important as it is not only soil fertility that determines high yields but also soil potential.

It is very important not to apply too much or too little fertilizer. Too much is expensive and too little has a negative effect on yields. In this regard, a reliable chemical or seed company will assist and inform producers.

Do not waste your fertilizer by attempting to improve poor quality or dead soil. Such soils require much more care and soil improvement before a fertilizer programme may be initiated.

Nitrogen N may become a problem if too much is applied on certain crops because it may stimulate vigorous plant growth but may also hamper yields. For instance, flower and fruit drop of tomatoes and beans may occur. On the other hand, applying too little nitrogen may produce damaging results such as reducing the leaf area, slowing growth and substantially reducing yields. It is therefore important to control the application of nitrogen N carefully.

It is important to not only know your crop, but also your cultivar. Each crop and cultivar may differ in it's response to different fertilizer applications and as a result, yields may be seriously impacted. It is important to understand that the foundation of any growing plant begins with its underground root system and therefore all necessary support to the developing plant should be given from the start. The aim is to cultivate a healthy, strong seedling which is able to face any environmental challenges it might encounter.

Achieving good results starts with soil. Special care and attention should be given to achieve and maintain fertile soil at all times. However, soil improvement can be expensive, but when it is done correctly, plants produce higher and better quality yields. This is the producer's reward. The basis for fertilizer recommendations are found in soil analysis tests which indicate the status of nutrients in the soil.

The following factors should be considered long before fertilization takes place:
- The pH at the root zone.
- The soil moisture level and the methods of irrigation.
- Salt content and the nitrogen N source.
- Temperature, light, humidity, soil types and planting dates. These could all have a considerable influence on yields and quality of crops.

There is no single recipe for fertilization and nobody can categorically inform producers what the nutrients and pH status of their soil is. Over time, experienced producers develop an understanding of the fertility of their soil. However, most producers still make use of a reliable soil analysis. This is to make sure that the best maximum or minimum fertilizer amounts are applied, for the best yields and quality, throughout the growing season.

The interpretation of a soil analysis should be done by a fertilizer expert. A report is produced to assist the producer to apply the correct combination and amount of granular fertilizer. For instance, a fertigation program may be implemented through a dripper irrigation system which administers the correct kinds and amounts of fertilizer. Fertilizer is injected into the main water line and flows out of the drippers into the soil next to the plants.

The required N topdressing is applied after the plants have emerged. N fertilizer dissolves quickly after irrigation. If possible, a better option is to grub the N fertilizer slightly into the soil, using a hand grub or a grub behind a tractor. To some extent, this would also assist in controlling the young weeds.

After applying the right kind of nitrogen N fertilizer, irrigate the land as soon as possible. Be aware that too much water may cause the N fertilizer to rapidly dissolve and drain beyond the reach of the plant roots.

The main nutrients are the macroelements that may be found in the mixed types of fertilizer. They are nitrogen N, phosphorus P and potassium K. N is much more movable in soil than P and K. N is mostly found in the mixed granular fertilizers which are to be applied in the open-field. Applications of N fertilizer after establishing crops is almost always necessary. P and K play an important role in different soil types, due to the movement of these elements. N drains quickly, especially in sandy soils, but in loam and clay types of soil, the N may be captured for longer periods. In the case of urea, the uptake of N is slow and may take as long as 2 weeks before it is available to the plant.

Soil temperature and the water content of soil have a considerable influence on the release of N. The release in cool or cold soils is much slower than in warmer soils.

When N is supplied through the water dripper system, it is recommended that the elements should be between 500 and 1 000 parts per million. When applied at the rate of 10mm water p/Ha then 100 000 litre of water is necessary to irrigate 1Ha. For example, 10kg of urea and 40kg of KNO_3 is applied in 100 000 litre of irrigation water.

No fertilizer program would be complete without giving attention to **microelements.** Some vegetable crops are more sensitive to elements such as calcium Ca, boron B, molybdenum Mo, zinc Zn etc.

Foliar sprays could be used to correct the microelements. However, they should be given at the right stage of plant development. Phosphorus P moves at a much slower rate in clay types of soil (even sandy soils) and therefore top or sidedressings are sometimes of no use. It is better to apply P before establishing the crop.

Calcium Ca and magnesium Mg plays a minor role in acid soil, since calcium is poorly transported downwards. Adequate levels are needed at the tip of the root system. If a dripper system is in place, soluble N fertilizer could be supplied by adding it into the water tank.

Sometimes only a fraction of an element may be necessary to bring the soil to the correct fertility levels. Fertilizer is expensive therefore do not over-fertilize. All fertilizer are salts and salts may build up in the soil. A wide range of nutrients and fertilizers are available on the market which may be confusing to many producers. It is important to know that fertilizer mixtures not only contain plant nutrients but also fillers and conditioners. Interpreting the label that is on the fertilizer bags can sometimes also be confusing.

Commercial fertilizer mixtures are generally described in the following format: 2:3:2:(22) 0.5Zn. This particular fertilizer is commonly used, as it is normally the cheapest balanced mixture. It is mostly used when no analysis recommendations are available. This means that the 2:3:2:(22) 0.5Zn mixtures contain 2 parts of nitrogen N, 3 parts of phosphorous P and 2 parts of potassium K. The 22 in brackets is the available plant nutrients. Be aware, that a concentration of 14 is too low, 22 is average and 40 is quite high. Zn means that the mixture also contains 0.5% zinc, because most soil in South Africa have a shortage of Zn.

Nitrogen N is essential for leaf development, phosphor P for **root development** and potassium K for strong healthy **flowers, fruit and pods.**

To calculate how much of each element is present in a mixed fertilizer such as 2:3:2(30) Zn, proceed as follows: N+P+K = 2+3+2 = 7. This is the total amount of the 3 elements.
Therefore **N** = 2x30 = 60÷7 = 8.6%, **P** = 3x30 = 90÷7 =12.8% and **K** = 2x30 = 60÷7 = 8.6%.
Therefore, the 2:3:2 (30) 0.5Zn mixture contains 30% nutrients (N+P+K) of which 8.6kg is N, 12.8kg is P, 8.6kg is K and 0.5 Zn.

The most commonly used and cheapest fertilizer type is the 2:3:2: (22) 0.5 Zn mixture.

Different crops require different kinds and amounts of nutrients.

For instance, it may sometimes be better to use a fertilizer such as 4:1:1:(20) for leafy crops such as spinach, as it contains more (N). For root development crops such as carrots and turnips use 2:3:2:(22) 0.5 Zn and for fruit, flower and pod development (tomatoes, egg fruit, cucumbers) use 3:1:5(30) SR. However, this combination is quite expensive and is seldom used. SR means that the nitrogen (N) is released slowly in the soil.

Fruit and the flowers of certain crops only begin to develop approximately 6 weeks after the seedlings are transplanted. For example, tomatoes produce ripened fruit 8-9 weeks after transplanting. More K is then needed for fruit and fruit cell development when the fruit is still fairly young, especially when fertilizer is applied through the dripper system.

Granular fertilizers are mainly used for open-field production (fertilization) and soluble fertilizers are mainly used for drip and hydroponic systems (fertigation).

It is important to know what kinds of nutrients certain crops require and the easiest way to determine this is to recognize which nutrients the crop has removed from the soil during its growing period. Those nutrients that the crop has removed from the soil should then be replaced back into the soil to prevent it becoming infertile.

Calculate how much nutrients are removed from the soil per 1 ton of plant material during the entire growing period of the crop. Scientists multiply 1 ton of ground plant material for the specific crop to give the amount of nutrients that needs to be replaced back into the soil.

For example, if a yield of 30 ton p/Ha is expected from a specific crop, then the withdrawn nutrients factor for a crop such as lettuce is 3.4kg N per ton of plant material. Therefore N = 3.4kg x 30 ton (the 30 ton is the expected yield) = 102kg, P = 0.5kg x 30 = 15kg and K = 2.4kg x 30 = 72kg.

It is clear that lettuce does not require much P and therefore the above formula could be used. The **optimal planting times** for crops have a major influence on the yields and quality of some vegetable crops.

When crops are established outside (before or after) their optimal planting times, the yield and quality may be affected. Also diseases and insects may become a major problem. However, the advantage of establishing crops earlier or later in the season is that producers may be on the market earlier or later and may receive much better retail prices. Also, take note that certain cultivars, especially the F1 hybrid cultivars, may be adapted or semi-adapted to withstand certain conditions.

When producers over supply nutrients in the soil, the plant only uses nutrients and water it requires. Plants will not take up the excessive nutrients, because they only take up enough nutrients needed for healthy plant growth.

Commercial farmers should always research and gather as much information as possible.. There is modern technology available to producers, especially for large enterprises producing maize, cotton, soyabeans etc. The internet may thus benefit producers.

15.2 INCORRECT FERTILIZING

Incorrect fertilizing is one of the biggest problems in South Africa and across the world. It is not only unnecessarily high levels of nitrogen N which is often applied, but also other nutrients.

Many producers are unaware of this while others don't care and continue to apply fertilizer year-after-year resulting in soil which is spoiled and poisoned due to too much salts.

Poor yields and quality may be the result. Producers should be aware of what is going on in their soil. Proper and detailed soil analysis needs to take place and calculations should be made correctly.

15.3 PLANT NUTRIENTS AND REQUIREMENTS

TABLE 4: THE NUTRIENT REQUIREMENTS FOR CROPS

HEAVY FEEDERS	MEDUIM FEEDERS	LIGHT FEEDERS
Cole Crops	Onions	Beans
Potatoes	Carrots	Peas
Cucumbers	Beetroot	Most Herbs
Tomatoes	Garlic	Maize
Pumpkins/Squash	Spinach	
	Mellons	
	Sweet Potatoes	
	Lettuce (more N required as this is a water loving crop)	

The correct plant nutrients give the plant a strong root system as well as a healthy, green leaf canopy, bright flowers, firm fruit, tough roots, firm bulb and tuber skins.

It is a well-known that when plants are healthy, the plant grows vigorously, develops tough cells and is more resistant to insects and diseases. Insects find it more difficult to penetrate tough, healthy plants skins.

Plants differ in their nutritional requirements. More than 13 nutrient elements are available as can be seen in the table below.

TABLE 5: MACRO AND MICRO NUTRIENT REQUIREMENTS OF PLANTS

MACRO NUTRIENTS		MICRO NUTRIENTS	
Nitrogen	N	Iron	Fe
Phosphorus	P	Copper	Cu
Potassium	K	Chlorine	Cl
		Manganese	Ma
		Molybdenum	Mo
		Magnesium	Mg
		Boron	B
		Zinc	Zn
		Calcium	Ca
		Sulphur	S

It is well-known that Mo, Zn and B shortages are usually present in acid soil and also when the pH of the soil is high.

When plants show symptoms of shortages of nutrients (elements), the usage of foliar sprays are sometimes very effective. It is however important to apply it at the correct stage of the plants development.

TABLE 6: RELATIVE TOLERANCE OF VEGETABLE CROPS TO THE ELEMENT BORON

TOLERANT	SEMI TOLERANT	SENSITIVE
Beet	Pumpkin/Squash	Beans
Carrot	Musk melons	Cucumbers
Spinach	Cole crops	Garlic
Tomatoes	Capsicums	
	Lettuce, maize	
	Peas, potatoes	
	Sweet Potatoes	

15.4 ORGANIC AND INORGANIC FERTILIZER

A combination of organic and inorganic fertilizers are used for vegetable crops. The most important inorganic fertilizers are nitrogen N, phosphorus P and potassium K. Inorganic fertilizer contain much more N, P and K than organic fertilizer. Organic fertilizer (animal manure) contains N, P and K in small quantities while chicken manure contains much more N than cattle manure.

15.4.1 STRAIGHT FERTILIZER
Many kinds of fertilizer and fertilizer mixtures are available. It can become confusing when attempting to choose the best one, especially without having a soil analysis available.

Single or straight fertilizer only supplies 1 nutrient, for example, phosphorus P. However, in the case of super phosphate which contains more than just P, the chemical composition is mono-calcium phosphate and calcium sulphate.

15.4.1.1 Phosphorus fertilizer
There are 2 types of phosphorus P fertilizers, namely, rock and super phosphate. Rock phosphate is known as langafos and is cheaper than super phosphate. It also takes longer to dissolve.

15.4.1.2 Super or langafos on acidic soil and supers on alkaline soil
Single and double super phosphates are available. Double super phosphate contains more phosphorous K than single supers but is more expensive. Super phosphate is a phosphorus carrier and other phosphorus carriers are double supers, calmafos and lagafos. They differ, according to the quantity of phosphorus that is presented in the formula. Super phosphate has 8.3% P and other supers 9.0% 13.2% and 19.6%, P respectively.

15.4.1.3 Potassium
Some producers use K_2SO on neutral and alkaline soil and KCL on acidic soil. Each fertilizer has its own characteristics and should be used optimally. There are also a large variety of inorganic fertilizers available.

Table 7 gives an indication of the straight groups of N, P and K fertilizers.

TABLE 7: DESCRIPTION OF STRAIGHT NUTRIENT FORMULA FERTILIZER

NUTRIENT ELEMENT	FORMULA	DESCRIPTION
N Fertilizer	NH_2SO_4	Ammonium sulphate
	$NH\,NO_3$	Ammonium nitrate
	LAN/KAN or LAN	Limestone ammonium nitrate
	Urea	
P Fertilizer	Super phosphate (8.3 P)	
	Double Supers phosphate	
	MAP	Mono ammonium phosphate
	DAP	Di-ammonium phosphate
K Fertilizer	KCL	Potassium chloride
	K_2SO_4	Potassium sulphate
	KNO_3	Potassium nitrate

15.4.2 MIXED OR COMPOUND FERTILIZER

Inorganic fertilizer is usually supplied in 50kg bags. Smaller quantities are also available but are usually more expensive.

Fertilizer mixtures contain N, P and K in specific ratios.

Several mixtures are available on the market, such as 1:0:1 (47), 2:3:0 (21)+0.75 Zn, 2:3:2 (22)+0.5Zn, 2:3:2 (30)+0.5%Zn, 2:3:2 (39), 2:3:4 (30), 3:2:0 (25), 3:2:1 (25), 4:1:0 (30), 5:1:5 (42), and so on. This is not a complete list.

15.4.3 NITROGEN FERTILIZER CARRIERS

A variety of nitrogen N carriers exist, such as limestone ammonium nitrate that is commonly used for top or sidedressings and is known as LAN/KAN or LAN and has 28% nitrogen N. In comparison, ammonium sulphate has 21% N and urea 46% N.

15.4.4 POTASSIUM FERTILIZER CARRIERS

The potassium K carriers are potassium sulphate (40% K), Potash magnesia (21% K) and potassium nitrate (38% K).

15.4.5 MICRO OR TRACE ELEMENTS

Plants only need small amounts of trace elements. However, be aware that some crops cannot grow properly without them. Examples include zinc oxide which supplies zinc Zn and copper sulphate which supplies the mineral copper Cu.

15.4.6 APPLYING MICRO OR TRACE ELEMENTS

Trace element fertilizers are difficult to apply to established plants, as only very small quantities are necessary.

If too much of these elements are applied, it may become toxic to the plants and soil.

15.4.6.1 The 2 ways to provide microelements
- Mix the trace elements fertilizer with water and spray the plants by using a spray pump.
- Mix the trace elements with another fertilizer and add it to the soil.

15.5 APPLYING THE CORRECT TYPE OF LIME

When the soil-water pH for certain crops drops too low, lime should be applied. The amount and type of lime to be used should be calculated after a soil analysis test is performed. Liming of acid soil is necessary and maintains soil fertility to some extent. Fertilizer which contains ammonia, especially when using topdressings such as LAN/KAN, may decrease the pH. However, several applications are required to achieve this.

Research and trials in the Mpumalanga-Highveld region have indicated that most soil in this region is too acid. It is therefore necessary to apply at least 1 ton of lime stone every year to achieve the optimum soil pH level.

It is important to identify which elements are present in the soil and in what proportions. This information will serve as a guide in order to apply the correct type of lime before establishing crops. The critical component of lime is the alkalinity which is generated when the acidity reacts with carbonate. Once a pH of 4.5-5 has been reached, the toxic levels of aluminium is eliminated which allows the roots of plants to exploit the reserves of sub-soil moisture.

It is important to know that the pH affects the availability of nutrients to plants and should be corrected by using the right kind of lime. The proportion of calcium Ca and magnesium Mg in the soil will determine which of the four types of lime should be applied.

15.5.1 THE FOUR TYPES OF LIME
- Calsitic lime contains only calcium Ca.
- Polemic lime contains calcium and magnesium Mg.
- Slaked lime contains calcium Ca and is more soluble.
- Gypsum is used when the pH of the soil is too low.

To raise the pH, gypsum is applied. It is however a slow process and may only be effective after the second year of application. Therefore producers should continuously apply lime to raise the pH.

It is important to know that the coarseness of **lime influences how well it is taken-up by plants.** Therefore, the finer the lime is, the faster it will be taken-up by the plant. Be careful when lime is supplied in bulk. It is usually dumped next to the production land and some of the powder may blow away when the slightest wind occurs.

Research has shown that dolomite and calsitic lime become available to plants at the same time. The decision of which lime to use should be determined after a proper soil analysis has been performed and a specialist has been consulted. Remember, it is quite simple to reduce the pH of the soil-water content, however, to raise it, is difficult. Therefore, producers use different nitrogen N fertilizers as a topdressing to increase or decrease the pH to a small extent. Contact fertilizer experts in this regard.

In some cases in South Africa, yields could be increased by spending money on lime rather than on fertilizer. Make sure that the ideal pH range is correct for the specific crop before establishing anything. Be aware that the choice of fertilizer affects the soil pH.

Some fertilizers are residually acidic, while others are residually neutral and have little or no effect. Some fertilizers tend to increase the pH such as LAN/KAN, while others such as ammonium sulphate would acidify the soil and decrease the pH.

Use ammonium nitrate as a top-dressing when the soil pH is high. Be aware that excessive lime dressings could induce the deficiency of trace elements such as magnesium Mg and manganese Ma.

Acid-forming fertilizer should be avoided in acid soil and basic fertilizer should be avoided in alkaline soil. To raise the pH, gypsum should be worked into the soil according to the soil analysis. Remember, raising the pH is a slow, long-term process and when gypsum is used this will only be active after the second or third year.

It is important when the pH of soil is low producers should know which elements are present and in what proportions, so that the **correct treatment could be applied.** Reducing the pH could also be done by adding sulphur, aluminium sulphate or sulphuric acid. This is very expensive however and is not a practical option applying on large areas. Soil with a pH range higher than 7.5 should rather be avoided for most vegetable crops.

15.6 TOP OR SIDEDRESSING

Topdressings for almost all vegetable crops is very important and has a major influence on quality and yields, especially for the broad leafed crops.

It is sometimes not necessary to apply topdressings when the soil analysis report indicates that the soil is fertile enough. A light N application at the correct stage of plant development may be beneficial for most crops, as long as it is not over applied. Otherwise, some crops such as onions, garlic, green beans, tomatoes and capsicums may abort flowers.

Topdressings should be compatible with crops. Development of top growth needs to develop at a reasonably rate, but not too vigorously. Therefore, make sure that correctly measured applications of fertilizer are made and that other cultivation practices are also adhered to. If this is done correctly then the producer will achieve healthy, top quality crops and high yields.

Be careful in areas which are warm and have high humidity. N applications may be used-up very quickly by the plants through the process of transpiration or vigorous plant growth may be promoted that causes the plants to overshadow one another. Such conditions are very conducive to disease, infections and early development of produce.

Note that 150kg ammonium sulphate is equivalent to 100kg ammonium nitrate and 75kg of urea. When N inputs are too low, root growth may not develop normally. The key to the successful management of N is to walk the "thin line" between applying too much and too little. This is not an easy task, as N losses are also affected by soil conditions and changes in weather conditions. The rate of most N inputs are difficult to estimate correctly. However, when topdressing nitrogen N is applied correctly, it often has a remarkable influence on plant growth.

In the application of N, it should be understood that it is highly soluble. It undergoes a transformation and can easily be lost in the soil when it drains away, especially in sandy soil when over-watering or heavy rains occur.

Most crops react optimally when 75% of the nitrate N is transformed through the plant in the nitrate condition and 25% in the ammonium condition. 100% ammonium can be toxic to plants (Adriaanse 1990).

LAN/KAN consists of 50% nitrate and 50% ammonium and dissolves quickly in the soil. However, urea has to undergo a transformation, as it is 100% urea and therefore is not directly available to plants. Only after 4-6 days (as in the case of LAN/KAN) is it available to the plants after it has been applied. N losses may often affect plant growth yield and quality.

When N is lost in a nitrate condition then there is a chance that ground water may be polluted. In the process of boosting the plant growth, if there is an over supply of nutrients, these nutrients will be lost as the plant only takes-up what it needs. Fertilizer is soluble in water and can leach out quickly past the depth of the roots. This occurs especially when heavy irrigation or continuous rainfall takes place.

Depending on the soil structure and fertility of the soil, 1 to 3 N topdressings may be necessary. In the case of transplanted seedlings, the first dressing is usually 2.5-3 weeks after transplantation. On the other hand, when seeds are sown in the land, then the first dressing takes place about 4-5 weeks after the seed has emerged. The second application is 6-7 weeks thereafter, usually at a rate of 100-150kg p/Ha N for low to fairly fertile soil. These amounts also depend on soil fertility, structure, temperature, humidity and plant density.

For fruit-bearing, long season crops such as tomatoes and capsicums, it may be better to use 1:0:1 fertilizer or urea. It could then be supplied through the dripper system. It is recommended to slightly work N fertilizer into the soil by using a hand grub or mechanically, with a tractor and 3 toothed grub. When raked or grubbed, the soil should be irrigated as soon as possible.

Experienced producers look at the plant colour of certain crops before applying N. When the plants are dark or slightly dark green, it may be an indication of enough N. When the plants are pale or slightly yellowish then N should be applied immediately. This yellowish colouring usually shows in lands that are uneven. Too much N may lead to flower drop or vigorous plant growth (tomatoes, pumpkins, capsicums, beans etc). In the case of onions and garlic, thick necks may result.

15.7 AN EXAMPLE OF AN AVERAGE TOMATO FERTILIZER PROGRAM

Different crops have different nutrient requirements.

The different stages of development of plants, often require more than one nutrient to be administered. If these requirements are met, higher yields and better quality is the result.

Tomato plants may need the following minimum amounts of N, P and K elements through the dripper system: 130kg N, 20kg P and 168kg K (according to Haifa requirements and the World-wide Fertilization Organization). In the table below, is an example of an average fertilizer application program for tomatoes.

TABLE 8: AN AVERAGE STRAIGHT FERTILIZER PROGRAM FOR TOMATOES

NUTRIENT ELEMENTS	TIME OF APPLICATION	% APPLIED	AMOUNT OF ELEMENTS APPLIED
N	2-3 Days before planting	50%	65kg
	4 Weeks after planting	25%	32.5kg
	8 Weeks after planting	25%	32.5kg
P	Before planting	100%	20kg
K	Before planting	50%	84kg
	4 Weeks after planting	17%	28kg
	6 Weeks after planting	17%	28kg
	8 Weeks after transplanting	17%	28kg

The fertilizer strength described above is LAN/KAN N 28%, Super phosphate P 8.3%, KCL K 48%.

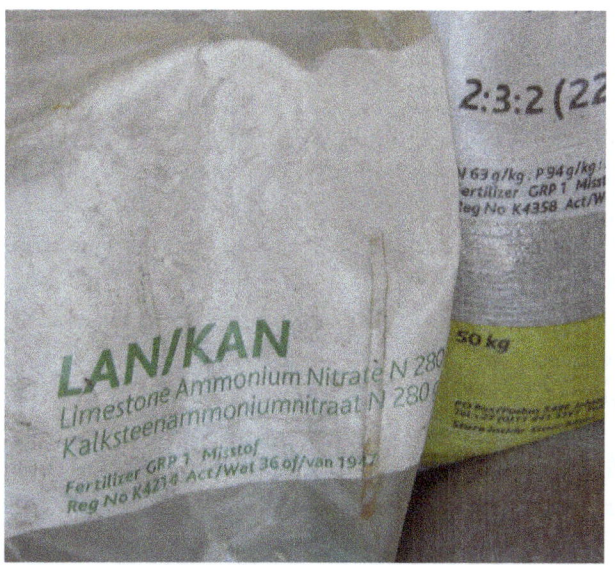

102 The 50kg bag of fertilizer at the back is the commonly used mixed fertilizer 2:3:2(22) Zn. In front, is a bag of LAN/KAN mostly used for top or side dressings of crops.

103 A fertilizer spreader with a swinging action is mounted on a tractor. It is very important to calibrate it correctly before using it.

15.8 FERTILIZER FOR INDETERMINATE OPEN-FIELD TOMATO CULTIVARS

According to a soil analysis, broadcast the amount of fertilizer evenly upon the entire land. If a dripper system is going to be used, N could be applied as a topdressing through the dripper system. A fertilization program through the dripper system is recommended. Consult fertilizer experts to advise on how to achieve best yields.

104 The person(s) applying fertilizer by hand, should be trained to do so, so that they can apply it evenly over the entire land.

105 A modern fertilizer spreader is very effective, provided that it is calibrated correctly.

15.9 APPLYING FERTILIZER OVER LARGE AREAS

If a fertilizer spreader implement is used to spread the fertilizer evenly throughout the land, it is important to calibrate it correctly. Research has indicated that when a fertilizer spreader is not calibrated correctly, approximately 40% of the fertilizer is effectively lost. Commercial producers make use of various methods, such as different types of mechanical fertilizer spreaders, some of which are computerised.

Fertilizer is extremely expensive and should be applied correctly. Too much, too little or the incorrect type of fertilizer applied unevenly, may be harmful to plants. In this case, usually a plant colour change will be noticed.

Fertilizer is applied in different ways. Small-scale farmers broadcast or spread it all over the entire land by hand. Soluble fertilizer is also applied through a dripper system and is known as fertigation. Producers usually apply the fertilizer mixes at a rate of 1 000-1 500kg p/Ha on fairly poor kinds of soil. If compost or kraal manure is going to be used, then fertilizer is still necessary, but usually 85% of the normal application will suffice. However, many advanced producers apply the full amount recommended before applying compost or manure.

Before applying fertilizer, the soil should be prepared properly. The fertilizer should be worked to at least a 15-20cm depth into the soil, ideally using a rotovator, tiller or harrow. In sandy soil, only a disk is required to create a fine and smooth seedbed.

15.10 APPLYING FERTILIZER OVER SMALLER AREAS

15.10.1 MEASURING FERTILIZER BY HAND
The beginner, small-scale and upcoming producers are sometimes inexperienced and do not always know how much fertilizer should be applied on their land.

15.10.1.1 Training labourers to apply fertilizer
Train the labourers as follows:
- One reasonably big handful of granular fertilizer weighs ± 50g and two hands 100g.
- When applying 100g (or two big handfuls of a granular fertilizer) spread it evenly over one square meter of the soil. This equals 1 000kg p/Ha.
- Mark one square meter on a clean soil surface. A fork or spade could be laid down in order to assist with estimating 1 meter square.
- Apply 1 big handful of mixed granular fertilizer and spread it evenly over the square meter on the soil. This equals approximately 500kg p/Ha.
- Carefully look at the density of the granular fertilizer that is exposed in the square meter of soil and apply the same quantity/density on the prepared land, as evenly as possible.
- When the area is 10 x 1 meters or 10 square meters, then 10 handfuls are applied (500g equals 500kg granular fertilizer p/Ha).

15.10.2 APPLYING THE FERTILIZER
- Generally, when a soil analysis is not available and fertilizer is to be applied on fairly fertile soil, most producers tend to use the cheaper, balanced 2:3:2:(22) 0.5 Zn granular mixed fertilizer. Apply 1 000-1 500kg p/Ha (2 handfulls per square meter equals 1 000kg p/Ha). However, the amount of fertilizer applied depends on whether the crop is a heavy or light feeder, as well as the type of soil, rainfall, soil fertility and soil condition.
- The fertilizer should be worked into the soil to a depth of approximately 18-20cm using a spade, rotovator, tiller, harrow or disk.
- When irrigating the lands, the fertilizer eventually penetrates deeper into the soil, as the water draws the nutrients downwards. This however depends on the quantity of water applied.
- To apply fertilizer over large areas it is sensible to use twine and lay it out on the prepared land.
- The twine lines should be approximately 3 meters from one another over the full length of the land.
- Fertilizer is then spread evenly by hand between the 3m lines.
- The applicator then moves to the middle of the line and applies the fertilizer between the 2 lines at the same density that was applied in the 1 square meter demonstration block.
- When reaching the half-way mark of applying fertilizer, half of the fertilizer should have been used. If fertilizer is still left-over at the half way mark, spread this evenly on the surface that has already being covered.

- Experienced producers know how much to apply and how to spread it over the production land to meet the quantity requirements without using twine.
- Remember 1Ha is a large land area (100 x 100m). This is 2 rugby fields or 10 000 m². 10 Bags x 50kg granular fertilizer equals 500kg p/Ha.

15.11 FOLIAR SPRAY FERTILIZER

Foliar nutrition applications are important and essential for the sustainable production of many crops.

According to experts of foliar nutrient applications, foliar sprays are a method that should be considered for plants that have a shortage of one or another element. Foliar spray is an effective way to improve yields and quality.

Applications of foliar fertilizer spray is not dependant on the physical condition of soil. Foliar sprays cannot be a substitute to remedy some nutrient problems. Many kinds of foliar fertilizers are available. Some combine multiple microelements such as sink Zn, copper Cu, iron Zn, boron Bo, molybdenum Mo and manganese Ma. Some of these sprays are very effective and important to apply at the correct stage of plant development.

When plants show any kind of nutrient deficiency at an early stage, it can be corrected by using foliar sprays. However, be aware that when a shortage of elements becomes evident on a plant, it may sometimes already be too late to correct the situation. Foliar sprays also sometimes contain hormones that stimulates plant growth.

15.11.1 REASONS FOR MICROELEMENT SHORTAGES
According to some researchers, only 10% of all microelements in the majority of soils, are available to plants. This is due to the following reasons:
- Soil type.
- Unbalanced elements in the soil.
- The pH binds some elements. Also climate conditions such as drought, drowning and cold conditions.

15.11.2 THE CONDITIONS UNDER WHICH FOLIAR SPRAYS MAY BE APPLIED
- On sandy soil when the elements are drained away.
- When the pH is too low or too high and some elements such as phosphorus P are bound. Similarly, if there was incorrect applications of lime. Too much or poor applications of lime as well as mixing lime in the soil may bind certain elements such as zinc Zn and manganese Ma.
- Too much potassium that may dominate manganese Mg.
- Hard layered pans below the soil that disrupts soil drainage.
- Any disease or nematodes which affects the root system.

It is also known that a high phosphate P content in the soil will cause a zinc Zn shortage. A product that binds elements on the surface of the soil where seeds are planted, is in the pipeline. It could be available even for small seeds. The result would be that small amounts of certain elements are released and are available for the germination of the plant.

Research is a high priority, especially for commercial maize, corn and soybean producers. However, even seed treatment against nematodes and rhizobium bacteria for potted plants such as beans, peas and soybeans are being researched.

Some of the foliar sprays work on plants in two ways. They supply nutrients through the leaves as well as downwards to the roots.

Most of the stomata of vegetable plants are underneath the leaves and a proper foliar spraying is therefore necessary.

It is important that the pH of the spraying water should be correct and that an adjuvant should be added. This keeps the plants wet for as long as possible and allows the plant to take-up the elements effectively. It is sometimes better to use stickers with a silicon base, especially for crops that have a waxy layer, such as for onion and garlic. If a oil based sticker is repeatedly used, it may dissolve the waxy layer. Some fungal chemical sprays contain enough zinc Zn, magnesium Mg and copper Cu. Plants should be sprayed early in the morning or late in the afternoon. If the humidity is high the penetration of foliar spray is more effective.

A sticker such as pH Green does not contain oil and is also adapted to the pH range. It is possible that when foliar sprays are applied on the leaves of plants, it may improve yields of crops, especially when the nutrient demands have reached the optimal stage. At this stage the plants will grow vigorously. However, this is not to say that the yields will increase. In some cases the yields may even decline. Plants that look nicer mean nothing to the producer unless the yield increases.

For example, when comparing beans which grow vigorously and look impressive to lower growing beans, it has been noticed that the lower growing beans yield much better.

106 Four litre per hour arrow drippers are inserted at the correct spacing in greenhouses. In double rows and on ridges these drippers are mostly used for the production of high-value crops such as tomatoes, peppers and cucumbers.

15.11.3 FERTILIZER THROUGH A DRIPPER SYSTEM IN SOIL

If soluble fertilizer is going to be applied throughout the dripper system in soil, then producers should make sure to consult with a fertilizer expert in addition to conducting soil analysis tests so that a weekly fertilizer nutrient program may be implemented that is suitable for the crop and soil type.

All the required nutrients will be applied through the dripper system. A water pump is installed and all the fertilizer is then dissolved in the tank in a 1 to 1 solution. The correct size irrigation motor must be installed. The pump is connected to a series of solenoid valves. The amount of solenoids depend on the size of the area to be irrigated. Contact irrigation experts for advice in this regard. The electrical water pump/motor and a small timer switch is used. Most producers prefer a small Hunter 6-8 station timer/computer that is connected to the solenoid valves.

The timing of the irrigation is implemented according to the water capacity of the soil which is measured with a tension meter or data logger. Small amounts of water-mixed fertilizer may be used to irrigate at intervals of 2 times per day. When the plants are fully grown and when very hot days occur, approximately 1.5-2 litres of water per dripper per day should be applied for crops such as tomato, squash, cucumbers and peppers. These crops are heavy water users.

A dripper opening of 2mm supplies 2 litres of water per hour while a 4mm opening supplies 4 litres per hour. Tension meters or data logger should be monitored and producers should irrigate accordingly. If the soil has already been fertilized according to a soil analysis, then only water is irrigated through the system. When a topdressing is required, it is only necessary to mix a soluble N fertilizer such as calcium nitrate in the water tank.

15.12 CHEMICAL SPRAYING THROUGH IRRIGATION WATER

Chemical spraying of plants is a method where chemicals are incorporated into the irrigation water, such as overhead, sprinkler, micro and drip irrigations systems.

This means that regular applications of nutrients and chemical sprays according to the plants requirements may be given through the irrigation system.

Chemical spray (chemi-spray) is a handy method that may be used economically. It requires a high level of management knowledge as well as an understanding of the basic principles of fertilization and irrigation including the ability to handle practical problems related to a well designed irrigation system.

When fertilizer is applied through the water, N can be applied at any time in the season. It is important that the concentration of the nutrients in the irrigation water should not be less than 50 parts per million. Therefore, when irrigating 1mm water, then 10 000 litres is necessary to irrigate 1 Ha. To achieve this 0.5kg of urea is necessary or 4kg KNO^3.

15.12.1 CHEMICAL SPRAY CLASSIFICATIONS FOR WATER-BASED APPLICATIONS
- Fertilizer spray - Fertigation
- Weed spray - Herbigation
- Fungal spray - Fumigation
- Insect spray - Insectigation
- Nematode spray - Nemagation

15.12.2 ADVANTAGES OF CHEMICAL SPRAY
- Fertilizer can be introduced next to the roots of the plants, mainly with drip irrigation.
- Logging of fertilizer, especially N, can be controlled.
- It is a tool to ease management and could also be part of an effective computerized control system.
- Changes in application methods can be adopted with minimum trouble.
- Saves time, labour, fuel and application costs in the long-term.
- Less tractor compaction of the soil and damaging of plants.
- Very effective on soil with a low fertility status.

15.12.3 DISADVANTAGES OF CHEMICAL SPRAY
- At the beginning, infrastructure is expensive. For example, water capture tanks, electrical motors and dosage injectors (dozetrons).
- Needs a well designed system. It is important that the irrigation water should be spread evenly over the entire land.
- Application is difficult when irrigation water is not necessary.
- Some products cause clogging of the system.

TABLE 9: SYMPTOMS OF ELEMENT DEFICIENCIES

ELEMENTS AND REQUIREMENTS	PLANT SYMPTOMS AND DEFICIENCIES
NITROGEN (N) Is part of a large numbers of organic components that includes amino acids, proteins and chlorophyll.	Sclerotic leaves which are light green and yellow begins on the lower leaves. Ripening is retarded and stunted plants are usually the result.
COPPER (Cu) Synthesis and stability of chlorophyll.	Dieback of shoot tips and brown spots on terminal leaves. Dull colour.
PHOSPHORUS (P) Photosynthesis and respiration. Components of cell membranes. Also part of necessary organic compounds including sugars.	Plant development is restricted, stunted and maturity is delayed after emergence. Some cole crops, carrots and sweet corn, show purple veins especially on the older leaves.
CLORINE (Ci) Photosynthesis. Water balance.	Wilting. Stubby roots.
POTASIUM (K) Important for protein metabolism. Activates certain enzymes, promotes growth, assists in regulating water related systems.	Wilting and then early abscission. Decreased yield. Poor ripening of fruit. Stunted plants. Mottled, spotted and curled leaves as well as browning of leaf margins.
IRON (Fe) Chlorophyll synthesis.	Upper or young leaves turn light yellow and sometimes exhibit green veins.
CALSIUM (CA) Component of the cell wall related to the building of cell walls.	Leaves do not emerge and unfold. Tip burn on young leaves.
MANGANESE (Mn) Enzyme complex. Involved in photosynthesis	Sclerosis of older leaves and leaf drop.
MAGNESIUM (MG) Component of the cell wall related to the building of cell walls.	Primary veins remain green, sclerosis of older leaves. Possible leaf drop. Brassica leaves may become brilliant orange, yellow or purple.
MOLYBDENUM (MO) Component of nitrate. More destructive in low pH soil below 4.5. Mo deficiency is also referred as Whiptail on certain crops.	Twisting and narrowing of leaves.
SULPHUR (S) Component of protein and vitamins. Activates enzymes for N fixation. Germination and growth of pollen. Translocation of sugars.	New leaves are uniform and golden yellow. Rosette effects. Plants are thin stemmed.
ZINK (ZN) Hormone synthesis and chlorophyll formation.	Intervenal sclerosis. Mottled leaves.
BORON (B) Cell division and development. Cell wall stability. Fruit and tubers are flecked.	Rosette effect. Thick, curled, brittle leaves that crack easily. Corky areas on stems and roots split. Hollow stems in cole crop.

CHAPTER 16

SOWING SEED AND THINNING SEEDLINGS

16.1 SOWING SEED

Vegetable seed such as beetroot, carrots, spinach, water, musk melons, pumpkins, squash, maize, beans and lettuce are usually sown directly in the production land.

It is cheaper to manage the stand by lowering the rate of seedlings. This can only be successful if soil preparation, advanced weed control, sprinkler irrigation, care for the young seedlings, disease and insect control and other cultural practises are carried out effectively. It is recommended that the producer uses good quality seed to ensure that the stand or plant population is satisfactory.

There are a few methods for sowing seeds by hand in furrows. One method of sowing seed by hand in furrows, is to hold some seeds between the thumb and fore finger and then to rub the these fingers together in order to drop them as evenly as possible in the furrow. Alternatively, seeds may be sown directly from a paper bag. Some producers mix the seed with mealie meal to ease the hand sowing process.

Constantly monitor the seedlings while they emerge to make sure that diseases and insects do not attack them.

Only one reliable labourer should be trained to sow seed, so that it is clear whom should take responsibility when mistakes are made, such as sowing too densely, sowing too sparsely or skipping rows.

16.2 THINNING SEEDLING PLANTS IN THE PRODUCTION LAND

When using a seed planter that is calibrated correctly, thinning is usually not necessary or reduced to the minimum. On the other hand, when sowing small seeds by hand such as beetroot, spinach, onion, lettuce and carrots, it is always necessary to thin the plants to the required spacing.

Thinning seedlings is generally a laborious and difficult task, especially if the seeds are sown too densely and weed control becomes a problem.

Sowing seeds by hand requires experience. Labourers should be trained to sow the seed correctly and to the best of their ability. It is important that the manager him/herself practises sowing small seeds to suit the spacing requirements of the crop or cultivar.

It is necessary that beginner producers should practice before they sow seed. They should take care not to sow too densely or too sparsely in order to avoid the unnecessary and uncomfortable labour of thinning the seedling plants afterwards. Seedling rates should be controlled as accurately as possible.

Seed is expensive and labour can be costly when sowing too densely. Weeds also emerge much faster in and between the rows which makes thinning very challenging.

If sowing beetroot and carrot seed by hand and thinning is neglected or delayed too long, the result is that the production is mostly unmarketable. Most of the globes or roots may develop without their characteristic sweetness and may even become woody when the plants are still young.

It is important to have a good reliable seed germination count to maintain a full stand. Sometimes thinning can be combined with weeding, but it is an entirely different and also uncomfortable job. When thinning seedlings, the seedlings that are removed may be transplanted if so desired (but not carrot seedlings). The first thinning of seedlings should be carried out when the plants are 3-4cm high and the second thinning when they are 7cm tall.

Usually weeds grow and develop faster than the seedlings and should be removed before thinning the seedlings. The best that the producer can do, is to sow the seed neither too sparsely nor too densely and to make sure that the germination of the seed is good. This reduces the uncomfortable task of thinning and keeps thinning to a minimum. **Producers use herbicides to control weeds. This is especially effective when seedlings are produced in seedbeds.** *See Chapter 17.2 for using herbicide-based weed control.*

CHAPTER 17

WEED CONTROL

17.1 INTRODUCTION

Weed control is very consuming and the major reason for failure is when weeds are controlled too late or not at all. Generally, weed plants consume considerably more nutrients than seedlings, as they normally outnumber the seedlings to a large extent.

Recently researchers have claimed that certain herbicide applications used to control weeds may cause cancer to humans and should be avoided. This should be investigated by producers and they should avoid these kinds of herbicides. Producers should therefore keep themselves up to date with any new developments in this space.

Producers should plan for labour in a timeous manner and not allow weeds to overgrow the seedlings. It is important to control weeds when they are still young. Vegetable production is more prone to weed competition than other crops, as vegetable seedlings are tender and develop slower than weeds. The important thing to understand about weeds is that they compete for water and nutrients. They should not be allowed to compete with the crop for **water, nutrients** and **sunlight.**

When applying nitrogen N fertilizer, weeds will use more N than is desirable and then store the N in their leaves. In this way the crop is robbed of this much needed element. It is best to control weeds before they break through the soil or while they are small. At these earlier stages, shallow cultivation will control the weeds and not injure the seedling roots too much. Good crop foliage may help to control weeds. For example, sweet potato vines can cover the land quickly and dominate most weeds, especially the lower growing weeds.

Weeds use much more water, sometimes 2-3 times more than vegetable crops. When weeds are controlled effectively, less irrigation water is necessary. A thick stand of weeds can draw nutrients from the soil at approximately 96kg nitrogen N, 5kg phosphorus P and 41kg potassium K p/Ha, over a period of 5 months. The critical period for controlling weeds is approximately 5-6 weeks after sowing and 3-4 weeks after transplanting seedlings.

If weed control is carried out too late and the weeds have already overgrown the seedlings, the damage caused by weed competition has already occurred. Producers should understand the way the weed plants reproduce, so that they may be able to control them.

17.1.1 WEED REPRODUCTION
Weeds reproduce as follows:
- Some weeds produce a large number of seeds.
- Weeds may grow from small pieces of their stems which break-off from the mother plant.
- Some weeds have a root system that can regenerate and grow again.
- Some weeds can produce seeds over a very short period.
- Weeds can often establish themselves on hard, compact soil.

It is best to control weeds when the soil is dry, otherwise when the soil is wet, they can grow extremely quickly. Weeds serve as a haven for diseases, insects, bacteria and fungi. Certain weeds may host insects and diseases through the winter and can cause serious problems for crops. It is known that nematodes, trips, red spider mites, aphids, leaf miners and a number of other insects feed over the winter and then reproduce on or in the weeds.

Commercial farmers use a tractor and a grub or disk to control weeds between the rows. This is provided the spacing is wide enough for the implements and tractor wheels to fit in the rows. Spacing should be adapted for the implements and tractor wheels. Most tractor wheels are 1.2m wide. Weeds **in the rows** near the plants are usually removed by hand using hand grubs. Annual weed plants produce a lot of seeds and they mature quickly and can be transferred by water, wind, humans, implements and animals.

Certain weed seeds pass through the digestive system of cattle and sheep without any germination damage. This is why it is not wise to use sheep or cattle manure to establish vegetables or for soil improvement. Otherwise a considerable number of weeds may be permanently introduced into your land. Therefore, it is advisable to instead make good compost from this manure.

When a high growing crop is established, it may help to overshadow weeds to a certain extent. Crop rotation using cover crops such as sun hemp, black mustard, lucerne, eragrostis, cow candy, lupines and so on, is an economical way to improve soil. However, note that some cover crops such as sun hemp is nematode (eelworm) sensitive.

107 A small production land completely overgrown with weeds. When sowing seed, weeds may emerge quickly and overgrow the seedlings. Do not sow seed directly in a land where weeds were a problem the previous season or where it is known that nutty weeds are a problem.

108 Some sweet potato cultivars may cover the land reasonably quickly and smother any weeds. This is provided that weed control was implemented properly when the plants were still young. Afterwards, no more weed control should be necessary. The large weeds should be pulled-out by hand.

109 Controlling weeds in rows using a hoe. Mechanical implements are used when weeds need to be controlled on large lands between the rows. The determinate tomato plants are to be trellised using the double wire method to a height of about 1.6m.

17.2 WEED CONTROL WHEN USING HERBICIDES

It has been found that some herbicides cause cancer and according to researchers should be withdrawn from the market. Keep yourself up to date with this information (June 2015).

A herbicide is a chemical that is used to inhibit plant growth or to terminate plants. One of the advantages of herbicides is their selectivity. A selective herbicide is toxic to some weed plant species but not others. When such a herbicide is applied to a variety of weed species, only some will be killed. Some weeds may be affected slightly while others not at all. Non-selective herbicides terminate all plant growth.

Herbicides are divided into two main groups, namely systemic and contact herbicides:

- Systemic herbicides are absorbed by the plant and translocated to the leaves, stems or roots of plants.
- Contact herbicides terminate the insects that feeds on the plants.
- Herbicides are applied as a pre-plant, pre-emergence or post-emergence treatment. These terms may be confusing, as it is not always clear whether reference is being made to the crop or the weed.
- Pre-emergence and post-emergence usually refers to the weeds. Pre-plant applications occur directly before establishing the crop. They are usually incorporated into the soil by using a suitable implement.

Pre-emergence herbicides are applied on the soil surface and needs to leach into the soil by rain or applying light irrigations. The young roots of the weed plants absorb the herbicide. However, most herbicides are applied after establishing the crop and before the emergence of weeds. Some herbicides could still be applied after the emergence of certain crops. These are mainly systemic herbicides that have a period of residual activity. The application rate of most herbicides is determined by the clay content of the soil. Disturbing the treated soil through tillage may drastically shorten the period of residual activity. In some circumstances it may be necessary to apply a post-emergence herbicide at a later stage.

Post-emergence herbicides are applied after emergence. Some herbicides may be applied after emergence of weeds but normally before emergence. Herbicides are most often effective when used together with good cultural practices in a proper weed management program.

When used correctly, the benefit of herbicides are that they often save **time and labour.** The weeds that are to be controlled should be correctly identified and the application performed accordingly. There are numerous kinds of weeds and it can sometimes be difficult and confusing to identify or classify them.

The use of herbicides is often expensive and mostly limited to larger farming producers. However, it can save considerable time and labour costs in the longer run.

The danger of overusing herbicide and other chemicals is when their carbon chains break down in the soil, causing toxic compounds to form which destroy the micro-organisms in the soil. Certain plants are very sensitive to herbicides and when they are sprayed with too much chemicals (especially when windy conditions occur) it may cause damage, especially for crops such as tomatoes, potatoes and cucurbits.

Farmers sometimes do not understand why their plants all of a sudden become infected with a disease or virus. They forget that they may have over-sprayed with herbicide. When using herbicides, it is important to study the label carefully and apply them correctly.

Take note of the following aspects concerning herbicides:
- The pH of the spraying water must be correct.
- Spray at the right time of the day. Do not spray in the middle of the day when it is still hot.
- Spray when there is little or no wind, such as early in the morning.
- Use a sticker that allows the remedy to remain on the plant for longer periods.

17.2.1 TYPES OF HERBICIDES
17.2.1.1 CONTACT HERBICIDES
This chemical must make contact with the weed to kill it.

17.2.1.2 SYSTEMIC HERBICIDES
This chemical must be taken-up by the plant through the roots, leaves and stems.

17.2.1.3 BEFORE PLANT HERBICIDES
This chemical is applied before the crop is established.

17.2.1.4 AFTER PLANT CHEMICALS
This herbicide is applied after the crop is established.

17.2.1.5 PRE-EMERGENCE HERBICIDES
This herbicide must be applied before the germinating weed seedlings appear above the soil.

17.2.1.6 POST-EMERGENCE HERBICIDES
To be applied after the weeds appear above the soil.

Selective herbicides kill or stunt weeds with little or no harm to turf grass. Non-selective herbicides kill or damage all plants and may, under certain situations, act selectively. If the dosage is excessive, even a selective herbicide may become toxic.

17.2.2 HERBICIDE ACTION

Herbicides are specific in their action. They should be selected carefully by the producer in order to choose the best one for their purpose. Doing so, partly prevents the damaging of the production land which may influence the yield and avoid injuring crops being grown.

Contact herbicides are the most effective against annual weeds and only kills the parts of the plant on which the chemical is deposited.

Systemic herbicides are absorbed either by roots or foliar sprays and are then translocated within the plant system to all tissues of the plant.

Systemic herbicides may be effective on both annual and perennial weeds, but they are particularly good at controlling perennial weeds.

No post-emergence herbicides exist for broad leafed weeds. Avoid using herbicides continuously. It could become harmful (toxic) to the soil in the long term.

CHAPTER 18

CROP ROTATION AND SOIL IMPROVEMENT

110 Rotational crops on a small plot. Remember to draw a plan of the layout of the plants because in the next season the producer may forget where the crops were situated.

111 A clean, well-maintained plot of land free from weeds. Here, different types of sweet potato cultivars were established on ridges which are 1m from one another. The bleached plants on the left are an orange flesh cultivar.

18.1 INTRODUCTION

The purpose of crop rotation is to keep the soil fertile by retaining and improving the levels of organic matter in the soil.

Crop rotation provides a way to prevent the same family of crops from being produced on the same soil year-after-year. The implementation of this practice helps to prevent diseases, insects and weeds which are related to a specific crop family, from building-up and multiplying in the soil.

It is recommended, that for most crops, a rotation program of at least 3 years be followed, for the same family of crops. Plants that are reincorporated into the soil, provide different nutrients back into the soil. However, it can be difficult for small vegetable producers to rotate crops, due to the limited land at their disposal as well as the short-term crops that are cultivated.

18.1.1 THE FOUR GROUPS OF VEGETABLES CROPS
- Fruit bearing crops: tomatoes, brinjals, capsicums and cucurbits.
- Bulbs and rooted crops: onions, garlic, carrots, beetroot and sweet potatoes.
- Leafy crops: spinach, lettuce and cole crops.
- Legumes: beans and peas.

Some producers rotate their land so that they may establish a cash crop for a few years, and thereafter, they make use of a cover crop. This is usually on farms that have limited lands.

Sweet corn is a good rotational crop that hosts few insects and diseases which may affect other vegetables. Usually small-scale farmers rotate their crops by relying on their memory. However, they sometimes forget which crops that they have previously rotated.

To make the most of crop rotation, producers should use a computer to draw up a rotation plan and keep records. Such a system will indicate where the locations of previous crops were established. Rotating crops is key to preventing diseases such as early blight, powdery mildew and other diseases. It also assists in improving the soil as it leads to advantageous changes in tillage operations and improved root depth for different crops.

If possible, it is recommended to make use of winter cover crops in the winter, summer cover crops in the summer or a yearlong green manure crop, such as Lucerne. These cover crops could be cut repeatedly before using it in the rotation program. This process also allows the land to rest from time-to-time, which is necessary to improve the soil.

112 These knots on the roots should not be confused with nematodes. They are the Rhizobia bacteria that correct nitrogen N levels in the soil. If the knots are pressed open, a pink/purple substance is noticed.

113 Sun hemp is an excellent leguminous cover crop that has considerable foliage that can be incorporated into the soil when the pods are still young. However, it is very sensitive to Fusarium and nematodes.

18.2 CROP ROTATION AND CROP NUTRIENT REQUIREMENTS

After harvesting crops, only the leftover crops and weeds will be incorporated into the soil. However, this is not nearly enough nutrients for the next crop. Some crops supply very limited plant material, which will only partially improve soil fertility.

If only fertilizer was used, it is recommended that a 3 year rotational production period is implemented. After this period, it is time to consider soil improvement.

18.2.1 LIGHT NUTRIENT USERS - ROOTED CROPS
Rooted crops are light feeders that use small amounts of nutrients. This includes sweet potatoes, carrots, beetroot and turnips, but not potatoes.

18.2.2 MILD NUTRIENT USERS - BEAN FAMILY
The bean family includes peas, beans and many other leguminous crops. Beans are mild or medium nutrient users. Take note that crops which bear pods, add a natural organism called rhizobium, to the soil. This is beneficial for the follow-up crop.

18.2.3 HEAVY NUTRIENT USERS - LEAF AND FRUIT BEARING CROPS
Tomatoes and cabbage are heavy feeders and withdraw large amounts of nutrients from the soil.

TABLE 10: AN EXAMPLE OF A 4-CROP ROTATIONAL SYSTEM THAT IS BASED ON A 3 YEAR ROTATIONAL PERIOD *(PLOT A: CARROTS, PLOT B: BEETROOT, PLOT C: CABBAGE AND PLOT D: BEANS)*

A	Carrots	B	Beetroot	C	Cabbage
B	Beetroot	C	Cabbage	D	Beans
C	Cabbage	D	Beans	A	Carrots
D	Beans	A	Carrots	B	Beetroot

18.3 REASONS FOR CROP ROTATION

18.3.1 INSECT AND DISEASE CONTROL
Crop rotation is one of the most effective ways to manage certain insects and diseases. When crops that belong to the same family are cultivated continuously on the same soil, it may lead to an increase in insect and disease populations which may become a major problem.

18.3.2 INCREASED SOIL FERTILITY
When producing the same crop repeatedly in the same soil, it withdraws the same nutrient elements from the soil. Some of these crops do not utilise minor elements which may lead to some nutritional imbalances in the soil. This is why different crops should be used to increase overall soil fertility such as deep-rooted versus shallow-rooted, heavy feeders versus weak feeders and nitrogen N fixers versus non-nitrogen fixers.

18.3.3 IMPROVED SOIL STRUCTURE
It is important to select the best rotational crop in your area which matches your crop establishment periods and seasonal weather conditions (whether it is cool or hot weather conditions). It is recommended to incorporate rotational crops that have considerable foliage. The more juicy green material is incorporated into the soil, the more beneficial organisms will be developed and reproduced in the soil. Green manure will provide large quantities of organic matter into the soil which will improve the soil structure. Soil fertility produces favourable effects as it incorporates beneficial bacteria into the soil.

18.3.4 PREVENTING WATER AND SOIL LOSSES
The higher the crop coverage, the more water will penetrate into the soil. Therefore, less water is necessary which prevents erosion and loss of top soil.

18.3.5 WHEN CROP ROTATION BECOMES UNNECESSARY
If soil is analyzed correctly and fertilizer is applied on good, fertile soil using a good compost, the question then arises: why should the same crop not be cultivated year-after-year? If wheat and sugar case is grown for many years on the same soil, why can this not be done for all vegetable crops?

The reason is that if soil-borne diseases such as bacterial wilt and bacterial cancer are present in the soil, this soil should never be used for certain crops because it is permanently infected with these bacterial spores. This is why producers in the Western Cape changed to indeterminate high-value tomato hydroponic systems in the 60's. They were forced to change, due to these and other soil-based bacteria and diseases.

Most tomato diseases requrie ideal conditions to develop. Healthy growing plants have thick outer walls which to a great extent, prevents the diseases from entering the plant tissue. Unless the plant is damaged by labourers or insects, the diseases will find it difficult to enter the plant cells or tissue.

As an experiment, the author established indeterminate high-value tomato cultivars in a shade net house in fertile soil which had been analysed. Fertilizer was then applied on the ridges for 3 seasons and no major problems were found. However, 1 out of 200 plants were infected with a Mosaic virus and also with Fusarium. These plants were removed as soon as they appeared and excellent yields were achieved even after the third establishment of the same cultivars. The production in this shade net house was established on the same ridges every year. Excellent yields of more than 160 ton were harvested.

It is important to control insects, diseases and weeds effectively and to avoid spores and insects which could survive over winter. Plants should be removed or ploughed into the soil as soon as the last harvesting takes place. Some producers establish indeterminate high-value crops in shade net houses and tunnels for many seasons, and then later change to hydroponic systems.

18.4 SOIL IMPROVEMENT

18.4.1 GREEN MANURE

The incorporation of green plant material into the soil is known as cover crops or green manure. This plant material could be obtained from any source, even from weeds. One of the most difficult challenges facing a vegetable grower is to maintain the organic content matter of the soil.

18.4.1.1 BENEFITS OF GREEN MANURE
- Increases water infiltration and holding capacity.
- Improves soil aeration.
- Improves soil fertility and structure.
- Adds nutrients and organic matter to soil.
- Reduces soil erosion caused by wind and water.
- Increases soil biodiversity, by stimulating the growth of beneficial microbes and other soil organisms.
- Ideal for organic crop rotation.

It is a practical way to improve the soil structure and restore productivity to unused or over-worked soil. The primary intention of cover crops is to provide food to the millions of microbes which can only do their work in favourable soil conditions. This benefits all plants and eventually benefits the producer.

Green manure crops should also be fertilized when producing them. This allows them to produce vigorous, juicy top-growth that takes-up especially N. Large quantities of organic matter is then incorporated into the soil. This serves as an excellent, readily-available source of nutritional elements for the follow-up crop.

When selecting crops for soil improvement, the following should be considered:
- Adaptation of the crop to the climate and soil.
- The time at which the crop needs to be planted.
- The amount of plant material that is produced.
- The best time when the crop has to be incorporated into the soil.
- The ease of incorporating the plant material into the soil.

Most of the nutritional elements taken-up by the rotational crop are stored within it, and would be available for the follow-up crop after the plant material has decomposed in the soil. Green manure should be worked into the soil at the correct stage of crop development, usually after the plants have flowered and when they are still young and juicy.

The greens should be ploughed into moist soil for at least one month in warm soil conditions and for 6 weeks in cool soil conditions before planting the production crop. When hot conditions occur, it gives the microbes a chance to decompose the plant material at a higher rate. In winter, when the soil is cool, the decomposition process is much slower. Keep the soil moist at all times in order to cultivate the most favourable conditions for the microbes.

Most vegetables are grown on farms that have been intensively cropped which translates to one or more cultivated crops grown each year on the same land. Note that the crop residues are not sufficient to replace the organic matter which is lost annually after harvesting. The harvested crops put very little organic matter back into the soil, because when they are harvested, they are usually quite old. The quality of the green material is therefore not very effective for this purpose.

Green manure crops are best when using leguminous crops that are high in nitrogen N. This is because they fix N from the atmosphere in combination with the plant material which binds rhizobium on the roots.

Some of the crops that may be incorporated into the soil are for example non-leguminous crops such as Eragrostis curvula, finger grass, maize and babala and leguminous crops such as peas, cowpea, sun hemp, black mustard, lucerne (after harvesting), soybeans, lupines, velvet beans, clover, and so on. Working these crops into the soil is very advantageous for the producer.

18.4.2 LEGUMINOUS CROPS THAT FIX NITROGEN INTO THE SOIL

Rhizobium is classified as a soil bacteria that fixes N from the atmosphere and binds it on the roots of pod bearing plants. The rhizobium cells multiply and penetrate the young roots and in the process form nodules on the roots. This bacteria also fixes nitrogen N from the atmosphere which assists in the process of producing amniotic acids and proteins. *See photo 112.*

Only vegetable crops such as beans, broad beans and peas provide N rhizobuim when they are incorporated into the soil. This N will then become available for the follow-up crop.

Other crops such as trees and shrubs which bear pods, also provide N. It is well-known that the Karop-tree provides a huge amount of sweet pods which are very nutritious for sheep and cattle and can be included in their feed. These trees can also replace some nitrogen N into the soil. This is why some grasses grow well beneath these trees.

It is a good option to **inoculate** the seed of legume vegetable crops before planting, to stimulate the development of the rhizobium bacteria. The condition of the soil and the development of the plant itself, also plays an important role in this regard. A well-conditioned plant encourages rhizobium to development to a greater extent which causes significant amounts of N to remain in the soil and become available for follow-up crops.

18.4.3 BIO-FUMIGATION CROPS MAY BE INCORPORATED INTO THE SOIL

Bio-fumigation refers to the suppression of various soil borne pests and diseases by using naturally occurring compounds.

Cole and other crops such as black mustard, release natural gasses from their plant tissues when they are crushed or finely chopped. These gasses are immediately incorporated into the soil. The plants, in the process of their decomposition, release a gas when an enzyme comes into contact with water. This gas destroys certain insects and diseases which are present in the soil.

Highly prioritized research is available in this regard (2013).
Contact ARC Roodeplaat for more information.

CHAPTER 19

SANITATION AND SCANNING FOR INSECTS AND DISEASES

19.1 INTRODUCTION

Sanitation needs to be practised by humans throughout their lives in order to protect and defend themselves from infections and diseases such as bad smells, rotten teeth, viruses, bacteria, fungi, mites, insects (flying and crawling), and so on. If not addressed, these conditions, infections and diseases may lead to illness or death.

It is important that vegetable producers inspect their production lands regularly. They should be aware that insects and diseases can do tremendous damage to the plants. Damage may occur in as little as a few days, if not controlled effectively. Action must be taken as quickly as possible in order to prevent insects and diseases from multiplying out of control.

Some diseases may spread via millions of spores. A single diseased plant could spread fungi or bacteria throughout the entire land via wind, rain, overhead irrigation, human movement or implements. Viruses can be spread by humans, aphids, white flies and to a lesser degree, trips.

Aphids for instance, can multiply without males. They are usually found in colonies or begin to multiply in spots in parts of the land. It is therefore important to thoroughly scan the entire land and pay attention to these colonies to prevent them from spreading any further. The higher the plant population of crops, the more intensive the scanning should be. Otherwise, it becomes difficult to effectively control the insects and diseases. Labourers who smoke can also transmit diseases such as Tobacco Mosaic Virus (TMV)
from one plant to another.

Irrigation water from rivers or dams may carry diseases to the production land. Producers should therefore investigate this, and if necessary, install proper filters to prevent infections from being distributed via water used from polluted rivers and dams. Specialized information should help the producer to select the most suitable filters and water pumps.

If plants have been infected with bacteria or viruses, it is better to immediately remove these plants from the production land and destroy them by burning or burying them. It is recommended to seal the virus infected plants in a plastic bag and to leave them in the sun for a day. This will destroy all aphids and viruses.

All plant propagating material such as seedlings, tubers and vine cuttings, should be free from insects, bacteria, fungi and viral diseases. It is therefore recommended to obtain plant material and seedlings from reliable suppliers. In addition, the packing store must be cleaned and disinfected regularly.

Implements and labourers may also carry diseases, insects, bacteria or viruses from one land to another. Tomato poles that are reused for trellising plants may transfer diseases and viruses to the new production land. Diseases may also be carried by tractor wheels, equipment and tools. When producing vegetables, beginners may notice that they have little problem with diseases and insects. This is because it is the first year of production and they may be isolated from nearby crops. However, eventually diseases and insects will be introduced into their environment and they may build-up and multiply rapidly.

Very cold, freezing winter areas are not likely to be attacked by diseases, insects or viruses. It is necessary to look at these factors and sterilization should take place before establishing crops. This prevents diseases from being introduced from the beginning. When harvesting crops, the remains should be incorporated in the soil as soon as possible. Plant stems, leaves and roots may serve as disease, insect or virus reservoirs which will allow them to survive over winter periods.

19.1.1 HIDING PLACES OF INSECTS AND WHERE TO FIND THEM
- Underneath the leaves (aphids, red spider mites, white flies).
- In the flowers of certain crops, such as tomatoes and pumpkins/squash (trips).
- Some insects tunnel through the leaves of tomatoes, peppers or potatoes (leaf miners).
- Others eat into fruit (bollworms).
- Others bore into stems (stem borer).
- Some hide in the soil and mostly consume the stems of young plants or cut-off the plants at soil level at night. These types of insects hide during the day to avoid birds from consuming them (cutworms).
- Even the roots are attacked by microscopic nematodes (eelworm).

19.1.2 SCANNING
Producers should inspect their land by walking in a zigzag pattern from one end of the production land to the other. The next scanning should then be performed from the opposite end. **It is important to scan the production land regularly, especially when previous crops/lands were infected with diseases or insects.**

To identify insects and diseases, producers should look for them in the correct places, either where they are hiding or out in the open on the plants. As soon as insects/diseases/viruses are spotted, they should be controlled. It is important that producers should scan regularly to identify insects, diseases and viruses before an outbreak takes place. It is necessary to act fast in order to prevent major outbreaks which may get out of control and incur heavy losses.

When heavy infections occur due to certain insects and diseases, they may be difficult to control due to their high concentrations. When high concentrations occur, more chemicals and controlling applications will be necessary. If sprayed repeatedly and there is no/little effect, it may be that the insects have become immune due to the use of chemicals which have the same mode of action.

Over time, diseases and spores may build-up in packaging stores and therefore producers should sterilize stores regularly. Spacing of plants should also not be too dense in order to assist in controlling any outbreaks. *See Chapter 11 Spacing.*

It is important to have toilets as well as basins for labourers to wash their hands, before they work with plants.

SUMIPLEO® 500 EC

Unique chemistry for the effective control of insect pests, including resistant Lepidoptera strains.

SumiPleo® is unique in its IRAC product class, widely trusted as an effective alternative chemistry for the control of mainly insecticide resistant Lepidopteran pests in S.A. Good control of African bollworm, Potato tuber moth and *Tuta absoluta* in potatoes, African bollworm and *Tuta absoluta* in tomatoes, Diamond back moth in crucifereae crops e.g.: cabbage, broccoli, Brussel sprouts as well as African bollworm in lettuce, is now registered. Registration for control of the new Fall Army Worm pest (*Spodoptera frugiperda*) in sweet corn was also recently obtained. Strong suppression of both the Dipterous, American and Pea leaf miners (*Lyriomyza spp.*), can be expected in potato and tomatoes. Suppression of looper in tomatoes is also an additional control bonus obtained with **SumiPleo®** in a Lepidopteran focussed spray programme.

SMART CHOICE FOR INSECT CONTROL

Strong efficacy of this proprietary insecticide Pyridalyl offers outstanding properties to meet the demands of modern effective insect control, as an essential part of integrated pest management (IPM) and insecticide resistance management systems:

- Controls a wide range of Lepidopteran larvae and other pest species in vegetable and field crops.
- Effective against larger larvae, with application timing thus less critical.
- Also active against insecticide resistant populations with its unique biochemical mode of action.
- Allows no cross–resistance with other pest control chemicals.
- Safer to a wide range of beneficial arthropods like natural pest enemies and pollinators and also earthworms.
- Relatively safe to mammals and birds. (Toxicity class 3. Blue band).
- Safe for field operators with a convenient quick re-entry period.
- Shows excellent rain tolerance especially in cruciferous crops against diamond back moth.
- Applied at low rates ensures superb control with the advantage of a short withholding period in vegetables prior to harvest.
- Where scheduled sprayings are needed, rotational programmes with Bt materials like: DiPel® DF or Florbac® WG is a good strategy.
- Acts through both dermal exposure and ingestion, therefore requires thorough wetting of the crop.
- An ideal product to use in Integrated Pest Management (IPM) and Pest Resistance Management.

For more information to achieve maximum efficacy against insect attacks with our modern novel **SumiPleo®** chemistry, contact your nearest Philagro agent or product manager: Henk Terblanche - 082 829 4070.

SUMIPLEO® 500 EC
CONSULT LABEL FOR DETAIL USE RECOMMENDATIONS
Sumipleo® is the registered trademark of the Sumitomo Chemical Company, Tokyo, Japan. SumiPleo (Reg. nr. L8377 Law 36/1947), contains Pyridalyl, caution. DiPel® DF is the registered trademark of Valent BioSciences, Libertyville, USA. (Reg. nr., L6641 Law 36/1947), contains *Bacillus thuringiensis* var. kurstaki , caution. Florbac® WG is the registered trademark of Valent BioSciences, Libertyville, USA. (Reg. nr, L5531 Law 36/1947), contains *Bacillus thuringiensis* var. aizawai, caution.

Tindrum 17/115R

Philagro South Africa (Pty) Ltd
Reg no: 98/10658107

PostNet Suite #378, Private Bag X025,
Lynnwood Ridge 0040
Pretoria Tel: 012 348 8808
Somerset West Tel: (021) 851 4163

PHILAGRO

PRODUCTS THAT WORK FROM PEOPLE WHO CARE

www.philagro.co.za

CHAPTER 20

OVERVIEW OF INSECTS, DISEASES, VIRUSES AND PHYSICAL DISORDERS

20.1 INTRODUCTON

There is no such a thing as a safe chemical pesticide which is designed to kill pests and control diseases. Even when used to prevent losses of plants and crops producers should remember chemicals are poisonous.

Insects and diseases which attack crops, cause the biggest problems in the agricultural industry worldwide. This trend has persisted from ancient times for thousands of years. The damage that insects and plant diseases cause amount to considerable financial losses. Expensive chemicals, equipment, implements and labour are used to control insects and diseases. This is often a difficult and laborious task. Not only should the correct clothing be worn by the operator, but he/she should also use the correct apparatus. Cleaning of clothes and apparatus after use is essential. *See photo 158.*

One of the largest expenses when producing vegetable crops is control measures. According to one of the leading agricultural research companies in the world, without crop protection, as much as 42% of all crops before harvesting and a further 10% after harvesting, would be lost due to diseases, insects and weeds. Not much would be left for human consumption. Insects are the most adaptable organisms. The diversity of insects is also increasing.

Most insects are small and have a short lifespan. Some multiply and spread extremely quickly. To produce healthy vegetable crops it is necessary to follow an effective management program. There is no other shortcut to controlling insects and diseases.

Over-dosage and the irresponsible use of chemicals are responsible for the destruction of the biodiversity of many good insects and organisms which are important and have certain functions that are beneficial to the producer, the environment and other plants. For example, certain predators, parasites, pathogens and good soil bacteria attack and destroy harmful plants and insects. **The decline of soil fertility is responsible for poor plant health and in-turn expands the opportunities for insects and diseases to penetrate the thin skins of weakened plants.**

Producers should educate themselves in order to identify the most common insects/diseases which may become a problem. They should also research the most effective ways to control them.

The general public has become aware of healthy vegetables and are concerned about the usage of poisonous chemicals. More and more, the health conscious public is demanding quality, healthy, clean and organic produce and other agricultural products.

Plant diseases are caused by living organisms known as pathogens, such as fungi, bacteria and viruses. Development of new methods to control insects and diseases, especially environmentally friendly chemicals, remains a constant focus. Plant diseases may be managed by addressing one or more of the following factors:
- The host plant
- The disease and
- The environment

When the environment is unsuitable, some diseases will not develop. However, bacteria and fungi may survive in a dormant state for many years and wait until the environment is favourable for their development.

Sometimes there are resistant cultivars available, especially F1 hybrid cultivars. Applying biological control methods is very intensive and time consuming. Usually large enterprises do not use these methods.

To begin organic production is very expensive for the beginner or small-scale producers. This is due to the very high standards and regulations that are laid down. *See Chapter 10 for organic farming.*

20.1.1 OUTDATED CHEMICALS

In this manual, the control measures for insects and diseases are not going to be discussed. The reason for this is that by the time that any publication reaches its readers, the information is likely already outdated. Intensive research is continuously conducted to develop better, less poisonous chemicals to replace the older ones. The producer should therefore consult more recent information from reliable sources, such as chemical and seed companies.

Diseases and insects have the ability to adapt to changing conditions and some of them may become resistant or partially resistant to chemicals.

Spraying chemicals with the same mode of action, or using a group of chemicals continuously is not best practice. Producers should instead periodically change to another action chemical. This will help to prevent insect resistance caused by the repeated use of one or more chemicals or groups of chemicals.

Unwanted insects, diseases and viruses may appear from time to time due to the changes in cultural practices. These changes may favour their development or be introduced and imported from other countries.

20.1.2 VARIOUS WAYS THAT DISEASES SPREAD
- Wind, rain splash and overhead irrigation.
- Insects, vectors transmitted by means of seed, diseases and sometimes from seedlings.
- Runoff and flood irrigation.
- Boots, shoes and equipment used in production areas.
- Plant residues.

20.1.3 THE CAUSES OF INSECT OUTBREAKS
- Favourable weather conditions.
- Bad chemical control or incorrect chemical usage.
- Lack of natural enemies.
- Neglecting to scan for insects.
- Resistance or partial resistance of insects to chemicals.

All grades and standards of vegetables on the national markets have strict requirements for the amount of disease and insect related injury/damage. Damaged produce receives lower market prices or are rejected. It is for this reason, that successful vegetable producers consider spraying and other similar control measures, just as essential as cultivation or any other farming operations.

20.1.4 HOW INSECT AND DISEASE INFECTIONS MAY CAUSE HEAVY CROP LOSSES
- By reducing yields and lowering quality of products.
- By increasing cost of production and harvesting.
- By requiring materials and equipment to apply chemicals.

Losses should be prevented as much as possible as it affects the producers' income thereby lowering their living standards.

Strict rules and regulations are set-forth by the Department of Plant and Quality Control for the **import and export of vegetable seed and plant material into or out of a country**. This is done in order to prevent infected plant material and unwanted insects and diseases from entering a country.

Most bacteria and viruses cannot be controlled effectively once plants are infected. Virus infected plants should be destroyed to avoid further transmission of infections. The best way to select the most correct and available pesticide, is to gain access to the most up-to-date, relevant and reliable information.

20.2 THE MAIN PARTS OF PLANTS THAT INSECTS/DISEASES ATTACK

20.2.1 LEAVES
When white powdery growth appears on the upper leaves it is called **powdery mildew**. Leaf blight spots and mildew can be controlled with one of the many **copper** fungicides. Remember that most insects such as aphids, red spider mites and whiteflies hide underneath the leaves.

20.2.2 STEMS
Brown or black areas on the stem cause wilting or death of the upper-stem which is known as stem blight. If the inside of the stem is blocked due to stem blight, it also affects water transpiration. When the stem is cut, a brown colouration is visible which is caused by a fungus called Fusarium. Stem borers penetrate the stem, especially of maize while cutworms cut stems off, partially or completely, at soil level.

20.2.3 ROOTS
When plants wilt, the problem may be caused by root rot due to various fungi and bacteria. Root rot may occur in soil that is very wet. Root diseases and nematodes that attack the roots, are difficult and expensive to control. Spores can be introduced via wind, irrigation water, seedlings, implements and on the feet of workers.

20.2.4 FRUIT AND FLOWERS
Black or brown spots and large, sunken, rotting areas which develop on fruit may be caused by several fungi or bacteria. Insects such as bollworms bore into the fruit while trips hide in flowers and feed on pollen. Disease spores are noticed as leaf-spot and blight. Remember most insects such as aphids, red spider mites and whiteflies hide underneath the leaves.

20.2.5 A SELECTION OF MAJOR DISEASES TO TAKE NOTE OF
- Brown spots with a yellow ring around them on the leaves is Leaf Spot.
- Browning of small to large areas of the leaf is Leaf Blight.
- White powdery growth on the upper leaf surface is normally Powdery Mildew.
- Browning or black areas on the stem causing wilting or death of the upper-stem is called Stem Blight.

20.3 GMO CROPS

Biotechnology and Genetically Modified Organism (GMO) crops are currently playing an important role and will continue to play an important role in the future of vegetable production. This means that certain crops could become partially or totally resistant to certain diseases or insects. This will be beneficial to producers and would also save money and labour as well as achieve higher yields.

It has been reported that more than 85% of the maize, soybean and cotton production in SA has been established (2.85 million Ha) from GMO crops. According to the Maize Trust, an accumulative 12 million Ha of GMO maize was established from 2001-2010 which produces grain crops of 40 million metric tonnes.

This grain is being consumed annually by 50 million South Africans, 800 million feedlot chickens, 1.4 million feedlot cattle and 3 million feedlot pigs without adverse affect to humans, animals or the environment.

It has also been reported that some crops have already lost their resistance against insects, as the insects have become immune.

Droughts in Africa may force governments to make use of GMO maize, even if some countries have reservations about the long-term health implications of feeding GMO produce to their populations. Some benefits of GMO crops include higher yields, insect resistance, drought resistance and lower production costs.

A GMO potato cultivar (developed in 2009) is almost totally resistant to an insect called codling moth. The moth is a major insect problem worldwide, causing a great deal of damage to the potato industry on an annual basis. Many millions of rand is lost. However, this potato cultivar has not yet been released. This is due to the strict GMO laws as well as the resistance of, especially health conscious people, who prefer non-GMO crops which are organically produced.

Regulations require that products from GMO crops must be labelled accordingly such as for mealie/maize meal. Consumers may then buy non-GMO mealie/maize if they so choose. Some people believe that GMO crops are harmful to humans, but to-date, this is unproven.

20.4 GUIDES FOR CONTROLLING INSECTS AND DISEASES

Producers should understand that it is not always possible to provide up-to-date information about controlling insects and diseases. Chemical products change rapidly, evolving to better, less toxic chemical products which remain just as effective. It is also against the law to use unregistered chemicals for controlling insects and diseases.

The Seed and Plant Act No 46 of 1947 makes it clear that only registered chemicals may be used for a specific crop. The problem arises when there are no registered chemicals available for certain crops and diseases. Garlic, for instance, has no controlling chemical measures which may cause producers to rely on onion regulations or other non-registered chemicals (some of which may be used by certain experienced producers). Sometimes certain insects and diseases are quite effectively controlled by these non-registered chemicals.

The following guides are available from Departments and the Agricultural Research Council:
- A guide for the control of insects by Annette Nel and a guide for the control of plant diseases. The latter is available from Agricultural Information Services.
- An effective, colourful photo guide of insects, their descriptions and habits, as well as the damage they may cause is available from the Agricultural Research Council (ARC) Roodeplaat. The author, Dr D Visser, is well known for his intensive research of insects in South Africa. Dr Visser is a specialist entomologist, especially for potatoes and other vegetable crops.
- Dr Reinette Gouws has conducted research regarding diseases and viruses as well as other problems. This includes control and pre-control measures, scanning, hygiene and the bio-effectiveness of crops.

However, the guides for the control of insects and diseases mentioned above may confuse some producers due to the many choices that are available of active ingredients and even choices of **trade names of chemicals.**

The aim of these guides is to allow the producer to select the most correct chemical for the control of plant diseases and insects. The guides also provide guidance of the correct usage and effective application of chemicals. Producers often do not know how to choose the most effective chemical to control a specific insect or disease. Also note, that often there is no instant solution.

Many producers experiment and seek advice from other experienced producers who use unregistered chemical control measures. They find the controlling action of unregistered chemicals satisfactory. Take note, however, that when producers use unregistered chemicals, they do so at their own risk. There are many insect and disease controlling chemicals that are not registered for certain crops. For example, garlic producers use chemicals that are registered for onions. It is not too difficult to identify insects. Diseases are, however, more difficult to identify and control effectively.

20.4.1 PLANT RESISTANCE AGAINST DISEASES AND INSECTS

Diseases and insects may adapt themselves to become partially or fully resistant to chemical remedies. This often occurs when contact and stomach chemicals are not correctly applied onto the plant, especially underneath the leaves where most insects hide.

When using systematic chemicals it is necessary to **spray the growing plants properly** so that they can take-up the active ingredients of the remedy. It is important to spray underneath the leaves when contact or stomach chemicals are used. A follow-up control is usually necessary, especially when the target population is high. Be aware that when heavy outbreaks of insects and diseases occur, it may be impossible to control them.

Many classes of insecticides have adverse effects on insects. They are:
- Parathyroids
- Organophosphates
- Carbonates

Botanical compounds such as nicotine, pyrethrum, canola and mineral oil, organ chlorines, soaps, plant extracts from garlic and chillies as well as insect extracts are used. It is possible that the following generation of insects (if they were spraying with insecticides incorrectly) may become immune or partially immune to a chemical. This occurs especially when producers **repeatedly and continuously spray with one kind of a compound that has the same active ingredients or mode of action.**

20.5 INSECTS

As discussed earlier in this chapter, due to the fact that chemical formations continue to evolve so rapidly, they will not be dealt with in this manual.

20.5.1 THE MOST COMMON TYPES OF INSECTS
- Insects that include the larvae of leaf and flower eating beetles such as stinkbugs and leaf hoppers.
- Fruit damaging and sucking insects such as aphids, red spider mites, white flies, bragada bugs, etc.
- Soil insects such as nematodes (eelworm).

Plant insects with chewing mouth parts can cause considerable damage to fruit, flowers, leaves and roots. In some cases, due to the physical damage caused by these insects, secondary damage such as root rot and bacterial/fungal infections occur.

To develop a new chemical requires scientific research over many years. It is very costly to develop a new, low toxic remedy that is both environmentally, as well as human friendly. This is a difficult science, as less toxic chemicals are sometimes used to control insects that are difficult to control even with highly toxic chemicals.

Insects such as red spider mites, trips, white flies and aphids are difficult to control once an outbreak occurs. These insects are usually controlled by using systematic chemicals, as the toxic chemical penetrates the entire plant and eliminates insects on all parts of the plant. **The plant still continue to actively grow and the chemical should not affect it adversely.**

The insects and diseases that are described in this manual are only some of the main, most common types which affect vegetables crops. Only a basic description is provided.

20.5.1.1 ROOT-KNOT NEMATODE See photo 114.
Nematodes or eelworms are a serious problem on various crops. They are microscopically small and may reproduce rapidly when conditions are favourable. Infected roots exhibit typically knotty swellings when heavy infections occur. These knots block the uptake of water and nutrients. Plant growth is retarded and wilting of the plant takes place even after regular irrigation is applied.

Root nematodes are quite common and widely spread. They can be found in most soils. When the crop remains in the soil for a long time, it gives the nematodes time to build-up and develop. Producers should therefore begin their production with the smallest nematode population that is possible in their soil. There are various nematode species that attack vegetables. The most common one is the root knot eelworm.

It is preferable that producers make use of resistant cultivars, if available, when nematode problems are expected.

Pastures such as Eragrostis curvula (Ermelo type) may be established in a crop-rotation program. Land planted with E. curvula for a period of 3 years will be fairly free of nematodes. The most effective control measure is to make use of soil fumigation. It is, however, labour intensive and expensive.

Some producers use a fumigation chemical called EDB that is applied at least 14 days before establishing the crop. The soil must be moist and free of clots when EDB is applied with a hand injector. The holes used for fumigation must be closed immediately. When using a mechanical apparatus, the soil should be sealed by rolling or using sprinkler irrigation to prevent the fumes escaping too quickly.

The first seed treated with an eelworm controlling chemical is called **Avicta complete maize**. It is now available with maize seed, but will also possibly be available for certain vegetable seeds at a later time.

20.5.1.2 CUTWORMS See photo 127.
The eggs of cutworms hatch quickly and the caterpillars are a dull, greyish colour and hairless. The moths and caterpillars are active at night. Weed-free areas around the land prevent the moth from laying her eggs. Cutworms are mostly a problem when plants emerge or when transplanting seedlings. Do not underestimate this insect. It may do considerable damage to young plants in a production land. The dirty looking cutworms cut-off the young seedlings just above soil level, almost entirely from their stems. They then attack the next plant. The worms are usually found near the damaged plant below the soil surface. When scratching the soil next to the plant they become visible and curl up when they are disturbed. Producers use a chemical bait powder or spray chemical which is applied around the young plants. It is important to clean the land and adjacent areas to prevent the moths from laying their eggs. At least 8 weeks before establishing the crop, crop residues that are on the land, should be incorporated. This is so that the residue decomposes in the soil and destroys any cutworm eggs that may be unhatched. This prevents cutworms and other over-wintering insects from surviving.

20.5.1.3 APHIDS See photo 117.
Aphids are tiny pear shaped insects, 1-4mm in length, when fully grown. They may be green or black, with or without wings. They are soft-bodied with long thin legs and antennae. Aphids are a serious threat to vegetable crops and other plants. Aphids suck the sap from the new growth of young plants. They may also transmit viruses. Aphids mostly hide beneath leaves or gather on the growing points of plants. They produce a juicy, sweet, honeydew substance that attracts ants and other insects. A symbiotic relationship exists between the aphids and ants. In exchange for the honeydew, the ants protect the aphids against natural enemies, such as lady birds, spiders and praying mantises. Aphids do not lay eggs. They produce small, live babies that suck on plants, literally from the time they are born. They produce these babies without mating. Aphid numbers can grow extremely quickly. When they increase in numbers and become over-populated and conditions become unfavourable (with foliage declining), they grow wings and go in search of other growing plants. Aphids are weak flyers and therefore they need to move to the top of plants and wait for the wind to carry them to their next location, sometimes quite far from the original production area. Aphids should be controlled as soon as possible. This must be done correctly, or they may become resistant to chemicals.

20.5.1.4 TRIPS
Trips are found on many crops. They are very small insects. They feed on the leaves by gnawing on young tissue. They feed on the liquid of onions and garlic. They usually hide between the leaves of the basal-end of onions and garlic plants. When the leaves of garlic and onion are pulled opened they can be seen near the main stem, usually at the bottom of the leaves. Trips are active insects and move quickly when they are disturbed. Young

are orange or yellow, while adults are dark brown to black. The adults may give a painful bite when they land on your body. They also hide in flowers, where they consume the pollen and also damage fruit. When large numbers occur, they may cause flower drop of tomatoes, peppers and even cucurbits. The amount of trips decline in the winter months (June/July), due to the cold conditions. They breed slowly in winter but when the temperature rises their breeding numbers escalate. When onion plants become heavily infected, they become silvery. This occurs due to the scratching effect of the trips mouth parts which scrapes away the waxy layer from the onion and garlic leaves. The result is unprotected wounds which leaves the door wide-open for other diseases. When trips are controlled effectively while producing onion and garlic, most other disease problems are also restricted.

114 A plant that is infected with root-knot nematode (eelworm). Note the clearly visible knots on the roots.

115 Bollworm damage on tomato fruit. The young caterpillar (bottom) penetrates the fruit. A moth can lay many eggs at night, which may result in considerable damage.

116 A CMR beetle feeding on a tomato fruit. They prefer to consume flowering plants, especially garden beans and pumpkins.

20.5.1.5 DIAMONDBACK MOTH
The diamondback moth is the world's biggest problem for cole crops, especially cabbage. The larvae are light brown and after hatching, they turn green. They mature to a length of 10-12mm. When the wings of the moth are folded on its back, it forms a clearly visible diamond pattern. The larvae spins itself into a cocoon under the lower leaves or in weeds or debris. These insects are adapted to a wide range of climates. They reproduce rapidly and may also develop resistance to chemicals. The moth lays eggs on the plant, and when they hatch, small, light green caterpillars appear. They are mostly active on the undersides of leaves where they eat holes in the green tissue. When heavy infections occur, the leaves, with the exception of the veins, will be destroyed. The larvae consume the soft parts of the leaves between the veins.

20.5.1.6 BRAGADA BUGS
Bragada bugs are dark grey to black and have orange-yellow spots on their backs. The males and females move together while they are attached and facing-away from one another. The insects suck the sap from plants and are a problem on cole crops. They suck the young cole plant leaves, causing them to dry out. They may also damage the growing points of seedlings and young plants, causing cabbage and cauliflower to form numerous small heads that are of no use. Bragada bugs are very active in the early spring before the summer rains.

20.5.1.7 BOLLWORM See photo 115.
Bollworms are the larvae of night flying moths that lay their eggs on weeds or vegetable plants. The young caterpillars are dark in colour with small black spots on their backs. The colour of older caterpillars, varies from nearly black to brownish, green, yellowish or even pink. Their underside is characteristically striped white to beige running the full length of the body. On the sides are the typically dirty white stripes. They feed mostly inside the fruit.

PART 1: A COMPLETE GUIDE TO VEGETABLE PRODUCTION

117 A heavy infection of aphids feeding mainly on the underside of tomato leaves. They also attack most other crops.

118 A 'looper' caterpillar feeding on a tomato fruit.

119 A leaf eating lady bird feeding on a pumpkin leaf.

20.5.1.8 LOOPERS See photo 118.
The looper is a caterpillar that is smaller than the bollworm and has a loop between its front and rear legs.

20.5.1.9 RED SPIDER MITES See photo 126.
Red spider mites are tiny and reddish brown with 4 pairs of legs. They are not insects but related to spiders and have 8 legs. They do not fly. Like aphids, red spiders suck the juice from the plants. They mostly prefer to stay on the underside of leaves. However, during heavy infection, they can clearly be seen on both sides of the leaves. The fine spider web is clearly visible. Infected leaves lose their colour and turn light yellow. When heavy infections occur, the leaves become bronzed and dry out. In serious cases of infection, the entire plant will be affected, with a dry coloured appearance. Red spider mites prefer dry, hot conditions. They are very difficult to control. They may become resistant or partially resistant to chemicals. Some producers close their tunnels, so that the mites might burn to death.

20.5.1.10 LEAF MINERS See photo 120.
Leaf miners are small, yellow and black flies with fringe wings. The larvae feed between the upper and lower epidermis of the plant leaves, tunnelling through from the one to the other side. They usually do little damage when their population is low.

120 Leaf miner damage on tomato leaves.

121 A heavy infection of white flies typically hiding under a leaf.

122 Cucumber leaves covered with white flies debris. This happens when the population of flies is very high and their debris lands on the leaves.

20.5.1.11 PUMPKIN FLIES

Pumpkin flies are brown coloured flies with yellow bands or spots on their bodies. Adults sting young fruit, lay their eggs just below the skin and cause sunken brown spots. White maggots develop inside the fruit. They are usually controlled by using a bait substance, such as brown sugar syrup mixed with a poisonous chemical. Producers dip a bag in this substrate and splash it on the leaves. The sweet brown sugar substance attracts the flies and kills them.

123 A lady bird larvae feeding on a spinach leaf.

124 Spotted maize beetles, also called two-by-two's. They can become a serious problem when they consume the silk and kernels of the maize cobs. They also invade the flowers of crops in large numbers.

125 A stink or green bug sucking the tips of crops, causing them to wilt. The bug moves very quickly when it is disturbed.

126 Red spider mites on tomato leaves. It is a difficult insect to control and it should be controlled as soon as they are spotted.

127 A cutworm removed from a plant. The moth and worm are active at night. They are dirty looking worms as they inhabit the soil. As soon as they are disturbed, they curl up, as shown.

128 A corn ear (cob) worm prefers corn but also attacks ripe tomatoes.

20.5.1.12 WHITE FLIES See photo 121.

White flies hide underneath leaves and can be seen when the leaves are turned over. Their numbers build-up towards towards the end of the season and are difficult to control once the population is high. The reason for this, is that they are very light, and the slightest wind will blow them away from the target. They usually land on the other plants nearby. They are small, 1-2mm in length, and can transmit viruses. Whitefly larvae are wingless. They produce honeydew, a sweet extract that falls onto the leaves and fruit. When there are many of them, their debris collects on the bottom leaves, giving these leaves a filthy appearance. Note that whiteflies reproduce quickly when conditions are favourable, especially inside tunnels. It has been reported that a single whitefly, can produce up to 7 000 eggs in a matter of 20 days.

20.5.1.13 LADYBIRDS See photo 119 and photo 123.
These beetles are light brown while others are a brilliant red, with black spots on their bodies. The larvae are orange coloured with thick, hairy, black spines. The larvae and the adult ladybirds feed on leaves. The light brown ladybirds are the worst as they are plant eating. The bright coloured ladybirds are mostly the good ones that feed on aphids.

20.5.1.14 SPOTTED MAIZE BEETLE (Bont mielie kewer) See photo 124.
This beetle feeds on pollen and can invade flowers on some crops, especially maize pollen and silk.

20.5.1.15 STINK BUGS See photo 125.
Stink bugs are very active and prefer to consume the tips of growing points which cause the tips to wilt. The bugs produce a foul smell, when killed. They are easy to spot due to their large size.

20.5.1.16 CMR BEATLE See photo 116.
This beetle prefers to attack the flowers of plants, especially those of green beans, tomatoes and pumpkins. They also attack soft fruit, such as tomatoes.

20.6 DISEASES

A disease is classified, when a plant becomes infected by a fungus, bacteria or virus. As a result, the plant usually becomes stunted, changes its colour and does not develop as it should.

20.6.1 THE 3 PRIMARY DISEASE CLASSIFICATIONS
- Diseases that attack the roots.
- Diseases that attack the upper plant growth.
- Diseases that attack crops/produce that is stored in a pack house.

It is more difficult to identify diseases than it is to identify insects, especially in the beginning stages of disease development. Diseases are very destructive if proper management and control measures are not followed. Bacterial and viral diseases are amongst the most problematic. In crops such as tomatoes, cucurbits and capsicums, when a bacterial or viral infection occurs, it is sometimes impossible to control. Any disease should be stopped from spreading any further. This is why it is important to identify bacterial and viral diseases as early as possible, while they are in their growing stage, before they are able to spread to other plants. It is often better to remove and destroy all infected plants. Sometimes diseases are of a secondary nature and can be controlled by good management practices. The effect of diseases can result in poor quality produce. It is important to recognize factors that lead to disease development. Environmental conditions play a role in determining whether a disease may develop, as well as how rapidly and to what extent. Factors such as temperature, moisture, light, soil pH, water content, gases and nutrients all play a role in the development of plant diseases. It is the micro-environment that plays the key role in disease development as does the relative humidity and acidity around the roots.

20.6.1.1 The causes of disease outbreaks
- Favourable weather conditions, especially high humidity near the coast.
- Bad chemical control or incorrect chemical dosage.
- Neglecting to scan for diseases.
- Insects/diseases that becomes resistant or partially resistant to chemicals.

Cultivars that are resistant or partially resistant to certain diseases, are available. A disease such as early blight, usually attacks plants such as tomatoes, peppers and potatoes from the bottom of the plant leaves. They attack the stems and fruit, resulting in the leaves at the bottom of the plants dying. In the local language, this is why this condition is sometimes called "creeping-up sickness".

When not controlled properly, early blight and other diseases leave millions of spores. Note that experienced producers who have identified the presence of this diseases in past seasons, use this knowledge to anticipate its appearance in future planting seasons. Therefore, they pre-emptively perform crop rotation and implement chemical control measures before the disease appears.

Fungal diseases mostly develop on the outside of the plant and are usually quite simple to control when sprayed effectively. Viral diseases are mostly fatal. When the virus enters the plant by means of a vector, it seriously influences the growth of plants. The plants usually exhibit a rosette appearance when infected. It this case, it is recommended to completely remove infected plants from the land. Diseases attack plants and restrict them from producing optimally. Plants may be partially or completely destroyed in the process. Malformed fruit may also be produced. If producers do not effectively control diseases at the beginning stages, an outbreak can be expected.

Tomatoes may be attacked by as many as 30 diseases. Of these, five of them are major and should be easily identifiable by producers. Fungi which cause many plant diseases, are usually so small that they cannot be visually identified. They may be identified with the aid of an electron microscope or magnifying glass, depending on their size.

Some diseases attack only one plant species while others may attack a whole range of species. Most fungal diseases reproduce themselves via spores which are distributed by wind, humans or water. Others fungi have thick-walled spores that can survive for long periods of time on plant debris, in the soil or on the surface of seeds. This allows them to be carried into the next growing season. By following good agricultural practices and taking preventative measures, the producer may avoid a crop failure due to disease problems. Most vegetable diseases vary in their development and intensity depending on weather conditions. Diseases are at their strongest during wet conditions or when dew is present on plants. Also, when too much water is applied, it may cause the wilting of plants. This is due to a lack of oxygen and the drowning of plants which may cause tuber, bulb or root rots.

20.6.2 CHEMICAL REGULATIONS
Chemical regulations is a very wide and challenging field, therefore only the most important diseases will be discussed in this manual, even though numerous other minor plant diseases exists across the world. Also, as mentioned previously, chemical products that control diseases are not going to be discussed or recommended in this manual. All trade names of poisonous chemicals are required to be registered for specific crops. It is against the law to use unregistered chemicals. Many producers use unregistered chemicals that may provide better control measures, however, they do this at their own risk.

20.6.3 DISEASE DESCRIPTIONS
20.6.3.1 DAMPING OFF DISEASE
This disease is caused by various fungi which usually attack seedlings in the nursery or after transplanting them in the land.

20.6.3.1.1 Other causes of damping off disease
- Black leg on cole crops.
- Alternaria, Fusarium and Phythium species.
- Cutworms.
- Water logging and algae build up.
- Fertilizer burn after transplanting seedlings and when applying fertilizer (especially N).

This fungus multiplies and attacks young seedling stems. The stems shrink and the plants usually fall over. Over-watering of seedlings and poor drainage may allow algae to build-up on top of the seed trays. This is favourable for fungal diseases. High temperatures and dense plant populations encourage this disease. Sterilize seed trays before planting seed. Seed should be treated with a fungicide seed dressing such as Thiram. Otherwise, seeds may carry the fungus on their surface to the next production land. All certified seed is treated with a chemical seed dressing.

20.6.3.2 PURPLE BLOTCH - ALTENARIA
The first symptoms of purple blotch or leaf spot is usually small, **purple,** coloured spots, especially during moist conditions. These spots then enlarge to approximately 1cm in diameter. The centre of the spots are usually purple or grey-brown and surrounded by a clear yellow ring. Black spores may be present at the centre of the spots. Several spots then emerge and lead to the death of the plant. High humidity encourages the disease. It is also a seed borne disease.

20.6.3.3 EARLY BLIGHT. See photo 129.
Early blight produces brown-to-black irregularly shaped spots, with concentric ring patterns surrounded by yellowing of the leaf tissue. The symptoms **mostly begin on the older, bottom leaves.** This is a destructive disease occurring mainly on tomatoes, capsicums and potatoes. It is a worldwide problem. This disease should not be ignored as it may cause considerable damage without remedial action. If producers are not familiar with early blight, especially beginner producers, they should take notice when the bottom leaves begin to develop brown patterns. At this stage, they should begin implementing control measures immediately. The disease usually begins on the older, bottom leaves of plants and crawls steadily upwards. Producers should continuously scan for this disease, as it is a very under-estimated. Most commercial producers control this disease by implementing preventative measures, especially on tomatoes and potatoes.

This disease thrives in hot, dry weather conditions and usually spreads rapidly. Tomatoes and potatoes should be sprayed regularly, preferably on a weekly basis, but this also depends on the weather conditions. The disease can be controlled with one of the many copper chemicals, such as copper oxychloride, commonly known as Cupravit. Dimildex is also used in combination with Dithane that includes Dithane M45 and Mancozeb.

20.6.3.4 LATE BLIGHT
This fungus is indicated by black lesions that begin on the stems of plants. It is visible as brown, sunken areas on tubers and fruit of certain plants. Late blight occurs in warm, wet, misty, rainy weather conditions or when heavy dew conditions exist. The disease is not common on tomatoes plants and it develops at a later stage of plant development.

20.6.3.5 POWDERY MILDEW See photo 135.
This disease **favours high temperatures, warm, dry conditions and moisture in the form of dew.** It is primarily a dry climate disease but can be destructive in the subtropical areas, especially in winter. White spots appear on the underside of the younger leaves. The symptoms are white powdery growths on the leaves, especially on the lower leaves. It may cause severe infections on leaves which usually turn brown from the edges inward and then die. The disease appears as white patches mainly on the upper/top surface of leaves.

Powdery mildew causes widespread yield losses and is **difficult to control** once the disease has spread throughout the land, usually later in the season. Systemic chemicals, combined with other chemicals, such as Lebaycid which controls pumpkin flies on the cucurbits families, should be used. **Powdery mildew also grows within the plant leaves and is latent inside the plants tissues. This may infect the crop and be visible after 21 days.** Begin chemical applications at the first signs and repeat spraying every 7-10 days. It is important to ensure total chemical coverage of all parts of the plants which are above ground level. The disease results in defoliation of the plant and recovery is slow.

Powdery mildew is a fungal disease that may become resistant to chemicals. Producers should from time to time change chemicals with those that have different modes of action. **Young plants** are not usually infected, however older plants, especially when under stress, are susceptible. In fact, it would be very unusual to grow cucurbit families through to maturity, without at least one appearance of this disease, especially later in the season and on the broad leafed pumpkins and squash. Note that disease resistant or partially disease resistant cultivars are available.

20.6.3.6. DOWNY MILDEW See photo 131 and photo 132.

Downy mildew favours high humidity, wet conditions and moderately high temperatures. It causes leaf blight and begins as a pale yellow-to-brown spot on the upper sides of leaves. The spots become larger and yellow and white areas appear. Typical mildew growth may be found underneath the leaves. Leaves will quickly die, beginning with the older ones. The leaves die from the edges, inwards. Yellows spots on the upper leaf surface are followed by the development of brown spots on the underside of affected leaves. Thousands of spores are produced and are spread to other plants via air, splashing water and human movement. Onion and garlic plants are all vulnerable to downy mildew, especially when trips are present. Trips are commonly found on these crops. They remove the waxy protective layer on the plant with the gnawing of their mouth parts. This enables diseases to infect the plants.

129 Early blight on tomatoes. The disease usually begins on the bottom leaves of the plant. Most producers preventatively control this disease if the disease has presented itself previously.

130 Early blight on tomato fruit. Note the white and black sunken spots and larger dark areas.

131 Downy mildew on tomato leaves.

20.6.3.7 BLACK ROT See photo 134.

Black rot is one of the more serious bacterial diseases that can infect cole crops, especially cabbage and cauliflower. It favours warmer and, to a lesser degree, humid, cool conditions. It does not spread in colder weather conditions below 18°C. The disease is usually introduced by infected seeds or seedlings. The lower stem turns brown and rots. As the lesions expand, the tissue turns brown and the small veins become black. Black rot spreads rapidly when warm, humid weather conditions occur. Once in the soil, the bacteria spreads by rain, irrigation water and wind. The bacteria enters the plants through wounds. If infected seedlings are produced in greenhouses, under cool conditions below 15 -18°C, the plants may not show any symptoms. When infected seedlings are transplanted into the land and temperatures rise to 25-35°C during periods of high humidity (80-100%), the younger plants may become stunted and will eventually wilt and die. The single most important preventative measure for this disease, is to use certified seed. In addition to this, a 3-year crop rotation program should be followed, once the disease has infected previously used lands.

20.6.3.8 BROWN RUST See photo 137.

Brown rust occurs mostly late in the growing season when humidity is moderate and when high temperatures occur. This fungal disease causes small white spots to develop on the surface of leaves. The spots later develop into bright orange, open pustules which are slightly raised. Yellows spots on the upper leaf surface are followed by the development of rusty brown, bright orange pustules. These are slightly raised on the underside of affected leaves. The pustules commonly occur in concentric circles. The spores which are produced are spread from plant to plant by air currents and water splashes. The disease favours dampness and high humidity. In these conditions it spreads rapidly. The disease may not appear every year on plants which were infected the previous season, as its development depends on ideal temperatures and humidity conditions to exist. The spores, however, remain active in the soil for many years. It is difficult to control this disease when it is in the later stages of development, due to the thousands of spores which are released. The disease usually develops later in the growing season, on crops such as garlic and onions. It is often not worthwhile controlling it, as it may begin a few weeks before harvesting.

132 Downy mildew on tomato leaves.

133 Phytophthora root rot on spinach.

134 Black rot on cabbage plants.

20.6.3.9 BACTERIAL CANCER See photo 141.

Bacterial cancer causes small, white spots with slightly raised brown centres to develop on the fruit. The entire plant wilts rapidly, even in moist soil. The disease begins on the lower leaves and moves upwards. The pith of the stems near the ground level, begins to rot. The disease occurs mostly in hot areas. Infected plants must be removed and burned once spotted. If not, they infect the soil and the infection can remain in the soil for as long as 10 years, where they will continue to infect tomatoes, potatoes and the capsicum families. In the 1970s, the disease was so destructive, that when crops were produced under protection in the soil, this disease could not be controlled properly. The reason for this is that no chemical products could control the disease once it spread in the soil. Producers in the Western Cape who produced tomatoes in soil under protection, were forced to produce tomatoes in hydroponic systems.

In 1972 Professor Brown, chairman of the organisation "Vegetables under Protection", travelled to overseas countries to study hydroponic production in greenhouses. The aim was to find a non-soil medium to grow high-value crops in. He started to research the production of crops in hydroponic systems. Consequently, Professor Brown is known as the father of hydroponics in South Africa. Thereafter, greenhouses were changed to produce high-value crops in hydroponic systems.

135 Powdery mildew underneath a cabbage leaf.

136 Immature garlic cloves which have germinated and grown inside the garlic bulb emerge and develop new leaves. This happens when high temperatures and too much irrigation or rain occurs near the end of the growing season. Usually only a small number of plants are involved.

137 A heavy infection of a common disease called Fusarium. It is a soil disease on tomato plants but which also attacks other crops. The symptoms show early in the growing season. When heavy infections occur it is not worthwhile to trellis the plants.

20.6.3.10 BACTERIAL SPOT AND STIP See photo 142.
These diseases are similar, and the important factor is that their control measures are the same. Tomatoes, spinach and capsicums are usually attacked. The symptoms on tomato fruit are brown, approximately circular, rough spots with margins of about 2-3mm in diameter. The spots eventually enlarge and become irregular in shape. Small circular spots turn brown and appear on the leaves. Infected leaves usually dry out and may easily fall off. Begin a spraying program for tomato, spinach and peppers, when the first sign of this disease occurs. Chemicals that include copper hydroxide are quite inexpensive and are a good option. The disease will, however, not go away as bacterial diseases cannot be fully controlled. They can only be retarded so that the harm they can do, is limited.

20.6.3.11 FUSARIUM WILT OR ROOT ROT See photo 137.
This disease is caused by a fungus that enters the plant through the roots. The top growth of plants may exhibit wilting symptoms, become yellow, and later, leaves will die. Some, but not all leaves, begin to turn yellow. When a cross-section of a stem of an infected plant is inspected, a clear brown or partially brown inner stem underneath the skin is visible. The stems of the cucurbit families and tomatoes are also brownish when they are cut-open lengthwise. In the case of water and musk melons, the roots turn a brownish colour and die. This causes the plants to wilt even if sufficient water is available.

When sweet potato vines and roots are scratched with your nail, a brown stripe becomes visible. This indicates the presence of the disease. Parts of the sweet potato vines become yellowish, but not all the top vines of the plant show this symptom. It seems that this disease is not a huge problem. Infections are usually wide-spread across the entire land and it requires favourable conditions to emerge. However, be careful when the disease builds-up in the soil. Avoid establishing related crops again. The conditions are favourable for the disease when temperatures rise above 20°C.

138 Botrytis or grey mould on tomatoes.

139 Cercospora leaf spot or leaf spot on spinach leaves.

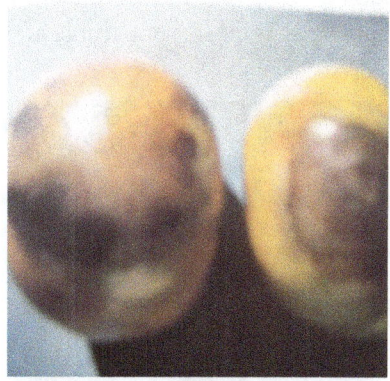

140 Common rot of tomato fruit occurs when the fruit is near or touching the soil and if combined with high humidity.

20.6.3.12 ANTHRACNOSE See photo 143.
Anthracnose is a fungal disease favoured by cool and wet weather. Lower leaves of plants develop small, water soaked spots that darken and became larger as the plant matures. The centres of the infected leaves dry and fall out, leaving openings with black margins. Affected leaves tend to die fast and then fall off. The disease affects younger leaves. The fungus is spread by splashing water and could become serious during long-lasting wet weather and humid conditions. Some weed species may also host this disease.

20.6.3.13 LEAF SPOT (Cercospora leaf spot) See photo 139.
This disease is caused by a fungus that penetrates the leaves. It causes round spots which multiply rapidly which give the leaves a rough appearance. The disease attacks young leaves when warm conditions occur. It seldom attacks in cold conditions. The disease can be easily controlled by using copper chemicals while the plants are still young. The disease is seed borne and commonly attacks beetroot and spinach.

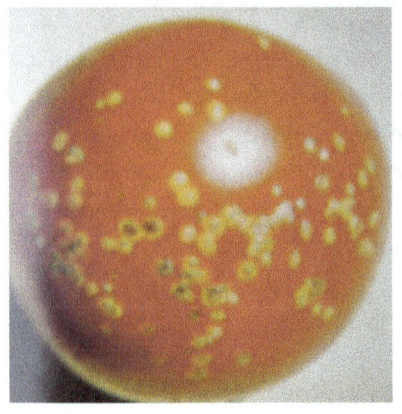

141 Bacterial cancer on a tomato fruit.

142 Bacterial spot on tomato fruit.

143 Anthracnose or ripe rot on tomato fruit.

20.6.3.14 BOTRYTIS OR GREY MOULD See photo138
This disease is also called grey mould, due to the brown spots which appear on the leaves causing a grey fungi to begin to grow. All parts of a tomato plant can be infected which could lead to fruit rot. This condition is favoured by moist conditions and high humidity. The disease is destructive mainly inside tunnels, where high humidity and favourable conditions exist. When the disease is wide spread, it is very difficult to control. Onion and garlic are vulnerable, especially when trips remove the waxy layer from the leaves, as this exposes wounds which are easily penetrated.

20.6.3.15 BACTERIAL WILT
This disease is caused by bacteria and is usually present in the soil. Characteristically infected plants wilt during the day and partially recover at night. A simple test to identify this disease is as follows: cut an infected stem laterally and place it in a glass of clean water. The stem will release a mass of bacteria spores and a milky substance is clearly noticeable after a few minutes.

144 Brown rust on garlic leaves.

145 Fusarium wilt on tomato plants.

146 Mosaic viruses are the most common problem on tomato, cucurbit and capsicum crops, especially in winter in the warmer areas. Tolerant cultivars, especially the F1 hybrid cultivars, are sometimes available and should be considered in problem areas.

20.6.3.16 PHYTOPHTHORA AND PHYTIUM
Phytophthora is a soil disease and attacks the root system of plants. This disease may build-up in a few days, especially in clay-type soil. It is caused by a fungus and can be devastating when favourable conditions occur. The disease favours long periods of high humidity and medium to high temperatures. The most common symptom is wilting. Usually, the lower stems of plants, just above the soil level, are infected. Affected seedling

roots are clearly visible because they become brownish and disintegrate. Young seedling roots from seed trays should always be healthy and white. The disease attacks the developing roots as soon as they appear, causing the white roots to change to a brownish colour and disintegrate. The disease is then carried to the stems of the plants, at which time the leaves begin to show signs of yellowing. However, usually not all the leaves show this symptom - it is mostly shown on one side of the plant. The disease does not favour aeration in soil, especially in clayish types of soil. Therefore, rather make proper, raised beds for sensitive crops such as carrots, beetroot, spinach and lettuce. Make use of ridges for tomatoes, peppers, chillies, paprika, potatoes and sweet potatoes.

20.6.3.17 WHITE BULB ROT

This disease may survive in the soil for more than 20 years and attacks mainly the onion families and a few other crops. The fungi remain dormant in the soil, waiting for onion roots to develop. At this stage, they enter the bulbs through the roots. The disease prefers cooler, wet soil and a temperature range of 14-18°C. In some parts of South Africa, white bulb rot is a huge problem, as previous productions had infected the soil with millions of fungi spores. Some soil is so heavily infected, that producers should totally avoid these areas for the production of onion and garlic. It is not worthwhile to attempt to control this disease.

20.7 VIRUSES

A virus is a parasite that lives inside the tissue of plants. It may hamper its development and sometimes develops very quickly and is destructive. Various kinds of viruses occur across the world which attack vegetables and other plants.

Once a plant is infected, the plant virus cannot be killed without destroying the plant. To kill the virus, the infected plant should be removed and be buried or burned to prevent the virus from infecting other plants. The fact that viruses affect the growth tempo of plants, is the most important symptom of a virus infection which can be used to recognize it. Typical virus symptoms which occur on the leaves, are mosaic patterns or patterns of light to dark green and sometimes yellowing on the leaves and vines. Malformation of tomato, cucumber and pumpkin/squash fruit is another symptom.

Viruses are mainly transmitted by aphids and, to a much lesser degree, by trips. Labourers who smoke, may transfer tomato mosaic virus from one plant to another. Disease/virus tolerant or resistant crops and cultivars are available. They are sometimes essential in certain areas when producing tomatoes, potatoes, cucurbits (pumpkin, squash and cucumbers) and capsicums (peppers, chillies, paprika).

Tomatoes, potatoes, pumpkin/squash and peppers are much more vulnerable to a variety of viral diseases. Tomato Spotted Wilt virus and Tobacco Mosaic virus are the most common, especially on tomatoes, cucurbits and capsicum plants. They may do considerable damage in certain warm-winter production areas. The latest development in the breeding and research of resistant or tolerant cultivars (2009) includes bacterial spot, tomato spotted wilt virus and also some resistance to root-knot nematode.

Research and breeding programs always remain a high priority in the production of tolerant, partially resistant or fully resistant crop cultivars. Currently in development are tomato cultivars which can set fruit at low temperatures and cultivars that have multiple disease resistances. These will soon be released to the industry. Remember, disease resistance is lost or partially lost as soon as vegetable plants become stressed. When re-propagating virus or bacterial infected mother plant material (such as cuttings, tubers, bulbs or cloves) all sister plants will be infected. When any part of the plant is used for propagating material which is intended to be transplanted, then these sister plants will also be infected.

Virus infections are a substantial problem in areas where producers establish summer-orientated crops in warmer, winter areas. Such crops are tomatoes, peppers, pumpkins/squash. When these crops are produced in these areas, they may become infected very quickly. These infections can be destructive to the extent that it is impossible to control them.

20.7.1 FACTORS TO CONSIDER WHEN CONTROLLING VIRUSES
- Other infected crops near the production land may infect the new production land.
- Leftover plants and plant material that survive from the previous season should be removed from the land because they may be a source of viruses.
- Weeds play a role in the survival of viruses.
- Some viruses survive on plant debris for long periods.
- Infected seed could be a source of infection.
- Infected plants should be removed immediately when they are noticed.
- Humans may carry viruses from one production land to another.
- Resistant cultivars should be considered, if available.
- Cuttings, such as sweet potato vines and garlic cloves, may carry virus diseases.

20.7.2 PLANT VIRUS DESCRIPTIONS
20.7.2.1 TOMATO MOSAIC VIRUS (TMV) See photo 146.
TMV is one of the most common plant viruses known. The symptoms are first noticed on the youngest leaves of plants. A mosaic pattern appears on leaves which causes them to become bubbly. Infected leaves curl upwards and the plant remains small. Fruit of infected plants have a knobbly appearance and usually remain small. The virus can be transmitted when handling plants during trellising and pruning. Damage caused by this virus is extremely severe, particularly in areas where tomatoes, peppers and cucurbits are produced throughout the year. This virus prefers to lie dormant in the tissues of plants, especially in the capsicum families. This means that the virus will affect the plant slowly and mostly in the later stages of the plant's development. That is also why pepper plant infections (besides nematodes), may slowly decline over time while a reduction in yields is sometimes severe at the end of the growing season. **Take note that the TMV virus has similar effects as the PVY virus.**

20.7.2.2 POTATO VIRUS Y (PVY)
PVY causes plant leaves to show a light and dark green discoloration as in the case of TMV. The leaves become yellow and dark green, stripes are visible next to the leaf veins. The leaves usually begin to crinkle. The plants remain small, fruit set is poor and becomes malformed even at an early stage. This is one of the most common viruses and appears in all the production areas in South Africa. This virus is transmitted mainly by aphids and to a lesser degree, by whiteflies. Insects, especially aphids, should be controlled effectively to avoid outbreaks of infections. Besides potatoes and tomatoes, the virus mostly attacks the capsicum families, especially sweet peppers. The virus infection damage is exhibited more clearly when the plants become mature.

Remember, capsicum families, especially peppers, when they are trellised in shade net houses or in tunnels, can remain in production for 6 months or longer, if insects and diseases are controlled effectively. The capsicum families are perennial plants and can be grown through to the next season. They can be pruned to produce fruit the following season when temperatures are favourable. However, because of the lower yields and virus infections, this is not practiced by farmers.

20.7.2.3 VERTICILUIM WILT
This viral disease is soil borne and infects the entire plant. The plant wilts as if suffering from drought stress.

20.8 PHYSICAL DISORDERS

Physical disorders are factors that are not related to diseases. Disorders prevent the plant from growing properly and may spoil the appearance of both the plant or fruit. Physical disorders are caused by:
- Insect, disease and animal damage.
- Sunburn and **high or low** temperatures.
- Drought and drowning.
- Poor drainage conditions and the pH of the soil.
- Poor pollination of female flowers.
- Herbicide damage and soil compaction.

- Uneven water applications and hail.
- Air pollution.
- Fertilization toxicity and deficiency.
- High salt content of soil.

20.8.1 PHYSICAL DISORDER DESCRIPTIONS
20.8.1.1 CAT FACE See photo 147.
Cat face is malformed fruit with enlarged scars and holes. It is usually found on the blossom end of tomato and pepper fruit. The disorder occurs more commonly on cultivars that provide larger fruit and is almost absent on the small fruited cultivars. It may also be caused by trips, when plants are still in the flowering stage.

20.8.1.2 FRUIT CRACKS See photo 149.
Fruit cracking may be a serious problem, especially on tomato fruit. Huge losses could be the result. It also occurs in greenhouses and hydroponic systems, when indeterminate tomato cultivars are produced for long periods. Cracking refers to the strength and stretching ability of the fruit skin. Rapid growing, high N and low potassium K levels as well as a lack of availability of certain nutrients make the fruit more susceptible to cracking. Cracking can be expected and is often related to peppers, potatoes, sweet potatoes, carrots, cabbage, pumpkins, water and musk melons.

20.8.1.2.1 Two types of growth cracks that occur on tomato fruit
- **Radial cracking** splits from the fruit shoulder (from the stem towards the blossom end).
- **Concentric cracking** splits in circular patterns on the shoulder around the stem of fruit.

It is possible that both types of crackling may occur on the same fruit. Some cultivars have partial or good resistance against cracking. Cracks may appear at any stage of the ripening of fruit. Cracking occurs when the inside of the fruit expands faster than the epidermis which causes the epidermis to split.

20.8.1.2.2 Other reasons for cracking
- When plants absorb too much water, usually after water shortages or stress.
- A sudden drop in temperature, when the plants lack water combined with a shortage of Ca.

If the cracks begin early when the fruit is still young, the scars will become deeper as the fruit develops and matures. The cracks may become infected with secondary pathogens causing ugly black scars. Cracking is more of a problem when fruit begins to ripen than in the early stages of fruit development.

20.8.1.2.3 Control management
Some control measures for cracking include selecting tolerant cultivars, reducing fluctuations in soil moisture and maintaining good foliage covering of fruit. Fruit can be covered by tomato leaves or leaves of pepper plants. Producers should do their best to retain foliage especially when hot, sunny conditions occur. A wide fluctuation in air temperature may also cause cracking. Producers should avoid applying too much N. It is important that water management should be properly scheduled. Rain and high humidity induce cracking because less water is lost due to transpiration. Cracks on sweet peppers could sometimes be a cultivar characteristic. In plants with a shallow root system or where shallow irrigation is practiced, fruit may crack sooner. It has been noticed that smaller tomato fruit cultivars as well as the thick walled roma and salladette cultivars are less prone to cracking.

20.8.1.3 BLOSSOM END ROT (BER) See photo 147.
BER disorder occurs at the blossom-end of the fruit. It begins with a light water soaked lesion that enlarges and turns black. BER is always associated with a calcium Ca deficiency. Calcium is a poor mobile element, which moves slowly in plants. BER shows early symptoms on young fruit, especially if water stress takes place. Underdeveloped root systems or the use of too much N will reduce Ca uptake. Rapidly growing plants are more susceptible to BER. Conditions such as high humidity, temperatures and moisture extremes promote this disorder. Cultivars differ in their tolerance to this disorder. Control of the humidity in greenhouses (not

possible in a shade net houses) and fertilization through the irrigation system may inhibit BER to a certain extent. A Ca-based foliar spray may also be helpful, when applied at the correct stage of plant development.

20.8.1.4 SUNSCALD OR SUNBURN OF FRUIT See photo 148.

This disorder may become very serious and for many producers it is difficult to avoid, especially in warmer areas. When fruit (tomatoes, peppers, green type pumpkins) become sunburnt, the fruit wall tissue becomes sunken, wrinkled and turns white. It is mainly the ultra violet sun beams that cause this damage. Sunscald even occurs in shade net houses and in tunnels in the warmer areas, especially when the foliage (the plant canopy) is not dense enough. Symptoms of affected fruit usually occur on the side or top half of the fruit as the sun bakes on the western side of the fruit. Often the affected area becomes infected by secondary pathogens. This disorder is not a huge problem where tomatoes and peppers are cultivated under protection. However, in warmer areas, it is still a problem inside greenhouses, especially when the plants become stressed. The problem can be reduced when cultivars with good foliage cover are established. When harvesting pepper fruit, care should be taken not to damage the plants too much. This will ensure that the remaining fruit is not exposed to too much sun. Be aware that the green cucurbit pumpkin types are very vulnerable to sunburn, especially the cultivar Green Hubbard.

20.8.1.5 ZIPPING See photo 148.

Zipping is thin scars running from the stem scar, down the length of the tomato fruit to the blossom end side. This scar looks like a zip (thus the name) and the scars on one fruit can be numerous. In severe cases, a hole opens in addition to the zipper scar. Zipping should not be confused with spider track. This disorder never becomes a serious problem and the fruit could still be marketable as second grade produce when they are mixed with other fruit.

147 Blossom end rot (BER) on the underside of tomato fruit.

148 Sunscald or sunburn.

149 Radial and concentric cracking of tomato fruit.

150 Cat face (top), zipping (bottom-left) and radial cracking (bottom-right).

151 Tomato fruit with the calyx still attached (top) and without (bottom).

152 A crop affected by both sunburn and blossom end rot (BER). This crop is very sensitive to sunburn.

20.8.1.6 SPIDER TRACK

Spider track on tomato and pepper fruit is generally more numerous than zipping and is confined to the shoulder area of the fruit. The cause is not known. Cultivars that are affected by this disorder but it is seldom serious.

20.8.1.7 PUFFINESS

It is not possible to notice puffiness until the fruit is cut. Fruit that is severely puffy will appear to be flat-sided. When fruit is cut open, its cavities are hollow and the weight of the fruit is very low in relation to its size. This problem is caused by factors that affect fruit set. This includes pollination, fertilization, or seed development. The most common causes are too low or too high temperatures during fruit set. Nutritional imbalances (high N and low potassium) and insufficient light are also factors.

20.8.1.8 RAIN CHECK

The cause of rain check is not known. This disorder does not occur in greenhouses, but could occur in shade net houses. This condition occurs especially during long, wet spells and heavy rains. Rain checks are not very common but spoil the appearance of the tomato fruit. Numerous tiny, concentric cracks develop on the shoulder of the tomato. Pepper fruit is also susceptible to this disorder. Sometimes these tiny cracks combine into larger, longer cracks or scars. Green fruit is most susceptible.

20.9.1.9 BLOSSOM DROP

The main cause of blossom drop is a poor uptake of calcium Ca. This occurs mainly when soil has poor lime content and when the plant grows too vigorously. Other causes include excessive potassium P and irregular water applications. Balanced fertilization and sufficient calcium Ca in the soil should be maintained. Factors such as temperature extremes, high relative humidity or excessive wind, could inhibit pollination. Insects, such as trips, consume pollen inside the flowers.

153 A physical disorder on cucumber fruit called tapering which has many causes.

154 Herbicide damage on tomato plants.

20.8.1.10 MALFORMED FRUIT

In crops such as pumpkins, malformed fruit can be expected due to incomplete pollination of the female flower by insects. Malformed fruit is mostly due to too low or high temperatures and is also caused by a lack of moisture or a nutrient deficiency, especially an N shortage.

20.8.1.11 FLOWER FRUIT DROP AND POOR FRUIT SET

Too much N as well as too low or too high temperatures may cause poor fruit set. The reason, is that pollen is not normally released when temperatures fall below 9-10°C or when temperatures exceed 32°C. Too high concentrations of N may also cause abortion or flower drop in the fruit of tomatoes and other crops like beans, pumpkins, cucumbers and peppers.

CHAPTER 21

INTRODUCTION TO INSECT AND DISEASE CONTROL

21.1 INTRODUCTION

Intensive and costly research takes place all over the world in order to produce non-poisonous or soft chemicals that eradicate insects/diseases effectively and but leave humans and other animals and insects (such as bees, spiders, predators, animals, fish and birds) unaffected and unharmed. However, in practice this is very difficult to achieve. It is the responsibility of the users to make sure that they follow the rules and regulations that is stipulated on the specifications of the container.

Research is intensive and ongoing to identify new and effective insect/disease control measures. Poisonous, soft or non-poisonous chemical products change quite rapidly and new improved chemicals are released periodically. Some chemicals are also withdrawn from the markets if they are identified as too dangerous or toxic.

155 The correct protective clothing for labourers' that spray chemicals is very important. The mist blower should be calibrated correctly by noting the droplet size and the spraying force.

21.2 CHEMICAL GUIDELINES FOR CONTROLLING INSECTS AND DISEASES

There are numerous kinds of chemical products on the market. Producers are required to continuously educate themselves on how to use chemicals correctly for nutritious, healthy vegetable production.

If a disease or insect cannot be identified correctly, then it cannot be controlled. Many producers use unregistered chemicals for disease/insect control according to their personal experience of what works best for them. Some chemicals are periodically banned due to their poisonous active ingredients or their harmfulness to animals, fish, birds and humans. These types of chemicals' do not comply with regulations. examples are the well-known DDT and the very toxic Temik that has been used for centuries. Improved, more effective and less poisonous chemicals may become available and replace the old ones.

The problem is that some of the new types of lower strength chemicals (less poisonous) are not always as effective as the previous, stronger (more poisonous) chemicals that have been withdrawn from the market.

Chemical consultants may sometimes recommend two or more chemicals to control insects or diseases in the hope that one of them will be effective. This may be expensive as chemicals are very expensive and are usually not supplied in small quantities.

Most diseases vary in their development and intensify as a result of weather conditions, especially when wet weather and high humidity occurs, usually near the coast when morning dew is present.

Diseases can be challenging to identify correctly and to control effectively. Knowledge of how to control diseases and insects is important for successful vegetable production. The life cycles of common insects should be known in order to correctly apply chemical measures. Sometimes preventative action should be taken, especially when conditions are favourable and the probability of insects and diseases are high. For example, early blight which may infect tomatoes and potatoes.

When producers are unsure of correctly identifying if they **cannot identify the insects or diseases correctly** then consult Chemical dealers, Seed Companies, Local dealers, Neighbour, Literature and the internet to find the best chemical control measure.

21.3 NATURAL METHODS FOR CONTROLLING INSECTS AND DISEASES

Natural and organic methods may help prevent and eradicate pests/diseases that the plants may already have. Here are some of the aspects to look at:

- **USE QUALITY, CERTIFIED DISEASE-FREE SEED**
  Establish seeds and seedlings plants that are free from insects and diseases. If this is not done, diseases may develop when the next crops are established and insect eggs or larvae may be carried into the land when transplanting seedlings. For example, if vegetative plant material such as sweet potato vines are infected with diseases and viruses, then the new production land may inherit these unwanted conditions.
- **LOOK AFTER PLANT HEALTH**
  Healthy, well-developed plants are more resistant to insect/disease related infections as the plants outer skin cells are thicker and hardier. It is more difficult for some insects and diseases to penetrate hardier skin. When plants become stressed, such as with water and nutrient shortages, they are much more vulnerable to diseases.
- **INSPECT THE PRODUCTION LAND REGULARLY**
  If insects or diseases are noticed during inspections, then they should be controlled as soon as possible. Some insects and diseases multiply and spread very quickly and therefore waiting too long before controlling them may cause damage. They may also be chemically sprayed for a second (or more) time if necessary. Insects and diseases may also become immune or partly immune if repeatedly sprayed with the same chemical and with the same mode of action. Virus and bacteria infected plants should be removed before they spread to other plants.
- **USE BIO-FUMIGATION CROPS IN A ROTATING CROP SYSTEM**
  See Chapter 18.4.3 for bio-fumigation.
- **INTRODUCE PREDATORY INSECTS**
  There are a number of insects that consume other insects. For example, ladybirds, praying mantises, spiders and wasps.
- **ESTABLISH RESISTANT CULTIVARS AND PERFORM SELF-EVALUATIONS**
  Scientists are constantly developing and improving cultivars from all over the world. Crops may have one or more resistant abilities such as resistance to insects, diseases and viruses. Producers should investigate and consider these cultivars.
- **PLOUGH IN WINTER**
  Winter ploughing may bury insects in the soil down where they may smother and die.
- **ESTABLISH SOME CROP EARLY IN THE SEASON**
  Generally, summer orientated crops are established as early as possible after the cool, dry winter season and when the temperatures begin to rise.

If establishing crops such as cucurbits, cole drops, tomatoes and so on early in the season, then fewer insects/diseases will attack them.

If establishing tomatoes/peppers early in the summer rainfall areas, then there is an opportunity to avoid the extremely hot temperatures. For instance, when establishing tomato seedlings in mid-September in the Gauteng area, they will begin flowering at the end of October when it is still fairly cool. However, when establishing seedlings in mid-October, they usually begin to flower at the beginning of December when temperatures are very high. This may influence blossom drop due to poor pollination. Pollen may burn inside the flowers which may cause flower and fruit drop. Ideal temperatures for these crops are 18° at night and 25° in the day. F1 hybrid cultivars are available that are more resistant to temperatures and may overcome this problem to a large extent.

Be aware that pumpkin flies are more active later in the season. When establishing cucurbit families early in the season, minimum damage can be expected as the flies only begin to rapidly multiply at the end of October, in the summer-orientated areas. Flies and most other insects are minimized after very cold winters. When the fruit develops to maturity, its skin becomes tough and thicker and the flies are not able to penetrate the fruit and lay their eggs.

Some insects and diseases prefer warm wet seasons. Others prefer cool dry seasons that are favourable for them to multiply quickly. The live cycle of most insects is usually very quick as they generally multiply quite rapidly.

CHAPTER 22

GUIDELINES FOR INSECT AND DISEASE CONTROL

22.1 INTRODUCTION

Chemicals are necessary to destroy diseases/insects or to prevent their development. However, to achieve this successfully then chemicals should be applied correctly. The correct chemical application required to control insects and diseases is sometimes difficult to achieve, especially when there is a build-up of chemicals over time in the production land. An efficient spraying program is the very basis of the producer's effort to control insects and diseases.

It is very important to use the correct dosage that has been prescribed by the manufacturers. Different degrees of chemical applications are available and the correct application method and amount is the most important factor when using chemicals. Research has indicated that more than 60% of poor chemical control (applications) is due to the use of ineffective apparatus or that chemicals are sprayed in unfavourable conditions.

22.2 COMMON REASONS FOR FAILURE TO CONTROL INSECTS AND DISEASES

156 A knapsack-type chemical spray pump is uncomfortable for labourers to carry as a fully loaded tank may weigh in the region of 15kg. The white attachment in front is to avoid over spraying.

157 A chemical spraying unit being calibrated. Producers should make sure they understand how to calibrate the pressure lever units correctly.

158 A power operated, driven mist blower is very useful and effective for spraying, especially in greenhouses and for small-scale producers.

159 A tractor with a hydraulic spray pump spraying 4 potato rows at a time. Commercial producers can manage 8 or more rows with modern sprayers.

160 Chemical spraying using a powered mist blower. The blower is very effective when calibrated correctly.

Failure to control insects/diseases is mostly blamed on chemicals, but the cause is usually elsewhere:
- Waiting too long before spraying or when insects/diseases are too large or too numerous.
- Spraying plants when it is windy.
- Neglecting to clean the spraying tank after using herbicides.
- Incorrectly calibrating the spraying nozzles of the apparatus.
- Making use of the incorrect type of chemicals.
- Spraying with the incorrect droplet sizes-either too large or too small.
- Spraying at the incorrect time of day. Chemicals are sensitive and may dissolve quickly in hot, humid conditions.
- Spraying during rainy days.
- Using expired chemicals and incorrect doses of chemicals.
- Uneven coverage especially when using contact or stomach chemicals.
- Incorrect pH of the water used for spraying. It may be necessary to apply stickers and softeners in the chemical water. It is important to adjust the pH before adding chemicals in the spraying tank.
- Making use of unregistered chemicals.

22.3 ADVICE FOR EFFECTIVE INSECT, DISEASE AND VIRUS CONTROL

- Scan the production land regularly. This will help determine the correct timing for the next application to take place. Do not neglect to scan regularly because you may miss problems.
- Many insects and diseases appear at the same time in the season, year after year. Make sure that you are prepared ahead of time.
- One proper spraying application during the early stages of the development of insects and diseases, would normally eliminate the need for another application to take place.
- Less toxic chemicals may be used when insects are still young and beginning to reproduce.
- Spray only when the wind speed does not exceed 4-5km per hour.
- It is claimed that for a wind speed of only 5km per hour, a 0.5mm droplet will drift 1.2m, when a spray pump is held 0.5 m above the soil. A 0.25mm droplet may drift 2.7m while a 0.1mm droplet may drift as far as 9.3m away from the target.
- Do not spray when the temperature is high.
- Overhead irrigation (sprinklers) and rain may wash the chemicals from the plants. Wait at least 24 hours after spraying chemicals before applying follow-up irrigation.
- Use spraying apparatus with effective spray volumes, nozzles and pressure.
- Proper spraying coverage of plants is essential. Some insects/diseases hide underneath the plant leaves.
- Increase spraying as the season prolongs and when the plant size increases.
- Use proper chemicals and know the insects and diseases to be controlled. Choose only recommended chemicals and rates of application.

22.4 SPRAYING EQUIPMENT USED FOR INSECT AND DISEASE CONTROL

The law is strict and complicated concerning the usage of chemicals. Many regulations are in place. If labourers are not trained in the correct usage of chemicals according to regulations and if someone is poisoned as a result of negligence, even accidentally, the farmer or producer will be held accountable. Therefore, it is important to properly train labourers and keep them informed and updated about the rules and regulations.

Some chemicals are very poisonous and may affect a person's health in the long term. Taking proper precautions such as masks to avoid inhaling chemicals or protective clothing to avoid chemicals coming into contact with the operator's body, should not be neglected. For example, it is not acceptable for an ordinary hand lever spraying apparatus which is fixed on the operators back, to leak. Poison may build-up in the lungs, kidneys and tissues and symptoms of ill health may take years to materialise.

22.5 SAFEGUARDING AGAINST POISONING

The safeguarding period is the period between the final application of the chemical being used and the harvesting time. This period is usually prescribed by the manufacturers and is important to adhere to. This period allows the chemicals to break down before harvesting the crop for human consumption.

Do not harvest the edible crop before the safeguard period has expired as this may affect the consumer, as the crop may carry small amounts of poison when harvesting. Producers are trusted by the public and consumers to make sure their crops are clean.

An example is for cabbage production which is controlled continuously using chemicals in summer. This is due to the high insect and disease pressure during the warm months. The safeguarding period should be followed carefully.

It is important to read the instructions that are provided on the labels of chemicals. Especially the safeguard instructions when using chemicals near harvesting time. For example, when 7 days is allowed before harvesting, then the producer must wait at least 7 days before harvesting the crop. The safeguarding period of chemicals differ, therefore be sure to check the labels of each chemical product for the minimum requirements.

It is the duty of the employer to provide and maintain a safe and healthy working environment. Labourers should be informed, trained and educated about the SABS standards when working with chemicals.

22.5.1 THE BASIC RULES WHEN USING CHEMICALS
- Wear gloves when measuring and handling chemicals.
- Wash off any spillage on the body using soap and water.
- Clothing should cover the entire body.
- Keep-out of the spraying drift.
- Do not smoke, eat or drink when spraying crops.

22.5.2 THE CORRECT PROTECTIVE CLOTHING WHEN SPRAYING CROPS
- Overalls that cover the whole body.
- Rubber shoes and goggles.
- Gloves.
- Masks and the correct type of filters.

See photo 158.

Chemicals should be stored in a room that has a smooth cement floor, good ventilation and a light. The room should have a shower and a washing basin. It should be locked at all times and the windows should be properly burglar proof. ***See Chapter 22.12 for how to build a Chemical store.***

22.6 TYPES OF SPRAYING EQUIPMENT FOR INSECT AND DISEASE CONTROL

A large selection of spraying equipment is available on the market including hand-pump operated knapsacks, mist blowers, motor driven, power and jet propelled sprayers as well as aeroplanes. The choice of spray pump should depend on the size of the production land.

- **KNAPSACK SPRAYERS:**
  One of the most widely used small sprayers is the hand-lever operated knapsack. It is carried on the back of the person operating the sprayer. Small-scale producers usually have 2 knapsacks (one for chemicals and another one for herbicides). ***See photo 156.***

- **POWER-OPERATED AND HYDRAULIC SPRAYERS:**
  Various power-operated sprayers are available and range in size from small, portable, engine-driven pump units to large self-propelled sprayers that have a series of nozzles mounted along a boom which are pulled behind a tractor and can spray 6m across at a time. *See photo 159.*
- **THE POWERED-OPERATED MIST BLOWER:**
  Ideal for production on a reasonable scale. It produces droplets in the range of 50 to 100mm. A mist blower sprays more effectively and applies the same quantity of pesticide for one-tenth of the volume. Mist blowers are popular for use in tunnels or in shade net houses. *See photo 155.*

22.7 CHEMICAL SPRAYING VOLUMES AND DROPLET SIZES

22.7.1 INSECTICIDE TIMING
To be effective, chemicals remain on plant foliage for a period of time as a thin layer. The period of time for which chemicals may remain on the plants range from hours to weeks. It depends on the break-down action of the chemical sprayed on the plant.

22.7.2.1 CHEMICALS ARE CLASSIFIED INTO THREE GROUPS
22.7.2.1.1 CONTACT AND STOMACH CHEMICALS
Eliminates insects when chemicals are sprayed directly onto them. Stomach insecticides should be applied over the entire plant, especially where the insects usually hide underneath the leaves.

22.7.2.1.2 FUNGAL CHEMICALS
Destroys fungal diseases and should be applied over the entire plant.

22.7.2.1.3 SYSTEMIC CHEMICALS
Penetrates into the leaf stem or roots of the plant and then dissolves into the sap stream. This introduces the active chemical to the part of the plant where the insects or diseases are feeding or hiding. It is important that the plants are sprayed at the optimum stage of their development as the systemic chemical control is more effective when the plants are still in their growing phase.

22.7.2 SPRAYING VOLUMES
Spraying volumes may be classified into the following groups:
- High volume: More than 600-700 litres p/Ha
- Medium volume: 200-500 litres p/Ha
- Low volume: 60-200 litres p/Ha
- Very low volume: 10-50 litres p/Ha
- Ultra-low volume: Less than 10 litres p/Ha

When using low volume apparatus then less of the active ingredients are wasted. The location where the chemicals are sprayed is known as the target. When using the correct dosage, the active ingredients should reach the target effectively.

22.7.2.1 LOW SPRAYING VOLUMES ARE EFFECTIVE BECAUSE
- It saves time when refilling the water/chemical tanks.
- Less water, containers and implements need to be transported to the land.
- Modern, low-volume applicators are designed to provide better spraying cover and better penetration.
- It is much more cost effective.

22.7.3 DROPLET SIZES
It is important that chemicals adequately cover the target areas of plants. An effective coverage depends on the droplet size and not on the spraying volume.

The droplet size of an aerosol spray can is about 50 microns.
- Mist sprays   70-100 microns
- Fine sprays   130-250 microns
- Medium sprays   200-450 microns
- Coarse sprays   450 microns and above

The micron count and number of droplets per cm² is as follows:
- 10 micron   190 droplets per cm²
- 100 micron   19 droplets per cm²
- 200 micron   2.4 droplets per cm²
- 400 micron   0.3 droplets per cm²

As the droplet numbers decrease, the droplet size increases. Therefore, the smaller the droplets, the better the chance is that they will stick to the target due to the higher numbers of droplets.

22.7.3.1 Smaller droplets are more vulnerable to outside conditions
- Small droplets may easily drift away from the target even if the slightest wind occurs.
- Small droplets evaporate easier before they reach the soil and may harm the environment.
- When small droplets are sprayed too far or too high from the plants, it will affect the lifespan and the drift distance of the droplet.
- When the soil and vegetation is warmed by the sun, warm air rises from the soil. The movement of the hot air from the bottom takes place. This may produce the unstable movement of air and the droplets could drift away from the target.

22.7.3.2 When a droplet finds the target, the following may happen
- The droplet may roll off.
- The droplet may bounce off.
- The droplet may be washed off if it rains shortly after application.

22.7.3.3 POINTERS FOR EFFECTIVE SPRAYING OF PLANTS
- Do not spray when the wind speed exceeds 5km p/hour.
- Plants that are wet with dew, rain or irrigation water should not be sprayed.
- When a systemic chemical is used, make sure that the entire plant including the lower parts are thoroughly covered to make sure that the plant takes up the systemic chemical effectively.

22.8 PH OF CHEMICAL WATER

Chemicals dissolve in water much more effectively at optimal pH levels. Fungicides are more effective when the pH of the water is neutral while insecticides, herbicides and foliar sprays are more effective when the water pH range is between 4.5 and 5.5. When necessary, Adjuvants should be added to the insecticide spraying water in order to increase the life span of the insecticide.

22.9 RESISTANCE OF INSECTS AGAINST CHEMICALS

Chemicals belong to different groups which have different toxic active ingredients. The groups are known for their mode of action.

Continuously using the same chemical to control insects may kill the susceptible ones, but leave those with natural resistance unaffected. The process is called selection pressure and leads to the resistance of most insects. To avoid resistance, producers should control the insects during the outbreak stage when they reproducing or when the insects are still young.

When using stomach, or contact chemicals it is important to spray plants thoroughly so that all the parts of the plants are covered, especially under the leaves. Producers should regularly scan for insects over their entire land. This is important because when insects reproduce quickly and are spotted too late, the chemical control measures will then have to be continuous and the chance of chemical immunity is greater. This is especially true for the hardier insects such as spider mites and aphids. When insects such as aphids become resistant to chemicals, they also produce resistant or partly resistant offspring.

22.10 BIOLOGICAL CONTROL OF INSECTS AND DISEASES

Biological control means that insects feed on other insects or lay eggs in their bodies that then emerge to become predators like their parents. The larvae of some predators feed inside the body of the host insect, eventually hatching and killing them in the process. Other predators suck the life out of their hosts (such as aphids) and only leave their skeleton behind. This usually occurs beneath the leaves of the plants.

Biological control is an insect management system that is dependant on the assistance of friendly insects and is integrated with other management practises, which may include cultural, physical and chemical measures.

The application of chemical control should be managed using softer, less toxic chemicals that do less damage to the beneficial insects as well as the environment. Learning to live with a small number of harmful insects is valuable. It is important to monitor these harmful insects and to control them. This is not always a simple task, because if you kill all the harmful insects, then there is no food for the beneficial insects. Therefore, using biological control measures implies that there must always be a population of harmful insects that is large enough to maintain the predatory insect populations.

Biological control is not possible for large commercial production areas in South Africa, however it may be a good option for small-scale farmers. Organic farmers do their best to introduce natural insects and maintain them. This keeps the balance between the harmful and harmless insects. Toxic chemicals may not be used by organic farmers while biological control chemicals and fertilizer are expensive and sometimes less effective. By understanding the environment and not disturbing the balance of it, producers can succeed.

The use of resistant or partly resistant cultivars for some crops is helpful and sometimes a must. More than 90% of crops such as maize, cotton and soybeans are GMO modified. This means that these crops are resistant to certain insects or diseases.

When plants are healthy they should never be stressed due to water shortages. Certain insects, such as red spider mites and trips thrive under hot and dry conditions. Therefore, irrigation is important, because water stress allows these insects to penetrate the plants more easily.

Natural predators of harmful insects are usually wasps, birds, ladybirds and different kinds of spiders. For example, there are good ladybirds which only consume aphids and there are bad ones which consume plants. Usually the bright red and black spotted ladybirds are the good ones. Different types of spider species attack and destroy insects. They are rarely seen because they hide underneath the leaves of plants or reside in the soil and are only active at night.

It is advisable to monitor these natural insect predators in order to give them a chance to assist in controlling or partly controlling any harmful insects. The biggest disadvantage of biological control is that it takes time, sometimes as long as 3 to 4 years, to establish these friendly insects. The friendly insects need to be given an opportunity to multiply their population before they are able to control other insects.

It is important that producers who are interested in making use of biological control measures, should carefully choose the best chemical control measures that does not harm the friendly insects. To assist them in this, producers should contact reliable chemical companies.

22.11 SPRAYING PROGRAM GUIDELINES

- Train the operators to use the apparatus correctly. They should know the different nozzle sizes, how to adjust the spraying mist as well as the speed that should be maintained during the spraying process.
- Know the insects/diseases. Be able to scan and identify their feeding places.
- Choose recommended chemicals and decide the spray volume (high or low) which is to be used.
- Mix the chemicals shortly before spraying the plants in order to maintain the stability of the mixture.
- The pH level of the spraying water is very important and should be correct.
- Make use of adjuvants, water softeners and stickers when necessary.
- Use penetrations for systemic chemicals and stickers for contact chemicals.
- Inspect the spraying nozzles regularly.
- Do not mix chemicals of different kinds together unless it is recommended by the manufacturer.
- Do not spray during hot, sunny days or when plants are wet.
- Do not spray when wind speeds are more than 5km/hour.
- Spraying equipment should be in good working order and should be calibrated correctly.
- Rotate between the different groups of chemicals to prevent resistance.
- Clean the equipment after every application. Left-over chemicals may influence the next crop the equipment is used for.
- The effectiveness of some chemicals may be influenced when they are stored too long. Some chemicals can be stored for long periods and others not. Read the label.
- Always use chemicals according to the recommendations on the label.
- Use protective clothing and follow the safety precautions.

22.12 BUILDING A CHEMICAL STORE

The entrance to the store is important. A 3-5m concrete strip should surround the store. The building should be constructed using bricks and a closed-off roof so that the store is sealed and kept cool. Use steel to inhibit the chances of burning. A stronger and better constructed building may also prevent burglars from breaking into the building.

The floor of the store room should be at least 20cm above soil level in order to prevent rain water from entering the room. The floor should be smooth in order to avoid chemicals falling onto the floor and accumulating in indentations. The door should be made of a non-absorbing material and should be strong to make breaking into the store difficult. The store should always be locked. Unauthorised persons should not be allowed to enter the store.

The store should be well ventilated by using a small extractor fan. A light should be installed to make it easy to identify the chemicals properly. Large windows could be installed on the shady side of the building in order to improve the light and to minimize temperatures.

22.12.1 WARNING SIGNS
A warning sign should be placed on the outside of the building indicating that poisonous chemicals are in storage. A warning sign on the door should display the cross and skull warning as well as a no smoking sign. A fire extinguisher should also be installed next to the door.

22.12.2 EQUIPMENT INSIDE THE STORE
Herbicides should be stored separately from other chemicals. A basin with clean running water as well as a shower should be installed inside the building. A bucket filled with sand should be in the store to cover any spills that might occur on the floor. Chemicals should be stored on racks/shelves.

GreenZone
THE WAY TO GROW

Tel: 011 868 1141   Email: info@greenzone.co.za   Website: www.greenzone.co.za

Greenhouse products for total crop management

CHAPTER 23

PRODUCING VEGETABLES UNDER PROTECTION

23.1 PRODUCING VEGETABLE CROPS UNDER PROTECTION IN SOIL

Most vegetable crops may not be economically produced in tunnels and shade net houses. One reason for this could be that the plant spacing and plant population is approximately the same inside these structures as outside in the open-field. Producers should do their own research, evaluations and experiments to assist them with understanding the economic viability of these choices.

Some leafy crops are limited to production in tunnels and shade net houses due to considerations around expenses and maintenance. These leafy crops are usually spinach, baby cabbage, baby cauliflower, chinese cabbage, lettuce, spring onions, runner beans and also certain herbs such as parsley, celery, basil, etc.

Take note that when crops are trellised in tunnels or shade net houses and planted in the soil, the yields and quality are higher due to the fact that the trellising allows the crops to grow up to a height of 2.5m. Also, the plant spacing and humidity is much higher and therefore can carry a higher plant population compared to the open-field. Examples include high-value indeterminate crops such as tomatoes, cucumbers and pepper plants. Plant population may increase up to 26 000 plants p/Ha for under-cover determinate tomatoes in comparison with the open-field where the plant population may only be 16 000 to 18 000 plants p/Ha.

Other advantages include less wind and sunburn, better protection against hail and storms, less insects and diseases (also easier to find them), more comfortable to harvest and cleaner fruit. The cost of shade net houses and tunnels production is much higher due to the structure and trellising costs. However, the yield is more than double the determinate tomato production in the open-field.

The estimated cost for a 1Ha fully installed shade net house is approximately R1 434 64 (Table 11) and for a 1Ha tunnel is approximately R355 692 (Table 14).

Different types and sizes of Greenhouses are used in South Africa and of these the 10 x 30 m UV treated plastic tunnel with open doors and flaps is the most commonly used by commercial producers. It is the ideal size and creates the ideal conditions. When erecting larger tunnels, then temperature, aeration and humidity control is not very effective. When using smaller tunnels, then plant population is lower.

Crops which are under protection in shade net houses and tunnels and which are established directly in soil, accrue more benefits when compared to open-field production.

High-value crops such as tomatoes, peppers, cucumbers and to a lesser degree squash are compatible with production in tunnels or in shade net houses. High-value crops help to cool tunnels off, especially the taller plants. This is because the high transpiration rate of the plants helps reduce the humidity, which would otherwise need to be removed from the tunnel.

Crops such as lettuce, spinach, cole crop families, determinate tomatoes, box peppers and certain herbs grow more or less the same in open-field production as in tunnels or shade net houses. Producers often prefer high-value crops in soil in shade net houses because it is cheaper and also because tunnels are more expensive than shade net houses. Tunnels provide good protection, much higher yields and better quality produce.

Be aware that the markets are sometimes limited and could easily be oversupplied with certain exotic crops.

23.1.1 ADVANTAGES AND DISADVANTAGES OF SHADE NET HOUSES
23.1.1.1 ADVANTAGES OF SHADE NET HOUSES
- Offers protection against storms, hail, heavy rain and also animals, birds and larger insects.
- Allows wind movement and thus has a cooling affect.
- Some shade nets may filter up to 70% of the sun's harmful UV rays. These UV rays are responsible for sun scald and physiological disorders on tomatoes and pepper fruit.
- Shade net does not absorb water and does not rot. It also does not burn easily.
- Shade net is strong due to its interlock knitting construction. The net stretches when hail accumulates on it and returns to its normal position once the hail has melted.
- Monkeys and other animals cannot tear the net due to its natural strength.
- Shade net has a life span of 10 to 15 years compared to plastic covered tunnels that have a life span of 4 to 5 years, especially in warmer areas.
- Shade net saves water due to the protection it offers from the sun and may reduce wind speed by 25%.

23.1.1.2 DISADVANTAGES OF SHADE NET HOUSES
- It is not possible to control temperatures and humidity.
- It is not possible to keep small insects out.
- Theft can still take place as the net may be cut to enter the House.
- When producing high-value crops, such as indeterminate tomatoes, fruit burst and blossom end rot (BER) may occur more easily. To avoid this, plastic covering strips should be established on the ridges inside shade net houses, so that the irrigation water can be better controlled to prevent these conditions from occurring, especially in high rainfall areas.

23.1.2 SHADE NETTING FOR SHADE NET HOUSES
40% white, silver or black netting as well as a combination of black and white are available from the trade. White netting provides 24% shade and filters 80% UV radiation and transmits 60% light. Black provides 40 % shade and filters 40% UV radiation and transmits 60% light.

Consult manufacturers and suppliers for the best type of netting for your area. They also issue producers with a manual on how to construct a shade net house.

23.1.2.1 TYPES OF SHADE NETTING TO COVER SHADE NET HOUSES
20% Shade Net is used to cover and protect fruit trees and some vegetable crops against hail.
30% Shade Net is used for cut flowers and seedling production in cooler regions.
40% Shade Net is used to provide maximum shade and is recommended for vegetables production.

161 This shade net house is 3.7m high and is mainly used for the production of high-value crops that are to be trellised 2.5m high. The net is buried and tensioned using the soil.

162 This shade net house is 2.5m high and is mainly used for leafy crops. The net is on top of the soil and fastened with wire which is a cheaper option.

163 Indeterminate tomato production in a hydroponics shade net house. The irrigation fertigation tanks are installed inside the shade net house.

164 Tomato plants that are 2 months old established in a fully temperature controlled hydroponic tunnel. The plants are pruned using the 1 stem horizontal method up to the 2.5m wire.

165 A electrical box installation with a small 8 station Hunter computer and a 5 000 litre tank. The solenoid valves are connected below the box and are set to open and close the water flow to various tunnels automatically.

166 Tomato production in soil in a shade net house. They are pruned using the one stem method and trellised up to the 2.5m horizontal wire. This is similar to the OBS hydroponic system.

23.2 CONSTRUCTING A SHADE NET HOUSE

For high-value crops like indeterminate tomatoes and pepper cultivars, a shade net house should be erected to a height of 3.7m. This allows plants to be trellised up to a height of 2.5m. *See photo 161.* However, when leafy crops are to be produced, the house only needs to be constructed to a height of 2.5m. This height could also be used for high-value crops, especially in the cooler areas. There is approximately a 20% cost saving of a 2.5m house when compared with a 3.7m house. *See Table 11 for costs to erect a 1Ha shade net house.*

23.2.1 POLES FOR A 3.7M HIGH SHADE NET HOUSE
- Use standard 4.2m treated gum or tantalized poles.
- Poles are usually spaced 3 x 6 m square apart from each other.
- Plant the poles 50cm deep.
- The higher the shade net house is erected above the top of the crops, the better the air circulation. A height of 3.7m is recommended for indeterminate tomato production.
- Indeterminate tomatoes need to be trellised to a 2.5m high horizontal wire.
- Anchor poles should be at least 110/125mm thick and the inside poles should be at least 100mm in diameter.
- Place a piece of shade net or plastic on top of the 4.2 m poles to reduce friction of the netting over the top of the poles, due to the natural movement/expansion/contraction of the netting.
- Shade net houses for leafy vegetable crops may be constructed to a 2.5m height as the plants are not trellised.

23.2.2 CABLES AND WIRES FOR SHADE NET HOUSES
- Use 5 or 6mm galvanized stay wire or 5mm cable to anchor the outer poles.
- Use 4 or 5mm galvanized stay wire across the 3 x 6m poles, along the length of the rows. The wire could be stapled with a U nail at the top of each pole.
- Supporting cross wires within each internal block may be used to further brace the structure if heavy hail storms are experienced.

23.2.3 FITTING THE SHADE NET
- Use 40% shade net for vegetables, white in warmer areas and black in cooler summer areas.
- For smaller or custom structures, the producer could provide the measurements of the required shade netting to a shade net company to create. This allows the net to be rolled off the roof of the house, if necessary.
- Make sure the area is clean and lay the shade net out to its full length next to the structure.
- Thread 2.25mm oval, high-tensile steel wire through the eyelets.

- The shade net together with the wires running through the eyelets is then laid on top of the structure between 3m (or 6m) rows.
- Stretch the shade net evenly across the structure and secure the four corners to keep the net in place.
- Using 2.5mm polyethylene lacing cord, secure one 3m wide end to the perimeter stay wire, then stretch the net evenly and secure it.
- The steel wires running through the eyelets can now be attached to the 3m poles and tensioned.
- Fasten the net to the cross cables at regular intervals of 1m or 1.5m to avoid wind flapping. This will prevent wear and tear of the shade net.
- Do not lay wires across the top of the net in order to stop any flapping unless there are supporting wires directly underneath. This is done so that the shade net is gripped between the top and bottom wires.
- The wires carrying the net can be clipped to each other at one metre intervals. The clips should be able to open, especially if the net is carrying a heavy load of hail.
- For protection against prevailing winds, all four sides should be closed with shade netting at a 45° angle. Use the anchor stay wires to support the net.

23.2.4 ANCHORS
- Anchors are one of the most vital parts of a well-constructed shade net house. The use of good, strong anchors will increase the lifespan of the net.
- Stay wires must be the correct specification and should be securely fastened to the poles and anchors.
- The structure must support the weight of all the plants that are going to be trellised up to the 2.5m horizontal wire. Galvanized steel wire is better than steel cables as they stretch less which avoids the need for constant re-tensioning.
- The stay wire or anchor cable should be attached around the pole and over the point at which the horizontal wires or cables (bearing the netting) are attached.
- Never put the stay wire into the soil when attaching it to the anchor. Instead use a steel rod (minimum 12mm). The steel rod must be above the soil level.
- The anchor should be buried to at least the height of the anchor pole away from the base of the pole. The optimum angle for the anchor cable is 45°. The closer the anchor is to the pole, the less effective it is in transmitting the horizontal force from the pole to the anchor.
- 1 meter fence or Y poles are a good option for anchors. Drive them into the soil at an angle of 45º.
- Anchors at the top of a slope will carry more load and will need extra bracing with poles or struts. Remember that corner anchors need to be stronger than other anchors.

23.2.4.1 TYPES OF ANCHORS
23.2.4.1.1 STANDARD FENCE POLES
Standard fence poles are the most popular provided that there are no rocks below the soil level. 1 meter iron fence poles or high tension Y poles are driven into the soil at an angle of 45º in the opposite direction of the 45-degree net.

23.2.4.1.2 MOTOR VEHICLE TYRES
Old truck tyres are the best. The tyres are buried 1m x 1m deep. A steel bar fits inside the tyre and is attached to a steel rod with an eye or hook onto which the stay wire or guy grips are then secured.

23.2.4.1.3 CONCRETE BLOCKS
Concrete blocks which are a cubic meter in volume could be used but is expensive.

23.2.4.1.4 DUCKBILL ANCHORS
The duckbill anchor is expensive but easy to use. The anchor is driven into the soil using a hammer and the rod is driven into the soil. Once the anchor is 1 meter deep, the drive rod is then removed.

CHAPTER 23: PRODUCING VEGETABLES UNDER PROTECTION

167 Leafy crops established in a Gravel Film Technique (GFT) System in a shade net house.

168 Indeterminate tomato production in a half-moon tunnel produced in soil using a drip system.

169 An indeterminate tomato cultivar for hydroponic production in a Gravel Flow Technique (GFT) system.

170 A shade net house structure should be well constructed when producing indeterminate tomatoes or the structure may collapse as shown above. The plants are trellised up to the 2.5m horizontal wire and pruned to the one stem method. The weight of the plants and fruit hanging on a 40m double-row weigh approximately 1 000kg.

171 Indeterminate high-value tomato production in a shade net house in soil using an arrow drip system and pruned to the 2.5m horizontal wire.

172 Ridges made in a shade net house and installed with an arrow dripper system. The ridges are covered with a white plastic mulch to control the water supply. If over-watering occurs, for instance due to rainfall, it may cause the fruit to burst and BER may become a problem.

173 The water inlet side of a GFT hydroponic system. The nutrient water flows underneath the gravel, back to the catchment tank and is continuously circulated.

174 Spinach established in a GFT hydroponic system.

175 Pepper plants in a GFT system.

176 A hydroponic commercial unit near Hartbeestpoort Dam producing cucumbers throughout the season. It is fully temperature controlled and makes use of warm water boilers to heat the tunnels in winter. All the flaps are opened during summer. Extractor fans are also an option to remove the humidity. The humidity is responsible for cooling the plants inside the tunnels.

177 Cucumber plants are trellised up to a 2.5m wire. This is a multi-span installation as the tunnel has no inner/side walls. This is an advantage as more space is available when no side walls are used.

23.3 PRODUCTION OF HIGH-VALUE TOMATO OR PEPPERS IN A SHADE NET HOUSE DIRECTLY IN SOIL

It is expensive to begin producing high-value indeterminate tomatoes and peppers due to the costs related to seedlings, trellising of plants, carrying poles, water tanks, labour and wires inside the house. However, if implemented correctly, the yield could be more than 3 times higher when compared with producing determinate tomatoes in the open-field.

The plant population of indeterminate tomato and pepper cultivars are approximately 26 000 plants p/Ha and the determinate open-field population is usually less than 18 000 plants p/Ha.

Trials have shown that some crops can be produced profitably and successfully in shade net house, in soil, in a way that is similar to hydroponic production. These crops include fresh market and cherry tomatoes, peppers and to a lesser degree zucchini squash. Then, to an even lesser degree, eggplant and melons could be grown, but it is not recommended (for instance, with melons there may be pollination problems causing the fruit not to develop).

23.3.1 SOIL PREPARATION IN SHADE NET HOUSES

It is relatively simple to construct a shade net house thereby saving a considerable amount of money. However, it is recommended that shade net houses be built smaller than 0.3 Ha, as wind storms and hail may damage larger structures.

Note that diseases can be controlled more effectively in smaller structures. More than one structure is useful because when one is infected with a disease or insects the others are usually left unaffected. Similarly, shade net houses may be isolated from one another enabling the producer to avoid diseases and insects being transferred from one shade net house to another.

It is advisable that the ridges in shade net houses should be directed to one side in the unit. This allows the rain water to drain out of the unit through the foot paths. If possible the ridges should be in the direction of the length of the shade net house.

The first step in producing successful indeterminate high-value tomatoes or peppers is to select a suitable cultivar for this purpose in your region. The soil may then be prepared in the following manner:

- Select preferably a sandy loam soil if possible. The slope should not be too steep.
- Mark the land area where the shade net house is to be erected.
- Decide the size of the house. Not larger than 0.3 Ha for reasons previously explained.
- Clean and level the land.
- Take soil samples and send it in for a soil analysis test at least 6 weeks prior to establishing the crop.
- Let a professional fertilizer expert interpret your soil analysis samples and provide you with information and advice. Accordingly, apply the correct amount of lime and fertilizer for the specific crop.
- When the pH of the soil-water content is too low, incorporate the recommended lime into the soil at least 6 weeks prior to establishing the crop. The pH should be in the range of 5.8 to 6.8 for most crops.
- Prepare the production land. Aim for a fine tilt.
- Apply fertilizer to the entire land according to the soil analysis test. Next, make suitable ridges at the required spacing.
- Create well-prepared ridges 1.8m from one another and at least 25cm high and 60cm wide.
- To improve soil fertility, apply good compost on the ridges and if possible, apply it at a density of 80 tons p/Ha or 8 big spades per square meter. Work the compost into the soil to a depth of 15 to 20cm into the ridges.
- Some experts claim that when using good compost on ridges at a high density (at least 80 ton per Ha) year after year, then crop rotation is not necessary. This is because well-prepared compost may assist in preventing certain diseases from developing. Also, it is difficult for insects (the vectors that carry viruses) and diseases to penetrate healthy, actively growing seedlings and plants, as they have thicker and tougher outer walls. In a 3 year experiment, only 1 out of 200 plants where infected with viruses in a shade net house. When using a drip system in a shade net house, ridges should be covered with plastic mulch to regulate the flow of the irrigation water through the dripper system. During heavy rains, nutrients may leach, fruit may burst and blossom end rot (BER) may be a factor. Covering ridges with black plastic liners and controlling the irrigation water through the dripper system may prevent this, especially for tomato and pepper production.
- Apply good compost and straighten the ridges before the shade net house is erected.
- Be aware that considerable trampling of soil may occur during the construction phase of the shade net house. Some producers first erect the shade net house and then make the ridges. However, poles and side netting may get in the way if a tractor is used to do soil preparation and create the ridges.
- Certain tomato cultivars, especially the small fruited cultivars, are less prone to fruit bursting and BER and therefore produce more fruit per plant.

23.3.2 SHADE NET HOUSE INSTALLATION
Also, see Table 11.

- Shade net houses should be divided into sections and irrigated accordingly. They should also be adapted to the water flow pressure of the water pump delivery system.
- Usually a 5 000 litre tank is installed to supply the arrow dripper system. Multiple tanks may be connected to one another to irrigate larger areas.
- Install an electrical box and make use of a time switch. Alternatively, install a small 6 or 8 station Hunter computer in the electrical box.
- The timer is set to irrigate the plants several times per day automatically.
- Connect the solenoid valves next to each other, for each block, underneath the electrical box.
- Fit the main irrigation line from the tank to the end of the last sister line. Usually a 50cm PVC pipe is the norm.
- Connect the sister line to each ridge and lay along the full length of the ridge. Usually 25cm PVC pipes are used for the sister lines with a stopper at the end of the line.
- Carefully punch the correct sized hole (to avoid leakages) every 85cm in the sister line in order to fit a connector.
- Fit a compensating valve on the connector and a 4 way manifold on the compensating valve.
- Fit 50cm spaghetti tubes on the 4 outlets of the manifolds.
- On the end of each spaghetti tube, fit a 2 or a 4 litre per hour arrow dripper.

23.3.3 ESTABLISHING SEEDLINGS IN A SHADE NET HOUSE
- To achieve 26 000 plants p/Ha, the spacing of the double rows should be 23cm in the rows and between the 2 rows on the ridges, the spacing should be 40cm.
- It is recommended to transplant tomato seedlings as early as possible in the summer orientated areas, when temperatures are suitable. Temperatures may become too high from December to February in the summer orientated regions.
- The seedling plants should then be established on well-made and fertilized ridges.
- The sister lines are laid in the middle of the ridges on the plastic liners. Use wire hooks to support the sister lines and to hold the lines straight and in position. They will move around if not supported properly.
- Space the 4 arrow drippers 23cm from one another using twine and straighten the planting line 20cm on both sides of the sister lines. The dripper manifolds are then placed 85cm from one another, the arrow drippers 23cm in-line from each other and 40cm from each other between the double rows on the ridge.
- Punch suitable holes in the plastic mulch liners at the correct spacing. Use a sharpened 32mm PVC pipe to make the holes. Plant the seedlings in the holes and place the arrow dripper next to the seedling plant.

23.3.4 IRRIGATION IN GREEN HOUSES
- A single indeterminate tomato plant may use up to 2 litres of water on a hot day, in soil, when fully grown. Adapt the water cycles and timing of irrigations to suit the plant under these circumstances.
- Water is only supplied through the dripper system.
- Only soluble nitrogen N fertilizer are to be applied and mixed into the tank when top dressings are required. Granular fertilizer should already have been applied to the soil according to a soil analysis.
- A small Hunter 6 or 8 station computer is commonly used and producers set the timing for 2 or 3 times a day to supply the correct amount of water to the drippers. As the plant develops the irrigation timing should also be adjusted.
- The duration of irrigation timing to supply water depends on the size of the plants and temperature. Be aware that a single fully grown tomato plant may use (transpire) 2 litres of water on a hot day.
- Most producers make use of tension meters or data loggers to determine the soil water capacity.
- The irrigation timing and the duration of timing is very important and is adapted to the amount of water that the plant would transpire throughout the day.
- Over or under irrigation should be avoided as both could negatively influence the yield.

23.4 PRODUCING CROPS IN TUNNELS AND IN SOIL

High-value crops such as tomatoes, peppers and cucumbers may also be produced in tunnels.

23.4.1 POINTERS FOR PRODUCING CROPS IN TUNNELS IN SOIL
- Tunnels without heat control in the cold winter spells (areas), would only provide a 2-3°C reduction in temperature compared with the outside temperatures. Long and cold overnight temperatures are exchanged from the outside to the inside of the tunnel.
- Plants transpire considerable amounts of moisture, which should not be allowed to build-up in tunnels. When plants become large, they release more moisture into the air and are therefore responsible for cooling the tunnels off. Transpiration and moisture exchange to the outside is important especially when hot summer days occur.
- The correct kind of plastic UV treated covering for tunnels is important and should effectively reflect the light. It is for this reason that the plant population can be higher inside tunnels.
- Air circulation is important when temperatures are high. It is therefore important to install the correct kind of fans.
- The spacing of plants inside tunnels is important.
- The soil should be suitable for the production of crops.

A permanent tunnel should be built on fertile, sandy loam soil with a pH in the range of 5.8 to 6.8. Non-shaded and well-drained soil is also a requirement.

Typically, a 10 x 30m tunnel accommodates six double rows of indeterminate tomatoes, peppers or cucumbers. However, most cucumber producers make use of 8 single rows spaced 1m from each other. Based on a recent soil test, all the necessary phosphorus P and the majority of potassium K should be applied before establishing the crop.

Ridges will help with soil warming, drainage and a greater volume of soil for rooting the crops. Nitrogen requirements for tomatoes depend on the soil quality (i.e. organic matter) and previous cropping. Generally, soluble N fertilizer should be applied via the drip system, starting 2.5 weeks after transplanting the seedlings. Since tunnels are manually ventilated (by opening doors and flaps), they should be accommodated in an accessible location, preferably to catch the wind direction.

Measure the soil moisture using a tension meter, data logger or smart phone. The humidity inside tunnels is much higher than outside as plants transpire and release considerable amounts of moisture into the air. This must be controlled constantly by opening the doors and flaps at both ends of the tunnel. In hot areas, 2 extractor fans in front of the tunnel at each side of the doors are often necessary. Too high humidity inside tunnels is harmful for plants.

Some producers use shade net to cover intersections at both sides of the tunnels. The biggest problem inside a tunnel is to control the temperature and humidity. The doors and flaps of tunnels should not be closed too early in the afternoon or too late in the early mornings when cooler conditions predominate. This is because the temperature and humidity would build up inside the tunnel. If the doors and flaps are not opened in the early morning, possibly due to negligence, temperature and humidity would build-up to such an extent that the plants would burn to death in a matter of hours, especially in hot conditions.

When the humidity builds up in tunnels then the stomata of the leaves close and may only recover after 24 hours after conditions become favourable again. Producers should therefore open and close the tunnels every day, at the correct times in the day, not too early and not too late. This allows the temperature and humidity to be maintained at the best possible levels. The doors and flaps could be left open during the nights in warmer areas, but in winter they should be closed early in the afternoon to retain the captured heat. Open the tunnels again the next morning when the sun is shining. Heat should be built-up inside the tunnel before the doors and flaps are opened/closed. If doors and flaps are left open and unattended, monkeys and baboons may help themselves, even if the crops are not yet fully ripe.

English or greenhouse cucumbers are not suitable for production in shade net houses although they produce excellent fruit. This is because during windy conditions, the fruit scratches against the leaves and stems of the plants, producing small marks on the young fruit. These marks enlarge as the fruit develops and become unattractive. In less windy areas this could be a minor problem and it may be worthwhile for producers to experiment with this. Due to the fact that irrigation is controlled in tunnels, the result is a uniform application of water which should reduce fruit cracking and other physiological problems such as Blossom End Rot.

Through experimentation, growers may identify crops that could be profitably produced in Green houses. Some exotic crops reach high prices if they are scarce, especially certain herbs such as celery, parsley, basil, spring onions, leeks, chinese cabbage, broccoli, baby cabbage and baby cauliflower etc. Producers should make sure to select the correct cultivar for their area, as this is one of the best ways to become successful, profitable and marketable.

All the above-mentioned crops also grow well in the Gravel Flow Technique (GFT) System. ***See photo 130.***

23.4.2 TRELLISING AND PRUNING OF CROPS IN TUNNELS AND SHADE NET HOUSES

The object of pruning and training is to achieve a balance between the vine and the fruit growth. Only high-value crops such as tomatoes, peppers and cucumbers are trellised. This means that the main stem of the plants are twisted around a horizontal string that is fastened up to a 2.5 m horizontal holding wire. When pruning and training plants, it is best to orientate the tomato or pepper plant to face in a North-South direction in summer, in order to enable the plants to utilize the morning and afternoon sun better.

23.4.3 INDETERMINATE TOMATO PRODUCTION IN GREENHOUSES

- Usually the plants are trellised up to the 2.5m high horizontal wire.
- A string is connected from the bottom stems of each plant to the horizontal wire.
- Only the main stem is trellised up to the 2.5m horizontal wire. In the case of cherry tomatoes, 2-3 stems could be trellised up and down from the 2.5m wire due to their vigorous growth.
- While tomato plants are still young, they must be pruned of all their side shoots which develop between the leaf and the main stem axles.
- The main stem is then twisted around the string as the plant grows higher. Do this regularly until the main stem reaches the top of the horizontal wire, after which the growing point is removed to prevent further plant growth.
- After the tomato plant reaches the 2.5m horizontal wire, usually 9 trusses develop on the main stem.
- New cultivars with shorter internodes are also available which produce more trusses up to the 2.5m horizontal wire and therefore higher yields.
- The harvesting time from when the first tomato fruit is picked to the last one, depends on the temperatures, how well the recommended cultivation practises are carried out as well as how successfully insects and disease are controlled.
- The last fruit for tomatoes are ready to be harvested approximately 8 weeks after the first harvesting.
- Be aware that when transplanting tomato seedlings in ideal conditions, the first harvesting is about 3 months after transplanting. Some cultivars could be harvested a week or two earlier or later but this depends on the cultivar choice and weather conditions.

The first, bottom-most tomato fruit trusses, up to 3 levels up, which develop on the main stem of most fresh market tomato cultivars, provide the largest and best quality fruit. The smaller cultivars and cherry types bear much more fruit on the lower stems. The first truss of the cherry types carries more fruit and they are also larger. Some cherry tomato cultivars bear more than 60 tomatoes on a single truss, especially in cooler summer areas.

178 Note the side shoot on the left on the main stem. It must be removed while it is shorter than 1-2cm. Only one stem is to be trellised up to the 2.5m horizontal wire, while all the other side shoots should be removed.

179 Young indeterminate tomato plants in a shade net house. The vertical twine is already secured to the 2.5m horizontal wires. The stems are to be twisted around the twine up to the top wire, after which the growing points are to be removed to prevent further growth.

23.5 PEPPER PRODUCTION IN GREENHOUSES

Also see Chapter 31.3 for Production of peppers under protection.
- Pepper plants grow as a single stem forming about 10 leaves and then a terminal flower is formed, after which the stem is then split into 2-3 stems. When harvesting this fruit it is extremely brittle and may easily break clean off of the main stem.
- A flower develops between every stem and the stems also repeatedly split into 2 side stems after every division.
- It is recommended that the first and second flowers from the bottom be removed. This is done to stimulate the root system as well as plant growth. If 2 flowers are formed in the split stem, it is recommended that one flower should be removed as soon as possible.
- The fruit on the lower stem should be harvested as soon as possible in order to stimulate the young plant so that it may develop faster and reach full maturity and maximum yields sooner.
- Remove malformed fruit during the early stages of their development or as soon as they appear.
- Note that capsicum crops are perennial.
- Plants usually do not grow above 1.8 meter depending on the cultivar. Some coloured cultivars do not grow much higher than 1.4m meter.
- Be careful when day temperatures fall below 10ºC as seed-set may not take place at flowering time and the fruit may become malformed.
- Remove all suckers up to the sucker below the first flower cluster allowing two stems per plant. Prune when the suckers are less than 4mm long.
- Do not prune the plants if they are wet in order to avoid diseases.
- Harvesting is a time consuming process and is usually done once or twice a week.

23.6 CUCUMBER PRODUCTION IN GREENHOUSES

- English cucumber cultivars are only produced and trained in tunnels or hydroponic systems.
- English cucumber fruit is long, slender, seedless and thin skinned and more difficult to produce than tomatoes. This is due to the lack of disease resistance and insect control measures.
- Tunnel cucumbers are trellised and pruned in a similar way as tomatoes and the vines may reach great lengths.
- Experienced growers remove all growing points long before the last harvest and are still able to harvest quality fruit.
- The main stem must be twisted around the vertical string up to the 2.5m horizontal wire as the plant grows higher.
- According to specialists, it is better to remove all developing young fruit up to 80-90cm from the bottom of the soil. This allows the plant and remaining fruit to develop fast and allows the root system a chance to become strong. The small cucumber fruit develop quickly and use considerable amounts of energy, therefore not always allowing the plant sufficient time to develop in a strong and healthy way.
- Remove all side shoots as soon as they appear. Only one main stem up to the 2.5m horizontal wire should be the norm. This will ensure good quality fruit and a longer plant life.
- Remove the growing points when the plant reaches the top of the 2.5m horizontal wire (Umbrella method). Side shoots will develop after removing the growing point.
- Select 2 strong shoots and remove the rest. Allow the side shoots to grow over the 2.5m horizontal wire and downwards to the ground. When the shoots reach 70cm above soil level, the growing points must be removed to prevent the plant from growing any further.
- Remove underdeveloped, damaged and crooked fruit as soon as they appear.
- From the day of transplantation to the first harvesting of the fruit is only 35-40 days. From flowering to first harvesting it usually only takes 12-15 days.
- The market requirements for cucumber fruit is usually ± 30cm long or 410-450 gram.
- Due to this fruit developing so quickly, it should be systematically harvested every 2-3 days.
- Each Greenhouse cucumber requires shrink-wrapping for the market. Therefore producers need access to a shrink-wrapping machine.

23.6.1 ISRAELI CUCUMBERS AND GHERKINS
The Israeli type of cucumber is a much smaller fruit and is packaged the same way as baby marrows. Baby cucumbers (Israeli) and Gherkins produce much more fruit (4-5 times more than the English or Greenhouse cucumbers). Gherkins are also popular for canning as they are small and good looking, with a rough, thorny, dark, green skin.

It is possible to grow Israeli small cucumbers in a shade net house without scratch marks on the fruit as smaller cucumber fruit does not scratch easily. Cultivar choice for small cucumbers is important as certain cultivars tend to be too short to meet the market packaging requirements.

23.7 TEMPERATURE CONTROL AND VENTILATION UNITS FOR TUNNELS

During colder winter conditions, a tunnel may only provide a 3 to 4% temperature difference inside the tunnel when compared with the outside conditions. The coldest time of the day is usually between 4h00 and 5h00am. Tunnels do not offer more advantages when heavy frost occurs.

The humidity inside tunnels should be controlled at all times. When hot conditions occur, producers use extractor fans or an expensive wet-wall system which works with huge fans that extract air through the wet wall. This system uses considerable amounts of energy and water.

Warm season crops are sensitive to direct frost and temperatures below 1°C. Tomatoes and peppers are classified as summer-orientated, semi-cool crops while greenhouse cucumbers prefer temperatures a bit higher. Ideal temperatures are 16 to 18°C at night and 28 to 30°C during the day.

High growing or indeterminate plants are normally trellised up to a 2.5m horizontal wire inside tunnels. The structure should be high enough to fit the 2.5 horizontal wires for trellising the plants.

Plants have the ability to cool tunnels off due to transpiration while also releasing moisture that needs to be controlled by using some form of ventilation. Small plants inside tunnels do not have the ability to cool the tunnels off via transpiration.

If plants have been established later in the season when the days are longer with higher day and night temperatures, the quality and yields will decline due to these higher temperatures. It is recommended to establish high-value crops as early as possible in the summer rainfall areas and/or later in the season. This avoids high temperatures that may cause bursting of fruit, blossom end drop (BER) and poor fruit-set, especially for tomatoes, peppers and cucumbers. Beginner producers and even existing commercial producers should understand this and do their homework. It is also advisable to evaluate of different kinds of cultivars to identify those that are more adaptable to specific conditions. Successful producers continue to ask questions and seek new information in order to challenge themselves to achieve higher yields and better quality.

23.7.1 TEMPERATURE CONTROL IN TUNNELS
Hydroponic producers make use of temperature control in tunnels when producing high-value crops. It must be emphasised that heating or cooling tunnels is expensive, especially when heating tunnels. This may cause the enterprise to be unprofitable, especially for small-scale farmers. No matter what type of energy is used to heat tunnels, it is still expensive. It is hoped that solar heating could change this in the near future. The cheapest way to heat tunnels is to make use of a Boiler heating system, especially the self or automatic loading systems that make use of coal as its energy source.

23.7.1.1 Boiler systems work in two ways:
- Blowing hot air through large plastic tubes about 25cm in diameter which are vented with holes along the length of the tubes from the beginning to the end of the tunnel.
- Pumping hot water through pipelines to a radiator and using a fan to blow the hot air into the tunnel.

The hot air system for heating tunnels is the most popular and inexpensive, as it does not use water, piping and radiators. However, a small heating boiler can only heat 2 to 3 tunnels at a time and is quite expensive. The number of tunnels should be carefully considered. There is little value in erecting only one or two tunnels and expecting to make good profits. Some producers establish their crops earlier or later in the season when temperatures become higher. It is then only necessary to heat the tunnel for a few weeks therefore saving on energy costs. The producer should however know their crops well, what the expected yields and market prices are and should plan production accordingly.

Tunnel heating provides a warmer growing environment so that the growing season may be extended earlier or later in the season, possibly throughout the winter in warmer areas. Tunnels also protect crops from insects, diseases, extreme radiation and weather conditions.

The object is to increase the temperature inside the tunnels through the use of white plastic covers. The plastic covers used are specially designed for vegetable crops and is the best absorber of heat. The covers may easily raise temperatures 2 to 3°C. This is an advantage when cooler conditions occur and will help to market vegetables earlier or later in the season when compared to outside lands. Tunnel production could be extended to an additional three weeks if proper planning is performed and the best cultivars are used.

The addition of row covers (mulching) in the tunnels is a vital component of warm-season vegetable production. It is generally known that by using row covers, an additional 2 to 3°C protection is generated so as to counter frost. Research has shown that some crops can be profitably grown in tunnels provided that the temperature not too high or too low. This includes summer-orientated crops such as: zucchini squash, chillies, peppers, cucumbers, runner beans, cherry and fresh market tomatoes. Through experimentation, growers may identify other crops that may be grown profitably in tunnels.

Due to climate differences in different regions, tunnels require somewhat different designs and constructions. The underlying characteristic in the design of tunnels is ventilation. A passive ventilated roof system is when the area of the roof-vent opening is 20% of the size of the floor space area. The sides of the tunnel that open (doors and flaps) should be in the same direction as the wind in warm areas, while the north-south direction is best for making use of the light intensity in cooler areas.

23.7.2 TUNNEL VENTILATION

Ventilation is vital for plants, especially in the warmer areas and should be achieved by opening the doors and flaps of tunnels. Sometimes it is necessary to open a tunnel for the entire day and only close it when cold conditions occur. This keeps the inside of the tunnel slightly warmer. Extractor fans are used in tunnels, especially for hydroponic systems in order to remove excess hot air and humidity (moisture) mostly in the summer orientated areas. *See photo 182.*

Tunnels that are more effective but also expensive, such as those with an open roof and roll up sides. In many cases, shade net is used on the sides of the structures to allow better ventilation. For most crops in tunnels, there is little need to ventilate during the first month of their growth period, especially early or late in the season. This is because temperatures are mostly moderate during these times of the year. Ventilation becomes necessary when the temperature rise inside the tunnels.

An option to consider when tunnel plastic is damaged, is to use shade netting rather than plastic sheeting. A 40% density white shade net house is able to reduce the air temperature by 2-3°C in hot conditions. The reduction in light intensity is beneficial for some crops such as lettuce. The reduction in temperature can be achieved using shade net, sprinkler irrigation or mist units inside the tunnels.

23.8 NUTRIENT FLOW TECHNIQUE (NFT) HYDROPONIC SYSTEMS IN GREENHOUSES

See photo 167 and photo 173.

Leafy crops are mainly produced in this system, but some producers also establish indeterminate tomatoes and peppers in this system and other modified NFT systems.

Indeterminate F1 tomato and pepper cultivar plants are trellised by constructing a trellising framework to allow the high-value crops to reach a height of 2m.

Soluble fertilizer solutions are circulated through troughs or cheaper special plastic liners or hydro liners that are specially designed for the NFT system. The liners are available from Gundle Plastics. The circulated water is mixed with a special fertigation solution that includes 16 Nutrients. The water is then returned (drained) back to the water reservoir which is usually a 5 000 litre tank. This is called a closed recycle system. The disadvantage of the Gravel Film Technique (GFT) system is that when disease pathogens are introduced into the system they can multiply easily and rapidly and infect all the plants in the system. Once the system is infected with fungal or other pathogens it is extremely difficult to eradicate.

23.9 MULCHING IN GREENHOUSES

It is advisable to cover the ridges in shade net houses with plastic strips and apply drip irrigation which can be controlled. If too much rain water occurs, especially at the end of the growing season, the fruit, especially tomatoes and to a lesser degree peppers, are likely to crack. This is because the calcium Ca has drained away. Blossom End Rot (BER) may then also occur.

When producing crops in shade net houses and when it rains in the hydroponic systems, the producer needs to adjust the nutrient application. This is especially the case when irrigating through arrow drippers and when using mulching plastic strips in shade net houses. Irrigation must be controlled to avoid over-watering when rain occurs. If mistakes are made due to the impact of too much rain and too little nutrient elements, it will affect the growth development of the plants at a later stage. When the new growth becomes pale green, it should be corrected by adding N, as soon as possible.

When early tomato production takes place, it is recommended that black, clear or white (infrared transmitting) plastic mulch should be used to increase the soil temperature, reduce weed emergence and prevent soil evaporation. For the most effective transfer of heat, the black plastic mulch should have good contact with the surface of the ridges. Clear plastic will increase the soil temperature significantly more than black plastic. Weeds however can emerge under the clear plastic film due to the penetration of light. White plastic or a combination thereof, can significantly lower soil temperatures and is ideal to be used for late summer productions.

Organic mulches such as straw, hay, grass or compost can be used for strip mulching to cover the rows. Organic mulches create a favourable environment for many beneficial insects. Organic mulches may be applied to increase the soil temperature. However, some organic mulch (straw or hay) may lower soil temperature which may not be effective when needing to warm the soil in the Spring. These mulches will be incorporated into the soil after harvesting the crop and will be beneficial in improving soil fertility. All mulches, including compost, increase soil temperatures but not as effectively as black plastic or other plastic mulches. Raised beds will significantly enhance soil warming and drainage. There is also a larger volume of loose soil for the roots to penetrate.

23.10 ESTIMATED COST OF ERECTING A 1HA SHADE NET HOUSE

The estimated cost of erecting a 1Ha shade net house as indicated in the table below, is only a guide. Certain labour costs are not included, such as for the measuring and the layout of the area, digging holes to plant poles, erecting the house and shade net, transport, wires, installation etc.

Normally, producers avoid erecting shade net houses over large areas for vegetable production due to various reasons such as cost and insect/disease control. Instead they erect smaller shade net houses less than 0.3 Ha.

The table below estimates that a 1 Ha shade net house would cost approximately R335 639 to erect and be ready for production. The production cost for 1Ha is then estimated in the following table and amounts to R170 630. The total yield for tomato production in a shade net house would amount approximately to R506 269. It is then possible that a producer could break even for the first season.

A 0.3Ha house would cost approximately R118 500.

TABLE 11: ESTIMATED COST OF ERECTING A 1HA SHADE NET HOUSE (2014)

Area: 1 Ha = (100 x 100m = 10 000 m²).			
Crops: High-value tomatoes or peppers (cucumbers are not suitable in shade net houses due to wind damage (gherkins or baby cucumbers are an option).			
Type of shade net house.	Flat roof with 45° side ends (slopes).		
Height of construction. Standard 4.2m poles to trellis/train crops. Shade net houses for leafy crops and herbs may be constructed 3m high and are cheaper to construct.	3.7m		
Spacing between double rows from center to center.	1.8m		
Spacing in rows: Double rows, 4 plants per 4 way manifolds.	Manifolds with 4 outlets spaced 85cm from one another.		
Length of rows.	100m (45° slope allow 100m rows)		
Number of plants 1 double row at 100m.	468		
Number of rows p/Ha.	55		
Amount of plants p/Ha.	25 740 (plant population)		
Plants per square meter.	2.5-2.6		
INSTRUCTIONS TO ERECT THE HOUSE	**QUANTITY**	**UNIT PRICE**	**AMOUNT R**
Clean the surface of the soil area. Remove rubbish, grass, rocks, trees and bushes. Level with a slight slope to one end.		R2 000	
Rip, plough disk and rotovate the soil. Make ridges 100m long x 0.35 to 40m wide and 1.8 from each other. Decide if the fertilizer is to be applied before or after creating the ridges.		R3 000	
40% white or black (or other colours) Shade Net.	6 864 run meter	R21/run meter	R144 144
Lacing twine 2mm (UV protected). 1kg rolls = 480m in length to bind the shade net together.	5 rolls	R140	R700

Item	Quantity	Unit Price	Total
4.2m 110-125mm tantalized poles spaced 3 x 6m from each other.	462	R120	R55 440
4.2m 125-150mm outside tantalized (anchor) poles.	66	R140	R9 240
6m 100-125mm treated poles to carry the horizontal straining wires at 2.5m high.	100	R140	R14 000
Crosby clamps to bind the cables or use cheaper binding wire.	50	R0.60	R30
Iron 1.8 HD, Y standard iron fence poles (anchors) cut to 60cm pieces. Make furrows 30cm deep. Drive poles 60cm deep into the furrows at the required spacing. Tension the wires and fasten the net. Wires 400m x 120m @ R0.80 p/meter = R384. Duckbill anchors, concrete blocks or car tires are an option but are expensive. A cheaper method is to fasten the net at soil level and tension it. Use 4.5mm wire and fix the net on the wire.	120 48 000m	R450 0.80	R5 400 R384
4.5mm wire. Use 16m per anchor. Fold double and twist around to tension poles: 520m x R0.80 per meter.	420	R0.80 per m	R336
Strain wires (non-stretch) are used on top of the poles to keep the net in position. Pull laterally and across every pole = 6 700m.	6 700m	R0.50	R3 350
Double horizontal wires to trellis plants to the 2.5m horizontal height. Space double wires 1m from each other at 2.5m height. 2 x 55 rows = 110m x 100 = 11 000m x R0.80 per m = R8 800.	1100m	R0.80 p/m	R8 800
INSTALLATION INSIDE SHADE NET HOUSES			
PVC black pipe 50mm for the main water supply. Pipe line and pump are not included.	425m	R4.90	R2 082
25mm PVC black pipes for the sister lines.	5 500m	R2.80	R15 400
Disc filters with back wash action. Establish in the main pipelines. It is very important to establish the correct filters.	3	R1 000	R3 000
4 000 litre tanks, 6 tanks required at R4 500 p/tank.	6	R4 500	R27 000
Compensating valves (8 litres per minute), 85cm from each other. Connectors to be fitted to connect the 4 way manuals.	6 435	R2.50	R16 087
4-way outlet manifolds to fit on compensating valves.	6 435	R0.60	R3 860
Spaghetti tubing 3mm. Fit on arrow drippers. 14 500 meters required, rolls are 300m long.	48 rolls	R480	R23 040
Arrow drippers, 4 or 8 litres p/hour. 8 litres p/hour may be a better option as they do not block easy.	25 740	R0.80	R20 592
Electrical motor (water pump) one per 0.3Ha.	3	R3 500	R10 500

Fittings: elbows, t-pieces, end-stop plugs, clamps etc.	120	R4	R440
TOTAL			R355 693

TABLE 12: PROJECTED PRODUCTION COST TO PRODUCE INDETERMINATE TOMATOES OR PEPPERS IN A 1HA SHADE NET HOUSE IN SOIL (26 000 PLANTS P/HA) (2014)

ITEM	COST P/HA
SOIL PREPARATION: Irrigate the House before soil preparation: plough, disk, rotovate.	R2 600
SOIL SAMPLES: Take soil samples for an analysis test. Each combined representative sample costs approximately R250 x 3 = R750.	R750
Analyze the results. Make use of a reliable fertilizer expert at approximately R250 per sample. At least 3 samples for larger units required.	R750
Fertilizer: Apply lime if the pH of the soil is too low. Broadcast fertilizer. approximately 1 300kg p/Ha 2.3.4 (30) Zn @ R290 p/50kg bag. Require 26 bags x 50kg p/bag = 26 x R290 = R7 540 (2014). 2 Top dressings LAN/KAN @ 150kg p/Ha = 6 bags x R240 per bag = R1 440. Application of fertilizer: mechanical, tractor and spreader R350	R9 330
PLANT MATERIAL: 26 000 F1 hybrid high-value indeterminate tomato seedlings from Nurseries = R3 per seedlings = 3 x 26 000 = R78 000.	R78 000
TRANSPLANTATION OF SEEDLINGS AND LABOUR: 1 labourer transplants approximately 1 000 seedlings p/day. 18 labourers transplant 18 000 seedlings p/day = 18 x R100 p/day = R1 800.	R1 800
TRELLISING: Use twine and wire hooks at the bottom to fasten the string. Trellis only one stem of the plants up to the 2.5m horizontal wire by winding the plants around the strings. Fasten it on the bottom by using a peg and a releasable knot above. Prune at the same time in a 1.5-month period for 6 labourers over 5 days = 6 x 5 = 30 x R100. Tomato hooks and 3m twine is also available but quite expensive.	R350 / R3 000
WEED CONTROL AND LABOUR: Control weeds between rows using hand tools: 2 times in the season, 6 labourers working 6 days = 6 x 2 = 12 x R100. Control weeds using herbicides before establishing the seedlings.	R1 200
IRRIGATION: Irrigate through the drippers 2 times per day for 16 weeks. Note that electricity/fuel/gas becomes costly and to pump these amounts of water is also expensive.	R7 000
HARVESTING AND LABOUR: 10 labourers harvest and transport 120 tons of tomato fruit to the store at R100 per labourer p/day. One labourer harvests approximately 600kg p/day and 200 labourers, 120 tons p/day = R100 x 200 = R20 000. Ripening stages of fruit could be uneven especially for the indeterminate tomato cultivars, as they are harvested many times and over long periods.	R20 000
SORTING, CLASSING AND PACKING: To sort, class and pack 120 ton tomatoes: 1 labourer sorts and packs approximately 1 000kg fruit per day, then 120 labourers = 120 x R100 = R12 000. 4 tons are rejected due to issues of size, cracking, blossom end rot, mechanical damage, physical disorders etc.	R12 000
PACKAGING MATERIAL: 5kg carton boxes to pack 120 ton tomatoes = 24 000 boxes @ R0.80 p/box.	R19 200
TRANSPORT: 120 ton tomatoes to the market/s: If 100km from the market and using a 10 ton truck and if travelling to the market 10 times = 2 000km @ R7 p/km = R14 000 plus labour for the truck driver and assistant (for 10 deliveries approximately R300).	R14 000 / R300
TOTAL	R170 280

PART 1: A COMPLETE GUIDE TO VEGETABLE PRODUCTION

TABLE 13: ESTIMATED YIELDS AND PROFITS FOR TOMATOES AND PEPPERS IN SOIL IN A 1HA SHADE NET HOUSE

TOMATOES	A reasonable yield is 5kg per plant or 128 tons p/Ha. Advanced producers yield up to 8kg per plant or more in a 5-6 month growing period. 25 740 Tomato plants p/Ha x 5kg = 128 700kg. Minus unmarketable fruit (1 000kg) = 127 700kg. Estimated retail price p/kg (good quality fruit) = R4.50. Total R127 700 x R4.50 p/kg = R574 650 p/Ha.	R574 650 p/Ha
PEPPERS	12 marketable pepper fruit per plant x 25 740 plants p/Ha = 12 x 25 700 fruit = 308 880 less 3 000 unmarketable fruit (due to small size, cracks, blossom end rot, sunburn etc.) = 305 880 fruit @ R1.80 per fruit = R550 980 p/Ha. Experienced producers harvest up to 18 fruit in an 8 month growing period, provided that virus carrying vectors are effectively controlled.	R550 980 p/Ha

180 Building a small shade net house 3.7m high to produce high-value tomatoes or peppers.

181 Removing side shoots on high-value indeterminate tomato plants and trellising them by twisting them around the vertical string up to the 2.5m horizontal wire. The growing point is removed to prevent the plant from growing any further.

23.11 ESTIMATED COSTS, YIELDS AND PROFITS WHEN PRODUCING HIGH-VALUE CROPS IN TUNNELS DIRECTLY IN SOIL (2014)

NB: All the information mentioned in the table below provides an estimate for input costs to produce the specific crop. The cost of erecting a tunnel is excluded. Costs continuously change and differ from supplier to supplier.

The table below will assist the producer to make good decisions when he/she decides to produce indeterminate tomatoes, peppers or cucumbers in tunnels and in soil. Some beginner and small-scale producers attempt to begin producing high-value crops in 1 or 2 tunnels. They usually have limited financial resources and tend to lose considerable amounts of money and eventually fail.

Vegetable production in tunnels is intensive and it is not profitable to produce high-value crops in 1 or 2 tunnels. Generally, producers should make use of 6 to 8 tunnels for one production system.

TABLE 14: ESTIMATED COSTS AND YIELDS IN TUNNELS FOR PRODUCTION IN SOIL OVER 1HA (2014)

SIZE OF STANDARD TUNNELS:	One 10 x 30m tunnel occupies 0.03Ha or 300 m² and 33 tunnels = 1 Ha.		
VENTILATION	Natural ventilation, open doors and open flaps on front side of the tunnels.		
COVER	Plastic, 200 micron UV protected plastic.		
CROPS-PRODUCTION IN SOIL	Mainly high-value crops: tomato, peppers or cucumbers established on ridges and in soil.		
SPACING OF PLANTS INSIDE THE TUNNEL	6 double rows, 1.5m from centre to centre of the rows.		
LENGTH OF DOUBLE ROWS	28m. Allow 1m spacing from the open door ends.		
NUMBER OF PLANTS FOR DOUBLE ROWS	128.		
NUMBER OF PLANTS IN 1 TUNNEL	768.		
NUMBER OF PLANTS P/HA	Plant population for tomatoes, peppers and cucumbers: 25 000 plants p/Ha and in Winter 22 000.		
PLANTS PER SQUARE METER	2.5 per m² in summer and 2.2 m² in winter.		
DESCRIPTION	QUANTITY	UNIT PRICE	AMOUNT
Cost of a 10 x 30m tunnel with open doors and open flaps (2014) = R45 000. 33 tunnels equal 1Ha when erected by a supplier.	33	R45 000	R1 320 000
INSTALLATION OF TUNNELS			
Level and clean the land properly.			R3 900
If desired, do soil preparation before erecting the tunnels.			R2 900
Take at least 5 soil samples for a soil analysis test. Let a fertilizer expert interpret the results.	5	R300	R1 500 R700
Apply granular fertilizer according to a soil analysis. Lime should be incorporated at least 6 weeks prior to planting time, when necessary.			R6 500
Create ridges. Apply fertilizer by hand before making ridges.			R1 500
If available, apply good compost on the ridges, at least 80 tons p/Ha or 8 full spades per square meter. Work it into the ridges (15-20cm deep) after erecting the tunnels.			R8 000
Install at least 4 x 5 000 litre plastic tanks.	4	R4 500	R18 000
Install at least 4 electrical motors. Allow specialist to do calculations and diagrams.	4	R2 900	R11 600
Solenoid valves. 1 valve for approximately 4 tunnels.	8	R300	R2 400
Install 4 electrical boxes. Connect all components and a small Hunter computer with 8 stations and	4	R600	R2 400

connect the motors and solenoid valves. Tunnel suppliers supply boxes that are fully installed.			
Install PVC pipes. The main pipe line is usually 50mm. About 300m of pipe is required.	300m	R5	R1 500
25mm PVC sister lines inside tunnels for double rows = 5 500m.	5 500m	R2.50	R13 500
Attachments to connect 4 way manifolds.	6 336	R0.80	R5 058
4-way manifolds to connect the spaghetti tubes.	6 336	R0.30	R1 900
Spaghetti tubes 0.5m long. 12 000m is required = 300m roll = R480 per roll.	40 rolls	R480 per roll	R19 200
Arrow drippers to connect onto the spaghetti tubes.	25 000	R0.60	R12 500
50mm to 25mm adapter for sister lines in tunnels. 25mm fittings from the tanks to all 33 tunnels. Inside the tunnels 198 end stops.	198 198	R7 R2	R1 386 R1 000 R396
TOTAL COST 1HA = 33 TUNNELS			**R1 435 840**

TABLE 15: POSSIBLE YIELDS AND PROFITS IN TUNNELS FOR PRODUCTION IN SOIL P/HA

TOMATOES	Reasonable yield per plant = 6kg in a 5 to 6 month growing period. Advanced producers yield up to 9kg per plant or more. If 780 tomato plants per tunnel and 6kg per plant is achieved: 780 x 6kg = 4 680kg. Minus 300kg per tunnel for unmarketable fruit that is too small, cracked, has BER, infected, suffered mechanical damage etc. Total yield for 1 tunnel = 4 680kg minus 300kg unmarketable = 4 360 x R5 per kg = R21 800 per tunnel. For 33tunnels x R21 800 = R719 400	**R719 400**
PEPPERS	17 marketable pepper fruit per plant over a 6 month growing period = 800 plants p/tunnel = 13 600 fruit p/tunnel. For 33 tunnels p/Ha = 448 800 fruit minus 5 000 unmarketable = 443 800 fruit @ R1.60 per fruit p/Ha = R716 480.	**R716 480**
CUCUMBERS	18 Fruit p/plant and 780 plants per tunnel = 14 040 fruit per tunnel and for 33 tunnels per Ha = 463 320 fruit @ R1.50 per fruit = R694 800 p/Ha using the umbrella trellising method over a 5 month growing period.	**R694 800**

CHAPTER 23: PRODUCING VEGETABLES UNDER PROTECTION

TABLE 16: PROBLEMS RELATED TO TOMATO PRODUCTION IN GREENHOUSES

PROBLEM	POSSIBLE CAUSE	SOLUTION
Flowers are falling off.	Temperatures too cold or too warm.	Proper ventilation and temperature management when producing in tunnels.
Flowers form together.	Too cool.	Proper temp management in tunnels.
Fruit-cat face.	Pollination disorder.	Humidity too high or temperature too low.
Cupping or rolling of leaves.	If upper leaves roll, scan for aphids. They produce sticky juice and attract ants. Some early cultivars roll leaves during heavy fruit load. May also be water stress.	Control aphids using chemicals. Irrigation management.
Poor fruit set.	Temperatures are too high or low or humidity is too high. Flowers not vibrating enough to release pollen.	Temperature management. Shake tomato stakes to improve pollen release before 12h00. Use a mist blower.
Border rows have fruit with holes.	Worms.	Chemicals should be applied every 14 days if worms are visible.
Stem lesions causing the plant to wilt.	Disease.	Have plants diagnosed.
Fruit fails to ripen.	Temperature.	Late tomatoes may not ripen because of low light and temperatures.
Black spots on the bottom of fruit.	Blossom end rot.	Deficiency of Calcium.
Fruit cracking.	Irregular watering.	Mulch and water uniformly.

182 A 30x10m tunnel fitted with the correct size extractor fan on both ends of the tunnel to extract heat and humidity inside the tunnel, especially when very hot days occur.

183 Excellent quality and high yielding cucumber plants in a hydroponic 10x30m tunnel. The plants are growing fast and the first cucumber fruit is harvested 4 weeks after transplanting healthy strong seedlings. The fruit is harvested at least 2 times per week.

CHAPTER 24

HARVESTING, HANDLING AND PACKAGING

24.1 WHEN TO HARVEST

Enormous losses occur each year due to carelessness when harvesting and handling vegetable crops. Much of this can be prevented through teaching labourers:

- the proper methods of harvesting
- handling products in the field
- storage
- transportation to market.

It is therefore essential that labourers should be properly trained in the harvesting, handling, sorting, classing and transportation of perishable crops. Some crops such as onions, potatoes, sweet potatoes and garlic allow some leeway when harvesting and others such as sweet corn, spinach, green beans and leaf lettuce pass the stage of maturing quickly and should be marketed as soon as possible. Producers should harvest indeterminate tomatoes 2-3 times a week, especially when hot weather prevails.

General rules cannot be laid down, but the aim should be to harvest the product at an optimum stage of maturity that **will provide the housewife with prime products for the table.** High temperatures hasten maturity of most crops and it is sometimes necessary to harvest more frequently.

Delaying harvesting may increase the total yield, but it may also allow heavy losses from rotting, insect/disease damage, sun scald, rain, hail and frost. Both quality and yield will decline. Most growers do not harvest their products at the proper time. Harvesting is often delayed so that vegetable crops become larger/heavier, as the producer aims to earn better prices. However, this sometimes results in the rotting, bursting of fruit, over ripe fruit, bean pods becoming stringy, tomatoes/peppers/pumpkins could be sunburnt, maize corn kernels harden.

Crops should be handled with care after harvesting. Crops that are not packed in proper containers may be damaged. Such damages may not be apparent at the time of harvesting and packaging. Damage may show at the market and this may lower the crop quality and prices accordingly.

24.2 POST HARVESTING

Most vegetable crops are perishable, some crops rotting rapidly after harvesting. It is important to transfer them from the production area to the packing store and the market as soon as possible. Due to high costs small-scale farmers do not use cold storage. This limits the storage time of perishable crops. Normally small-scale farmers depend heavily on the fast sale of their crops. Post-harvest quality control begins in the field with clean hands and sanitary conditions are required at all times. When handling certain crops careful supervision and proper instructions regarding the harvesting process must be given. Labourers are essential to the success of any hand harvesting of crops. Pack house problems and complaints are often the result of poor instruction/supervision of labourers.

24.2.1 CAUSES OF POST HARVESTING DAMAGE
Vegetables lose quality in many ways. The problem between the field and the table can be summarised as follows:

24.2.1.1 THE NATURAL AGEING PROCESS OF RIPENING
Over-ripened and too old crops often appear at the market. This is mainly caused by external factors like high

temperatures. This is the most important factor and without temperature control, the shelf life of crops is sometimes limited.

24.2.1.2 WATER LOSSES

Many vegetables contain between 85-95% water. Drier air and low humidity cause the rapid loss of moisture in fresh vegetables. Garlic producers pack their bags 10% heavier in anticipation of water loss. When the weight of bags falls under the prescribed weight at National markets, the product is downgraded and lower prices result.

24.2.1.3 MECHANICAL DAMAGE

Cuts, bruises and other damage from implements are responsible for many losses. These cuts and bruises encourage disease infection during storage, especially in warm pack houses.

184 Packing different coloured pepper cultivars according to supermarket requirements in a clean, disinfected pack house.

185 Top: Greenhouse English cucumbers are always wrapped in plastic covering in a neat and clean way. The different pepper colours at the bottom are red, green and yellow.

24.3 GRADING

Grading refers to the classification of crops into groups based on quality, condition and size. The factors upon which grading are based depend on the cultivar, size, maturity, colour, shape, freshness, firmness and defects. Defects include decay breakdown, rot spots, insect/disease, mechanical damage, clay on the surface of the product, colour changes (too old) and cracks on fruit and sunscald.

24.4 PACKAGING

Remember when packing your product that it should comply with market requirements. For your own benefit, use a colourful logo on your container so that consumers take note of your quality product and remember it in future. It is important to pack your product clearly classified in the different required classes. For example: class 1, 2 and 3. Be aware that class 1 should be of the best quality, good looking, fresh and eye catching. Producers should attach logos on the containers of their best class. Make sure that the logo is artistic, eye catching and colourful.

Remember most consumers, especially housewives, buy vegetables visually and when your product is eye catching, good quality, neat, clean and have a colourful logo, this is a good advertisement. Class 1 receives higher prices at the markets. Therefore it makes sense not to advertise your logo on lower class produce because

186 Tomatoes packed in cheap open carton boxes. Note the well-known ZZ2 logo. These are popular because consumers know that ZZ2 tomatoes are quality products and can be bought with confidence.

187 It is clear that the green pumpkin is more popular than the white. However, do not over produce one over the other.

this may also be remembered by the consumer. Keep in mind that your product is inspected at the market and when a class or grade fails to comply with the market requirements then it is downgraded to a lower grade and at the expense of the producer. Producers could add value to their products by packaging and marketing it themselves. Packaging is not simply a container for goods because it can also enhance product quality, improve shelf-life and protect the product from tampering. It should be hygienic and preferably re-usable.

Packaging could prolong the shelf life of a quality product, however, **you cannot pack a poor quality product in any container and expect its quality to improve**. The moment a product enters the distribution chain at the Supermarket the packaging requirements become considerably stricter. Shrink-wrap plastic covering is used mostly to keep bacteria out of the pre-packed container as well as preventing the product from drying out.

188 Different grades of tomatoes. They must be sorted into different sizes and packed accordingly. Usually grade 3 is mixed and of poor quality. Therefore prices are low. Product of this quality may not be acceptable to the markets or could be downgraded.

189 Beans, ginger, peppers, garlic and chillies in a Fruit and Veg store. The peppers are yellow, green and red, blocky with thick walls. Confirm the packaging, sorting and class requirements for each crop before delivery.

The objective is to get rid of as much oxygen as possible from inside the container. Using micro perforated plastic allows gas and moisture to escape from the container. When some crops, such as sweet potatoes and onions are stored with damage from the land even for short periods, fungi may start to develop and may affect healthy bulbs, tubers, globes or roots. Those in contact with the damaged products may become infected quickly. This occurs especially during hot spells and high humidity.

24.5 COOLING AND DETERIORATION

The importance of cooling to longer maintain the quality of perishable crops is obvious. Producers that are in a position to cool their perishable products in cold rooms can market it over an extended period by keeping it fresh longer. Marketing time can be extended considerably. The temperatures should not be too cold as cold conditions may damage the produce. However, when temperatures are too high, the product may ripen too soon and lose moisture, wilt or go bad. When crops are harvested and kept in warm stores and then introduced into cold rooms the surrounding moisture can condensate on the fruit and this may promote fungal and bacterial rotting.

24.6 STORAGE AND THE PACK HOUSE

STORAGE OF VEGETABLES
Fresh vegetables are stored under various conditions:
- Little or no temperature and humidity control.
- Use of ventilation in cooler outside temperatures or during the cool season of the year.
- Use of artificial refrigeration either in cold storage or by means of ice.

Both temperature and humidity are important for vegetable preservation and fresh consumption. Humidity may be increased by sprinkling water on the floor or using wet bags that absorb water. An accurate thermometer should be used to check the temperature in the store.

24.7 PLANNING THE PACK HOUSE

Important factors to consider when planning a pack house:
- the type of building
- inner and outer spacing
- traffic flow inside and outside
- layout
- receiving and dispatching of different crops

The pack house should be large enough for when the post-harvest processes take place. Depending on the size of the operation a pack house can be anything from an old converted garage to a large industrial building of hundreds of square meters. Whatever the size the principle of management and layout inside the store will be the same.

Farmers tend to concentrate all their effort on production and often overlook the importance of the pack house. Equipment should be laid out so that it is accessible at all times. Once you have organized the pack house observe the movement of people for a while. **Make sure that labourers don't constantly cross and bump into each other.**

Labourers should remain at their sorting/packaging tables and should not have to walk great distances to fetch or to load the product. The manager should be able to walk into the pack house and within a few seconds know exactly what is going on. This applies whether the pack house is small or large. Clean and sterilize the pack house at least once a week. Give the floor a good wash and disinfect it using Jik, formalin or spore kill.

The work surfaces should be thoroughly cleaned and disinfected every day. Many producers do not pay much attention to this but when packing certain crops sensitive to bacteria and fungi, disinfection should be a high priority.

Remember infection on your product may show only later and rotting or deterioration may become a problem. This usually happens after the product has been delivered to the market causing your crop to be rejected or downgraded. The result is the producer is the loser.

One of the biggest problems in pack houses is the tempo of product flow. Proper product flow minimizes handling, reduces damage and saves time. It will be helpful to design the inside and outside of the pack house on paper with a view to facilitating efficient product flow. The sorting and grading area should be next to the packaging area where the packaged products are ready to be transported to the markets. Outside the pack house locate the toilets and washing areas as near as possible to the working area and ensure they are clean.

In overseas countries, when the producer creates a pleasant, clean working environment in the pack house this tends to create good relations with labourers. Some large farmers have music, radio and also air conditioning to make the environment pleasant. This should be controlled however because different ethnic groups may differ in taste when listening to music

190 Wrapping Greenhouse or English cucumbers using a wrapping machine. The cucumbers were sorted to the correct sizes. All the oversized, crooked, defective, coloured fruit with various problems were rejected and sold separately.

CHAPTER 25

MANAGEMENT AND MARKETING

25.1 MANAGEMENT

Good management can be summarized as attention to detail. One way of achieving good management is by working with experienced people.

25.1.1 THE TWO MAIN FACTORS FOR GOOD MANAGEMENT
25.1.1.1. ORGANIZATION OF WORKERS
- Purchases: What, how much and when to buy seed, fertilizer, tools, equipment appliances. etc.
- Labour: How many and when needed.
- Production: When and what to cultivate, plant and harvest.
- Sales: When and what quantities to be sold.
- Machines: Which appliances/machines are required for production and their state of readiness/repair, condition of implements, tools.

25.1.1.2 ORGANIZATION OF WORK
This is the analysis of each job. Study each job and find the quickest and easiest way to get it done and then implement. When even simple jobs are not performed with the greatest efficiency/speed then such jobs will unnecessarily consume time and energy. Time and energy unnecessarily wasted by employees consume money in two main ways:
- The job takes longer and cost more
- The extra time wasted could have to be spent on other jobs.

Labourers, when paid daily wages, have little to gain from efficient and quick work and a lot to gain by stretching the work out as long as possible. If labourers are paid for contract work then labourers are incentivised and will normally perform better and more efficiently in minimum time.

25.2 MARKETING

A general definition for marketing your crop is the importance of choosing the correct crop and cultivar, supplying the crop at the correct price at the right time to the right consumer who has a specific need for your product.

The following aspects are important in achieving the best prices on the market:
- Produce the best quality that the consumer requires.
- Pack the product in the way that the consumer prefers.
- Build a reputation and advertise your quality produce with a logo on your class 1 produce.

In producing excellent quality for the market it is the ideal of all producers to achieve the above aspects. Many producers do not meet all the standards and then fail. The following are the main reasons:
- Producing the wrong cultivar. (Producing white pumpkins instead of the popular green one.)
- Poor soil preparation.
- Poor irrigation
- Incorrect application of fertilizer.
- Incorrect planting time.
- Too large production lands.
- Use of soil that is infertile.

- Too much clay or too sandy.
- Failure to properly control insects/diseases and weeds.
- Harvesting crops too early or too late.
- Dull colour of produce such as sweet potatoes, beetroot and carrots.

If you have not established a market for your crop you cannot produce it first in the hope of getting a market. This is a recipe for financial failure.

It is important to decide where you are going to sell your produce and if there is a market. Study the market requirements, statistics, price differences throughout the year, delivering periods etc. The success of marketing lies in the ability of the farming industry and the marketing sector working together so that the farmer makes a profit. A lot of vegetable produce is rejected at the market because these basic principles were not followed and the producer is the ultimate loser. The fresh market system is highly competitive and the role players that follow the correct basic principles should survive in the long term.

Producers today need to have the knowledge and technology to facilitate access to different cultivation and marketing methods for their produce. They should constantly seek additional information to broaden their personal knowledge. The fresh market system, production and marketing methods change all the time and therefore producers must evaluate and adapt to these changing conditions and keep themselves informed.

Profits can be made when producing vegetables but this depends on competent harvesting, handling and marketing and production practices.

The producer is responsible for the appearance of his product when it reaches the market. He is constantly reminded to produce quality vegetables if he wants to achieve success. Counter to this however are oversupplies at the market. Some producers battle while others manage to cover costs and even make some money. Why do some have success and others don't?

It is important for farmers to visit the market whenever there is an opportunity. Market agents encourage their producers to visit them.

If a farmer is going to entrust his produce to somebody in a distant market then he/she should also make the effort to meet that person and see the set-up there. Serious commercial producers should keep in close touch with their market agent. The agents get daily feedback on prices, market conditions and plan their market accordingly. This has also changed, due to technology all relevant information can be found on the Internet, E-mail and on a smart phone. If producers stay in touch with their market agent and visit the market frequently they will learn to adapt their production accordingly. The other requirements are continuity of supply to the market and matching the supply to the market requirements.

Consult market agents and together plan volumes and rates of supply to the market. Produce the best cultivar which the market and consumers prefer. Go and have a look at the fresh market and see firsthand how the market operates if your intention is to sell your vegetables at this particular market. It is important that the producer meets some of the buyers at the market and this gives you the opportunity to promote your produce while listening to their requirements. If you are serious about your marketing then you should make sure that your customers are satisfied.

It is also an ideal place to see what your competitors are up to. After having a closer look at your competitors' packaging, grading and presentation you can also improve your own standards. In that way you keep up with the latest development of new cultivars, new trends, newest packaging and more. Some of the larger farming organisations visit their market agents regularly because it is part of their strategy. There must be a good reason for this and the farmer who does not make an effort to visit the market is destined to lose out in the long run. Sometimes large farming establishments have a market monopoly and their produce gets preference at that market.

Many farmers have contracts with large organizations and consequently when there is a surplus, you can expect your produce to fetch lower prices.

It is important that the farmer has a good relationship with his market agents. This person sells your produce and plays an important part in your financial well-being. Producers should also meet the members of the market management team.

Supplying fresh produce to the market has three keys to success:
Quality, Continuity, the **relationship of trust** between the producer and agent t (built up over a period of time.)

Market agents may give you advice on products that are popular during certain periods of the season. The consumers might prefer green pumpkins instead of white ones. Grow larger volumes of green pumpkin and smaller volumes of the white one.

Get to know the markets. Certain vegetable crops and cultivars are popular during a particular time of the year. If possible, you can then produce crops out of season to achieve better prices and be absorbed by the market. However, this is easier said than done, as the climate and other factors do not always allow you to grow out of season.Markets are filled with examples of producers who aim for quality and continuity of supply. Producers who work closely with their sales agents will eventually enjoy the kinds of return that make it all worthwhile. It is sometimes necessary for the commercial producer to supply his vegetables to the market instead trying to sell it locally or at other markets. The difference in prices, however, is not much.

Finding local markets is sometimes not worthwhile when small quantities have to be delivered and sold. Commercial farmers should visit the market agent after they send their products to the National markets. Choose an agent and deal with him/her because it is better to know the market agent personally. Don't be afraid to phone him every day in order to find out about your product. A commercial carrot producer, Mr Rugani, near Carletonville sometimes supplies the market with 100 ton of carrots at one time. They are already pre-packed and he is well-known. Market agents therefore like dealing with him and also communicate with and look after him.

The market agents will let you know if there is a shortage of certain vegetables on the market, but he will not advise you if there is enough or if the market is overstocked. The producer should phone, use his smart phone or E-mail the market agent and find out how much of his product was sold the previous day. All the market data is processed every day so the agent can advise the producer almost immediately about what the position is.

Remember the market agents gather every morning and they discuss the prices of the different crops and what these crops are going to be sold for. Prices between agents do not differ very much. Some producers, especially when they are far from the markets, sell their products to hawkers at slightly lower than the market prices. They are paid cash and save transport and sometimes packaging costs. It could be important to establish your entire crop at the same time. The produce will be ready to be marketed at once. However, plan in such a way that you don't have to harvest all your crops at the same time. Some F1 hybrid cultivars are uniform and the more there is to harvested the more labour is needed. This can push your costs upwards and overstock the market.

In South Africa there has been a long tradition of producers jumping on the bandwagon after a spate of high prices. Farmers, who for instance do not normally grow tomatoes, suddenly begin to produce tomatoes and others who might be regular suppliers, start increasing their production. When farmers study the historical data they may realize that what goes up must come down. When they plant certain crops after the boom they are probably going to experience a downward spiral.

PART 2

GUIDELINES FOR CROP-SPECIFIC VEGETABLE PRODUCTION

PART 2 CONSISTS OF CHAPTERS 26-38 AND INCLUDES THE FOLLOWING FOR EACH CROP:
- Production requirements for vegetable crops in the open field and partly undercover protection.
- Estimated production costs, yields and profits for all the above crops.
- Designing a cash flow and business plan for all the above crops.
- Planting times, cultivar choices, seed and seedlings.
- Producing seedlings in seedling trays and in the soil.
- Climate and temperature requirements.
- Soil analysis, pH of soil and interpretation of soil analysis tests.
- Soil preparation and improvement.
- Step-by-step planning instructions.
- Seed planting and sowing methods as well as transplanting of seedlings.
- Spacing, crop rotation, mulching and cover crops.
- Overhead sprinkler, pivot, cannon, wheel and drip irrigation systems.
- Establishing crops under protection in greenhouses, shade net houses and hydroponics.
- Fertilization: Granular, soluble and fertilizer through drippers.
- Weed control: Mechanical, by hand and by using herbicides.
- Scanning for insects and diseases.
- Trellising and pruning.
- Yields and possible profits of crops.
- Handling, harvesting and packaging.
- Cleaning, sorting, grading, classification and storage.
- Descriptions of diseases, insects, viruses and physical disorders.
- Guidelines for insect and disease control.
- The system is designed to enable the producer to change to an Open Bag hydroponic system whenever he/she wishes. The soil was fertilized according to a soil analysis. Granular fertilizer and 80 ton p/Ha or 8kg per square meter of good compost were band placed on ridges. The above plants are trellised and pruned to the one-stem method up to a 2.5m horizontal wire. This is similar to the hydroponic system and yields more than 160 ton p/Ha (2010).

CHAPTER 26

PRODUCTION GUIDELINES FOR TOMATOES

191 Indeterminate tomato production (high growers). Trellis and prune to the one-stem method in a shade net house. Excellent quality and high yields have been achieved. The soil was fertilized according to a soil analysis and compost was applied at a rate of 80 ton p/Ha or 8 big spades of compost per square meter on the ridges. The fertilizer have been applied before making the ridges. Double rows were established with arrow drippers next to each plant.

192 A newly bred indeterminate F1 hybrid tomato cultivar that has a good appearance with a very good shape and colour. It is high-yielding with a good taste and lots of internal jelly.

193 A high-yield indeterminate F1 small fruit tomato cultivar. Certain producers prefer small fruit cultivars as they are more resistant to fruit burst and Blossom End Rot.

194 Typical of a long shelf life cultivar which has thick walls and little jelly. A priority for breeders' is more jelly for better taste, more seed cavities and longer shelf life abilities.

CHAPTER 26: PRODUCTION GUIDELINES FOR TOMATOES

195 A cherry plum tomato cultivar.

196 A cluster truss tomato cultivar in good condition. The whole truss is retailed and usually has only 8 tomatoes per truss.

197 An oval red cherry tomato cultivar. Note the heavily loaded trusses. Some cultivars bear 70 fruit per truss mainly on the lower trusses.

198 A Hybrid Roma or bush tomato cultivar. It is not trellised and is a wide-spacing tomato cultivar. It has become important to produce this crop for various reasons.

199 A 10 Ha determinate fresh market open-field tomato cultivar established near Hartbeespoort Dam, South Africa. The cultivar is trellised with the double-wire method and high yields can be expected.

200 A Saladette F1 indeterminate cultivar popular because of its long shelf life and good colour. Acceptably high yields are achievable.

201 Indeterminate tomato production in a shade net house. Plants have been trellised up to the 2.5m horizontal supporting wire and a dripper system has been utilised. When necessary, fertilizer are applied as top-dressings.

202 At the front is a very short growing tomato F1 hybrid cultivar that has an exceptionally high yield. At the back, is an indeterminate F1 hybrid cultivar that is trellised up to the 2m horizontal wire and which is using a drip system on well-prepared ridges.

203 Indeterminate tomato production in a hydroponics temperature-controlled tunnel trellised up to the 2.5m horizontal wire. Note the wet wall at the back, the water trough and the large fan units all used to cool the tunnel off.

26.1 INTRODUCTION

In South Africa, tomatoes are one of the most popular vegetable crops along with potatoes. The crop is classified as a fruit, but in the 1920's it was also classified as a vegetable crop. Tomatoes are the second largest production vegetable crop, after potatoes and overall the fifth most produced crop besides maize, corn, soybeans and wheat.

Commercial tomato production requires a high level of management, a large labour force at peak times, high capital inputs and attention to detail. Labour requirements, training/trellising of plants, disease control, harvesting, grading, packaging and transportation is usually very intense. There are several hundred tomato cultivars that exhibit different growing behaviours. High or low growers, different fruit sizes, shapes, colours and cultivars that are adapted to different temperatures and regions.

The crop belongs to the same family as potatoes, peppers, chillies, gooseberries and eggfruit. Tomatoes are very sensitive to a variety of insects and diseases that attack the plant, roots, leaves, stems, flowers and fruit. Due to the difficulties of controlling, especially diseases, as well as the expensive trellising methods usually required, make this one of most expensive and difficult vegetable crops to produce. The production of tomatoes therefore require more attention than most other vegetable crops.

Annual production of tomatoes in SA is nearly a million tons (2011). Approximately 50% of the total tomato production is supplied by the organisation known as ZZ2 Enterprises. They are also the largest tomato producers in the Southern Hemisphere (2011). The manager, Mr Berty van Zyl began with less than 1Ha in the 1950s and at that time only harvested 30 tons p/Ha. Since then, he has continuously improved his operations and presently produces more than 60 tons p/Ha. ZZ2 Enterprises is situated in Limpopo, Moketsi (Low Veldt), South Africa. Production in this area takes place throughout the year. Exceptionally high yields are achieved due to the fact that several expert managers are involved, each with their own field of expertise.

Soil management as well as the continuous improvement of soil by means of good composting and rotation of crops is extremely important. Production costs are reduced as growers become more efficient in the methods of production. For this reason, it is suggested that beginner producers should begin at a small-scale, not larger than 0.25Ha and then gradually become more and more efficient. Starting production on a large scale and then running out of cash with bad cultivation practises is not recommended. Producers should be aware of the price changes of products, training, labour, water and fertilizer in order to be successful in this trade.

Tomatoes may be eaten raw, used in salads, cooked, fried, sundried, frozen, processed, included in jam, chutney and also canned. Different types and cultivars of tomatoes are used for different purposes. Many types of tomatoes are available, such as fresh market, processed, roma, salladette and the various cherry varieties including red, yellow, pink, orange, plum, bottlenecks as well as the larger cherry cocktail varieties.

Cherry tomatoes produce numerous small fruit, sometimes as many as 70 fruit per single truss or more. Small cherry tomato fruit may produce approximately 40 small cherry fruit p/kg, while in comparison, medium fresh market tomato fruit may produce 8 to 9 fruit p/kg. The quantity of market tomatoes produced per Ha may equal that of cherry tomatoes.

Tomatoes are a good source of vitamin A and C as well as essential minerals. Surprisingly, they contain more vitamins than oranges. It has been reported that tomatoes do not contain harmful acids for humans and may prevent certain types of cancers. However, some consumers suffer from high blood acid and should therefore avoid, especially raw tomato fruit such as in fresh salads. When tomatoes are cooked, they releases many of these acids.

The arrival of long shelf life, high-value tomato cultivars has changed the nature of the markets in that producers and markets now demand long shelf life tomato cultivars.

26.2 DIFFERENT TYPES OF GROWING BEHAVIOURS

26.2.1 DETERMINATE GROWERS *See photo 199.*
Plants grow approximately 0.8 to 1.9m high and then stop. There are excellent low-growing cultivars available which are usually harvested earlier but produce lower yields. Higher growing cultivar plants develop vigorously and could reach 2m or higher. Numerous F1 hybrid cultivars are available all with their specific requirements and characteristics.

26.2.2 INDETERMINATE GROWERS *See photo 2 and photo 201.*
Indeterminate tomato plants do not stop growing, unless the main growth point on the plant is removed. Plants usually reach a height of 2.5meter in 2.5 months. Thereafter, the main growth point is removed as the plants are trellised. If the plants are left to grow continuously, they could reach to a height of 6-7m in a 6-month growing period.

Indeterminate tomatoes, cucumbers and peppers are known as the high-value crops. This is because of their indeterminate growth characteristic, special training methods, long shelf live, good quality, high yields, long growing periods as well as their numerous harvesting periods. Indeterminate tomato cultivars are mainly produced in hydroponic systems. However, they have also become popular to grow in soil, inside protective structures and even in the open field.

Production of determinate tomatoes for the fresh market is steadily declining as the production of indeterminate long shelf live cultivars have become more popular. Harvesting more than 80% first-class fruit from indeterminate cultivars is possible, while it is about 50% for determinate growers. Many producers establish indeterminate, long shelf life, high value tomato cultivars directly in the soil, in the open field, in shade net houses and/or in tunnels. They do this by using the one or two trellis stem pruning method.

Almost all the indeterminate tomato cultivars are F1 hybrids. The seed and the seedlings are very expensive. The saladette F1 hybrid indeterminate cultivars are high growers and are trellised. Determinate open-field cultivars with similar characteristics to the saladette are the well-known roma and block types. Saladette and the roma types have smaller fruit with thicker walls, less inner compartments and little jelly. Improvement of saladette cultivars is a high priority. *See photo 200.*

26.2.3 BUSH OR ROMA TYPES *See photo 198.*
Bush or roma type tomatoes do not need to be trellised. They are established on ridges that have good soil drainage abilities. The spacing of non-trellised plants is normally wide. This allows labourers and implements to operate in the land without damaging the plants. Excellent F1 hybrid cultivars and block types that have high yields are available and quite popular.

26.3 CULTIVAR SELECTION AND GRAFTING OF SEEDLINGS

It is expensive to produce determinate tomato cultivars, due to the high trellising, training and labour input costs. The trellising costs are usually only for the first season, as nearly all trellising material is reused for the next production season.

Selecting the best tomato cultivar for your region is sometimes a difficult process as often there are numerous cultivars available. This is especially the case for new, improved cultivars, such as the F1 hybrid cultivars. Producers should investigate and verify that they are selecting the best cultivar/s for their region.

Breeding programs for tomatoes and other crops are a high priority and new cultivars, which have been tested and evaluated continuously, are released periodically. These cultivars have special characteristics such as:
- Disease, insect and virus resistant or partly resistant.
- Better quality fruit, fruit sizes, internal fruit jelly, better outside and inner colour, better shapes, more inner

seed cavities and longer shelf live.
- For indeterminate plants, shorter internodes which produce more tomato trusses over a given distance. These cultivars therefore produce higher yields when trellised up to a 2.5m horizontal wire.
- Adaptability to higher and lower temperatures.
- Stronger root systems and nematode tolerance or resistance.

26.3.1 GRAFTED SEEDLINGS

The under-stem of a cultivar that has a strong root system and which may be resistant to nematodes and other soil diseases, is grafted onto another cultivar of the producers choosing. An example of such a strong and resistant under-stem could be a wild tomato species.

The roots of the under-stem then develop stronger and become more active, thereby transporting water and nutrition to the top/upper growth in a more effective manner. This encourages more vigorous growth of plants, larger fruit, shorter ripening periods, thicker and tougher cell walls and less insect/disease damage.

Grafting of plants is well-known, especially in overseas countries where it has been done for many years. Commercial producers should take note of this and investigate it for their own benefit.

26.4 ESTIMATED PRODUCTION INPUT COSTS AND YIELDS FOR DETERMINATE CULTIVARS

The table in this section should provide producers with a good estimate to follow in order to determine the costs and possible yields which are to be expected during production. Production costs for producing vegetables may vary from region to region across the country and should therefore be adapted accordingly. Therefore, producers should do their own research and enquiries in order to create an accurate budget and a cash flow plan.

Financial lenders may require a business plan before they will consider lending money. The table indicates the estimated capital that is needed to produce 1Ha of determinate tomato cultivars. The production cost of determinate tomatoes is high, but indeterminate production is even higher. This is because they are usually produced under protection.

Overhead costs are not included in the table. The trellising cost of determinate tomato cultivars is high but only for the first season. Most of the trellising materials, such as trellising poles and wires, are re-usable and therefore the costs would be reduced by at least 30% for the following season.

The costs as indicated in the table are based on a determinate, open-field production and a yield of 64 tons p/Ha or 3.5kg per plant over a 3.7 month growing period when using a plant population of 18 500 p/Ha.

TABLE 17: ESTIMATED PRODUCTION COSTS AND YIELDS FOR DETERMINATE TOMATO GROWERS IN THE OPEN-FIELD FOR 18 500 PLANTS P/HA (2014)

ITEM	COST P/HA
Clean and level the production area.	R650
SOIL PREPARATION: Irrigate the land before beginning any preparation. Plough, disk, rotovate and irrigate the area to full capacity (approximately 25mm) as part of soil preparation.	R2 600
SOIL SAMPLES: Take soil samples for an analysis test. Each combined, representative sample costs approximately R250 x 3 = R750.	R750
Analyse the results by making use of a reliable fertilizer expert (approximately R250 per sample and there are at least 3 samples for larger areas).	R750

Fertilizer: Apply lime if the pH of the soil is too low. Approximate quantity of broadcast fertilizer is 1 300kg p/Ha of 2.3.4 (30) Zn @ R290 p/50kg bag. 26 bags x 50kg p/bag = 26 x R290 = R7 540 (2014) 2 Top dressings LAN/KAN @ 150kg p/Ha = 6 bags @ R240 per bag = R1 440. Application of fertilizer: Tractor and spreader R350.	R9 330 R350
PLANT MATERIAL: Open pollinated, determinate cultivar seedlings from Nurseries: If plants are spaced 1.3m x 40cm apart within rows = 19 225 plants per Ha. Determinate tomato seedlings @ R400 (per 1000 seedlings) x 19 = R7 600. **INDETERMINATE F1 HYBRID SEEDLINGS:** R1.70-R2.90 per seedling. Grafted seedlings are R3.80 or more per seedling. **INDETERMINATE, SINGLE-ROW, OPEN-FIELD PRODUCTION:** 12 500 Plants p/Ha @ R3 per seedling = R25 000 p/Ha (2014). **INDETERMINATE, DOUBLE-ROWS IN SHADE NET HOUSES WITH IN SOIL:** For high-value crops, 26 000 seedlings are needed and could cost up to R50 000.	R7 600
TRANSPLANTING OF SEEDLINGS AND LABOUR: 1 labourer transplants approximately 1 000 seedlings p/day. 18 laborers transplant 18 000 seedlings p/day = 18 x R100 p/day = R1 800 Seedling planters could transplant 3 000 or more seedlings p/hour.	R1 800
TRELLISING: Poles for determinate growers are approximately 1.8m high. Drive them 0.5m deep and 4m apart from each other in rows which are 1.5m apart. Therefore, approximately 2 075 poles may be necessary @ R6 per pole = R16 600. If 2 labourers plant approximately 200 poles p/day, then 16 labourers may plant 1 320 poles p/day = 14 x R100 = R1 600.	R16 600 R1 600
WIRE: If using 1.25mm galvanized wire for the double-wire method: 66 rows x 100m per row = 13 200m x 4 double wires = 52 800m, R300 per 1 000m rolls implies approx. 52 rolls x R300 = R15 600. Labour: Tension wires in rows. 4-5 double wire layers = 66 x 100m rows. 2 labourers' fix and tension 15 double rows per day, working 6 days: 2 labourers x 6 days x R100 = R 1 200. If using non-slacking nylon twine or cheaper baler twine for single rows, then double twine method is much cheaper. For shade net houses, the double-row, single-stem method can be implemented using tomato hooks and twisting the plant around the twine.	R15 600 R1 200
WEED CONTROL AND LABOUR: Its recommended to control weeds between rows mechanically. Using hand tools, twice per season, 6 labourers working 6 days = 6 x 2 = 12 x R100 = R1 200. When controlling weeds and using herbicides before establishing the seedlings, the cost may be less at approximately R1 000.	R1 200
IRRIGATION: Since 1ml water p/Ha = 10 000 litres of water p/Ha, then irrigating 30mm per week for 2.5 months or 10 weeks (always deduct rainfall), the following water would be necessary: 10 000 litres p/Ha x 30 = 300 000 litres p/Ha x 10 weeks = 3 000 000 litres p/Ha. Note that electricity/fuel/gas costs may become high when applying and pumping this amount of water, especially in low rainfall areas. Using sun energy may be an option.	R7 000

HARVESTING AND LABOUR: 10 Labourers harvest and transport 60 tons determinate growers' fruit at R1000 per labourer p/day. One labourer harvests approximately 600kg p/day and 100 labourers harvest 60 tons. Therefore, 100 labourers x R100 = R10 000. Note that the ripening stages of fruit may be variable, especially for the indeterminate tomato cultivars, therefor they may have to be harvested many times over longer periods.	R10 000
SORTING, CLASSING AND PACKING: To sort, class and pack 60 tons of tomatoes: 1 labourer sorts and packs approximately 1 000kg fruit per day, therefore 60 labourers are required to pack 60 tons: 60 labourers x R100 = R6 000. 4 tons are rejected due to small size, cracks, blossom-end-rot, mechanical damage, physical disorders etc.	R6 000
PACKAGING MATERIAL: 5kg carton boxes to pack 55 tons of tomatoes = 11 000 boxes @ R0.80 p/box = R8 000.	R8 000
TRANSPORT: To transport 55 tons of tomatoes to a market 100km away, using a 5 ton truck that travels to the market 10 times = 2 000km @ R5 p/km = R11 500 + R1 500 (approximate labour cost for a truck driver and assistant for 10 deliveries) = R13 000.	R13 000
PREPARING THE LAND AFTER HARVESTING: Cleaning after harvesting, remove the poles, hooks and wires and ploughing all the materials and debris into the soil. Also establishing a crop to produce green manure.	R2 500
ESTIMATE PRODUCTION COST FOR THE FIRST SEASON P/HA	R106 530

TABLE 18: ESTIMATE YIELDS AND PROFITS FOR DETERMINATE GROWERS: Harvesting 64 tons p/Ha or 3.5kg p/plant, less 9.6 tons or 15% unmarketable fruit = 64 – 9.6 tons = approx. 54 tons.

Total marketable fruit = 54 tons/Ha 50% (27 tons) Class 1 @ R5.50 p/kg = R148 500 30% (16 tons) Class 2 @ R3.50 p/kg = R57 600 20% (11 tons) Class 3 @ R1.50 p/kg = R16 500	
TOTAL (54 TONS P/HA) LESS PRODUCTION COSTS ESTIMATED PROFIT FOR DETERMINATE TOMATO GROWERS	R222 600 R106 530 R116 070

26.4.1 EXPECTED YIELDS FOR DIFFERENT SYSTEMS OF TOMATO PRODUCTION
- Bush or non-trellising tomato cultivars may yield up to 30 tons p/Ha and Hybrids even more.
- Determinate, open-field production for 16 000 to 18 000 plants p/Ha, produce average yields of 40 tons p/Ha while experienced producers could yield up to 80 tons p/Ha or more. *See photo 202.*
- Indeterminate, single-row, non-pruning production may yield up 70 tons p/Ha (12 500 plants p/Ha or 5.5kg per plant). *See photo 193.*
- Double-row, single-stem trellising up to a 2.5m horizontal wire, produced in soil or in shade net houses yield approximately 160 tons p/Ha or more (26 000 plants p/Ha or 5.5kg p/plant over 5.5 months of the growing season). *See photo 203.*

Note that class 1 tomato fruit production output of a determinate cultivar is only approximately 50%. Indeterminate tomato production in shade net houses or tunnels, when following the correct procedures, may be 80% or more.

Some producers make use of smaller fruit cultivars sizes because they are more resistant to blossom-end-rot (BER) and fruit cracking. Be aware that the cherry types of tomatoes have very thin skins and therefore cracking of the fruit could become a problem, especially at the end of the growing season.

The cost of tractors, implements, maintenance, general equipment (wheel barrows, watering cans, spades, forks, hoes, crates, twine, pegs etc.), irrigation systems, water, water pumps, electricity/petrol/diesel, spraying chemicals, housing, toilets, communication (telephone, computers, fax, mail), clothing for spraying chemicals and so on, are not included, nor are maintenance costs of tractors and implements.

Certain labour costs are also not included while salaries are based on 8 working hours at R100 per day. Labour costs are rising rapidly due to union activity and labour cost differ all over the Country. To reduce costs, permanent workers should also supervise, manage and train other workers as well as operate tractors, trucks and motorbikes.

26.5 QUICK STEP-BY-STEP PRODUCTION PLANNING GUIDE FOR TOMATOES AND OTHER CROPS

TABLE 19: QUICK STEP-BY-STEP PRODUCTION PLANNING GUIDE FOR TOMATOES AND OTHER CROPS

ACTION	ACTION PERIOD
FINANCIAL PLANNING Identify a suitable market for the specific crop that you intend to produce. Decide how big the production area is going to be. Determine how much it's going to cost and perform financial planning accordingly. If a loan is necessary, create a business plan and/or cash flow plan that indicates how much cash is needed during specific periods until marketing. Make decisions that allow you to produce economically and to know the limits of your abilities. Do market research. Be aware that when markets are over-supplied you may be negatively affected.	3 months prior to establishing the crop
SEED AND CULTIVAR CHOICES Select the best cultivar for your region. Order your seeds well before you plan to establish your chosen cultivar or ask the nursery to order it. This is the time when most producers prefer to establish their crops and the demand for popular cultivar seed is high. Therefore, sometimes seed companies or dealers may run-out of seed stock shortly before planting time when optimal temperature conditions exist. The producer may then be persuaded by an adviser to use the second-best cultivar. Note that F1 hybrid seed is expensive but is usually uniform, stable and may produce better yields as many are resistant or partly resistant to diseases/insects.	3 months prior to establishing the crop
SOIL SAMPLING: Take soil samples for a soil analysis test and hand them in for testing. It is important to make use of a reliable fertilizer expert to interpret the soil samples and inform the producer. The advice will include the quantities and the correct types of fertilizer to be used as well as whether lime is to be incorporated or not.	2-3 months prior to establishing the crop
SOIL PREPARATION: *Also see Chapter 8.2 and Chapter 26.6.5 for soil preparation.* Select medium texture, sandy loam soil if possible. Clean and level the land. Determine if a hard soil layer is present and if so, break it up by using a ripper. Irrigate soil when it is too dry or wait for rain. Do not cultivate dry soil. Good land preparation is the first step to successful production. Early soil preparation is important because it allows the breakdown of organic matter, which improves moisture conservation and controls weeds/diseases/insects. Most of all, it creates a good, smooth, aerated seedbed.	3 weeks prior to transplanting the seedlings

Create well-prepared ridges for tomatoes, capsicum families and sweet potatoes. Create seedbeds for beetroot, carrots and to a lesser degree lettuce. If the pH of the soil is too low according to a soil analysis test, apply the correct type of lime 5 to 6 weeks prior to establishing the crop and plough it in. *See Chapter 15.5 page 85 for the correct type of lime.* The soil should be fumigated (if nematodes (eelworm) are expected to be a problem) at least 14 days prior to planting.	
SEEDLINGS: Obtain seedlings from nurseries or produce your own seedlings in seed trays. Take the necessary steps to ensure that healthy, disease-free seedling plants are provided by the nursery. When necessary, the seedlings should be hardened off, especially when warm, sunny or hot conditions are expected. Select the date during which the seedlings should be ready. Inform the nursery manager when you need your seedlings for transplanting. Tomato seedlings are normally ready to be transplanted within 5 weeks. Seedlings that have been produced in soil seedbeds take as long as 6 to 8 weeks to be ready. It is important to plant the seedlings at the required spacing according to the specific cultivar.	5 weeks prior to planting seed in seed trays or 6 weeks prior to supplying seed to a nursery
FERTILIZING: *Also see Chapter 15 Fertilizer.* The correct type of lime (according to a soil analysis) should be applied and worked into the soil at least 5 weeks prior to establishing the crop. Irrigate soil when it is too dry. Do not cultivate dry soil except when the soil is sandy. Apply fertilizer according to the soil analysis results. Use a calibrated fertilizer spreader. Small areas can be spread by hand. *See Chapter 15.10 for how to apply fertilizer on a small-scale.* Fertilizer should be worked into the soil with a disk or rotovator to a depth of 15 to 20cm before transplanting the seedlings, after which a smooth seedbed should be produced. If a soil analysis is not available, then apply a recommended amount of fertilizer amount. This applies to fairly fertile soil on smaller areas and for small-scale farmers. Note that too much N may cause vigorous growth, flower-drop, cracking of fruit and fruit abortion.	2 to 3 days prior to transplanting the seedlings in the production land
SOW OR PLANT SEED DIRECTLY INTO THE PRODUCTION LAND Certain crops such as onions are sown in soil seedbeds to produce seedlings. Certain crops such as carrots, beetroot, spinach, beans, maize and garlic cloves are sown in furrows and in rows in the production land by hand. Certain seed is planted directly in the prepared land by using seed planters.	At transplanting
IRRIGATION: The soil should be moist when transplanting seedlings or when sowing or planting seed. Irrigate 15mm after establishing seedlings, thereafter 25mm for the first week and 35mm every week thereafter for most crops. Accurate rainfall measuring apparatus should be used. Tension water meters and data loggers are reliable and may be considered. Take note that it is not advisable to over or under irrigate plants.	Directly after transplanting the seedlings
INSPECT: Inspection plants on a regular basis, especially when they are immature, for insects and diseases and apply effective control measures when necessary.	Inspect at least weekly

WEED CONTROL: Control weeds as they become visible. Do not allow weeds to overgrow the young plants.	When visible. Begin early.
TRELLISING AND BOXING: Trellis tomatoes, runner beans and box peppers. *See Chapter 26.6.10 for trellising.*	As plants grow
TOP DRESSINGS: Apply 2 to 3 top or side dressings when necessary, especially for crops that grow for long periods (3 and 5 weeks in fairly fertile soil after transplanting seedlings or just after emergence). Do not apply too much N as it may lead to poor flower-set and the abortion of fruit.	3 and 5 to 6 weeks after transplanting
Sort and pack your produce according to the market requirements.	After harvesting
Store in a cool or ventilated place.	After packaging
MARKETING AND TRANSPORTATION: Constantly study the markets to evaluate seasonal trends. This includes prices, as well as supply and demand consumer related trends. Harvest at the stage of ripeness that the market requires. For example, the market may prefer colour transitioning tomato fruit instead of red and ripe fruit. *See Chapter 24 for harvesting and handling of crops.*	Local/national supermarket etc.
Clean the land and plough in debris. Establish a rotation crop.	After harvesting

26.6 TOMATO PRODUCTION GUIDELINES AND REQUIREMENTS

26.6.1 SOIL REQUIREMENTS
Also see Chapter 5.3 Soil Requirements.

Tomatoes can be produced in many different soil types. The best soil, is a deep, well-drained, sandy loam to loam soil. Clayish soil with a very fine structure may be a problem. This is especially the case during long rainy periods. The plants may drown and it may also be difficult to harvest the fruit due to the muddy conditions.

26.6.1.1 SOIL PH
Tomatoes are tolerant of high acidity soil. Liming is recommended when the pH falls below 5.0.

26.6.2 CLIMATE REQUIREMENTS AND OPTIMAL PLANTING AND SOWING PERIODS
Also see Chapter 3 Climate.

Tomato production is classified as a cool weather crop and is sensitive to both cold and very hot climatic conditions. The ideal temperature for tomatoes is a minimum of 16°C at night and a maximum of 28°C during the day. The growing period has an impact on yields.

Determinate tomato seedlings could be established in mid-September and the fruit may be harvested in December to January. This is results in more or less 80 to 100 days for the first harvest of the fruit. In cooler conditions harvesting may take up to130 days. When it is very hot this may influence fruit set and abortion of fruit.

High and low temperatures, especially when occurring over lengthy periods, has a negative effect on crops and may cause flower drop and fruit set. Cold conditions lengthen the growing period and high temperatures decrease it. The earliest time when seed should be sown or seedlings transplanted in a prepared land, is when the soil and air temperatures meet at least the minimum temperature requirements. Hot and dry winds may also cause flower-drop, while good irrigation management may lower temperatures, raise the humidity and prevent flower drop. Also, the occurrence of hot conditions and the appearance of insects/diseases may influence planting dates in some regions.

Diseases increase rapidly mostly towards the end of the growing season. Dry, warm conditions worsen the problem of bacterial diseases. Fungal disease may be a problem during wet, humid and cloudy conditions, especially when production is near the coast. A severe infection of leaf related diseases usually lead to leaf drop and the sunburn of fruit.

Humidity inside a tunnel could be an advantage, however a single fully-developed tomato plant could transpire 2 litres of water per day on a hot sunny day. Humidity inside tunnels should be controlled by opening the ends/sides fully where warm areas occur. This is essential when producing high-value crops. It is sometimes also necessary to make use of extractor fans to protect the plants against humidity and temperatures which are too high inside the tunnel. Higher yields and better quality can be expected in well controlled tunnels.

26.6.2.1 OPTIMAL PLANTING TIMES FOR TOMATOES
Areas with light/little frost: Sept-Nov
Areas with heavy frost: Oct-Nov
Frost-free areas: Feb-March and July-August
Cooler winter rainfall areas: Sep-Nov

26.6.3 TOMATO CULTIVARS
All over the world, tomato breeding programs for producing new and improved cultivars have been given much attention, in the past and recently. This is why hundreds of cultivars are available and each cultivar has its own specific characteristics and requirements. It is important that producers select the best cultivar for their region and in doing so should communicate with other producers in their region, as well as with local dealers and reliable seed companies.

It is advisable that producers establish more than one cultivar and evaluate them. Most commercial producers evaluate cultivars on many criteria such as: high yields, disease/insect resistance, tolerance adaptability to temperatures, better fruit shapes, better outer and internal colours, longer shelf life, more inner cavities, better and sweeter taste, good quality, shiny, eye catching and so on. Ideally the characteristics of the fruit should match the cultivar requirements for the various market sizes.

Breeding programs are extremely expensive. It is recommended to regularly evaluate new cultivars and when a cultivar is identified that provides even a small improvement, the current/standard cultivars should be systematically replaced with the new one. Large producers use evaluations to improve yields and quality on a consistent basis. In the past and even presently, producers may lose considerable money when they neglect to evaluate different cultivars against their current/standard cultivars. Any added expense it worthwhile, not only for tomatoes but for all crops.

Tomatoes are divided into the following groups:
1. Bushy roma and blocky types: Non-trellising. Excellent Hybrid cultivars are available.
2. Determinate growers: Low growers which stop growing at certain heights.
3. Indeterminate growers: High growers which do not stop growing.
4. Saladette: Indeterminate grower with roma shape.
5. Cherry types: Both indeterminate and determinate growers.
6. Cocktail types: Indeterminate grower.

Some minor tomato cultivar fruit are jointed, which means that they exhibit an attachment called a calyx, which is to be found on the fruit shoulder. ***See photo on page 151***. Jointless fruit are free from a calyx. However, the calyx may serve as an indication of the age of fruit. Some market agents can identify the age of the fruit by this attachment but this has become an outdated technique. Cultivars are mainly jointless as the calyx may be a disadvantage due to the fact that it may puncture other fruit in certain packaging containers. Sometimes when picked, fruit of a non-calyx cultivar may have a calyx on the fruit and should be removed.

26.6.4 SEEDLING PRODUCTION IN SEED TRAYS
26.6.4.1 PRODUCTION OF SEEDLINGS
See Chapter 4.6 about seedlings obtained from nurseries and Chapter 4.11 for producing seedlings in seed trays.

26.6.4.2 SEEDBED PRODUCTION, ESTABLISHING SEED IN SEEDBEDS WITHIN SOIL
Also see Chapter 8.4 to produce seedlings in soil.

Production of tomato seedlings in soil seedbeds is seldom practiced today due to the practice of wide-spacing as well as the number of seedlings that are necessary to produce 1Ha, which is about 18 000 plants p/Ha. The spacing of long-life, indeterminate tomatoes produced in shade net houses is 26 000 plants p/Ha.

Almost all producers use seedlings that are produced from seed trays as well as nurseries. However, it may be an option to produce your own seedlings in soil. F1 hybrid seed are expensive and it is therefore not advisable to produce them in soil as losses of these types of seed should be avoided at all costs.

When preparing seedbeds to produce seedlings in soil, it is preferable to use a virgin soil and a sterilized, sandy loam type of soil, if possible. However, fertile soil that is prepared properly that exhibits fine, smooth characteristics, is also acceptable. Soil should also be disinfected when used repeatedly.

26.6.4.3 THE METHOD OF MAKING SEEDBEDS IN SOIL
Make use of a soil analysis. When a soil analysis is not available, the following mix is recommended: 60g Super Phosphate, 30g Potassium Sulphate and 90g Dolomite lime per m^2 (100g = 2 large, handful's). If soil is very acidic, more lime will be necessary.

Another option when a soil analysis is not available is to apply fertilizer, as follows: 120g 2.3.2 (22) Zn per m^2 as well as at least 8kg m^2 of good fertile compost. Fertilizer should be applied before creating the seedbeds. The soil should be moist but not overly wet before making the seedbeds.

Make the raised seedbeds 1m wide and less than 10m. The paths between the beds should be 50 to 60cm apart from each other. Add the surrounding, fertilized soil from the paths to the raised beds, so that they are not more than 6cm high. Create straight furrows across the seedbeds, 10 to 15cm apart from each other and 2-3cm deep. Sow the seed in the furrows approximately 3cm apart from each other.

An option is to create furrows 2-3cm deep in a 1m wide, well-prepared seedbed. Approximately 320 seeds would be necessary to establish a 10m long seedbed and 60 rows will cater for approximately 19 200 seeds. Approximately 10g of seed equals 10 000 seeds, depending on the cultivar due to the variation of seed sizes. Additional seed is required for the cherry types of tomatoes' as this seed is very small. Cover the furrows with smooth loose soil. When seedbeds are very cloddy, use smooth, sterilized sandy soil from elsewhere to cover the furrows, if possible. Remove all the clots from the covered furrows.

26.6.4.4 CARE FOR SEEDLINGS IN THE SEEDBED
Also see Chapter 8.7 Mulching.

Whenever hot days occur, irrigate the seedbeds twice-a-day or as necessary, until the seedlings have emerged. An option when hot conditions occur, is to cover the beds with grass. The grass keeps the soil cool and moist. The grass should be removed 5 to 6 days after sowing or as soon as the seedlings emerge.

Transplant seedlings when they are approximately 10cm tall or 6 to 8 weeks after sowing the seeds. Seedlings produced in soil seedbeds take longer to be ready for transplanting as compared with producing them in seed trays.

26.6.5 SOIL PREPARATION
Also see Chapter 8.2 Soil Preparation.

Good land preparation is the first-step for successful tomato production.

It is a good option to establish the rows in an east to west configuration, so as to make best-use of the sun for the plants. When the rows run north to south, then the one side of the plants receive morning sun while the other side receives afternoon sun. In humid areas, it takes longer for rain or morning dew to evaporate. In addition, the longer afternoon sun on the south-western side makes the fruit more vulnerable to sunburn.

It is necessary to irrigate the land when the soil becomes too dry. Determine if there is a hard layer below the soil, as may usually be found in new and sandy soil and eliminate it in order to promote good drainage and root penetration. It is recommended to plough the soil well before planting time, as it allows the soil time to settle. Early soil preparation is a good option for breaking down organic matter, improving moisture conservation and helping to control weeds. It sometimes also assists in controlling insects/diseases. It creates a good, smooth, aerated seedbed before beginning final land preparation later on.

It is a good option to plant on ridges (except for sandy soil), in order to improve the development of roots, as well as improve drainage and partly prevent soil diseases. Ridges should be made at least 30 to 40cm wide and flat on the top. Broadcast the fertilizer according to a soil analysis, before making ridges. Make ridges by using a disk (adjust the 2 disk blades) or a mechanical disk rigger.

Most producers establish seedlings of the non-trellising, roma types of tomatoes, at the bottom of well-prepared ridges. As the plants develop, they are trained to grow up towards the top of the ridges and then spread on top of the ridges.

On ridges, for tomatoes and peppers, drip irrigation and mulching with black plastic strips is a good option. *See Chapter 14.7 for drip irrigation.*

26.6.6 TRANSPLANTING SEEDLINGS
See Chapter 8.8 to transplant seedlings from seed trays and Chapter 8.5 to transplant seedlings from soil seedbeds.

26.6.7 WEED CONTROL
Also see Chapter 17 for Weed Control.
Tomato plants have a shallow root system, but can penetrate to a depth of 1m if established in well-prepared soil.

It is also necessary to make use of shallow weed control measures, especially when the fruit starts to mature.

26.6.8 FERTILIZATION
Also see Chapter 15.1 for commercial fertilizer and Chapter 15.4 for organic and inorganic fertilizer.

Fertilizing of tomatoes has been researched over many years and much information is available on the subject. According to experts, different tomato types and cultivars may differ in their specific feeding requirements. Contact the suppliers of these cultivars for the applicable information.

A reliable soil analysis should be a high priority. The amounts of macro-elements needed for tomatoes vary according to experts. It is important to consult a fertilizer expert to interpret the soil analysis so that the producer may apply the correct amount of fertilizer in the production land. Poor soil needs much more fertilizer in order to produce high yields and good quality. Very poor soil should however be avoided when producing tomatoes. The expense and time to improve poor soil is not justified.

When a soil analysis is not available, especially for small-scale farmers, then one needs to estimate the amount of fertilizer required for the specific planting area. It is almost always necessary to apply at least 1 000kg agricultural lime p/Ha even when the pH is fairly low. Also, it is generally known that on 1Ha of land, tomato plants remove approximately 5 to 6kg nitrogen N, 2 to 3kg phosphate P and 7 to 8kg potassium K. Basic fertilizer requirements for determinate open-field tomato production with a yield expectation of 50 tons p/Ha would be approximately 140kg N, 35kg P and 190 to 210kg K in a reasonably fertile soil.

A general guideline for small-scale producers, in fairly fertile soil, when a soil analysis is not available is 900kg p/Ha 2.3.4(30) Zn or 1 300kg of the common, cheaper 2.3.2 (22) Zn. This also depends however on the soil fertility. Small-scale farmers could use 2 handfuls of 2.3.4(30) Zn per m² or 100g. Two very big handfuls of fertilizer equal approximately 1000kg p/Ha. Most underprivileged farmers use the more balanced, cheaper fertilizer 2.3.2 (22) Zn (most soil have a shortage of Zn). One large handful of fertilizer is approximately 50g per m² which equals 500kg p/Ha.

If compost is going to be incorporated, at least 80 tons p/Ha or 8kg per m² should be applied on ridges in warmer areas. Don't waste your time applying too little compost in the warm summer areas.

Some producers use calcium Ca and boron B foliar sprays on a weekly basis when rain is continuous and when the fruit starts to set. This is to avoid fruit rotting, Blossom End Rot (BER) and bursting of fruit. BER and bursting of fruit may sometimes cause heavy damage.

26.6.8.1 TOP OR SIDE DRESSINGS
It is almost always necessary to apply top or side dressings. When soil are fairly fertile it is necessary to make use of 2 or sometimes 3 top dressings of N at 3, 5 and 7 weeks after transplanting the seedlings at a rate of 130kg LAN/KAN. Make use of the correct type of top dressing depending on the pH of the soil. *See Chapter 15.6 about top and side dressings and Chapter 15.5 about the correct types of lime.*

A second dressing or even a third one may sometimes be necessary, especially in sandy soil. In this case, use more or less 100kg p/Ha. Take note of the colour of the plants. If they become dark green it is an indication of too much N. However, too much N may cause flower and fruit drop that could affect the amount of fruit per plant. Too little N may cause a light green plant colour, especially at the growing points of the plants. If plants begin to turn light green, N should be applied immediately to the entire land.

When applying soluble fertilizer, then a pivot, overhead sprinkler or dripper system is used. However, be careful, as when too much Potassium K is applied, the elements are then pushed out of balance by other cat-ions which include calcium Ca and magnesium Mg which may result in the development of a Mg deficiency. The overall result will be large, lower leaves, which become pale in colour.

Plants absorb elements in the proportions they are available in the soil. Potassium K is often applied at a later stage of development through an irrigation or dripper system in order to obtain good sized, firm, quality tomato fruit. However, be wary of complex, weekly applications of expensive fertilizer through the dripper system. Certain dripper representatives may promote a weekly nutrient chemical program, but this is expensive and not always necessary.

A well fertilized land should do the trick and 2 to 3 N top dressings should be adequate. Look at the colour of the plants before applying too much N. Tomato fruit begin ripening from 3 to 3.5 months before the first fruit may be harvested. The plants require more potassium K for fruit and cell development. This applies especially to the indeterminate tomato types as the fruit ripens in series and is picked at intervals over longer periods.

26.6.9 IRRIGATION
Also see Chapter 14 for Irrigation.
Research has indicated that overhead irrigation does not assist in causing tomato diseases as long as the plants

are not wet after sunset. However, in areas with high humidity such as alongside the coast, plants are more vulnerable to diseases. Be watchful not to over or under irrigate, especially when the fruit begin to ripen. They may crack and become unmarketable.

Tomato roots penetrate at least 90-100cm deep into well-prepared soil. However, most roots may be found in the top 35cm of soil and should be irrigated accordingly.

26.6.9.1 WATER APPLICATION – AMOUNTS AND MEASUREMENTS

Always take rain water into consideration and adapt water application accordingly. Due to the long growing period of indeterminate tomato cultivars (5 to 6 months), the plants remove a large amount of water from the soil, especially during warmer conditions. Transpiration is high and the plants need more water when the first fruit begin to develop.

Determinate tomato cultivars with overhead sprinklers, require approximately 580mm water throughout the growing season. Begin by irrigating 20mm during the first 3 weeks after transplantation and thereafter 35-40mm for the rest of the growing season. Be careful not to over irrigate, especially later in the season and when rainfall occurs. Adapt irrigation by taking all these changeable factors into consideration.

When using a dripper system under plastic covering and on ridges, use reliable tension meters or data loggers to measure water application. Smartphones are also a good option. Most drippers provide 4 litres of water per hour. The timing of the drippers should be adjusted in order to meet the specific requirements. Too much or too little water may cause cracking, especially when warm conditions occur. Fluctuations of irrigation will promote blossom end rot (BER).

26.6.9.2 CRITICAL IRRIGATION TIMES
Immediately after transplanting the seedlings.
During the first 4 weeks.
During the flowering and fruit development stages.

26.6.10 TRELLISING AND SPACING METHODS
Trellising and spacing methods for determinate and indeterminate tomato cultivars in the open-field and in greenhouses.

26.6.10.1 TRELLISING TOMATO PLANTS
Also see Chapter 12.2 for trellising.

There are many different trellising methods used in South Africa for determinate and indeterminate cultivar production. There is always new technology being developed to save money. Trellising and labour costs are high. In this section, we will be dealing with some of the main trellising systems.

Producers should decide which system they intend to use and what type of tomatoes are going be cultivated. Determinate, indeterminate, saladettes or the roma non-trellising types. It may be a good option to include the non-trellising roma types.

High yields are achieved when producing indeterminate long-life tomato cultivars. Although the trellising cost of indeterminate tomato cultivars are high, many producers choose to change from determinate to indeterminate production. This is due to a range of factors including higher yields, good quality and the ease of management related to disease/insect control, trellising, pruning, harvesting, classing, sorting and packaging. It should be noted that more than 80% of the crops harvested may be of a class 1 quality, if the producer is able to care properly for their indeterminate F1 high-value cultivars, within suitable areas. Seed companies should be consulted for the best and most cost-effective trellising methods as well as for the different types of tomato cultivars available.

CHAPTER 26: PRODUCTION GUIDELINES FOR TOMATOES

204 Single pole and double wire trellising method for determinate and indeterminate tomato growers.

205 Single plant trellising method for indeterminate growers, trellised up to a height of 2m using a supporting wire.

206 An arrow dripper system installed in soil on ridges, in a 3.7m shade net house for the production of tomatoes or peppers. Establish 1 plant per dripper.

207 In front, a short Hybrid tomato cultivar which has an excellent yield. This cultivar will save money due to its low growing behaviour which requires less trellising material. At the back is a high-value tomato cultivar, trellised to a 2m high horizontal wire.

208 Trellising indeterminate tomato plants using non-pruning of plants. Note the double nylon twine to keep the plants upright. The spacing is 1.8m between the rows and 45cm in the row, which gives a population of 12 300 plants p/Ha when using single rows and cultivars that are suitable.

209 Tomato production in a shade net house, directly in soil. The plants are trellised to a height of 2.5m using the double-row and single-stem method. An arrow dripper is installed next to each plant. 80 ton p/Ha compost has been applied on the ridges which yielded more than 160ton of fruits p/Ha.

210 This production is installed with an arrow dripper system in a shade net house and trellised up to a height of 2.5m to the horizontal wire. More than 160 ton p/Ha were harvested.

211 In front is a determinate F1 hybrid cultivar which has an excellent yield up to a height of 1.8m. In the back are F1 hybrid cultivars trellised up to 2m.

212 Cherry tomatoes. Non-pruning with very good yields.

PART 2: GUIDELINES FOR CROP-SPECIFIC VEGETABLE PRODUCTION

213 The Indeterminate tomato plants have collapsed due to the heavy fruit load. The 2.5m-high horizontal supporting wires were only 3mm thick. However, this was an exceptionally good yield that was produced in a shade net house, in soil.

214 A F1 hybrid determinate, large fruit cultivar. The three fruit weigh exactly 1kg. The first, bottom fruit are usually the largest. This cultivar has many seed cavities and lots of jelly. The side walls were a bit too thin but the quality was still good. These fruit were harvested in the author's garden (Feb 2016).

215 This land was established with non-trellising roma type tomatoes. However, the producer neglected to cultivate it properly due to a shortage of water, poor weed control and a lack of covering which caused sunburn. As a result, the plants did not grow vigorously enough.

216 A flowering cherry tomato cultivar. This healthy plant bears abundant fruit and its yields are similar in weight (fruit per plant) as the fresh market tomato cultivars.

217 Top left, an oval/round tomato fruit. Top centre, a long shelf life fruit showing the thickness of the side walls and the seed cavities. Top right, a blocky cultivar fruit. Front left, a bottle-neck cherry fruit. Second from left, an oval, long cherry fruit, small bottle-neck cultivar and front right, an oval long cherry fruit almost the size of a cherry cocktail tomato cultivar.

26.6.10.2 ADVANTAGES OF TRELLISING TOMATO PLANTS
- Harvesting is earlier and potential for sunburn is less, especially under protective structures.
- Better yields and quality as well as good shelf life.
- Due to the indeterminate growth behaviour, insect and disease control is more effective and allows the labourers to be more effective.
- Plants become dry more quickly due to improved air flow which also reduces the potential of certain diseases.
- It is more comfortable to harvest the higher growing fruit and the fruit is more visible.
- Less damaged plants due to the wider spacing.

26.6.10.3 NON-TRELLISING TOMATO TYPES

The bush/roma and blocky tomato types are cultivated without trellising and are usually established on ridges. These types of tomatoes have a bushy growing pattern and are mainly used for processing, jam, canning and freezing.

26.6.10.4 DETERMINATE TRELLISED TOMATO CULTIVARS

Determinate tomatoes usually grow between 0.8 and 1.8m high. Be aware that shorter F1 hybrid cultivars are also available. Shorter growing cultivars save on trellising costs but yields are much lower.

The aim for breeders' is to provide determinate tomato cultivars that grow as low as possible. This saves money on trellising costs and labour and also results in reasonably good quality and yields. Producers should work with seed companies to determine the best cultivars to produce. Some cultivars are specific to certain areas.

26.6.10.5 DIFFERENT TRAINING (TRELLISING) SYSTEMS FOR DETERMINATE AND INDETERMINATE TOMATO GROWERS

Producers are often testing new methods to reduce the high trellising costs.

26.6.10.6 TRELLISING METHODS:
1. Bush or roma types: Non-trellising, short, bushy growers.
2. Determinate growers: Double-wire method.
3. Indeterminate trellised: Salladette cultivars.
4. Shade net/tunnel, indeterminate: Double-row, single-stems pruning method.
5. Single-pole method.
6. Minor and other trellising methods.

Certain producers use more cost-effective baler or nylon, non-stretching twine to trellis tomato plants as well as to box pepper plants.

26.6.10.6.1 PROCESSING OF BUSH OR ROMA-TYPES, NON-TRELLISING

Here we are dealing with bush/roma, pear or blocky shaped types of tomatoes. Non-trellising is used mainly for freezing, processing or canning, but some producers also use it for the fresh market. Some producers have contracts with processing factories.

The roma-types are cheaper to produce and the newest F1 hybrid cultivars are high yielders. They have a good external and internal colour and can also be supplied to the fresh market. The problem with the roma bushy types is that their internal jelly content is low. This is mostly due to their shape (some cultivars are blocky shaped), thick fruit walls and few internal chambers.

The saladette types are indeterminate growers. The pear-shaped roma could also be used successfully for processing. The yields are fairly high and the fruit has more jelly than the roma types. They are also bred to contain more internal chambers. *See photo 200.*

26.6.10.6.2 SPACING

Spacing of the bushy or roma types is normally not less than 1.5m between the rows and 30 to 40cm within the rows. For example, with plant spacing of 35cm within the rows and 1.5m between rows, which are produced on well-constructed ridges, the plant population will be 19 500 plants p/Ha. Spacing should be fairly wide between the rows to allow enough sunlight in and to allow labourers to move between the rows without damaging the plants.

26.6.10.7 TRELLISING OR STAKING – THE DOUBLE WIRE METHOD FOR DETERMINATE TOMATO PRODUCTION

The trellising methods for determinate tomato cultivars are:
1. Double-wire method

2. Mexican method
3. Philippines method
4. Taiwan method
5. Brazil method

26.6.10.7.1 THE DOUBLE-WIRE METHOD

This method is most commonly used by commercial producers. It makes use of a dripper or overhead irrigation system. The capital investment cost for the first season is high. The input cost for the second season is approximately 30% less, due to the fact that the trellising material is to be reused.

26.6.10.7.2 SPACING

Aim to establish determinate tomato seedlings in single rows, preferably on ridges. When making ridges they should be at least 30-40cm wide on top. A recommended spacing for transplanting seedlings is 30 to 50cm apart within the rows and 150cm between the rows themselves. This spacing results in a plant population of 18 800 plants p/Ha or 1.8 per m^2.

Commercial producers prefer 16 000 to 18 000 plants p/Ha in the summer-orientated areas. In the warmer winter areas, the plant population should be approximately 16 000 plants p/Ha. This is due to the limited sunlight and shorter days. For example, with spacing of 150cm between the rows and 45cm within the rows and the plant population would be 16 500 plants p/Ha. This is a good option for the warmer winter production areas.

Some producers prefer to use tractor wheels to cut into the soil to make seedbeds. The tractor wheels are usually 1.2m apart. The seedlings are then established 40cm apart within the rows which results in a plant population of 14 000 plants p/Ha.

Wide spacing makes it more convenient for tractors and implements to control weeds, insects and diseases, especially if a destructive disease was troublesome in the previous season. Weed, disease and insect control could be applied mechanically until the trellising poles are established approximately 3 weeks after transplanting the seedlings.

It is important to note that when rows are spaced wider, then the spacing of plants within the rows could be closer. This is particularly applicable when producing tomatoes and other crops in the summer-orientated areas where bright light is present. Closer spacing results in smaller tomato fruit due to the overlapping and competition between plants as they fight for light, water and nutrients. When the plant population is dense it is also more difficult to control insects, diseases and weeds.

Consumers prefer larger tomato fruit as these have more jelly and they achieve better prices. The markets are usually oversupplied with small and medium fruit.

SPACING OF POLES/STAKES IN THE LAND FOR THE DOUBLE-WIRE METHOD
- Drive 2m long, sharpened, treated wooden poles with a 7-10cm diameter, 50cm deep into the soil and set them 4-5m apart from one another within the rows using a hammer or pipe driver. *See photo 221.*
- Use thicker poles at the ends of the rows to serve as anchor poles. These poles will be placed under tension to keep the structure upright and prevent the plants from falling over.
- To make a pipe driver use a steel pipe ±12cm in diameter and 70cm long. Weld a lid on one side to seal the top and then fit the handles using thinner pipe of about 6cm and weld them onto the 12cm pipe. It is more efficient to use a pipe driver to drive the poles into the soil. *See photo 221.*
- The poles should be planted 2 weeks after transplanting the tomato seedlings in the land.
- Stretch and fasten 1.4-1.6mm galvanized wire around the end or outer side of the poles to establish the double wires.
- The first wire, closest to the ground, is spaced 30 to 40cm from the soil level. Thereafter, each wire is spaced 30cm apart to allow the plants to grow taller.
- The initial financial outlay for wires and poles is quite high. For example, if implementing a 1.5m distance

between rows and 4 double wires in the rows, then about 10 560m galvanized wire is required for 4 double-wire rows p/Ha and 1 650 poles are required when spaced 4m apart from each other.
- To remove the wires after harvesting, loosen them at both ends and simply roll them into a drum.

26.6.10.7.3 TRAINING OR TRELLISING FOR THE DOUBLE-WIRE METHOD
Determinate tomato cultivar plants stop growing when they have approximately 2-4 flower trusses and reach a height of 1-1.9m. However, this depends on the cultivar choice. Some cultivars reach a height of more than 1.8m if they grow vigorously in ideal conditions. The trellising costs will then be much higher.

Many producers still make use of the double-wire method which may be the most cost effective method in the longer run due to the reusing of materials. As the plants grow higher, the wires are placed under tension across their entire length. Check that the anchor poles are driven deep enough into the soil. The poles should be thick enough to carry the weight. The double wires should also be fastened to every pole in the row to spread the load.

As the plants grow higher, they are continuously trained between the double wires. To provide enough room for the plants, tomato hooks are attached to both sides of the plant, with the exception of the top pair of wires which should not be hooked. ***See photo 219.***

When using this method, it is not necessary to tie the plants and shoots. They only need to be trained between the pairs of wires. Choosing a lower growing tomato is a good option.

26.6.10.8 TRELLISING INDETERMINATE TOMATO PLANTS IN SOIL USING THE SINGLE POLE AND SINGLE ROW METHOD
See photo 199.

Broadcast fertilizer according to a soil analysis and then create ridges. If compost is going to be used, then also broadcast it on the 40cm wide ridges. Apply at least 80 ton p/Ha or 8 spades per square meter of good compost, if available.

26.6.10.8.1 TRELLISING INDETERMINATE TOMATO CULTIVARS
This method has become popular and producers are using it more frequently. Using the best F1 hybrid cultivars is important when using this method.

Single rows are to be established for this method. The poles should be 2.5m long and are driven into the soil 0.5m deep and are placed 4 to 5m apart from each other. Be aware that the poles will be carrying considerable weight and should therefore be strong and well anchored. Usually wire is attached at the top of these poles and tensioned. The dripper lines may be installed next to the plants.

26.6.10.8.2 SPACING
Do not establish the plants too close to each other due to their non-pruning and bushy growth tendency. When the plant population is too high, the fruit size may decline. According to experts, good spacing achieves the best quality. For example, 1.8m between the rows and 45cm within the rows results in a total of 122 100 plants p/Ha or 1.2 plants per m^2. Another example is 1.5m between rows and 45cm within the rows resulting in a plant population of 14 652.

Producers should experiment with plant spacing to achieve the best possible plant population for their region and climatic conditions. When producing tomatoes in the warmer winter areas, the plant population should be reduced. This is due to shorter days and less sunlight.

In order to trellis the higher growing plants, box them by using non-stressed nylon twine or the cheaper baler twine. Space the plants to the required distance from one another and as they grow higher, fasten them to the horizontal wires. It is recommended that all the leaves up to the first trusses are removed. It is not necessary to prune side shoots, due to the wider spacing. Cherry types also grow well using this method.

26.6.10.9 TRELLISING INDETERMINATE PLANTS IN SOIL AND IN PROTECTIVE STRUCTURES
26.6.10.9.1 INDETERMINATE CULTIVARS AND PRODUCTION IN THE OPEN FIELD

When producing indeterminate high-value tomato cultivars in soil, it is recommended to use ridges with a dripper system in soil, within protective structures such as tunnels or shade net houses. Similar to hydroponic production, producers could produce indeterminate tomatoes and peppers directly in soil within. A producer may use the Open Bag System (OBS) and later, may change to full hydroponic production from soil, without too much difficulty.

The benefit of establishing tomato plants directly in soil is because it is cheaper than hydroponic production. Less labour is required, higher yields and better quality (more class 1 fruit) can be expected. Overall it is easier to manage. This is because granular fertilization before establishing the crop, is not necessary. Instead, fertigation takes place through a dripper system using a timer and a soluble fertilizer directly administered via the irrigation water.

When using double rows and an arrow dripper system in tunnels or shade net houses, the quality and yields of indeterminate tomato cultivars produced in soil, are excellent. The plants are trellised the same way as for hydroponic systems. Some producers change from hydroponic systems to soil production inside tunnels or in shade net houses. The reason for this is due to various factors such as the high cost of expensive hydroponic mediums. This includes sawdust that has become expensive due to the greater demand. Instead, cheaper granular fertilizer are applied directly into the soil according to a soil analysis in a similar way as to open-field production. This fertilizer application should be adequate for the entire growing season, except for the N top dressings which are administered through the water tanks.

Producers can build shade net houses in the same way as for hydroponic systems in order to produce crops directly in soil. *See Chapter 23.2 on how to construct a shade net house and Chapter 23.8 on how NFT hydroponics systems are constructed.* The only difference between soil and hydroponic production is that hydroponic production needs an extra water tank which is repeatedly filled-up with expensive soluble fertilizer.

With regards to the dripper lines: arrow drippers remain the same as for the QBS system and are inserted 10 to 15cm adjacent to the individual plants. Generally speaking, arrow drippers are not utilized in South Africa. However, producers are aware of the benefit of constructing shade net houses or tunnels and installing high-value indeterminate crops such as tomatoes, peppers and cucumbers. This is done more or less the same way as for hydroponic production.

A problem when producing tomatoes and peppers in shade net houses within soil, is when there is too much rainfall during the middle and end growing periods of the plants. When this happens, it may lead to the bursting of fruit and blossom end rot (BER) which can become a considerable problem. Producers may lose a substantial amount of physically distorted fruit, especially in high rainfall areas. The fertilizer could also drain-away, causing a calcium Ca and other nutrient shortage. This may be corrected using foliar sprays, however this can be difficult and advice from an expert is advised. That is the reason why producers use white plastic mulching to cover the rows and only irrigate on a scheduled basis using a dripper system.

The quantities of Class 1 fruit are increased when good compost is incorporated while also improving the soil structure. Producers and experts claim that by using good compost, it may reduce soil-borne diseases such as Pythium, Fusarium and Phytophthora as well as viral infections. When such diseases occur, no rotation is possible for a couple of seasons. Proper disease and insect control should therefore be a high priority.

26.6.10.9.2 TRELLISING PLANTS:
- When indeterminate tomato cultivars are produced in soil in greenhouses (tunnels and in shade net houses),

the growing point is to be removed when the plants reach the 2.5m horizontal wire. This prevents the plant from growing further.
- All the side-shoots that develop should be removed before they reach a length of 4cm or as soon as they appear. This prevents the plant from becoming bushy. The removal of side-shoots begin 2 to 3 weeks after transplanting the seedlings.
- Single stems are trellised up to the 2.5m horizontal wire using a single vertical string and hook for each plant. The hooks are attached onto the lateral supporting wire and the plants are wound around the vertical strings as they grow higher in a similar way as for the hydroponic systems. *See photo 3.*
- Harvesting begins at about 2.5 to 3 months and continues on a weekly basis as fruit matures.

Indeterminate production results in higher yields, better weed, insect and disease control and overall, cleaner and better quality fruit.

It is not necessary to establish poles in tunnels and in shade net houses due to the fact that the horizontal wires and supporting poles are fixed on the framework of the shade net houses which support the trellised plants.

26.6.10.9.3 SPACING FOR THE DOUBLE-ROW ONE-STEM METHOD IN GREENHOUSES
Also see Chapter 14.7 for drip irrigation.

Spacing indeterminate tomatoes and peppers in tunnels and in shade net houses are the same as for hydroponic systems. Double rows are spaced 1.5m apart in a 10 x 30m tunnel and 1.8 To 2m apart from one another in a shade net house.

Details are as follows:
- Install a 32-40mm main PVC pipeline from the water tank pump to the tunnel or shade net house. Split and reduce this main pipeline into 25mm sister pipelines inside the tunnel/shade net house and place them in the middle of the double rows on the ridges.
- Insert 8 litres-an-hour compensating valves every 85cm from one another on the 25mm sister pipeline on the ridges.
- Insert at top of the compensating valves a 4-way manifold.
- Cut spaghetti tubes into 45 to 50cm lengths.
- Insert the tubes into the 4-way manifold.
- Connect arrow drippers on the open ends of the spaghetti tubes.
- Insert the drippers in the soil next to the plants at the required spacing.
- When producing tomatoes in shade net houses, space the compensating valves and the 4 way outlet connections 85cm apart from one another on the 25mm sister pipeslines in the rows. Space the lines 1.8m between the rows.
- Recommended spacing of rows for a 10 x 30m tunnel is 1.5m from one another, starting the first row 1.25m from the side of the tunnel and 1m from the door ends.
- The plant population inside the tunnel is 792 plants or 26 000 plants p/Ha (2.6 plants per m^2). In shade net houses the spacing in the rows is the same as for tunnels but between rows a spacing of 1.8m is recommended which results in a plant population of 26 432 plants p/Ha. However in cooler or warmer winter areas do not exceed more than 22 000 plants p/Ha as this will diminish the sunlight.

The purpose of the compensating valves is to prevent water from flowing back to the lower point of the pipelines, especially when the irrigation motor stops. This keeps the pipeline always filled with water. It is important that there are no leakages when compensating valves are installed.

New cultivars that have shorter internodes are a good option. The shorter internodes allow more trusses up to the 2.5m horizontal wire and therefore higher yields. If the plant population is too high, the plants compete for light and as a result the stem plant internodes could become longer creating less trusses up to the 2.5m horizontal wire. Fewer plants are grown when double instead of single stems are used. Less plants p/Ha reduces cost because the indeterminate long shelf high-value seed and seedling tomato plants are extremely expensive. Producers should do their own research regarding whether to use more than one stem per plant up to the 2.5m horizontal wire.

Producers should always make sure to apply cutworm bait as they cannot afford to lose any of the transplanted seedlings. Early blight diseases should be controlled before they appear and an effective insect/disease spraying program should be followed.

26.6.10.10 MINOR TRELLISING METHODS
26.6.10.10.1 SINGLE POLE METHOD
This method is not used by commercial farmers as it is labour intensive, time consuming and damage may be caused to the plants while training the foliage of the plant. This method could be used by small-scale farmers as it is a economical and simple way to trellis the plants. Note that the plants should not be tied directly onto the single poles. A loop is made and the loop is tied around the plants next to the poles. Make the loops large enough for the plant stems to expand. As the plant grows higher, more loops are then attached to the main stem.

26.6.10.10.2 SINGLE PLANT, VERTICAL TWINE TRELLISING METHOD FOR INDETERMINATE TOMATO CULTIVARS
Cultivar choice is important for this system and therefore cultivar options should be investigated with reliable seed companies. This method is cheaper than the double-wire method, as only a single steel wire is tensioned on top of each 2m high pole. The wire should be strong enough as it will carry substantial weight when the plants and fruit are fully developed. The anchor poles should also be well anchored and planted deeply to withstand the weight of the hanging plants.

Twine for each plant is fastened vertically to the top and bottom ends and as the plant grows higher, it will be twisted around the vertical strings. When the plant reaches the top of the 2m horizontal wire then the growing point must be removed to prevent the plant from growing any further.

26.6.10.10.3 DOUBLE TWINE TRELLISING METHOD FOR DETERMINATE CULTIVARS
Consult with seed companies when using this method. Take care as the wind may be a factor in windy areas and may impact the plants. Plants are going to be trained (boxed) between the double tensioned nylon (or baler) twine that is laterally 20 to 30cm apart from each other and fixed at the full length of the rows. The double twine keeps the plants in position and no hooks are necessary.

The poles in the rows should be spaced closer to each other at a distance of about 3 to 4m. Use non-slack nylon or the cheaper baler twine. The twine should be properly tensioned and firmly fastened to all the poles in the rows. Twine could also be used to keep the rows together by using it to fasten the double lines.

Make use of a dripper or overhead irrigation system.

26.6.11 CROP ROTATION
Also see Chapter 18 Crop Rotation and Soil Improvement.

Tomatoes and related crops such as peppers, chillies, paprika, potatoes, tobacco, ground nuts and gooseberries, should not be established in the same soil every year. Instead, they should be rotated with non-related crops for at least 3 years. Also, establish green manure crops to improve soil fertility.

26.6.12 MULCHING
Also see Chapter 8.7 Mulching.

Tomatoes grow more vigorously with better yields and quality when the soil is covered or mulched. An increasing number of tomato growers use plastic mulching. Various coloured and improved types of plastic liners are available. Fertilizer is applied according to a soil analysis on the ridges before laying the plastic liners. Drip irrigation is usually laid underneath the plastic mulches. Some producers punch holes into the plastic liners using a sharpened PVC plastic pipe to establish the seedlings.

218 In front, double-wire method. At the back, indeterminate single rows with trellising and non-pruning cultivars.

219 A tomato hook attached around a pole used for the double wire method.

220 Indeterminate cultivars established in soil using a dripper system. Note the marigolds in the rows which the producers claim reduces nematodes in the soil.

221 A pipe-driver used to drive the tomato poles into the soil.

222 A determinate tomato cultivar using a drip system. Note the black plastic liners.

223 A new determinate tomato cultivar displayed at a farmers' day.

Plastic mulching controls weeds and certain diseases, conserves moisture, increases the quantity and quality of the fruit, saves water and improves the early marketability of the fruit. The right colour liners keep the soil warmer. When using liners, be sure that the soil or ridges are in a good condition. They should preferably be level, free from sharp objects, stones or large clods to avoid punching holes through the plastic liners. It is recommended to use ultra-violet treated plastic to cover the ridges as it may be used repeatedly. Note however that UV plastic coverage is expensive.

26.6.13 HARVESTING AND PICKING
Also see Chapter 24 Harvesting, Handling and Packaging.

Determinate tomato production is harvested periodically due to the fact that not all the fruit ripens at the same time. Certain F1 hybrid determinate cultivars may only be harvested 2 to 3 times due to the evenly homorganic ripening process. For the first harvest, harvesting usually begins at 2.5 to 3.5 months. Be aware that early-harvesting cultivars are available but the yields are usually much lower.

Indeterminate tomato cultivars are harvested approximately 2.5 to 3 months after transplanting the seedlings, however, this depends on the type of cultivar. Many cultivars may be harvested continuously for about 2 months after the first fruit have ripened. This is due to the long indeterminate growing period of 5-6 months. When establishing tomato seedlings in cooler conditions, the period of harvesting may initially take longer and also be extended. Sometimes it may be as long as 4 months before the last harvesting takes place.

The best time to pick tomato fruit is early in the morning when the temperature is still quite cool. This is not always possible however. As soon as possible, transport the tomato picking crates to the pack house. Do not allow them to be exposed to the baking sun for long periods.

Harvest and post-harvesting handling is very important. When fruit is allowed to become completely ripe on the plant, to a degree the flavour might be improved, but the fruit may crack and other disorders or diseases may occur. Jointed calyx cultivars should be avoided as this may damage the fruit when they are packed closely together in containers or in boxes. The calyx may puncture the fruit especially those at the bottom of the containers. *See photo 151.*

Tomato fruit, when they are red and ripe are perishable. They should be picked and handled with care and labourers should be aware of this. Make sure that tomato fruit are not dropped into the picking crates as it may bruise the fruit. A 1m fall for a ripe, red tomato will bruise it and shorten its shelf life. Do not stack the fruit crates to high, especially with ripe fruit. The weight of the fruit on top may damage the ripe fruit at the bottom of the crates. Keep in mind that it is not sunlight that ripens the fruit but temperature.

Indeterminate tomato cultivars are picked at short intervals due to their indeterminate growing behaviour. This is labour intensive and considerable amounts of time may be spent walking through the rows, selecting and picking the fruit at the correct stage.

224 A colour chart indicating the harvesting stages of tomato fruit as well as the sizes to class them.

26.6.13.1 PICKING STAGES FOR TOMATO FRUIT WHICH ARE TO BE MARKETED

Indeterminate tomatoes are picked at short intervals over longer periods. Fresh market tomatoes are harvested at different stages of ripeness. It is important for the producer to harvest tomato fruit at the correct stage of ripeness. This largely depends on the distance to the market. There are 7 stages of maturity of determinate and indeterminate cultivars and these are indicated on the colour chart above.

There are several determinate F1 hybrid cultivars that have a very short harvesting period. This is to benefit the producer. Research has indicated that small fruit begin to ripen at the same time as the larger fruit. If cooling facilities are available, then the fruit could be picked when red.

If long shelf life tomato cultivars are picked when they are too green they may not develop to a full, red colour. Most producers pick fruit when they are at a light red stage. The fruit may become fully red in 2 to 3 days, especially in warm stores. It is important to train labourers to identify the correct stages of harvesting.

26.6.13.2 PICKING STAGES
Also see photo 224 above for colour chart.

26.6.13.2.1 GREEN STAGE
The fruit is still green and exhibits a very pale colour. Do not pick long shelf life tomatoes at the very green stage. They may not develop their colour fully when ripened.

26.6.13.2.2 MATURE GREEN STAGE
This stage is several days before the first light red colour begins to show. It's at this stage that it is ready to be picked. The seed cavities of the fruit are then filled with jelly. The fruit has a light red colour which starts at the blossom end. Fruit at this stage can be stored at 10-12°C for 7 to 14 days.

26.6.13.2.3 LIGHT RED BREAKER STAGE
Most producers harvest at this stage, especially when they are a reasonable distance from the market. The fruit usually ripens after 2 to 3 days in warmer conditions after harvesting. When temperatures are high the fruit will ripen faster. More picking times are necessary, because the fruit ripens periodically. Fruit could be stored at temperatures of 14 to 16°C for 7 to 8 days.

26.6.13.2.4 HALF-RIPE STAGE
When long, cool weather conditions occur, it is better to pick at the half-ripe stage. The fruit should be picked when they are light red and stored at temperatures of 10 to 14°C. They usually begin to ripen in 4 to 5 days.

26.6.13.2.5 RIPE STAGE
At this stage the majority of the fruit has a red colour and is firm without any sign of softening. Tomato fruit that are packed at this stage should be picked every day when warm conditions occur. This prevents the fruit reaching the consumer in an overripe or soft stage. Most long shelf life cultivars are exempted from this.

26.6.13.2.6 RED RIPE STAGE
The entire fruit is red and should be picked as soon as possible.

26.6.13.2.7 FULL RIPE STAGE
The fruit is fully red. The open-pollinated cultivars should be handled gently and marketed immediately. The long shelf life cultivars could be kept for longer.

26.6.14 SORTING, CLASSING AND MARKETING
Also see Chapter 25 Management and Marketing.

Producers should produce the preferred type of tomato cultivar and fruit that is most acceptable to the consumer. Fruit may be classified as fresh market, saladette, cherry, cocktail, novelty and speciality tomato fruit. The largest production is for fresh market fruit. Cultivar choice is important, however it depends on the needs of the market, the processing factories as well as the housewives.

CLASS OF TOMATOES ACCORDING TO MARKET REQUIREMENTS

Large	73-82mm	Medium	56-63mm
X Large	83-94mm	Medium plus	64-72mm
XXX Large	95mm and larger	Small	55mm and smaller

26.6.15 INSECTS
ROOT-KNOT NEMATODE: *See Chapter 20.5.2.1.*
CUTWORMS: *See Chapter 20.5.2.2.*
APHIDS: *See Chapter 20.5.2.3.*
LEAF MINER: *See Chapter 20.5.2.10.*
RED SPIDER MITE: *See Chapter 20.5.2.9.*
TRIPS: *See Chapter 20.5.2.4.*
BOLWORM: *See Chapter 20.5.2.7.*
WHITE FLIES: *Occurs especially in tunnels. See Chapter 20.5.2.12.*
CMR BEATLES: *See Chapter 20.5.2.16.*

24.6.16 DISEASES
Also see Chapter 20.6 Diseases.

DAMPING OFF: *See Chapter 20.6.3.1.*
EARLY BLIGHT: *See Chapter 20.6.3.3.*
LATE BLIGHT: *See Chapter 20.6.3.4.*
DOWNEY MILDEW: *See Chapter 20.6.3.6.*
POWDERY MILDEW: *See Chapter 20.6.3.6.*
ANTRACNOSES: *See Chapter 20.6.3.12.*
FUSARIUM WILT: *See Chapter 20.6.3.11.*
PHYTOPHTHORA WILT: *See Chapter 20.6.3.16.*

ALTENRIA LEAF SPOT: *See Chapter 20.6.3.2.*
BACTERIAL SPOT: *See Chapter 20.6.3.10.*
BACTERIAL CANCER: *See Chapter 20.6.3.9.*
BACTERIAL WILT: *See Chapter 20.6.3.15.*
BOTRYTIS OR GREY MOULD: *See Chapter 20.6.3.14.*

26.6.17 VIRAL DISEASES
Also see Chapter 20.8 Viruses.

Knowledge of viruses and diseases is essential when producing tomatoes. The most important virus is the Tobacco Mosaic Virus (TMV) and Potato Virus Y. The latest development (2012) is that some F1 hybrid cultivars are resistant or partly resistant to Bacterial Spot, Tomato Spotted Wilt Virus (TSWV) and Root-knot nematode.

TOBACCO MOSAIC VIRUS (TMV): *See Chapter 20.8.2.1.*
POTATO VIRUS Y (PVY): *See Chapter 20.8.2.2 Potato Virus.*

VERTICILUIM WILT
This viral disease is soil borne and infects the entire plant. The plant wilts as if suffering from drought stress. The plants should be removed from the land and be destroyed to prevent further infections.

225 Sunburn disorder.

226 Blossom End Rot (BER) disorder.

227 Lateral and concentric cracking disorder.

228 A Heavy cat face disorder.

229 Zipping disorder.

230 Tomato mosaic (TMV) virus infection on tomatoes.

26.6.18 PHYSIOLOGICAL DISORDERS
Fruit disorders and other physiological problems may cause severe losses for fruit-bearing crops such as tomatoes, peppers and pumpkins and field cucumbers.

The most common disorders should be known to producers.

BLOSSOM END ROT (BER): *See Chapter 20.9.1.3.*
CRACKED FRUIT: *See Chapter 20.9.1.2.*
CAT FACE: *See Chapter 20.9.1.1.*
MALFORMED FRUIT: *See Chapter 20.9.1.10.*
EXCESSIVE ABORTION OF FLOWERS, FRUIT AND POOR FRUIT SET: *See Chapter 20.9.1.9 Blossom Drop.*
SUNBURN OR SUN SCALD OF FRUIT: *See Chapter 20.9.1.4.*

CHAPTER 27

PRODUCTION GUIDELINES FOR CUCURBITS: PUMPKIN, SQUASH, WATERMELON AND MUSK MELON

231 A Green Hubbard cultivar in good condition. Overseas it is classified as a squash and in South Africa it is known as a pumpkin. Good foliage to cover the fruit is a bonus as it protects the green fruit pumpkins and squash from sunburn.

27.1 INTRODUCTION

There are five groups in the cucurbit family, namely:
- **Pumpkins**
- **Squash**
- **Cucumbers**
- **Watermelons and musk melons**

All the above cucurbits are alike. They have similar cultivation requirements and suffer from similar diseases/insects. All the groups have a trailing vine growing characteristic except certain squash and Hybrid cultivars which have a bushier growing characteristic. All cucurbits are sensitive to frost.

All groups are insect pollinated, primarily by bees. The flowers of pumpkins and squash are unisexual and monoecious. This means that the male and female flowers are situated on the same plant and on the same runner. Commercial producers often find it necessary to introduce bee hives to the production land, especially if bees are scarce in the area. This ensures maximum pollination and fruit-set. Verify that there are enough bees present near the production land when the plants are flowering. Bees need to visit a female flower several times (according to specialists at least 25 times) before full seed-set takes place.

Fruit size for pumpkins and squash are between 30 and 45kg. Currently, the world record for the largest pumpkin cultivar (Goliat van Gat) is 515kg (2015) and the South African record is 487kg (2016).

It is interesting to know that the watermelon does not have the highest internal water content at 92%. Asparagus is 93%, celery 94%, cucumber 96% and lettuce 97%.

PART 2: GUIDELINES FOR CROP-SPECIFIC VEGETABLE PRODUCTION

232 A new breed of squash/pumpkin. Plant breeders' are always striving to improve and develop new cultivars.

233 Newly bred yellow and green flesh melon cultivars displayed at a farmer's day, together with chillies, peppers and cabbage.

234 A gherkin cultivar belonging to the Cucumber family.

235 Pumpkin, tomato and pepper cultivars displayed at a farmer's day. The pumpkin on the right is a large Flat White Boer (van Niekerk type). This pumpkin cultivar can easily weigh up to 30kg.

236 An iron bark cultivar, yielding up to 80 tons p/Ha. The keeping quality is excellent – it can be kept for more than 5 months before marketing.

237 A world record pumpkin at 497kg (2008). The new record is more than 500kg (2010). The SA record is 487kg and is a Goliat van Gat cultivar, specially bred for its large growth ability.

238 A baby marrow cultivar established on well-maintained ridges using a drip system. The ridges are not covered with plastic mulch because of the fast harvesting of the young baby marrows due to the good fertilization and irrigation program.

239 A new iron bark cultivar displayed at a farmer's day. The keeping quality is very good. This Hybrid fruit is uniform and a high yielder.

240 An overdue baby marrow cultivar. The fruit is harvested when young due to the fast development of the fruit. They are normally picked every second or third day.

241 In front, baby marrow fruit is shown to labourers to indicate to them what the correct picking stages are. Some of the fruit still have flowers attached to them.

27.1.1 PROTECTION AGAINST SUNBURN

Pumpkin fruit has dark green skin and is susceptible to sunburn. The plants foliage, mainly the leaves, protect the fruit from the sun by providing shade. Therefore, it is important to allow the foliage to grow as densely as possible until the fruit matures. Fruit could also be covered with branches, straw, chaff and even pumpkin leaves. This is however labour intensive and not effective in windy conditions nor economical on large production lands.

It is recommended to make use of good management practises and control diseases such as Powdery and Downy mildew. It is not possible to produce cucurbits without eradicating these diseases in your production land, especially during the later stages of production.

27.2 CROP PRODUCTION

See Chapter 26.5 for a quick step-by-step planning guide for crop production.

27.3 ESTIMATED PRODUCTION COSTS FOR CURCURBITS

Production costs differ across the country for items such as seed, seedlings, fertilizer, soil preparation, electricity/diesel, chemicals, irrigation, water, labour and so on. Therefore costs are only estimated and producers should do their own research about production costs of all the various aspects. The table below estimates the cost and profits for pumpkin/squash production (2014). Overhead costs are not included.

Commercial and small-scale farmers should make use of a budget and cash flow plan. The tables below will guide the producer and help him/her compile a financial cash flow projection and business plan. The tables also indicate how much capital may be needed for the entire growing period. Financial lenders may need a budget before they consider extending loan facilities.

The yields and possible profits of pumpkins as indicated in the tables below are determined on 40 tons p/Ha for the Flat White Boer and Green Hubbard cultivars. Experienced producers achieve yields of up to 60 tons p/Ha or more. *See photo 216.*

The costs are based on production in the land. The following costs are excluded: tractors, implements, general equipment, irrigation systems, water, water pumps, electricity, spraying chemicals, housing, toilets, telephones, computers, clothing, wheel barrows, watering cans, spades, forks, hoes, crates, twine, pegs and so on. Maintenance, costs of spraying, water pumps and certain labour costs are also not included in the table.

TABLE 20: ESTIMATED PRODUCTION COSTS FOR PUMPKINS (FLAT WHITE BOER AND GREEN HUBBARDS) (2014)

ITEM	COSTS P/HA
SOIL PREPARATION: Clean and level the chosen production land. Rip, plough, disk, till or rotovate if necessary.	R2 000
SOIL SAMPLES: Take at least 3 separate soil samples for the analysis test. Ask for a farmers package. Each representative sample costs approximately R300: 3 x R300 = R900. Analyze the results using a reliable fertilizer expert = R300.	R900 R300
FERTILIZER: Apply lime according to a soil sample for pumpkins (spaced 50 x 1.50cm). Broadcast fertilizer Suggest 1 200kg p/Ha 2.3.4 (30) Zn @ R290 per 50kg bag. 24 bags x 50kg = 24 x R290 = R6 960 (2013). 2 top dressings LAN/KAN @ 100kg p/Ha = 2 bags x 50kg @ R240 p/bag = R480. Mechanical application of fertilizer (tractor and spreader = R350).	R7 440 R350
METHOD TO PLANT SEED: Need 3.5kg seed @ R300 p/kg F1 hybrid **cucurbit cultivar seeds and seedlings are expensive**	R1 050
PLANT SEED: By hand in the prepared land at the required spacing (labour).	R600
CHEMICALS: Labour to control insects and diseases throughout the season. If soil is to be fumigated against nematodes then add R4 000.	R1 200
WEED CONTROL: Mechanical (2 times between rows) = R650 2 labourers control weeds in rows for approximately 3 days: 2 labourers x 3 days x R100 per day = R600.	R650 R600
IRRIGATION WATER: Pump water to irrigation land: Diesel/electricity and labour (tractors, trailer, transport, fixing and shifting irrigation pipes etc). The cost of water depends on rainfall and low rainfall can be costly if it occurs (1mm water equal 10 000 litre p/Ha).	R5 800
HARVESTING AND TRANSPORT: 1 labourer harvests, loads and transports fruit to pack store at 3 000kg p/day. For 40 tons of fruit, 14 labourers required @ R100 per day = R1 400.	R1 400
SORTING: Labour to cut stems and sort/pack in bags as required by the market.	R1 500
BAGS: 40 000kg = 8 pumpkins per bag = 5 000 bags @ R0.70 per bag.	R3 500
TRANSPORT: 40 tons to be transported to the markets. If 100km from the markets, a 7 ton truck needs to travel 5 times @ R5 p/km = 10 x 100km x R5 = R5 000. Labour for truck driver and assistant = R800.	R5 000 R800
Clean and plough what remains into the land, remove and store irrigation system and establish a cover crop if necessary.	R1 200
ESTIMATE PRODUCTION COST FOR PUMPKINS/SQUASH	R34 290

TABLE 21: YIELD AND POSSIBLE PROFIT FOR PUMPKINS

POSSIBLE YIELD: 45 Ton p/Ha minus 5 ton unmarketable fruit = 40 Ton marketable. Average weight of 3.5kg per pumpkin = 11 400 pumpkins @ R9 per pumpkin.	R106 600
MINUS PRODUCTION COST	R34 290
POSSIBLE PROFITS	R72 310

27.4 PRODUCTION GUIDELINES FOR CURCURBITS

27.4.1 SOIL REQUIREMENTS
Also see Chapter 3 Climate for temperature requirements.

Cucurbits grow best in well-drained, sandy types of loam soil. When an early harvest is required then producers should make use of virgin sandy soil. The reason for this is that sandy soil warm-up faster which has the advantage of causing the watermelons to ripen earlier. This allows them to be ready for the December holidays.

27.4.2 PH OF THE SOIL
Pumpkins and squash do not grow well in acid soil. The pH for cucurbits should be between 5.5 and 6.8. Watermelon and cucumbers tolerate more acidic soil.

27.4.3 CLIMATE REQUIREMENTS
Cucurbits prefer length in duration, warm and dry weather conditions. Areas with a high humidity during the growing season could be aggravated by fungal diseases, especially Powdery and Downy mildews. This occurs usually before the end of the growing season.

Water and muskmelons are more sensitive to this disease and are dependent on an efficient disease spraying program, especially in humid areas. Long, rainy periods before or during harvesting may cause the fruit to rot if the underside of the fruit makes contact with the soil. Heavier, clay-types of soil are particularly susceptible to this problem. Producers should plan their planting times carefully as cucurbits plants do not bloom properly during long periods of rain, cloudy or misty weather conditions. Also, be aware that bees do not usually work effectively in these conditions.

As mentioned above, when producing cucurbits in the summer rainfall areas, the increase in humidity may sometimes lead to the serious diseases. Pumpkins, squash and cucumbers are less sensitive to fungal diseases than watermelons and muskmelons. Early plantings in winter rainfall areas cause more vulnerability to leaf diseases due to the high humidity and low temperatures. To a large extent, diseases may mostly be avoided when planting early in the summer rainfall areas.

In South Africa, most cucurbits do not have ideal temperature conditions throughout their growing period. This is due to the fact that soil and air temperatures are either too low or too high during the 3 to 4 month growing period. Some producers in the summer-orientated regions, attempt to reach the markets in the early or late stages of the season, especially for water and muck melons. The prices are normally also much better over these periods. Cucurbits and water melons could be delivered to the markets early in the season, due to warmer winter production areas and warm sandy soil types. Some pumpkins, especially the iron bark types and squash such as butternuts, may be stored for extended periods.

The earliest planting time, is the time when the soil temperature rises to 16°C and there is no longer a danger of frost. When soil temperatures rise to 16°C to 18°C, it is advisable to plant pre-germinated seeds. Seeds are germinated in warm hessian bags for 6 to 7 days. Establish the pre-germinated seed in the warmer, northern side of the ridges which face from east, to west. You may also make use of seedlings in seed trays with large cavities, as pumpkins roots develop quickly in trays.

The most important factors when producing cucurbits are soil, air temperature and humidity.

Knowledge of environmental factors enables the producer to determine the following:
- It is unfavourable if cucurbits are produced in regions where the humidity is high in combination with hot conditions, as this would certainly trigger fungal diseases. On the other hand, cold conditions may also affect plant development and cause lower yields.
- Early and late planting dates should be investigated and proper planning should be done.
- Assessing the suitability of land to produce the crop, is important.

27.4.3.1 AIR TEMPERATURES
Air temperatures have a significant impact on plant development, especially when the plants are flowering. Temperatures may prevent or encourage fruit abortion. Cucurbits are sensitive to frost which may damage the broad leaves of the crop. At 0°C the broad leaves could be damaged permanently. The optimum temperature for growth and development is 18°C to 28°C. Low temperatures have a negative effect on plant pollination and fruit set.

Relatively low temperatures and short daylight periods promote the formation of more female flowers than male flowers. On the other hand, when temperatures and the daylight period increase, more male flowers develop. Some F1 hybrid cultivars are the exception to this rule. Low temperatures have a negative effect on the pollination of flowers and fruit set. Pollination may still occur at 15°C but above this temperature the chances for pollination are better.

When young plants are exposed to low temperatures at the beginning of the season as well as low night temperatures later in the season, then the maturing of fruit will be delayed. Pumpkins, squash, watermelons and cucumbers grow optimally at temperatures of 24 to 30°C and muskmelons 22 to 24°C. Growth virtually stops when temperatures fall below 12°C. Plants may be badly injured and maturity could be delayed if temperatures fall below 3°C for several days.

Muskmelons are a popular crop but are not easy to cultivate. The fruit develops the best flavour when cultivated in hot, dry climatic conditions. They also require a longer growing season than pumpkins in order to mature. Watermelon requires a long growing season and fairly high temperatures. Early F1 hybrid cultivars are available, especially for the sweet baby and seedless types.

27.4.3.2 SOIL TEMPERATURES
Soil temperature has an effect on germination and plant development, also influencing germination and root growth. The minimum temperature for field germination is approximately 16°C and the maximum 30°C. Pumpkin, squash and cucumber seeds do not germinate very well if the soil temperature is too low. Germination of seeds is poor below 15°C and even more so below 10°C.

The longer the seed takes to emerge, the poorer the stand would be. Do not establish the cucurbits until the danger of frost has passed and the soil is thoroughly warmed. Producers often establish cucurbits very early in the season, directly after winter, due to the fact that they wish to be early to the market. However, this may be fatal as an unexpected late cold snap may damage or kill the young seedling plants.

Seeds of pumpkins, cucumbers and squashsshould emerge in 6 days. In order to be early to the market, some producers produce watermelons seedlings in seed trays, in heat controlled units and transplant them in the land as early as they can. It is important to transplant the seedlings when the climate conditions are suitable. ***See Chapter 3 Climate for temperature requirements.*** Cucurbits seedlings should not be kept in seed trays for too long, due to the fast development of brittle roots. Delaying transplantation for a few days too long, may cause the seedlings' roots to become root bound in the cavities of the seed trays. It is therefore sometimes better to plant seeds directly in the land or use seed trays or pots with larger cavities.

242 A newly-bred F1 hybrid butternut cultivar displayed at a farmers day. Note the thick necks and small seed cavities at the blossom end. This gives the fruit more flesh. This cultivar is much larger than the common Waltham butternut cultivar.

243 A plot of different watermelon cultivars displayed at a farmers day. Old and new cultivars, as well as small, yellow, red, baby and seedless cultivars were displayed.

27.4.4 CULTIVARS

Cultivar choice should be a high priority, especially for the kind of pumpkin or squash that the consumer and market prefers. It is important that beginner producers should perform their market research before producing cucurbits. For example, the market prefers a green skin type of pumpkin instead of a white one. Many improved F1 hybrid cultivars are available, but the seed is more expensive than the open pollinated cultivars.

THE FOUR PUMPKIN AND SQUASH SPECIES:
1. Curcubita Maxima
2. Curcubita Pepo
3. Curcubita Moshata
4. Curcubita Mixta

27.4.4.1 CURCUBITA MAXIMA SPECIES

Hubbard's are sometimes wrongly referred to as squash. They are actually true pumpkins. The Flat White Boer and Queensland Blue cultivars are classified as pumpkins, but Green Hubbard and Golden Hubbard are also known as pumpkins, however, this is not correct. The Flat White Boer and Queensland Blue types are open cultivars.

F1 hybrid pumpkin/squash cultivars are available but are expensive. There are numerous F1 hybrid cultivars to choose from. Producers should seek advice and identify what the market and housewives prefer.

Some F1 hybrid pumpkin cultivars have long runners and most cultivars are bred in this way. It is important that the first female fruit is set as close as possible to the main stem (crown), as this makes a difference to the yield. When the first fruit set starts near the crown, then there is more open ground in order to procure fruit, closer to the main stem. When pumpkin runners begin to grow into each other between the rows, fruit set declines drastically and newly developed fruit is aborted as the runners rank into each other.

27.4.4.2 CURCUBITA PEPO SPECIES
27.4.4.2.1 SQUASH
Little Gem, Rolet, Table Queen, Yellow and White Custard, Caserta, Long White Bush and Long Green Bush among others.

27.4.4.2.2 HUBBARD AND BABY MARROW (ZUCCHINI)
Classified and known as squash. The group is divided into two species according to their growth habits. They are the stem and runner types. Baby marrow grows bushy and produces fruit on the stem and is harvested when the fruit is still very young. If the fruit is left untouched, it could easily reach a weight of 2kg or more.

27.4.4.2.3 CROOK-NECK SQUASH
Also known as yellow squash.

27.4.4.2.4 STRAIGHT-NECK SQUASH
This type is popular because the fruit is straight and is more easily packaged into containers.

27.4.4.2.5 SCALLOP SQUASH
Known as summer squash.

There are a number of baby marrow cultivars to choose from with different shapes and colours. Some are dark green, others deep yellow with white stripes or patterns. Usually baby marrows are harvested when the fruit is still young. It takes 4 to 5 weeks from transplanting healthy seedlings to harvesting them. The open cultivar Caserta, is a good yielder and its fruit is light green, with light stripes and it is fairly resistant to mildew. The white silvery spots on the leaves of Caserta and other Hybrids F1 cultivars are an indication that the plant is partially tolerant to powdery mildew.

Most markets prefer more cylindrical, dark green baby marrows. Usually Hybrids F1 cultivars with dark green colours tend to be less productive, but the markets and housewives have the final say.

27.4.4.3 CURCUBITA MOSHATA
27.4.4.3.1 BUTTERNUT
A great variety of butternut cultivars are available and they have different shapes, patterns, colours and sizes. Many of these new cultivars differ from the well-known and original Waltham cultivar.

27.4.4.4 CURCUBITA MIXTA
27.4.4.4.1 TART TYPES
Green, white and yellow tart cultivars are also harvested as small baby marrow and look nice when they are used as decoration among other vegetables. When the tart shape cultivars are left to mature, they become as large as 39cm in diameter.

27.4.4.5 OPEN POLLINATE CULTIVARS OR OPEN CULTIVARS
The cultivars mentioned below are older and may be outdated.

PUMPKINS
Flat White Boer A (medium size)
Flat White Boer B (smaller size)
Flat White Boer van Niekerk (very large)
Queensland Blue

SQUASH
Waltham
Caserta
Long green bush
Green/yellow Hubbard
Lemon squash-Rolet
Table Queen and Table King
Patty Pan (yellow/green/white)

FIELD CUCUMBERS
Special rust resistant
Victory
Cherokee
Sweet slice

WATERMELONS
Klondike
Charleston grey
Congo

MUSKMELON
Honeydew Green Flesh
Lion Jumbo
Sweet Delight

All sweet and Sugar doll etc.
Hales Best

Watermelon differ in size, shape and internal colour. Red, orange and yellow and also a number of small round Hybrid seedless cultivars are available.

27.4.4.6 GREEN HOUSE OR ENGLISH CUCUMBERS
Develop fast and thrive best in cooler climates.

FOUR TYPES OF CUCUMBERS ARE AVAILABLE
1. **English cucumbers:** Tunnel produced and picked when 30cm long.
2. **Israeli cucumbers:** Picked and packed the same way as baby marrows.
3. **Open-field cucumbers:** Fruit is short and does not need trellising.
4. **Jerkins cucumbers:** Small fruit mainly used for pickling or canning.

27.4.5 SOIL PREPARATION
See Chapter 8.2 Soil Preparation and Chapter 5.3 Soil Requirements.

27.4.5.1 PUMPKINS AND SQUASH
It is important to clean the land inside and outside. Do this at least 8 weeks before establishing the crop to control cutworms. Incorporate crop residues and non-seed weeds so the crop residues and weeds can decompose. This also prevents the cutworm moths and pumpkin flies from laying their eggs on the plants in the lands.

Previously, if there is a history of nematodes (eelworm) in the soil, it is important to fumigate the soil. Cucurbits are very sensitive to nematodes which multiply quickly on the juicy roots. However, most producers do not fumigate due to the wide spacing and expense of fumigating the lands. They prefer to use new lands. Some producers use the chemical, EDB to fumigate the soil 3 to 4 weeks before establishing the crop.

Pumpkins and squash can be established in all types of soil, from sandy to the heavier clay soil. For an early maturing crop, sandy or sandy loam soil is preferable. For a late crop when a large yield is important, consider a clay loam type of soil. Cucurbits should not be grown in soil that is shallower than 35cm. This is due to the fact that the highest root concentration of the crop is in the top 30cm. Pumpkin and watermelon require deep, drained, loose, sandy loam soil that is fertilizer balanced as root penetration is at least 45-60cm deep if the soil is cultivated properly.

27.4.5.2 CUCUMBERS
Soil for cucumbers should have a high water holding capacity and good drainage abilities.

27.4.5.3 WATERMELONS AND MUSK MELONS
Water and musk melons are mostly established in sandy to sandy loam soil. Loose texture and good drainage is preferable. If early production is required, then the soil should warm up rapidly in spring.

27.4.6 PLANTING TIMES
27.4.6.1 OPTIMAL PLANTING TIMES FOR CUCURBITS
In summer-orientated areas where increased humidity occurs, later planting dates may give rise to serious problems such as leaf disease, mosaic virus infections and insects. For example, pumpkin flies may increase rapidly later in the season. This can be avoided by planting early in the season.

It is recommended to plant early in the season in the Gauteng Highveld. Cucurbits should be planted before the end of November and in the Lowveld and Winter rainfall areas, from August to mid-September. The latest planting date is determined by the length of the growing season and the time when winter weather is on hand.

1. Highveld, Gauteng (Mpumalanga, North-West and Free State): In heavy frost areas between October and November and in warmer areas between September and October as well as February.
2. The Midveld: In hot summers between February and March.
3. The Lowveld: In cooler and low frost areas between January and March as well as July and August.
4. Free State, Northern Cape and Central Karoo: In mild frost areas between September and November.
5. KwaZulu-Natal Midlands: Between August and November.
6. The Western Cape South Coast: In winter rainfall areas between July and September.

Seed companies need to be consulted for the ideal planting times for certain F1 hybrid cultivars in the different areas of South Africa.

27.4.7 SEED AND THE QUANTITY OF SEEDS PER HA
Pumpkin seed quantities: 3 000-12 000 p/kg.
Small seeds: Butternut, Lemon squash, Baby marrows.
Medium-large seeds: Queensland Blue and Flat White Boer B.
Large seeds: Hubbards, van Niekerk and Ceylon pumpkins.

Squash seed quantities: 6 000-20 000 p/kg.
Make sure to use treated seed that inhibits fungal infections.
Pumpkin: Ceylon's and Hubbard's 3 to 4.5kg.
Squash: Stem and rank 1.5 to 2kg.
Cucumber: 1kg Seed and 0.5kg for seedlings.
Watermelon: Small seeds, baby types: 1kg and large seeds: 1.8kg.
Musk melons: 1 to 2kg for seed and 0.3 to 0.7kg seedlings.

Remember, seed size differs from cultivar to cultivar.

27.4.8 PLANTING METHODS
When planting Cucurbit seeds, usually 2 to 3 F1 hybrid and 4 open-pollinated seeds are planted in close proximity of one another, at the required distances within the rows. When the seedlings emerge, the weakest should be thinned by cutting them with scissors. Only the 2 strongest seedlings should be left to grow. Cucurbit seedlings established in soil cannot be transplanted. A poor stand may result, if at least one extra seed is not planted in each plant position.

27.4.8.1 METHODS OF ESTABLISHING CURCUBITS
The most suitable method for establishing cucurbits depends on the following:
- Area to be planted.
- Soil type.
- Available irrigation system.
- Method of fertilizer application.

The land should be treated to control nematodes (eelworm), as this crop is very sensitive to nematodes. After transplanting the Cucurbit groups or once the seed emerges, it is important to protect these plants against cutworms or considerable damage may occur.

When establishing in smaller areas, the seed is usually planted by hand. Larger areas are planted using a modified maize planter or other vegetable planter. If planters are going to be used, it is not practical to drill the seed plate of the planter so that three or four seeds are planted at a time. The use of closer, in row spacing is an option, where two seeds are planted together and eventually thinned to one.

The land should be irrigated before establishing the seed. When the full-soil capacity method is to be used (where the soil is irrigated to full capacity), be careful that the land is not too wet or muddy. Irrigation should not be applied until the plants have emerged, as otherwise a hard crust may form on the top of the soil that may hamper emergence of the seed. Be careful that the top soil never becomes too dry when practicing this method.

Another method is to slightly irrigate the land, sometimes 2 times a day. Do this especially when high temperatures are prevalent to keep the top layer of the soil soft and cool until the seed emerges. However, be aware that a hard crust may form if the top layer is kept too moist.

27.4.8.2 MULCHING AND DRIPPERS
The soil surface could be mulched during hot conditions. This prevents the soil from drying out. Mulching prevents poor seed germination and keeps the soil cool, especially when the seed is planted at a shallow depth. It also prevents the top soil from warming-up and drying-out. Mulching using plastic liners and a dripper system has many advantages and is a good option for some crops. ***See Chapter 8.7 Mulching.***

27.4.8.3 TRANSPLANTING SEEDLINGS IN THE PREPARED LAND
The majority of cucurbits are planted directly in the soil. Cucurbit seedlings should not be left in seedling trays for too long. It is a good option to use seed trays with bigger cavities when pumpkins and squash seedlings are to be established. Bigger cavities give the seedlings more time in the seed trays. If problems of bad weather occur at planting time, the seedlings may then be kept for longer in the seed trays.

The faster a cultivar matures, the more it is setback when using old seedlings. When transplanting old seedlings, a reduction in yield may be expected. This is due to the roots being bound. It takes time to settle the seedling plants after they are transplanted. Seedlings should not become root bound as this would seriously influence the yield, disease tolerance and maturity, among other factors.

Producers often supply F1 hybrid musk and watermelon seeds to the nursery as they establish a part of the production land using the expensive F1 hybrid seed tray seedlings and the other part with seed. Seedlings from seed trays have the advantage of allowing the crop to be harvested approximately 3 weeks earlier when compared with planting seeds directly in the land. Make sure to use hardened seedlings if they are to be transplanted during warm conditions. ***See Chapter 4.16.*** Verify that the soil is prepared well and is moist before establishing the seed. Seedlings in seed trays should also be moist before transplanting, especially during hot days.

When transplanting seedlings, create a hole with a flat iron, planting stick, small spade or the corner of a spade at the required spacing. Create the holes deep enough so the plug of the seedling is slightly covered with soil. The soil should then be compacted slightly around the plug to remove air around the plug roots.

When hot conditions occur, consider applying a little water to each seedling after transplanting. Use a hose or watering can. This should also eliminate air pockets from the roots and promote downward root growth. If seeds are planted too deeply, then poor field germination may result. Apply cutworm bait after transplanting the seedlings or after the seed has emerged. It is important for this crop as damage can be very costly.

27.4.8.4 PLANTING METHODS
27.4.8.4.1 LEVEL GROUND METHOD (PLANTING DIRECTLY IN THE SOIL)
The soil should be irrigated if it is dry. Examine the land and avoid hard layers, otherwise use a ripper to break the hard layers up. Some small-scale producers create large, deep holes, add compost or manure to these holes and then plant the seed or seedlings into these holes.

With the direct planting method, it is important to irrigate the land thoroughly before planting the seed. However, the soil should not be too wet, thereby causing labourers to be tramping in mud. Broadcast the fertilizer and incorporate it into the soil to a depth of about 20cm over a level surface without any further cultivation. This method is suitable on sandy soil using sprinkler irrigation and in regions that have low humidity.

Establish large seeds of pumpkin/squash 3 to 4cm deep and cucumbers, musk melons and watermelons 2 to 3cm deep. Watermelon might not do so well when sprinkler irrigation is used. If rain occurs before the seeds emerge, a crust may sometimes form, which may become hard when the soil dries out. Make sure to keep the top soil layer damp and soft. A light, overhead irrigation application should ensure that the seedlings emerge into damp, soft soil.

When seeds are planted directly in the soil, 2 or preferably 3 seeds (1 to 2 when using F1 hybrid seed) are established in close proximity to each other, at the required spacing distances. The weakest emerging seedlings should be thinned-out after 2 or 3 weeks.

27.4.8.4.2 BAND PLACE METHOD
The fertilizer is banded along the row and then incorporated. If rain does not occur during the growing season, no further operations are necessary. Drip irrigation with or without plastic mulch may be used.

When large amounts of fertilizer are to be applied, half should be broadcast and the remainder banded. A band of fertilizer can be applied underneath the seed, in furrows. Be careful that the seed does not make contact with the granular fertilizer. The granular fertilizer should be mixed with the soil. Bands of fertilizer could also be worked into the soil 15-20cm deep.

27.4.8.4.3 BED AND PLASTIC MULCH METHOD
The fertilizer is applied in bands and beds are created above these bands. The beds are then covered with black plastic mulching.

The width of the beds is determined by the plastic. If it is 1.5m wide, then a 1 meter bed is covered effectively. Next, producers may use a sharp implement such as a sharpened 40 or 50mm PVC pipe (or something similar) and punch holes at the required spacing. The seedlings are then established in these holes. The soil should not be too sandy, as otherwise, when punching the holes, the surrounding soil may fall in. The soil should therefore be moist enough, so that the soil grains bind and do not interfere when punching the holes. The dripper irrigation system is installed underneath the plastic liners.

Another advantage of the plastic mulch method is that the undersides of the fruit are less prone to rotting, as they are protected by the plastic sheets that they are lying on. Plastic sheets also restrict weed development.

Producers should verify that the amount of water passing through the drippers does not cause water logging and over-irrigation. Make sure to use UV (ultra violet) sunlight protected plastic cover strips. They are expensive but can be used repeatedly.

27.4.8.4.4 WIDE SPACING METHOD
When using wide row spacing, producers only prepare the soil in the planting rows and leave the rest of the ground undisturbed. This cuts costs as fertilizer and soil preparation is then limited.

27.4.8.4.5 FURROW FLOOD METHOD
The method is not recommended, especially on sandy soil where fast drainage may occur and cause water wastage. It is an advantage to use this method when rain is expected or in heavy rainfall areas. Furrows for the flood water could be made by using a ridging plough. Fertilizer could be broadcast or a band of fertilizer could be placed so that the ridging plough incorporates it into the soil. It is however difficult to use a planter under these circumstances. The seed is planted on the side of the furrow just above the water level. The developing vines or runners are trained over the side and away from the water level.

27.4.9 SPACING
Also see Chapter 11 Spacing of Crops.

There is no ideal plant population for cucurbits as many different spacing's may be used with the various kinds of cultivars which are available. Different factors such as competition between plants, fruit size, temperature, long days and different areas may affect the plant maturity, the duration of the harvesting season as well as the amount of fruit per plant. Producers should experiment with spacing and cultivars to suit their own needs.

Some producers plant cucurbits (ranking) in square blocks as it is more comfortable controlling the weeds using mechanical methods. Establish 3.5 to 4m square blocks in the planting rows and plant 2 plants per hole. It is a good option and gives excellent yields.

Climate plays an important role. For example, pumpkins can be spaced closer together in the winter areas (Lowveld). Winter production in the Lowveld is problematic due to viruses. Areas with high infection potential should be avoided.

The spacing below is a general guide.

	IN ROWS SPACING cm	BETWEEN ROWS cm
Pumpkins	50-90	150 -175
Hubbard's	50-90	150 -175
Squash (baby marrows)	30-40	110 -150
Squash Rank (butternuts)	50-60	140 -175
Water melons	80-90	90-190
Musk melons	30-50	130-180
Cucumbers	40-60	110-130

Spacing depends on the cultivar, market, irrigation system and climate. Usually ranking Cucurbit rows are 1.8 to 2.5m apart from each other. Establish between one and six plants per meter in the rows. When implementing closer spacing for ranking cultivars, it usually results in a slightly reduced fruit size. It is a more concentrated fruit set and normally no second fruit set takes place. Closer spacing may also have the disadvantage of leaves and vines getting damaged due to the trampling of the labourers during picking, weed control and the application of fungicides/insecticides. Such damage often leads to the deterioration of the runners and allows diseases and rotting to set-in.

Spacing between drippers in some soil types, may also affect the spacing of the plants themselves. The plants shouldn't be spaced too far from the dripper openings, especially in sandy soil. They should be approximately 15 to 20cm apart. In clay soil, if spacing is too close, the dripper irrigation may cause waterlogging and result in the plants developing poorly. When producing musk melons and implementing a 1m distance between drippers, it is better to plant four plants around each dripper (0.5 x 0.5m).

Dripper lines expand during the heat of the day and shrink during night time. This often causes the dripper line positions to shift around. To address this, dripper lines should be slightly tensioned by using strips of inner tube or wire hooks. The longer the dripper lines the more they will shift from their original position.

When close in-row spacing is practiced, the foliage of the runner types usually become very dense in the early stages of vine development. In humid areas, this increases the risk of diseases. Therefore, it is recommended to establish slightly lower densities to reduce the foliage growth. To do so, lower levels of N and less irrigation should be applied, especially when producing musk melons. However, in areas with hot, dry summers, it is an advantage to have abundant foliage.

27.4.9.1 DRIP IRRIGATION FOR MUSK MELONS

The spacing between drippers as well as the soil type will influence the spacing of plants. On sandy soil, the plants should not be placed too far from the drippers, approximately 15 to 20cm. In clay types of soil, if plants are placed too close to the drippers, they may waterlogging and develop poorly.

If dripper openings are 60cm apart, a spacing of 30cm in rows is recommended. Musk melons should be established 15cm either side of the dripper, but it is better to plant four plants around each dripper (0.5 x 0.5 m).

27.4.10 DURATION BETWEEN TRANSPLANTATION AND MATURATION

Most cucurbits require a 4 to 5 month growth period, from planting the seed, until the final harvesting. However, cucumbers and certain squash crops are harvested as early as 2.5 to 3.5 months. When harvesting the young baby marrow fruit, they are picked from 7 to 8 weeks after planting the seed. When establishing healthy seedlings, the first picking is usually only after 6 weeks.

Pumpkins and squash can be grown all across the warm, winter areas, however in these areas, virus infections may be problematic. During early spring and summer, cucurbits may be grown almost anywhere in the Country.

THE GROWING SEASON (FROM PLANTING TO FIRST HARVESTING) IS AS FOLLOWS:

	Warm		**Cool**
Squash	50	to	70 days
Squash (Butternut)	90	to	110 days
Pumpkins	110	to	140 days
Watermelons	100	to	130 days
Musk melons (Sweet)	110	to	130 days

Temperature influences the length of the growing season. Early and late production will remain on the land for longer periods. The maturation time of New F1 hybrid cultivars is also normally shorter.

27.4.11 FERTILIZING
Also see Chapter 15 Fertilizer.

If squash are over-fertilized, it promotes leaf and stem growth at the expense of fruit development. When producing plants under plastic layers, the amount of nitrogen N for top or side dressing may be reduced as leaching is minimized. Nitrogen N soluble fertilizer may be applied through drippers and overhead irrigation. Side dressings could be applied alongside the plastic mulch. Less N fertilizer should be applied if legume crops were ploughed into the soil previously. Excess nitrogen N encourages a dense foliar canopy that may reduce the number of female flowers. However, a dense canopy is preferable as it protects the green fruit cultivars from sun burn.

If a soil analysis is not available the following recommendations may serve as a guideline: 25kg N, 50kg P and 50kg K p/Ha should be applied just before planting time and then be followed by a top dressing of 30kg N at the four-leaf stage of the young plants and also about 3 weeks after emergence. An additional top dressing of 30kg N and 50kg K p/Ha may be applied at 5 weeks after emergence or when the first male flowers appear. The application norm for Zucchinis and gem squash is 35kg N, 55kg P and 75kg K.

If a large quantity of fertilizer needs to be applied, rather broadcast it when the plants are spaced closely. Smaller quantities of fertilizer are used when the fertilizer is incorporated in or near the planting holes or furrows. When large yields are expected, a combination of broadcasting and strip application may be suitable. Half is broadcast and the rest applied in the rows and before planting the seed.

Some Hybrids F1 pumpkin cultivars are strong and vigorous growers, but it is important not to over fertilize, especially with N applications. N normally increases vegetative growth and a reduction in flower drop and fruit set may take place. If heavy rainfall occurs for long periods, the soluble N fertilizer drains away and should be replaced. Also, it is of little use applying N after the young female fruit have appeared. Foliar spray applications are a good option as the broad leaves of cucurbits take up the foliar applications very effectively.

Cucumbers and squash are fast-growing crops and should be well supplied with nutrients. The upper layers of the soil should be well-prepared and fertilized to obtain the best results. It is also important to verify that the calcium Ca level in the soil is sufficient. Ca is responsible for fruit firmness, better shelf-life and less rotting of fruit.

The question always arises as to whether fertilizer should be broadcast or applied in the planting holes or directly in furrows. For gem squash and baby marrow's that are spaced closely together, broadcasting is recommended. In the case of pumpkins, watermelons and musk melons which are spaced further apart from each other, strip or band applications is a good option. When producing crops in planting holes, certain small-scale producers apply fertilizer and kraal manure (or good compost) in the planting holes. This saves fertilizer and money. A good option when planting in furrows, is to apply half the fertilizer recommendation in the planting furrows and to broadcast the remaining half. When it is known that a high percentage of phosphates P will be "fixed" in the soil, broadcasting of P is not recommended.

Ca and Mg are important during fruit expansion, especially for musk melons. Weekly foliar sprays of these two elements during the early stages of plant development will ensure the best quality fruit.

27.4.11.1 MUSK MELONS
Fertilizing is crucial when producing melons. More often than not, lack of fertilizer is responsible for failure.

When a melon plant begins to set fruit, it has taken up less than 1% of the total N, P and K. This is the point at which the last N should be provided. Nutrient uptake reaches a peak and then declines fast. In the case of N and Ca the peak is reached approximately a month before harvesting, while K uptake is reached at the beginning of the crop harvesting period. At the time of peak nutrient uptake, as much as 30kg N and 70kg K is taken-up, per week. It is however difficult to maintain this level of nutrients through the dripper system. Fertilizer programs through dripper lines should aim to have surplus nutrients in the root zone when peak demand from the plants is anticipated. During this period, irrigation should be optimum.

Nutrients such as Ca can only be taken-up by the fine root hairs of plants. Any moisture stress or boron B deficiency will lead to root die-back and limited nutrient uptake. Foliar sprays using boron B and molybdenum Mo are important and a shortage at the early stage of plant development would lead to poor fruit set. Too much N would result in huge healthy vines with little fruit.

The best time to apply top dressing applications is when the vines start to run. It is important that the vines should grow vigorously. This can be done by supplying fertilizer to the soil or by using foliar sprays. Foliar applications are only effective if the young fruit is sprayed directly using Ca applications. If foliar sprays are applied at the correct stage of the fruit and plant development, it should prove very effective.

27.4.11.2 MICRO NUTRIENTS AND FOLIAR SPRAYS
It is not always possible to fulfil micro-nutrient requirements by spraying them on the leaves. As a general rule, it is better to correct all micro-nutrient deficiencies in the soil before stabilising the crop. However, this is not always possible as the chemical properties of the soil may influence the uptake of the micro-nutrient. Therefore, the only solution is to apply foliar feeding sprays to correct the situation.

27.4.11.3 TOP DRESSING FOR CURCURBITS
The first top dressing application of nitrogen N makes it possible for the plants to get off to a good start. It is important to monitor the colour and size of the leaves. They should be healthy and fairly large. If they do not look healthy and large according to expectations, then it is a possible that they have not received enough water. This assumes that there are no nematodes (eelworm) in the soil. Nematodes can cause wilting, discolouring and smaller leaves. Experienced producers are aware of this.

Nitrogen N leaches quickly when too much water is supplied. Irrigation should be in 1 or 2 mild doses. Usually a single top dressing of 100 to 150kg per Ha LAN/KAN that is broadcast, is sufficient. This should be applied 3 to 4 weeks after emergence. On sandy soil, sometimes 3 smaller top dressings are necessary. Pumpkins respond well to compost and manure and it is usually supplied in strip applications.

27.4.11.4 MOLYBDENUM DEFICIENCY

Pumpkins and squash are very sensitive to a shortage of Molybdenum. Producers should ensure this does not happen. Molybdenum is necessary in small quantities and is likely to occur in acid soil. The deficiency shows as light or yellow patches between the leaves, veins and toward the edges of leaves. Leaves become smaller and are sometimes badly affected. Symptoms begin later in the plants development stage. If the seedlings are treated with a sodium-misdate foliar spray, this condition may be avoided.

27.4.12 IRRIGATION
Also see Chapter 14 for Irrigation.

Cucurbits may be irrigated using overhead, drip or furrow irrigation. The amount of irrigation water for cucurbits should be reduced towards the ripening stage of the fruit. Too much water as the fruit begin to ripen, may cause internal rot, lower sugar content and cause cracking.

Wet conditions in clayish soil should be avoided and therefore it is important to schedule irrigation carefully. Over-irrigation in clay and loam soil may wash the calcium Ca out of the soil causing blossom end rot (BER). This may also lead to heavy foliage vine growth that promotes botrytis, root diseases, larger fruit and less internal fruit sugar. Irrigation is acceptable but this leads to smaller plants/fruit and earlier fruit set.

Light irrigation should be regularly provided during the early seedling stages. This is because the plant has little soil-water surrounding its roots and irrigation can be less frequent so as to allow deeper penetration of water in the soil. Avoid applying more water than the soil can hold at one time. The soil-water holding capacity in light textured soil should be about 60%. For clay types of soil, irrigate frequently and apply less water after each irrigation cycle. Clay soil tend to waterlog easily and needs to dry-out between irrigation cycles to allow more oxygen to the root system.

27.4.12.1 IRRIGATING PUMPKINS AND WATERMELONS
430mm water is the norm throughout the growing season. Apply 25mm the first week and there after, 30mm every week. Musk and watermelons require small amounts of water before the runner starts to develop due to the limited leaf area. If the soil is irrigated thoroughly before planting, then no early irrigation will be required, except on sandy soil.

Some producers claim that if they reduce the irrigation water after the first female flowers emerge, then this period of stress will ensure more female flowers and more fruit will result. This seems logical and once the fruit has set then normal water management may be resumed. It is effective to reduce irrigation when producing pumpkins, especially butternuts.

27.4.12.2 IRRIGATING SQUASH
During the entire growing season, apply 320mm water (25mm the first week and 30mm every week thereafter). If water is limited, the plants will wilt due to water stress.

27.4.12.3 IRRIGATING MELONS
Because of their desert origin, melons have a vigorous root system. Most of the water is taken up by the shallow roots which develop just below the soil surface. The root system is very efficient and makes use of the lightest rainfall. When using sprinklers, the roots spread at the same rate as the vines. Therefore, when correctly irrigated the plants will take up a lot of water and nutrients.

27.4.12.4 FURROW FLOOD IRRIGATION
Furrows are drawn at the required spacing by using a suitable implement. Fertilizer is to be applied in the

furrows and should be slightly mixed into the soil using a toothed implement. Fertilizer could also be broadcast over the land. The furrows are irrigated by using flood irrigation. When the water in the furrows has drained properly, the seeds or plants are established just above the irrigation water level.

27.4.12.5 DRIP IRRIGATION FOR MELONS

Problems with melons are due to incorrect application of water as well as incorrect spacing between the drippers. Over-irrigation leads to large, oxygen deprived zones in the soil. Very little root development takes place as a result. Calcium (Ca) is not easily available in soil and over-irrigation can lead to Ca deficiencies, except for certain plant development stages such as fruit-set. It is usually better to under-irrigate than to over-irrigate melons. However, under-irrigation tends to lead to smaller plants and earlier ripening of the fruit.

27.4.13 CROP ROTATION
Also see Chapter 18 Crop Rotation and Soil Improvement.

Wait at least 3 years before planting any cucurbits crops in the same soil. Soil fungi such as Fusarium can result in total crop failure. If possible this period should be extended to more than 3 years. Cucurbits may follow any unrelated vegetable in a rotational system except for potatoes, as the Fusarium species attacks both cucurbit and potatoes.

27.4.14 WEED CONTROL
Also see Chapter 17 for weed control.

Weed control should be practised in cucurbit lands when the runners (of the runner types) are still small. This is to avoid trampling and damaging the runners. After a heavy rain storm, it is sometimes necessary to break the hard crust of the top layer the soil as the soil dries-out. Keep out of the production land to avoid damaging the plants and also compacting of the soil. Cultivation should be halted when the plants are large enough or if they obstruct the passage of the implements.

Do not control weeds when the plants are wet as this may promote diseases. Cucurbits roots are very swallow and develop rapidly even on the sides. Their roots are mostly in the top 20cm of the top soil. Research has indicated that the roots spread in a loose type of soil almost as fast as the vines or runners. Because of this, weed control should be practiced at a very shallow level.

Mechanical weed control should be halted 5 to 6 weeks after transplantation or after the seed has emerged. After this period, the vines of the runner types develop very quickly and will cover the remaining open areas. This rapid vine growth will then control the remaining weeds. When planning, row-width spacing makes a difference, as producers must take their implements into consideration. A slight adjustment could make mechanical weed control a more comfortable and easier task. Some producers establish cucurbits in square blocks and then control weeds using mechanical implements along and across the area. This makes it easier and faster to control the weeds.

Planting within 3.5 to 4 meter squares and leaving 2 plants per hole gives excellent yields. No chemical herbicides are currently registered for cucurbits. This may change in the future.

27.4.15 WIND DAMAGE
Pumpkins and squash have very thin skins when they are still young. In windy areas, the vines and leaves may scratch and damage the young fruit, leaving marks on them. These permanent marks affect the price on the market, as they expand and later becomes brown. This spoils the appearance of the fruit. Some producers establish windbreaks such as Napier grass to reduce wind force.

27.4.16 POLLINATION AND ABORTION OF FRUIT
Pumpkin, squash, watermelons and musk melons are cross-pollinated. They are pollinated by wind and insects,

but bees are the primary pollinators. Cucurbit flowers open early in the morning and close at sun set. Pollination must take place before the flowers close. Large fruit develops only if enough pollen lands on the female flower stigma. If too little pollen lands on the stigma, the young fruit is aborted.

Fruit set could be very low when windy conditions occur. This is because the pollen may dry out and not germinate, especially if hot, dry winds occur over long periods. It is advisable to plan planting times to avoid this. Fruit set may also be affected when cool, rainy conditions occur, as bee activity is normally reduced. Temperatures are important for effective pollination of the crop and when day temperatures fall below 12°C for long periods then lower yields can be expected.

Natural abortion or fruit drop of pumpkin female flowers is normal. Usually 80% (including young fruit) are aborted. It is lower for cucumbers and squash. This condition is normal and occurs at the bud stage as the plant cannot mature all of the flowering fruit and therefore has to reject them. The load that a plant it can handle is normally 2 to 2.5 pumpkin fruit per plant. Other factors such as plant stress, disease (especially nematodes (eelworms)), temperature, wind storms and unpollinated female flowers may also cause the plant to lose too many female flowers.

Bees are necessary to pollinate crops. The producer should monitor the plants to see if there are enough bees visiting the plants flowers. Most bees are killed due to insecticides. Therefore, when using insecticide, it should be applied in the late afternoon or early morning, as these are the times that most bees are absent. Inadequate pollination may lead to an abortion of fruit and also malformed fruit. This is due to poor seed set, causing the fruit to become malformed.

According to specialists', bees should visit a single pumpkin female flower at least 20 times to pollinate the fruit successfully. Producers usually place 2 bee hives p/Ha, 150m from the production land for best results. Bees are sensitive to temperature and humidity. They do not work well when temperatures are 9°C or lower and when humidity is very low. Bees are attracted to any other flowering plants near the production land. Make sure that no other flowering crops or trees are nearby.

Seedless F1 hybrid watermelon cultivars need pollinator plants (male plants) to pollinate the diploid (female plants). Usually every 5th plant in the row (or next to the 5th plant) is adequate. Diploid fruit bearing cultivars can also be used as a pollinator for this purpose. The male plants are established a week later to ensure maximum pollination.

27.4.17 YIELDS

F1 hybrid cultivars are to be reckoned with, because leading Companies are affiliated to overseas Companies that spend enormous amounts of money providing the best cultivars for all producers. This applies especially in the breeding of tomatoes and cucurbits. This kind of research is expensive and is aimed to benefit all producers. Leaders in this research are Sakata May Ford, Hygrotech and Asgrow. On Farmers days, some of these leading Companies make an appearance.

POSSIBLE YIELDS

Pumpkins	35-45 ton p/Ha
Squash butternut	25-30 ton p/Ha
Squash baby marrow	20-25 ton p/Ha
Cucumber	25-30 ton p/Ha
Muskmelon	25-35 ton p/Ha
Watermelon	30-50 ton p/Ha

See photo 236. Iron bark F1 hybrid cultivar that yields 80 ton p/Ha.

As a general rule, the larger the fruit sizes are, the less fruit can be expected per plant. The larger pumpkins types usually produce only 1 to 2 mature fruit per plant.

27.4.18 HARVESTING AND STORAGE
Also see Chapter 24 Harvesting, Handling and Packaging.

27.4.18.1 PUMPKINS AND SQUASH
Pumpkins and some squash should be well matured before they are stored. They should not be harvested before the skin hardens. Any dry area is suitable for the storage of pumpkin and certain squash. They keep best on shelves where they can be placed next to each other with a small space between each fruit. The fruit should be cured (ripened) by keeping them inside a room at room temperature for a week or two. This also improves the shelf life of the fruit. Curing hardens the skin, heals wounds, reduces water content and improves the sweetness of the fruit.

Butternut and the iron bark types have a very long shelf-life and are less prone to diseases when compared with most other cucurbits. They are a good option for late summer production and may be stored until spring. When storing pumpkins, they should be inspected regularly and fruit that shows any sign of rotting, should be disposed of immediately, otherwise they can infect the surrounding fruit, sometimes very quickly.

Flat White Boer, Iron bark, Green and Orange Hubbard's may be stored after harvesting. The fruit should however be left on the land until the skin becomes hardened. Irrigation should be stopped when the skin starts to harden. Fruit of all cucurbits should be cut with a sharp knife, leaving a short stem on the fruit. Otherwise, when the stem is broken off directly from the fruit, it creates a large wound that may cause rotting, when in storage.

Green Hubbard's are usually harvested when the fruit still looks attractive and green. Do not store the fruit for too long a period.

Flat White Boer is normally stored and cured in the field. Be careful not to store fruit for too long a period, as it may change the white colour of the fruit to pale yellow. During the curing period, pumpkin should not be exposed to severe frost in the land, or the fruit may be chilled and damaged. Light frost however, should not be a problem. Butternut and iron bark pumpkins can be stored for up to 5 months or longer.

27.4.18.2 BUTTERNUTS
They could be harvested when they are small and then packed as baby marrows. If it is done this way, the plants will produce more fruit but this requires a fine balance. Producers should stop harvesting at a certain point to allow the plant to recover or else the plant may not be able to replace the loss and a poor yields may result.

27.4.18.3 MARROWS AND SQUASH
Baby marrows are ready to be harvested then their skin is soft and easy to scratch with a fingernail. Care should be taken not to damage their tender skins. Fruit should be harvested by cutting the stem with a sharp knife or by twisting it from the plant. Baby marrows are harvested when they are 10 to 12cm long, in-line with the market requirements and the packaging material.

When harvesting baby marrows, be aware the fruit develops very quickly. The young fruit is harvested as soon as 2 months after establishing the seed. When transplanting healthy seedlings, it takes only six weeks for the first harvest when the temperature is optimal and favourable conditions exist.

Harvesting old fashioned baby marrow cultivars, may bother the labourers due to the hairy, thorny nature of the stems which may puncture the skin of the labourer's arms. Cultivars are available that have very soft hair or exhibit little leaf stem hair.

27.4.18.4 CUCUMBERS
Open-field cucumbers are harvested when they are green before the skin changes its colour. A certain amount of experience is required to determine the correct stage of harvesting. The fruit nearer the stem will be harvested first. Size and smoothness are signs of the correct harvesting stage.

Cucumbers are stored for only short periods, normally 7 to 10 days. English cucumbers produced in tunnels are wrapped in a thin layer of plastic. This keeps them fresh for much longer, especially in cold rooms.

27.4.18.5 WATERMELON
Watermelons do not ripen after harvesting. Make sure that they are properly ripe before harvesting. When the fruit is left too long in the field they may become overripe and the skin changes its colour and becomes tinted with a yellowish colour. Overripe fruit are normally not crispy, have a mealie texture and are not juicy. This is undesirable. Practical experience is required to determine the optimum harvesting stage.

The following methods may assist producers to test the ripening stage of watermelon fruit:
- A cracking sound may be heard when pressure is applied to a ripe fruit. To test this, place the watermelon on your head and squeeze it.
- A method commonly used is to lift the fruit and tap it with your knuckles and when a dull sound is noted this indicates the fruit is ripe. A solid sound indicates immaturity.
- Cutting a small plug of flesh is a sure test but the fruit is then damaged.
- The skin of a ripe watermelon can be scratched easily with your fingernail.
- Place a watermelon level on the soil or on a table. Use a plastic straw and place it diagonally on top of the watermelon fruit. If the straw turns 180 degrees laterally on the fruit, then this indicates that the fruit is ripe. When the straw turns only a little bit or half way or not at all, the fruit is still green. Nobody can explain this.

Watermelon fruit is not stored under cool conditions and is rarely stored longer than 2-3 weeks.

27.4.18.6 MUSK MELON
When melon fruit reaches maturity, a slight crack develops around the stem where it is attached to the fruit. When the crack completely encircles where the stem is attached to the fruit, then some melon cultivars (especially the yellow types) will be ripe and ready to be harvested. Some Honeydew types do not exhibit this behaviour. Experienced producers are familiar with this.

The yellow melon types can be stored for 3 to 4 weeks and the green ones longer.

27.4.19 INSECTS
It is recommended to use systemic pesticides while the plants are flowering. Preferably spray in the afternoon or when cool and cloudy weather conditions occur as bees visit flowers in the morning and seldom in the afternoon.

Monitor production lands regularly. Pay attention to pumpkin flies that sting the fruit and lay eggs underneath the soft skin. This will damage the young fruit.

For more information about insects, read the following sections:
INSECTS: *See Chapter 20.5.*
ROOT-KNOT NEMATODE: *See Chapter 20.5.2.1.*
APHIDS: *See Chapter 20.5.2.3.*
LEAF MINER: *See Chapter 20.5.1.10.*
RED SPIDER MITE: *Chapter 20.5.2.9.*
BOLWORM: *See Chapter 20.5.2.7.*
LADYBIRDS: *See Chapter 20.5.2.13.*
CUTWORMS: *See Chapter 20.5.2.2.*
CMR BEATLES: *See Chapter 20.5.2.16.*
TRIPS: *See Chapter 20.5.2.4.*

Trips are neglected by many producers as the small insect hides inside the huge flowers and consumes the pollen inside the flowers. This may lead to excessive abortion of small, young fruit. This small insect is not visible if one does not inspect the flowers regularly. The easiest method for identifying trips is to inspect the male flowers by pulling them apart. Trips may reduce fruit set by consuming the pollen and by damaging young female flowers. They scratch the young fruit by rasping the surface with their mouth parts. These marks will then enlarge as the fruit become larger. Several chemicals are registered to help control trips, but these should be selected with care. This is because trips are most prevalent during the fruit set period.

PUMPKIN FLIES
See Chapter 20.5.2.11.

This brownish fly is a small wasp characterized by yellow spots on its body. Pumpkin flies lay their eggs underneath the soft skin of the young fruit. The eggs hatch and white maggots start to grow inside the young fruit. If the eggs do not hatch inside the fruit, the mark from the sting remains and makes a dent that will be visible at a later stage as a dry spot on the fruit. Therefore, the flies should be controlled before they are able to sting the young fruit.

Some producers establish cucurbits as early as possible in the summer orientated areas to avoid the high population of flies later in the season. A common way to control flies is to spatter the plants with sugar using poison bait. They splatter the solution using a bag but this practice is undesirable because it also kills bees that are attracted to the sugar.

27.4.20 DISEASES
Also see Chapter 20.6.

The success of production depends on an effective spraying program. Diseases may be prevented to a large extent by the following factors:
- Use disease-free seed and rotate crops.
- Implement good drainage and perform regular inspections.
- Implement early planting times.

DAMPING OFF
See Chapter 20.6.3.1.

FUSARIUM WILT
See Chapter 20.6.3.11.

PHYTOPHTHORA
See Chapter 20.6.3.16.

Do not underestimate this disease. It often appears when the plant dies back and the fruit rots, usually in the later summer months.

POWDERY MILDEW
Also see Chapter 20.6.3.5 Powdery Mildew.

Develops during dry conditions and may cause a great deal of damage. Control should begin as soon as white powdery spots appear on the lower side of the leaves. The disease then spreads to the top surfaces of the stems. Some producers use Sulphur. This is the cheapest way providing that the sulphur is applied during the cooler part of the day when the plants are dry otherwise the plants may burn when hot conditions occur.

DOWNY MILDEW
Also see Chapter 20.6.3.6 Downy Mildew.

Usually develops when cooler weather conditions occur. Control should begin as soon as wet weather conditions occur and should not be delayed. Some producers use Dithane M45 on a weekly basis until the weather clears.

ANTRACNOSES
Brown concentric rings develop on the fruit and later black rings develop within the brown concentric rings. This is one of the fungal diseases that attack the fruit when they are inside stores. It develops very rapidly once it has started.

FUSARIUM ROT
One of the most common fruit diseases which is caused by fungis. Many large, bleached, brown, oval spots develop and may produce a sticky droplet when the fungi penetrates the fruit around scars or damaged areas.

27.4.21 VIRUSES
Also see Chapter 20.8 Viruses.

MOSAIC VIRUS
Also see Chapter 20.8.2.1 Tomato Mosaic Virus.

The symptoms begin on the younger leaves which become uneven, begin to curl and remain small and underdeveloped. The fruit of the infected plants have a knobby appearance and usually remain small and exhibit a discharge. The virus is mainly spread by aphids and to a lesser degree by white flies. The damage caused by this virus is severe in areas where cucurbits are grown throughout the year. The virus spreads from the older to younger plants. It is sometimes not worthwhile to produce in winter in the warmer Lowveld areas.

VERTICILIUM WILT
This viral disease is soil born and infects the entire plant. It prefers to attack watermelon plants. The plant wilts as if suffering from drought stress.

244 *Fruit disorder. Tapering of cucumber fruit is a problem in Greenhouses.*

245 *A baby marrow fruit ready to be picked. They are usually picked when they are 10 to 12cm long. Some fruit still have the flowers on. The longer dark green cultivars are popular. This is a quick cash crop because the first harvesting is only 4 weeks after transplanting the seedlings.*

27.4.22 PHYSICAL DISORDERS
ABNORMAL OR MALFORMED FRUIT
Also see Chapter 20.9.1.10 Malformed Fruit.

This happens when poor pollination, a lack of moisture, water stress or a N deficiency takes place. If abnormal fruit is removed at an early stage, it allows new fruit to develop again.

CRACKING OF FRUIT
A thin outer skin and high sugar content of fruit leads to the cracking of fruit. The chances of cracking increases when warm soil conditions occur. Increased transpiration takes place, the water uptake becomes too high inside the fruit, which leads to cracking.

Water may build-up in the fruit of some cultivars that contain high sugar levels. The high water content in the fruit lowers sugar content and this causes cracking, especially of watermelons and musk melons. Too much water also causes the fruit bearing crop cells to expand to such an extent that the fruit might crack. Sometimes cracks begin to develop when the fruit are left too long on the vine after they have already matured. It may be a good idea to cut the first maturing fruit from the vines of the ranking types of pumpkins and leave them in the land until all the other fruit is ready. This may be beneficial to the growth of the other immature fruit.

SUN BURN OR SUNSCALD
The green or dark skin pumpkin cultivars are vulnerable to sunburn. The upper sides of the fruit turn white when they begin to mature. This happens because of sun burn and when the plant foliage is poor. The result is that the fruit begins to dry out, especially at an early stage of the plant development.

MALFORMED FRUIT
The fruit develops only on one side due to incomplete pollination. This also occurs due to a lack of moisture, fertilizer deficiency or insects.

ABORTION OF FRUIT
Abortion of young fruit happens naturally as the plants begins to settle and as a result only a few fruit will mature. For instance, pumpkins will only mature 1 to 2 fruit per plant. The rest of the female fruit are aborted at an early stage. Sometimes older fruit are aborted due to high temperatures, pumpkin flies and a lack of pollination.

CHAPTER 28

PRODUCTION GUIDELINES FOR CABBAGE, CAULIFLOWER, BROCCOLI, BRUSSELS SPROUTS, TURNIPS AND CHINESE CABBAGE

246 Various cole crops. Note the different cabbage head sizes and the strong healthy root systems. Note also the good quality and ideal leaf cover of the cauliflower plant. At the back, medium-sized cabbage heads. Plant population is 30 000 p/Ha and for baby cabbages 80 000 to 90 000 plants p/Ha. The cauliflower curds are large, as is the broccoli curd. Far left, is a F1 hybrid cultivar which is very large and firm. The cauliflower curds are excellent quality and are a brilliant white.

28.1 INTRODUCTION

The term "cole crops" is also known as the brassicas group. Cabbage is widely grown and fairly easy to produce. Cole crops which include cabbage, broccoli, cauliflower, brussels sprouts, turnips and chinese cabbage are hardy in cold conditions (except some chinese leafy cultivars) and can be produced in all areas of South Africa.

Cole crops are susceptible to pests and diseases, especially in the summer production areas or when established in humid areas. Insects and diseases multiply rapidly and are more active in warm summer conditions. As a result, many producers fail when producing summer cabbage, broccoli and cauliflower. Proper scouting (to identify insects/diseases) and a spraying program should be followed when summer production of cabbage, cauliflower and broccoli is considered.

The early and mid-season cultivars are better suited for fresh market sales when smaller heads of 1.4 to 2kg are desired. A number of excellent cauliflower and broccoli F1 hybrid cultivars are available. They range in maturity from 55 to 95 days for cauliflower and 55 to 75 days for broccoli. Take note that the longer the growing season of a cultivar is, the larger the heads will be, especially for cauliflower and cabbage if they have been established in clay type soil. *See photo 254.* Cauliflower is relatively difficult to grow due to the requirement of an optimal planting time in comparison with broccoli and cabbage.

28.2 ESTIMATED PRODUCTION COSTS AND PROFITS FOR COLE CROPS (CABBAGE)

The main questions a beginner producer or small-scale farmer wants answered is what the production cost and yields will be and how much profit can be generated. This is a common question, as commercial farming is all about profit and therefore the producer should design a cash flow and business plan.

Production costs differ all over the country due to the variable costs related to, for example, seed, seedlings, fertilizer, electricity/diesel, irrigation water, labour etc. Therefore, the costs should be adapted accordingly.

The tables below are an estimate of the costs, yields and profits for cabbage production (2014-15). Producers should also do their own research concerning production costs. Overhead costs are not included in the tables. The cost of producing 1 Ha of cabbage heads is indicated in the table and is based on an average yield of 60 Ton p/Ha. Experienced producers yield up to 80 ton p/Ha or more. 60 ton (or 33 000 medium sizes cabbage heads weighing approximately 2.5kg). An estimated 3 000 unmarketable heads should be deducted which leaves a massive quantity of heads to manage.

When small-scale producers establish cabbage in smaller quantities they should find a market for their overproduction. It is often not possible to market all the heads locally while it is not always worthwhile to transport huge amounts of heads to a market that is far away. *See Chapter 2 Planning and Record Keeping.* Additional labour is necessary during peak times such as for transplanting seedlings, weed control, harvesting, sorting, packaging and marketing.

Due to the effort of handling large numbers of cabbage heads at the same time, producers usually establish cultivars as follows:
- Establish the crop during different planting times/stages. Seedlings may be established every 3 to 4 weeks.
- Establish cultivars with different harvesting times. Early, medium or late cultivars can be established at the same time and therefore harvesting is extended over a longer period.

Producers should know what the plant populations are of the different sizes of cabbage heads, especially for baby cabbage. Plant populations are between 80 000 to 90 000 for small/mini cabbages. Cauliflower plants are usually 100 000 plants p/Ha except for the medium size heads, which are established at 30 000 plants p/Ha. If the producer can identify a market for baby cabbages, it may be a profitable option. Ethnic African people prefer larger size cabbage heads, while Caucasians prefer the smaller, medium size heads.

The cost of tractors, implements, general equipment (wheelbarrows, spades, forks, hoes, crates, twine, pegs etc), irrigation systems/equipment, water, water pumps, electricity, chemicals, housing, toilets, telephone, computer costs, overalls, clothing for spraying chemicals and so on, are not indicated in the table below. Neither are maintenance costs of tractors, implements, spraying equipment and tools. Some labour costs are also not included in the table below.

A budget and cash flow plan should be prepared long before establishing the crop. This will assist producers to manage their finances throughout the planting period as well as determine the amount of capital and credit required.

The table below will assist the producer to design a cash flow, business plan and a budget for cabbage. Financial lenders may require a budget and cash flow plan before they consider lending money.

PART 2: GUIDELINES FOR CROP-SPECIFIC VEGETABLE PRODUCTION

TABLE 22: ESTIMATED COSTS AND PROFITS FOR CABBAGE PRODUCTION (2014)

ITEM	COSTS P/HA
Clean and level the production land.	R900
Soil preparation: Rip, plough, disk and rotovate the production land.	R1 900
Soil analysis: A Farmers package is R300 per sample (2014). Interpretation of the analysis by a specialist. At least 3 overall samples p/Ha = R900. Interpretation of the soil samples = R200 per sample x 3 = R600.	R900 R600
Fertilizer: Apply the correct type of lime if the pH is too low according to a soil analysis. Broadcast fertilizer: 1 200kg p/Ha 2.3.2 (22) @ R290 per 50kg bag. Require 24 bags x 50kg p/bag = 24 x R290 = R6 960. 2 Top dressings LAN/KAN @ 150kg p/Ha = 6 bags x 50kg @ R240 p/bag = R1 440. Apply fertilizer mechanically using a tractor and spreader = R350. Note: fertilizer prices increase every year.	R8 750
Plant material: Seedlings from Nurseries (open pollinated cultivars): 33 200 seedlings p/Ha @ R0.35 per seedling = R11 620 00 (2014). The seed/seedlings for the Hybrid cultivars are more expensive. Produce your own seedlings in soil seedbeds as it is cheaper.	R11 620
Transplant seedlings: In prepared land, 1 labourer transplants approximately 1 400 seedlings per day and 25 labourers transplant 35 000 seedlings @ R100 = 25 x R100 = R2500. Mechanical seedling planters are also available. 2 labourers may transplant 3 000 seedlings per hour and more.	R2 500
Chemicals: To control insects/diseases and the necessary labour.	R1 700
Weed control: Mechanical weed control between rows = R500. 5 labourers control weeds in rows for 2 days, 2 x 5 x R100 = R1 000. Herbicides are cheaper and less labour intensive when applied at the right time.	R1 500
Irrigation: To irrigate 1mm of water p/Ha, 10 000 litres is required. Therefore, 30mm per week for 10 weeks = 30ml per week = 300 000 litre p/Ha. Then, for 10 weeks approximately 3 million litre p/Ha is required which costs approximately R6 000. Diesel/electricity, tractors, trailer, transport, labour, fixing and shifting irrigation pipes and so on should be factored-in. It can be costly to pump irrigation water, especially when low rainfall occurs.	R6 000
Harvesting: Transport 70 tons of cabbage to store. 1 labourer harvests and loads approximately 3 000kg cabbage heads per day. 12 labourers harvest 36 000kg per day = 2 day @ R100 per day = 12 x 2 x R100 = R2 400.	R2 400
Cabbage bags: Green potato bags for 70 000kg: 8 cabbages per bag = 8 750 bags = R0.50 per bag = R4 375.	R4 375
Sorting: Pack for market requirements. Usually in green potato bags including labour.	R1 300
Transport: To the markets 70 ton cabbages or 8 750 bag, including labour. If market is approximately 100km from the farm: Travel 7 times with a 10-ton truck, there and back: 2 x 7 x 100km = 1 400km @ R6.50 p/km = R9 100 plus labour for 10 ton truck driver and assistant = R800	R9 100 R800
Clean and plough residue into the harvested land. Remove and store irrigation system. Establish a rest crop if necessary.	R1 600
ESTIMATED PRODUCTION COST FOR CABBAGE P/HA	R55 945

TABLE 23: POSSIBLE YIELD AND PROFITS FOR CABBAGE

POSSIBLE YIELD AND INCOME: 60 ton heads p/Ha = 33 200 medium size heads at 2.5kg per head minus 3 000 unmarketable heads = 30 200 @ R5 per head = 30 200 x R5 = R151 000	R151 000
MINUS PRODUCTION COSTS:	R55 945
POSSIBLE PROFITS:	R95 055

28.3 INTRODUCTION TO COLE CROP PRODUCTION

28.3.1 CABBAGE

Cabbage is the main cole crop produced across the world and is related to cauliflower, broccoli, brussels sprouts, chinese cabbage, kale and turnips. Cabbage is fairly simple to grow and is popular amongst farmers. Cabbage and broccoli are less difficult to cultivate than cauliflower. Cabbage can be produced in many areas in South Africa while cauliflower fails to produce quality curds in some locations.

28.3.2 CAULIFLOWER

Cauliflower is more difficult to cultivate than cabbage and other cole crops. Cauliflower requires constant soil moisture and enough nutrients. It prefers cool growing conditions, especially in the later stage of development. New improved F1 hybrid cultivars adapted to summer and winter production are available. Producers should evaluate cultivars which may improve yields and quality. *See photo 248.*

28.3.3 BROCCOLI

Broccoli is not a difficult crop to produce. It requires less nitrogen N and matures much earlier. Summer-orientated F1 hybrid cultivars which can be established in warm areas, are available. Be careful to select the correct cultivar that is adapted to the temperature of your region. Note that the crop develops slower when cold or frosty conditions occur. High temperatures and uneven water supply may cause the curds to become loose which may cause them to go over to early bolting (flowering).

28.3.4 BRUSSELS SPROUTS

Brussels sprouts are not among the most popular vegetables, although their unique flavour is appreciated by many people who buy frozen packs. Their long growing season, susceptibility to attack by pests and unsatisfactory yields are the main reasons for the lack of attention given to this crop. However, new cultivars are available with better yields and firm sprouts.

247 A Savoy cabbage cultivar with typical dark green, glossy, crinkled leaves. Usually used for sauerkraut and salads due to its crispy and sweeter taste.

248 Various summer Hybrid cabbage cultivars displayed at a Farmers day.

249 A cauliflower cultivar. The curds are nicely covered with the outer leaves, protecting them from sunburn which would otherwise result in the colouring of the curds.

PART 2: GUIDELINES FOR CROP-SPECIFIC VEGETABLE PRODUCTION

250 A good summer production, medium sized, Hybrid cultivar.

251 A large F1 hybrid cauliflower cultivar. The curd is covered with outer leaves and is ready to be harvested.

252 A good quality, visually appealing, medium cauliflower head size. A bit yellowish but still acceptable.

253 An open-pollinated cauliflower cultivar. The curds of this cultivar mature unevenly, which prolongs the harvesting time. This is an advantage for producers.

254 A solid cauliflower F1 hybrid curd that weighs 9.5kg. This particular cultivar has a very long growing period. It is enough to feed more than 2 dozen people. This curd was grown by the author (1999).

255 A very solid summer produced cabbage cultivar.

256 A visually appealing cabbage field near Hartebeestpoort Dam, South Africa. The removable overhead and quick coupling irrigation system is visible at the back.

257 Some cabbage heads and cauliflower curds have huge outer leave coverage which may be used to make good compost or to feed livestock such as cattle and pigs.

258 A rare case of a cross between a cabbage and a brussels sprout cultivar. It is extremely difficult for plant breeders' to reproduce similar cultivars in a genetically true/pure way as the brassica family is highly susceptible to cross-breeding and self-pollination. Not only that, but it would take an extremely long time and many breeding selections. If it were reproducible, it would make producers very rich.

259 This F1 hybrid cauliflower cultivar has an excellent curd covering to protect it against sun light. As a result, the curd will stay brilliant white.

260 This flowering plant is a Pac-choi cultivar. It is a lettuce-like plant, from the chinese cabbage family. They all belong to the cole crop or brassica group. The flowers and stems are harvested and are very tasty and visually appealing in salads.

28.4 PRODUCTION GUIDELINES FOR COLE CROPS

28.4.1 SOIL REQUIREMENTS
Also see Chapter 5.3 Soil Requirements.

Cole crops can be grown in all soil types, but prefer a well-drained fertile loam type of soil. Heavy clay, clay loam and low-lying soil is acceptable but should be drained properly. Generally, coarse textured soil is favoured because it warms more quickly and provides earlier harvesting. In spring, the soil warms up faster and is also less likely to become waterlogged, especially when heavy rains occur. Heavier textured soil that has a greater water holding capacity and can be used to produce cabbage that matures later. Where possible, avoid very heavy clay soil.

The best results come from planting early cabbage cultivars in soil that are lighter in texture and late cultivars on heavier soil. Cole crops have a fairly shallow root system and have an effective feeding depth of 50 to 60cm. Performing deep soil cultivation before establishing cole crops will assist the movement of water downwards through the soil. Cole crops respond well to applications of compost, green or kraal manure.

Some soil form a hard crust when heavy rains occur or even when overhead irrigation is used. This may compact the top layer of some soil. It is important to break this crust whenever it occurs, in order to promote water penetration and aeration. Cole crops are very sensitive to poor soil aeration and sometimes tend to stop growing under these conditions. Cole crops develop much better when the soil surface is loose.

Cole crops are also sensitive to soil acidity. Soil with a low acid content often contain low levels of available aluminum and manganese which may affect plant growth and yields. High levels of aluminum may give rise to phosphorus P and molybdenum Mo shortages. The requirement for these two elements are high.

28.4.2 PH OF THE SOIL
A pH of 5.8 to 6.8 for cabbage and 6.6 to 6.8 for cauliflower, broccoli and brussels sprouts is preferable. Some crops are tolerant of high levels of soil-salt content. Cabbage may be produced at a pH as low as 5 with no reduction in the yield. Red cabbage cultivars however are sensitive to high pH levels.

The higher the salt content of the soil the more susceptible the crop will be to certain diseases. It has also been found that at a higher pH level, cabbage production can take place repeatedly. It is advisable to not produce cabbage in soil with a pH lower than Ph5 or above 7.5. If the pH is below 5.5 it is advisable to apply lime, but the amount and type of lime should be indicated by a soil analysis.

28.4.3 CLIMATIC REQUIREMENTS AND OPTIMUM PLANTING TIMES
Also see Chapter 3 Climate.

If producers do not establish ideal conditions for cole crops in the first 2 to 3 weeks, the crop may not reach its full potential. Selecting the correct climate is the key to successful production.

Bolting (premature flowering) of cauliflower and broccoli occurs when the plant begins to develop and suddenly slows down as a result of stressful environmental conditions. Examples include:
- High temperatures.
- Old seedlings from seed trays.
- Insufficient water and/or a shortage of N.
- Weed competition and dense plant populations.

Some cultivars are resistant to the above mentioned conditions.

28.4.3.1 CABBAGE
Cabbage prefers cool, moist climate conditions and can be grown throughout the year in most areas. They have a wide adaptability except when producing them in cold regions, during May to July. This is due to the heavy cold and frost during the long winter periods. When producing cabbage in the Lowveld or any other regions where the summers are very hot and/or where high humidity conditions occur, insects and certain diseases may multiply quickly and get out of control.

The optimum temperature for growth and development is between 16°C and 29°C but cole crops will produce in temperatures as low as 5°C. Cabbage is also resistant to frost and can survive minimum temperatures as low as minus 3°C without being damaged unless the heads are already fully developed, at which time they may burst or crack due to the pressure in the compacted heads. The plants are more susceptible to frost when there is a big difference between day and night temperatures.

Insects are much more active, multiply quickly and are more difficult to control in the summer conditions. Special attention should be given to insects and disease when warm conditions occur during the growing period. The best production quality is achieved when the crop matures during late autumn, winter or early spring. F1 hybrid cultivars are available, which may overcome these problems to a large extent.

Cole crops, especially cabbage, develop quickly under fairly warm conditions while the growing tempo declines when low temperatures occur. For example, the growing period may be 90 to 100 days during warm conditions and 130 to 140 days during cold conditions.

Sometimes old seedling stems harden. As a result, when they are transplanted, their hardened stems struggle to expand thereby remaining thin and leaving the plant with a thin underdeveloped stem.

28.4.3.2 CAULIFLOWER
Temperature requirements for cauliflower and cabbage are similar but cauliflower is less adaptable and more sensitive to high temperatures and also does not withstand very low temperatures. Younger plants usually recover completely when frost conditions occur. The summer-orientated cauliflower cultivars are usually intermediate types and do not need cool weather to initiate the curd formation. They will respond to a cool period by initiating curds after they have reached the mature vegetative stage.

Some cauliflower cultivars can grow throughout the winter and mature in the late winter. F1 hybrid cauliflower cultivars are available that can be established in a wide range of temperatures. Some cultivars have crucial sowing times and good quality crops can be produced when the plants become mature during autumn and winter periods.

Care should be taken not to transplant old cauliflower and broccoli seedlings. This has a negative influence on yields due to the early flowering (bolting) of the curds when they become too old.

28.4.3.3 BROCCOLI
This crop prefers cool, moist weather, but it is also produced in the cooler Highveld regions. When high temperatures occur, the broccoli heads may quickly turn yellow. This may happen particularly after harvesting and also before the crop is harvested. It is called 'brown' and some cultivars are more commonly affected, especially when under stressful conditions. Brown bud may become a problem when warm conditions occur and the curds start to mature. New improved F1 hybrid cultivars are available which are much more tolerant.

28.4.3.4 BRUSSELS SPROUTS
This crop is more susceptible to high temperatures than cauliflower. Good quality and yields are possible during the coolest period of the year as it can withstand lower temperatures than cauliflower and cabbage. Cultivar choice should be a high priority and planting times are critical for this crop.

28.4.3.5 OPTIMUM PLANTING TIMES FOR COLE CROPS

Areas with light to mild frost:	Feb-April and Aug-Oct
Areas with heavy frost:	Aug-Feb
Frost-free areas:	Feb-June
Cooler and winter rainfall areas:	Nov-April
Warmer areas (Limpopo):	Jan-May

28.4.4 CULTIVARS AND CULTIVAR CHOICES
There are many cabbage cultivars on the market, especially F1 hybrid cultivars that perform exceptionally well. To a lesser degree, broccoli and cauliflower are available and are suitable for South Africa's severe weather conditions.

Most leading seed companies spend a considerable amount of money and get involved in cultivar trials to evaluate and compare results between the standard and new cultivars as well as the F1 hybrid cultivars. Cultivars mentioned in this manual only serve as an indication of what is available. Some older or so called open cultivars and are still available in the trade. Because these open cultivars have been produced successfully for years, they have no breeder's rights associated with them which means that any person is able to produce seed from them for their own use. However, when trading with the seed, they must comply with the standard seed requirements.

Open cultivates are especially popular among small-scale farmers, home gardeners, underprivileged and upcoming producers. This is due to cultivars uneven maturing, long harvesting period and also the lower cost of seed and seedlings. Cabbage cultivars maturity times are classified as early, medium and late. Producers should be aware of this, for their own production planning.

28.4.4.1 CULTIVAR CHOICE
Considerations for selecting cabbage cultivars:
- The importance of producing the best cultivar for your region is important.
- Type of cole crops: Head, baby, novelty, savory or red heads.
- Head shapes: Round, oval, flat or pointed.
- Head sizes: Spacing should be adapted to the size of the cultivar.
- Climate: Winter or summer cultivars.
- High tolerance cultivars: These cultivars should be considered when black rot and other diseases are a particular problem in your area.

28.4.4.2 CABBAGE CULTIVARS

Different sizes and shapes are available from baby cabbages which are the size of a tennis ball, to huge heads, 50cm in diameter or more. Ethnic African consumers prefer a large cabbage head while most Caucasian consumers prefer smaller ones.

Baby cabbage cultivars differ in size, are cultivated earlier and their spacing is much closer. The small, medium cultivar Puma is popular and can be established all year round in most areas (2013). Purple head types are used to prepare sauerkraut. Savoy cabbage, originating from Italy, has crinkly leaves, is the most tender and has the sweetest taste of all cole crops.

28.4.4.3 CAULIFLOWER CULTIVARS

The type of cauliflower cultivars grown in South Africa have been developed with long outer leaves to offer the developing curd protection from the hot sunlight. Prior to 1970, open cauliflower cultivars were mostly unstable and matured very unevenly. However, uneven maturing is beneficial to some farmers. The marketing period is extended by longer harvesting periods.

Certain cauliflower cultivars are more heat tolerant than others while some are also affected by the length of the day. When winter-orientated cultivars are produced out of season, they may develop large outer leaves with under developed heads. On the other hand, when warm conditions occur, it may cause the heads to become ricey, hairy, discoloured or malformed. The curds may also become pink, purple, brown or green.

28.4.4.4 CABBAGE HEAD SHAPES

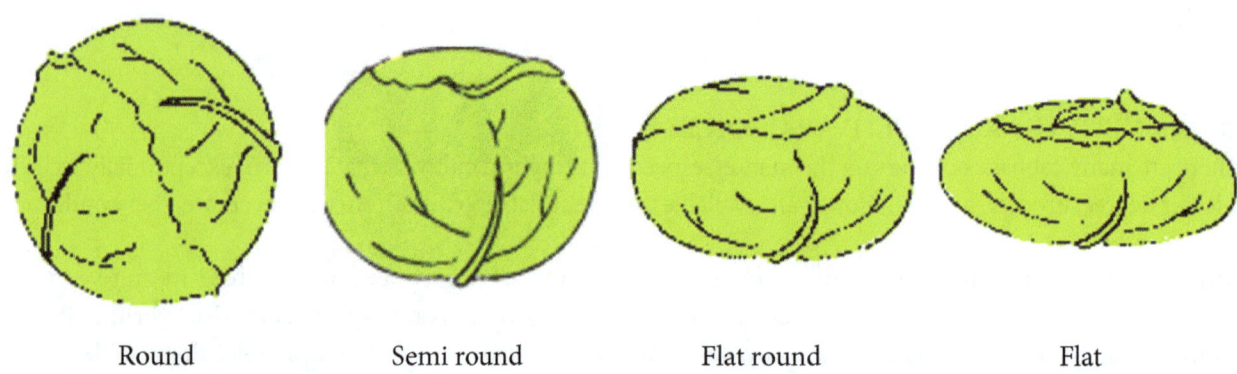

Round   Semi round   Flat round   Flat

261 Good size and quality brussels sprouts.

262 A good leafy kale cultivar. The leaves are harvested systematically and are especially popular amongst Portuguese people. Cultivar choice is important for this crop.

263 A Pak Choi chinese cabbage cultivar which belongs to the brassica family, however production in South Africa is limited. Numerous chinese cabbage cultivars are available, some taste like vinegar and others like mustard or pepper.

28.4.4.5 BROCCOLI CULTIVARS

Broccoli is grown for its large, thick, green stalks and for its curds that contain many flower buds. The edible head is a flower curd that develops many yellow flowering buds when they transition into the seed stage.

The first developed curd is the largest one and when the crop is continuously harvested, smaller side shoots will develop that can be marketed at a lower grade. This also lengthens the harvesting period. Some cultivars, especially the F1 hybrid cultivars, only produce one curd per plant.

28.4.4.6 BRUSSELS SPROUT CULTIVARS
See photo 267.

Brussels sprouts are a traditional winter crop and have a strong flavour. Brussels sprouts contain vitamin A, C, folic acid and fibre. It is believed that the sprouts may prevent colon cancer. Besides being extremely healthy, sprouts are very tasty and are usually used in stir-fries, slow cooking or soups.

28.4.4.7 TURNIP CULTIVARS
See photo 266.

The production of turnips is limited and is mainly used in soups. Two cultivars are commonly grown in South Africa, namely purple top and white globe.

28.4.4.8 CHINESE CABBAGE CULTIVARS

There are many kinds of chinese cabbage cultivars, which may be grouped. For example, there are loose heads, medium firm heads as well as flowering and edible stalk types. Some are excellent to use in salads, stir-fries, stews and soups. They may be used to replace certain herbs in salads as some of them may have a peppery, sour, salty and/or vinegar taste. The head type is usually very fleshy and juicy. **See photo 260 for a flowering chinese cabbage cultivar, photo 263 for medium loose heads and photo 264 for an extremely large head type cultivar.**

28.4.4.9 KOHLRABI CULTIVARS
See photo 265.

Kohlrabi is a swollen root and is similar to a turnip. The name originates from the German word, kohl that means cabbage and rabbi (turnip). There are 2 types and the roots are purple and green. The taste is similar to a broccoli stem, but sweeter and tastes more like a turnip. Kohlrabi may be eaten raw and when the root is sliced thinly, it is used in salads, stir-fries, steaming and soups.

264 A chinese cabbage cultivar. The plants are very juicy and can weigh up to 4kg or more. The crop is reasonably popular among the chinese people especially for making delicious soups and stir fries.

265 A purple kohlrabi cultivar. The edible part of kohlrabi looks like a turnip. It grows on top of the soil with the leaves growing over the surface of the globe. The leaves are also edible. The crop belongs to the cole crop family.

266 Turnips belong to the cole crop family. Above is a purple top globe cultivar.

267 A Brussels sprout plant is not commonly produced in South Africa. It is mainly produced for processing.

268 A good sized and quality broccoli cultivar. The plants usually grow vigorously while the curds begin to develop at a later stage. They remain small and then develop quickly. Be careful as some cultivars quickly transition over to flowering.

28.4.4.10 KALE CULTIVARS
See photo 262.

Kale is not a familiar crop but is a good vegetable source. It is easy to cultivate as the plant and stem develops in an upright position. It produces many broad leaves that are harvested from the bottom upwards as they develop. The focus should be on a cultivar that has large leaves.

There are numerous kinds of kale cultivars, from small to the taller growing types. Not all cultivars taste very good. Some Portuguese people use kale leaves instead of cabbage in their home gardens. They cook it in a similar way to cabbage and use it to make soup. Health conscious people make juice from the plant. The juice has healthy nutrients, especially vitamin A, B and Iron. Kale leaves are sold by health Organizations and Supermarkets. Some home owners establish a few plants in their back yard as an easy crop to care for and which is extremely healthy.

The leaf size of some cultivars are very large and are harvested continuously as the plant grows taller. The plant has a long growing season if the leaves are sensibly harvested. The leaves and stems are more nutritious than cabbage. When harvesting is completed, the plant ends-up with a long, thick stem, like an ostrich neck, with a few small remaining leaves at the top. It takes approximately 5 to 6 months after establishment to reach this stage.

28.4.5 SEEDLING PRODUCTION IN SEEDBEDS AND IN THE LAND
Most producers obtain seedlings from Nurseries. **See Chapter 4.6 about seedlings obtained from nurseries.** Some producers produce seedlings in seedbeds directly in the soil and are usually successful if they are familiar with this practise. **See Chapter 8.4 to produce seedlings in soil.**

About 280g cabbage seeds are necessary to produce enough seedlings for 1Ha. Approximately 3g seeds are used for a 10 m furrow row in a soil seedbed. Cauliflower and broccoli seedlings should be transplanted at the correct time. It is important to not wait too long before transplanting these seedlings.

Do not use seedlings that are weak or have other abnormalities. Some of the seedlings have no growing points and therefore they should be disposed of. They cannot develop to produce a healthy crop. Some insects, such as Bragada bugs can damage the seedlings quickly. The growing points of the seedlings are mostly attacked first. These insects are also called Two-by-two's as the adults couple together and try to move in opposite directions. They can do considerable damage to seedling plants.

28.4.6 SOIL PREPARATION AND TRANSPLANTING OF SEEDLINGS
Also see Chapter 5.3 Soil Requirements and Chapter 8.2 Soil Preparation.

When necessary, control root-knot nematode using a soil fumigant 2-3 weeks before transplanting the seedlings. This is usually not necessary for winter production. Cabbage can be grown in all types of soil, from sandy to heavier clay soil. Sandy to sandy loam soil is the best when an early crop is required. This is because sandy soil warms up more quickly and is beneficial to a strong root system, earlier plant development and earlier harvesting time.

When higher yields and large heads are the most important factor, then clay loam and silt soil should be considered. Late head producing F1 hybrid cultivars that are known to produce larger heads may also be considered. **See photo 254.** It is important to keep the soil loose, because when heavy rain occurs or after heavy irrigation, a hard top crust may develop as the soil dries out. This hampers water and oxygen penetration into the soil. In this situation, the water usually runs away to the lower end of the land.

If crust formation is a problem, the irrigation intervals should be shortened which will ensures that less water is lost. Cole crops are sensitive to poor aeration and stop growing in these conditions. It is therefore important to loosen the soil in between and within the rows themselves, as soon as a crust begins to form. Sometimes an N application can be applied, before breaking the crust.

28.4.6.1 TRANSPLANTING SEEDLINGS TO THE LAND
See Chapter 8.5 Transplanting seedlings from soil seedbeds and Chapter 8.8 Transplanting seedlings from seed trays.

If possible, transplant seedlings when cool and overcast conditions occur, or late in the afternoon. This provides the best results. Avoid transplanting old and leggy cole crop seedlings. Rather use shorter sturdy seedlings. Success of transplanting depends largely on weather conditions and treatment of the seedlings.

Irrigate the prepared land shortly before transplanting. The seedlings should be transplanted into damp soil. Make the holes deep and wide enough so that the plug of the seedlings has enough room to fit into the holes. The soil around the seedling should be slightly compacted to remove the air particles around the roots.

It is advisable to water seedlings when transplanting them in hot conditions. Using a watering can or hose, provide each seedling 0.3 litres of water. This would ensure a good plant population and avoid air pockets around the roots of the seedlings. Irrigate the land shortly after transplanting the seedlings.

28.4.7 AMOUNT OF SEED P/HA
Cabbage seedlings in seed trays or seedbeds: 0.2kg p/Ha.
Cauliflower transplanted from seed trays or seedbeds: 0.3kg p/Ha.
Broccoli transplanted from seed trays or seedbeds: 0.2 to 0.5kg p/Ha.

28.4.8 SPACING
Also see Chapter 11 Spacing of Crops.

Some producers and small-scale farmers do not always use ideal spacing between plants due to the limited area available. They often make the mistake of narrow spacing or establishing plant populations that are too high. This is understandable, but generally, closer spacing produces smaller heads. The total mass may remain the same when comparing high and optimum spacing, but a very high plant density should be avoided due to the large outer leaves of cabbage and cauliflower.

28.4.8.1 SPACING FOR CABBAGE
28.4.8.1 LARGE HEADS
A good option is 70cm between the rows and 55cm in the rows. The plant population would then be 28 400 plants p/Ha.

28.4.8.1.1 MEDIUM SIZE HEADS
50-60cm between rows and 45-50cm in rows. For a spacing of 60cm between rows and 50cm in the row, the plant population would be 33 200 p/Ha.

Medium size cabbage heads are the most popular and producers establish a plant population of 28 000 to 30 000 plants p/Ha. That results in a spacing of 55cm in the row and 60cm between the rows. This ensures a plant population of 30 000 plants p/Ha. Spacing of 55 x 65 results in a plant population of 27 700 plants p/Ha.

28.4.8.1.2 SMALL BABY SIZE HEADS
45-55cm between rows and 30cm in rows. A good option for medium size baby cabbage heads is 45cm between rows and 30cm in the row. This results in a population of 73 900 plants p/Ha.

Smaller tennis ball baby heads are spaced up to 80 000 to 90 000 plants p/Ha or more. If a producer has a market for small heads then this may be an advantage.

28.4.8.2.3 SPACING FOR CAULIFLOWER CULTIVARS
LONG SEASON LARGE CURDS
75-90cm between rows and 50-60cm in rows. A spacing of 80cm between rows and 55cm in the row results in a plant population of 22 600 p/Ha.

SHORT SEASON SMALL CURD CULTIVARS
60cm between rows and in rows 50cm. This results in a plant population of 33 200 plants p/Ha.

28.4.8.3 SPACING BROCCOLI
Spacing and plant population for this crop is very important because it may affect the curd size and quality. Wider spacing causes less competition for nutrients and light. Plant population can be 35 000 to 40 000 plants p/Ha for smaller head cultivars.

When warm weather conditions occur it is advisable to use closer spaced populations and in cooler conditions wider spacing.

28.4.8.4 SPACING BRUSSELS SPROUTS
60-70 between rows and 50-60 in rows. Brussels sprouts are mainly produced in mild winter conditions, therefore producers should keep the spacing lower, as there is less light and temperature intensity.

28.4.9 IRRIGATION
Also see Chapter 14 for Irrigation.

28.4.9.1 WATER REQUIREMENTS
Cole crops are sensitive to water stress and over irrigation in clayish soil types. Cabbage, cauliflower, broccoli and brussels sprouts require approximately 420mm water throughout the growing season. After transplantation, irrigate 25mm the first week and 35cm every week thereafter. The average water loss of a cabbage plant is 300ml per day for a young plant and 400ml to 500ml per day for an adult plant. If water stress occurs during the growing period, yields and quality can be affected.

If soil crusting is a problem, use shorter irrigation intervals. Experts that use overhead irrigation systems claimed that using this irrigation method improves plant and head development. The crop should be irrigated as soon as 50% of the field capacity is achieved. On heavier clay soil, a lower graduation should be the norm depending on the age of the plants, temperature and humidity.

Cabbage responds well when standard irrigation is applied, however some farmers irrigate less, as they believe the plants should be stressed to a certain extent. Afterwards, they then apply the ideal amount of water believing that some crops produce higher yields because of this. A permanent irrigation or centre pivot system will be helpful and improve yields.

Proper irrigation from the second week after transplanting the seedlings is critical. It is important that the young plants receive adequate water to ensure vegetative growth before the heads develop. When cabbage and cauliflower plants are adapted to less favourable conditions, their growth may slow down. The cells of hardened seedlings which have remained in seed trays for too long, may cause the stem to harden more and prevent proper transpiration.

The plants will adapt to changing conditions. For example, if receiving water stress, the plants may adapt and return to normal after a period of 24 hours. Any problems related to applying too little irrigation water may harden the crop but also lengthen the time it takes for the plant to get back to normal. This could result in seedling plants falling over, which will result in a yield loss.

Do not apply too much water once the cabbage heads have formed, as this may cause cracking or splitting of heads, especially during temperature fluctuations. Cauliflower and broccoli needs adequate, but not too much water, at the later stages of their development, especially when hot weather conditions occur. If cauliflower and broccoli plants become water stressed in the later stages of curd development, even for one day, it may harm the crop and influence the curd quality.

28.4.10 FERTILIZER
Also see Chapter 15 Fertilizer.

A soil analysis test for commercial and small-scale producers is advisable for each production land. Cole crops do well when organic matter is available as they are heavy nutritional feeders, especially of nitrogen N and potassium K. Farmers have learnt, that when corn is followed by cabbage, the cabbage produces smaller yields and vice versa. This is due to the fact that when the cabbage heads are harvested, there is a heavy depletion of nutrients from the soil.

Cole crops have a high nutritional requirement and a shortage of N, P and K are responsible for lower yields and quality. 30 ton cabbage plants p/Ha may remove 120kg N, 20kg P and 100kg K from the soil. Unavailable molybdenum Mo and a too low pH will affect plant growth. Applying small amounts of fertilizer through the growth period is beneficial and cost effective and will produce a good quality crop. Cabbage plants withdraw 3.4 N, 0.6 P and 3.6 K for 1 ton of plant material. Cauliflower, brussels sprouts and broccoli withdraw 8.5 N, 1.7 P and 9.2 K for 1 ton of plant material.

Suggested fertilizer for cole crops in fairly fertile soil is approximately 120kg N, 50kg P and 115kg K. When a soil analysis is not available and if a producer prefers a fertilizer mixture, the following amount of granular fertilizer is suggested: 1 300 to 1 500kg 2.3.2 (22) Zn and 150kg potassium chloride. This mixture should be broadcast and worked into the soil, not deeper than 20 to 30cm into the soil, just before transplanting the seedlings. Calcium is very important for all cole crops as it builds plant cells and is responsible for the heads becoming firm.

28.4.10.1 TOP DRESSINGS
Apply approximately 130 to 150kg LAN/KAN top dressing on fairly fertile soil, 3 to 4 weeks after transplantation and a second application 6 weeks after transplanting the seedlings. For the later developing cultivars, use the right type of lime (it depends on the pH of the soil). ***See Chapter 15.5 for using the correct type of lime.***

Always look for nitrogen N deficiencies when producing cole crops. This condition usually begins in the hollows of the land. When the leaves become pale in these areas, immediately apply N to the entire land. Be careful not to apply too much N later in the production season, especially for cabbage. The heads may crack as they become mature. If too much N is applied when producing brussels sprouts, the sprouts may become loose as they mature.

28.4.11 THE IMPORTANCE OF TRACE ELEMENTS
28.4.11.1 BORON
Cole crops, especially cauliflower, are sensitive to a shortage of boron B. The curds of cauliflower and broccoli turn brown or black and may become out of shape with corky lesions appearing on the stem and the midrib of older leaves. Cabbage and brussels sprouts are less sensitive to a boron deficiency. By the time boron deficiency symptoms are noticed in the field it is usually too late to correct it. If the plants are young it may be beneficial to use foliar sprays.

28.4.11.2 MOLYBDENUM
Cauliflower is more sensitive to molybdenum deficiency than other crops. The leaves become malformed, dark green and curly. Most leaf tissue does not develop and this why it is called whiptail. Other cole crops show similar symptoms, but to a lesser degree. Molybdenum deficiency is fairly common in acid soil that has a pH of 4.5 or lower.

28.4.12 CROP ROTATION
Also see Chapter 18 Crop Rotation and Soil Improvement.

Cabbage, cauliflower, broccoli, kale, chinese cabbage and turnips are related to cole or brassicas crops and should not be established in the same soil. A 3 year waiting period should be followed. To keep the soil fertile it is advisable to practice crop rotation using leguminous crops such as sun hemp, black mustard, velvet beans, green peas etc.

28.4.13 WEED CONTROL
The importance of effective weed control cannot be over-emphasized. Research has shown that the critical time of weeds in the summer-orientated areas exists at 15 to 45 days after preparing the soil. Plants competing with weeds may result in a 30% yield loss. Weed control should be performed at a shallow depth to prevent damaging the roots. Loosening the soil should stop as soon as it is damaging the roots. Herbicides are cheaper than labour and also less labour intensive for commercial farming, provided it is done correctly and at the right time before or after the planting or sowing of seed.

28.4.14 YIELDS AND DAYS UNTIL HARVESTING
Cabbage	60-95 ton/Ha
Cauliflower	15-20 ton/Ha
Broccoli	9-15 ton/Ha
Brussels sprouts	11-15 ton/Ha

28.4.15 DAYS FROM TRANSPLANTING TO HARVESTING
Cabbage	85-120 days
Cauliflower	90-120 days
Brussels sprouts	100-120 days
Broccoli	90-120 days

Broccoli matures between 90 and 100 days from sowing seed to maturity. It also depends on the cultivar. Curds tend to develop rapidly from small to the larger curds before harvesting. When transplanting broccoli seedlings from seed trays at an early stage, cultivars may mature in 60 days before the first harvesting.

28.4.16 HARVESTING, PACKAGING AND MARKETING
Also see Chapter 24 Harvesting, Handling and Packaging and Chapter 25 Management and Marketing.

28.4.16.1 HARVESTING CABBAGE

Some cabbage cultivars are vulnerable to head burst, splitting or cracking. It is important to harvest when the heads are firm before heads begin to burst. The heads should be cut with a sharp knife when they become firm. Some cultivar heads do not become firm all at the same time and therefore harvesting can be performed over a period of several weeks. F1 hybrid cabbage cultivars are more uniform and some cultivars become firm quickly. Do not leave cabbage heads, cauliflower or broccoli curds in the harvesting crates after harvesting.

28.4.16.2 HARVESTING CAULIFLOWER

Cauliflower produces one curd per plant and when they are harvested at a young stage, they are brilliant white. However, when exposed to sunlight they turn to a snowy white and later become brownish in colour with loose heads. The curd of the cauliflower should be brilliant to snowy white. The market demands this and the curd should be fully developed, compact and clean.

28.4.16.3 PROTECTION OF CAULIFLOWER HEADS BEFORE HARVESTING

It is important that the cauliflower curds should be well protected from the sun. Producers should choose more adaptable cultivars. The F1 hybrid cultivars have leaves that surround and protect the head until they become fully developed. Over maturity results in an open head that may become coloured. This is due to the sun which usually affects the curds directly and may result in an undesirable quality for the market and buyers. Discolouring lowers the grade and reduces prices substantially. Some producers protect the cauliflower curds by binding the longest leaves together over the curds and tying them with twine, tape or tying straps. This could be worth the effort to produce a healthy and large white curd.

28.4.16.4 HARVESTING BROCCOLI

Curds are picked when they are well firmed and a preferred size. This should happen before any signs of loosening and discolouring takes place. When producers prefer larger curds and they allow the curds to continue to expand but care should be taken that they do not become loose in hot conditions. They should be harvested before the curds become even slightly loose. As most cultivars mature evenly, all the curds are going to be ready for harvesting at the same time. Note that the curds can become brown quickly especially when high temperatures are a factor.

The primary curd of some cultivars is the most valuable and commands the highest prices. They are cut in the first 2 weeks of harvesting and are more or less 60 to 80% of the total crop. Most F1 hybrid cultivars produce only one main curd and seldom secondary ones. When the plant density is high, the curd size may also decline. However, the total yield may also be slightly higher because the primary curds are the largest proportion of the crop. The secondary curds develop fast and are stimulated once the primary curds have been removed. The secondary curds are therefore going to be harvested after the second or third weeks over a period of about 3 weeks. After this, they become too small to have any value.

Broccoli is a more perishable crop than other brassica or cole crops. Broccoli is very sensitive to climate and stress conditions. This fact is not always known by producers or the markets. Most of the product is pre-packed and sold directly to retail stores or consumers. The heads are generally placed in polystyrene containers and covered with a plastic film. It is important to cultivate this crop in such a way that it is not stressed in any way before harvesting occurs.

If the crop is harvested and the curds quickly change to a yellowish colour, this is usually an indication that favourable conditions did not occur and/or cultivation practices before harvesting were not practiced. As in the case of cabbage (that can be held back for a week or two before harvesting), there is no waiting period to harvest broccoli. The curds mature rapidly when they are still small and should be harvested every two or three days. Do not wait too long before harvesting. Broccoli curds should be harvested when the buds are firm and tightly closed. Curds should be tight and individual flowers should not show signs of yellowing. Curds of some cultivars develop quickly after the main curds are harvested and the smaller side curds may continue to be harvested for another week or two.

Curds are harvested when they have a good, bright green colour and before any signs of loosening and discolouring appear. This may happen rapidly, especially when hot weather conditions occur. Over-maturity reduces the quality of the heads. The curds must be cut before the flower buds show any signs of discolouration. When broccoli curds reach the proper harvesting stage they should be cut by including 15cm of the stem or more. The stem has a more favourable, sweeter taste than cabbage and also cooks perfectly soft. Some F1 hybrid cultivars produce only one curd and none or very little small side shoots. This is an advantage due to the shorter harvesting time required as only one curd needs to be harvested.

If broccoli and cauliflower are to be harvested on warm/hot days, the producers should begin doing so early in the morning when the curds are still cool. The harvest should also be kept it in a cool place before packaging. It is important for the soil to be moist as the curds will then stay juicy before harvesting.

28.4.16.5 HARVESTING BRUSSELS SPROUTS
New improved cultivars are available. The sprouts develop on the long, thick stalks of the plant from which they are picked by hand. Plants grow fairly tall (sometimes more than 70cm high). Brussels sprouts are a cool winter production crop and should be harvested when cool weather conditions occur. The sprouts develop in the leaf axils and mature alongside the stalk. The lowest sprouts mature first and should be harvested when they are firm. The lowest leaves should be removed before harvesting to allow the sprouts to mature properly. When harvesting late cultivars in warm/hot conditions, aphids are more active and loose sprouts may result.

28.4.17 STORAGE
Broccoli and cauliflower curds should be stored at low temperatures. This controls moisture loss and disease deterioration. It allows shelf life to be extended and longer transportation times to the market, allowing better market flexibility. The curds can also be stored with ice or ice water to extend their storage time.

28.4.17.1 PACKAGING
Cabbages are divided into 3 groups and packaged accordingly. They are:

28.4.17.2 SMALL, MEDIUM AND LARGE HEADS
The most popular cabbage heads are the medium size heads. They are approximately 2.5 to 3.5kg and the large heads 4 to 4.5kg. Local African people prefer large heads and often buy directly from producers. Large and medium size heads are usually packed in large, green mesh bags for the National markets. Retail shops retail the heads separately.

28.4.17.3 BABY AND NOVELTY GROUPS
Baby cabbage is popular in overseas countries and seems to be increasing in popularity in South Africa. Many retail shops prefer baby cabbages weighing in at about 0.5kg which are usually packed in pairs and in polystyrene containers. Larger heads of about 1.5kg and are wrapped in plastic films and sold separately. The middle and higher income communities usually prefer smaller heads as they can be prepared and consumed in one meal.

28.4.17.4 SAVOY RED AND SPITS HEAD GROUPS
Savoy cabbage heads are part of the novelty group. This market is limited however. The Germans know how to make excellent sauerkraut salads. It is not easy to make because it goes through secret processes to produce this healthy and tasty, sour cabbage salad product. The spits or pointed cabbage heads are well-known for their crunchy sweetness and juiciness. This makes an excellent crispy sweet cole salad. The well-known Cape spits and other similar salad cultivars have a short growing period and can be used in the early stages of development.

28.4.17.5 MARKETING CAULIFLOWER
The markets usually require clean, brilliant white cauliflower curds that are packaged in polystyrene containers and covered with a plastic film. It is important that the curds are carefully handled to reduce any marks or

damage to the curds. The quality and appearance of the cauliflower curds should be good and of a reasonable size according to the market and buyer requirements.

28.4.18 INSECTS
INSECTS: *See Chapter 20.5.*
ROOT-KNOT NEMATODE: *See Chapter 20.5.2.1.*
CUTWORMS: *See Chapter 20.5.2.2.*
APHIDS: *See Chapter 20.5.2.3.*
LEAF MINER: *See Chapter 20.5.2.10.*
RED SPIDER MITE: *See Chapter 20.5.2.9.*
TRIPS: *See Trips. Chapter 20.5.2.4.*
WHITE FLIES: *See Chapter 20.5.2.12.*
BRAGADA BUGS: *See Chapter 20.5.2.6.*
DIAMONDBACK MOTH: *See Chapter 20.5.2.5.*
LADY BIRDS: *See Ladybird. Chapter 20.5.2.13.*
BOLWORM: *Also See Chapter 20.5.2.7.* They drill holes in the ripening cabbage head and cause heavy losses. The attacks are usually sporadic.

28.4.19 DISEASES
Also see Chapter 20.6 Diseases.

DAMPING OFF: *See Chapter 20.6.3.1.*
DOWNY MILDEW: *See Chapter 20.6.3.6.*
FUSARIUM WILT: *See Chapter 20.6.3.11.*
PHYTOPHTHORA WILT: *See Chapter 20.6.3.16.*
BLACK ROT: *See Chapter 20.6.3.7.*

BACTERIAL SOFT ROT
Bacterial soft rot becomes more of a problem after harvesting cole crops than they are in the land. This disease could also affect crops on the land where they may develop a soft, watery rot. This pathogen needs a wound to enter the plant which may be as a result of sunburn, insect or mechanical damags.

BLACKLEG
This is a fungal disease that may occur during any stage of the plant development. Red, brown or black dots on the stem and leaves in the early growing stage may occur and affect the plant. The disease is more destructive in areas that have high rainfall and humidity. The disease can be identified at an early stage and should be controlled to avoid heavy yield losses.

28.4.20 PHYSIOLOGICAL DISORDERS
28.4.20.1 FACTORS THAT CAUSE PHYSIOLOGICAL DISORDERS ON COLE CROPS
- Cold temperatures, especially below 7°C. Also fluctuations in temperatures.
- The over fertilizing of seedlings.
- Incorrect sowing times and temperature differences.
- Seedlings which are too old at transplanting time.
- Temperatures, drought and diseases.
- Insects, especially during the young stage of plant development.

BROWN BUD (BROCCOLI AND CAULIFLOWER)
The crown or curds become brown due to high temperatures and sun burn. Broccoli plants should not be allowed to wilt due to water stress near harvesting time.

BLINDNESS (ALL COLE CROPS)

A percentage of the plants form no heads due to some injury. The plant produces numerous shoots due to insects feeding on it. This may damage the growing point of the plant when the plant is still small or at the beginning stages of head formation.

LEAFY HEADS (BROCCOLI AND CAULIFLOWER)

The presence of leaves on the curd of broccoli and cauliflower is high due to high temperature and vigorous plant growth. This is because of excess water and N applications.

LARGE, COARSE BUDS (BROCCOLI AND CAULIFLOWER)

Bud sizes of broccoli and cauliflower differ and depends on cultivars. The curds could also develop to large buds when they become mature. High temperature, delays in harvesting and the curds may develop to large or open buds.

HEAD ROT AND BROWN BUD

This is caused by several factors. Problems with calcium Ca uptake and rapid growth even if Ca levels are high. Curds rot because of bacteria that infect the plant when wet conditions occur. Flower buds may be aborted when dry conditions occur. Long periods of wet conditions may give rise to rapid plant growth and poor Ca uptake. Different cultivars which control the rate of maturity are the only practical way to prevent this disease.

HOLLOW STEM

This is commonly found in broccoli and sometimes in cauliflower. The symptoms are usually not visible from the outside. Small cracks develop on the stem and become larger as the crop develops. This problem has been linked to a boron B, nitrogen N and calcium Ca imbalance. Rapid plant growth may also cause hollow stem. Select less susceptible cultivars. Keep the curds cool with ice or ice water after harvesting to prevent bacterial activity and wilting.

TIPBURN

Tip burn may cause major losses when producing cabbage, cauliflower, broccoli and brussels sprouts. Tip burn is a breakdown of plant tissues near the center of the head due to a shortage of Ca. Symptoms usually appear on the leaf tips which turn brown and then black. High temperature and fluctuating soil moisture hampers the movement of calcium Ca into the leaves which leads to tip burn. Many cabbage and cauliflower cultivars are tolerant and resistant to tip burn. The best way to avoid tip burn is to use tolerant cultivars and to manage irrigation and fertilization well.

BOLTING
Also see Chapter 3.10 Bolting.

Bolting, also called buttoning of cauliflower, occurs on many other crops. It occurs when broccoli plants begin to grow and then slow down as a result of a stressful environment caused by conditions such as:
- Exposure to sub-optimal temperatures that force the plant to respond by transitioning to flowering.
- Temperatures for broccoli are for instance 15 to 25 C. It is important that the producer chooses adaptable cultivars. Bolting may also be caused by changes of temperatures, poor soil conditions and low nutrient levels.
- Insufficient water, nitrogen N and poor weed control may also trigger bolting.
- A too high plant population could also be a factor.
- Low temperatures early in the season may cause premature seed stalk development.

CHAPTER 29

PRODUCTION GUIDELINES FOR ONIONS

269 These are onion trials to research shelf life quality and bolting resistance of new cultivars against standard cultivars. This is a large bulb cultivar with a light yellowish outer skin and internal colour. Herbicides were used to control weeds and no further weed control was practiced. A chemical company also tested herbicide action on weeds which is why there are weeds evident on these test production lands.

29.1 INTRODUCTION

Onions are an important vegetable crop across the world. In terms of consumption, it is the 3rd largest vegetable crop, after potatoes and tomatoes.

Onions have a long growing season and need to be well-managed in order for success to be achieved with its production. After harvesting, onion bulbs are one of the few vegetables crops that can be kept in storage for long periods before marketing. If the bulbs are correctly dried after harvesting, they may be stored and sent to the market only when the prices are favourable.

Onion production is more difficult than other vegetable crops and requires considerable attention. Practical cultivation experience is valuable when producing this crop. If the basic cultivation practices and principles are followed correctly, a beginner producer should be able to successfully harvest at least 40 tons p/Ha.

It is important to establish the correct F1 hybrid cultivar at the correct time of the season. When producing certain cultivars out of season, the plant may develop a heavy leaf top growth and produce smaller bulbs. Some F1 hybrid cultivars may be established out of season in certain areas, but others have specific establishment times.

The bulb shapes of various cultivars vary from flat to round. Bulb formation is influenced by temperature and by day length. When bulb formation begins, the process cannot be stopped and the plants become dormant when temperatures rise. Onion production growth is very slow during the short, cool winter days. This minimizes disease and especially insect infection or damage.

29.2 BOLTING AND SPLITTING OF ONION BULBS

Bolting or premature development is when onion plants suddenly begin to go-over to producing seed stalks and begin flowering. Splitting is when bulbs in the production land suddenly begin to split. This sometimes occurs more than once.

When the plants are still young but reasonably mature, then splitting is not visible to the naked eye. Splitting is mostly visible on union bulbs above the soil surface. Low temperatures during crop development, the growth period, the day length, warm conditions, or planting at the wrong time of the season, may all trigger bolting. Some cultivars are tolerant to bolting and splitting, especially the Hybrids F1 cultivars.

When establishing F1 hybrid cultivars, it is important to make sure that they are established at the correct time of the season and within the correct regions. Always confirm which cultivars as best suited to your region, by talking with neighbours who may have experience in producing onion cultivars as well as with reliable seed companies.

270 Onion production using an overhead sprinkler system. The plant population is 550 000 p/Ha.

271 Onion seedlings that have been removed from a soil seedbed, ready to be transplanted.

29.3 ONION SET PRODUCTION

Set production is the method of harvesting onion cultivars out of season, which is generally done, in order to allow early marketing of produce. This method is not often practiced in South Africa. Those producers who do practice it, understand and are experienced with the various stages of set production within certain areas. Beginner producers should investigate this subject for a better understanding.

The production of sets is performed in 2 stages:
1. The seed is sown out of season for cultivars that are adapted to this system (July to September in the summer-orientated areas). This is done to produce small bulbs that are called sets. The small sets are then harvested in January. These sets remain small due to their out of season establishment and the day length.
2. The small sets that are harvested, are sorted into acceptable sizes and established in the production land for the following year, from mid-February to March. The sets are split into more than one bulb (from 1 to 4 bulbs). Approximately 50% of the sets split and these splits begin to develop into larger, mature bulbs which may be marketable from May-June.

The growing cycle for these small sets is between 85 and 95 days. The ideal size of a set bulb should be 20mm in diameter. Sets with a diameter of 25 to 35mm usually occur if they have not split. A relatively large percentage of the sets will begin to grow and mature normally. Some of the split onion bulbs may be dented on one side when harvesting, but the bulb shape usually recovers after a while. The quality and yield produced from sets is fairly good as compared to normal production. The advantage is that onions are produced out of season which usually results in achievement of better prices on the markets.

Production of sets is a specialized practice. Production and yields differ from season to season and it also depends on the weather conditions. Large, specialized producers use precision planters to plant seed. They usually do so 3 rows at a time and space the seed approximately 4cm from one another in the row. Modified set planters are used to plant the small sets and therefore the sets have to be carefully selected. Farmers in the Brits, Western Cape and Northern areas in South Africa, make use of sets planters to their advantage. Set producers usually plant both onion and sets, but open-field production is still their main operation.

29.3.1 SEED PRODUCTION
When producing onion seed, select true-to-type cultivar onion bulbs. These sorted onion bulbs are then stored until April in the northern parts of South Africa or in the summer-orientated areas and then established in a prepared land.

The bulbs are also split into 2 to 5 splits and each split produces a seed stalk. The stalks produce flowering heads and are pollinated by bees. The seed is harvested in September and then cleaned and stored in order for it to become fully dry. This seed is then supplied to the producers for establishment in their production lands. *See photo 40 for an onion seed production unit.*

29.3.2 MEDICAL USES
Raw onions, onion juice, fresh tomatoes and garlic cloves are excellent natural antibiotics. Cook 3 to 4 large onions in one litre of water for 35 minutes. Squeeze the cooked onions through a sieve and regularly drink the juice. According to health experts it assists in cleansing your blood.

29.4 ESTIMATED PRODUCTION COSTS AND PROFITS

The production costs to produce vegetables may differ across the country due to the variances in cost of seed, seedlings, fertilizer, electricity/diesel, irrigation water, labour and so forth. Therefore, costs should be adapted accordingly.

The table below provides estimated costs, yields and profits for onion production (2014). It will assist with designing a cash flow and business plan. Financial lenders require a business plan before they will consider lending money. Overhead costs are not included. Producers should do their own research on production costs when they produce vegetables.

The cost to produce 1 Ha onions bulbs is indicated in the table below and based on an average production yield of 55 tons p/Ha. Experienced producers achieve up to 70 tons p/Ha or more. 55 tons of onion bulbs is a considerable amount. It is labour intensive when cutting leaves and roots, sorting, classing, packing and supplying the product to the market. In order to manage these large amounts of onion bulbs, careful planning and organization is necessary.

The cost of tractors, implements, general equipment (wheel barrows, spades, forks, hoes, crates, twine, pegs etc.), irrigation equipment, water, water pumps, electricity, spraying chemicals, housing, toilets, telephone/computer costs, overalls, clothing for spraying chemicals and so on, are not included in the table below, nor the maintenance of tractors or implements. It is important to make sure that tractors are in good condition before production activity begins.

Certain labour costs are not included in the tables below. Permanent labourers should be used as often as possible, so that they may supervise and train other labourers and also be tractor/truck drivers. Producers should determine labour cost themselves, as costs differs from region to region. Labour is expensive and should therefore be well managed.

TABLE 24: ESTIMATED PRODUCTION COSTS FOR ONIONS (1HA) (2014)

ITEM	COSTS P/HA
Clean and level the production land.	R950
SOIL PREPARATION: Rip, plough, disk or rotovate/till the land.	R2 500
SOIL ANALYSIS: Farmers package is R300 per sample (2014). The interpretation of the analysis by a specialist is approximately R250 and therefore at least 3 samples p/Ha = R750.	R900 R750
Fertilizer: If the pH of the soil-water content is too low, apply agricultural lime. 1 200kg p/Ha 2.3.2 (30) @ R290 per 50kg bag. Require 24 bags = 24 x 290 = R6 960. 2 Top dressings LAN/KAN @ 150kg p/Ha = 6 bags x 50kg @ R240p/bag = R1 440. Mechanical application of fertilizer via tractor and spreader = R450.	R8 400 R450
SEED: Hybrid seed = R2 000p/kg (2014). F1 hybrid seed = R12.50 per 1000 seeds. Require 600 000 seeds p/Ha = 600 x R12.50 = R7 500. **OPEN POLLINATED SEED:** Require 4.5kg open pollinated seed to produce your own seedlings in seedbeds = R450 p/kg = R2 025. Cost to produce seedlings in a soil seedbed = R5 000. When planting F1 hybrid seed with a planter (Brits planter): Seed = R2 000 p/kg. Require 4.5kg = R9 000 p/Ha + planter and labour of R400 = R9 400. **ESTABLISH SEED USING A PLANTER: Open pollinated cultivars:** Open pollinated seed @ R450 p/kg. Require 4.5kg seed p/Ha = R2 025 + labour of R600 (planting seed with a seed planter). Total = R2 025 + R600 = R2 625. **ESTABLISH OPEN CULTIVAR SEEDLINGS FROM NURSERIES:** Open pollinated seedlings @ R60 per 1000 seedlings. Require 500 000 to 700 000 seedlings = 60 x 700 = R42 000. Transplant seedlings (labour) = R4 000. Total = R42 000 + R4 000 = R46 000. **HYBRID SEEDLINGS FROM NURSERIES:** R12.30 per 1 000 seeds. Require 500 000 to 600 000 seedlings p/Ha = 600 x R12.30 = R7 500. Nursery F1 hybrid seedlings R70 per 1 000 seedlings. Require 600 000 seedlings = 600 x R70 = R42 000 + R7 500 for F1 hybrid seed = R49 500. Transplant seedlings (labour) = R4 000. Total = R49 500 + R4 000 = R53 500.	R2 025 R5 000
TRANSPLANT SEEDLINGS: Transplant 600 000 seedlings in the land including labour = R4 000.	R4 000
CHEMICALS: Insect and disease control including labour.	R1 700
WEED CONTROL: Labour: 3 labourers control weeds in and between rows for 4 days, twice in the season. Therefore 6 labourers work for 12 days = 72 x R100 per day = R7 200.	R7 200
IRRIGATION: Diesel, electricity and labour costs as well as shifting the irrigation system, cleaning and repairing it. Water can be costly due to the long growing season as well as the lower rainfall in the summer orientated areas during winter.	R3 900
HARVESTING: Harvesting, drying, transporting and storing bulbs, including labour.	R2 800
ONION BAGS: For 52 tons when using 15kg bags = 3 466 bags @ R0.80 per bag.	R2 800
CLEANING: Cutting leaves and roots, sorting, classing and packing in bags including labour.	R3 800
TRANSPORT: Transport 52 tons of onion bulbs to the markets if 100km from the market: 10-ton truck travels 5 times = 5 x 2 (there and back) x 100km = 1 000km @ R5 p/km = R5 000 plus driver and assistant costs of R800.	R5 800

Clean the land and plough-in the debris. Establish a rotational crop to improve soil fertility if necessary.	R900
PRODUCTION COST P/HA	R53 875

TABLE 25: POSSIBLE YIELDS AND PROFITS FOR ONIONS

YIELDS: 55 tons minus 3 tons unmarketable onions = 52 ton p/Ha @ R30 per 15kg bag = 5 200 bags x R30 per 10kg bag = R156 000.	R156 000
MINUS PRODUCTION COST:	R53 875
POSSIBLE PROFITS P/HA Production costs in the Western Cape is much higher than those in other areas due to the higher labour costs and the expensive seedlings and seedling plants which are produced by specialized organizations. Order seedlings well ahead of time from these Organizations.	R102 125

29.5 PRODUCTION GUIDELINES FOR ONIONS

29.5.1 SOIL REQUIREMENTS
Also see Chapter 5.3 Soil Requirements.

Onions thrive in moist, fairly deep and well-drained types of soil. The ideal soil structure should hold sufficient moisture around the roots to allow proper expansion of the bulbs. However, as the bulbs become mature, there should not be too much water in the soil, otherwise the bulbs may rot. When producing onion bulbs in heavy clay type soil, they may be difficult to clean after harvesting. Gravel and coarse sandy soil are not suitable due to their fast drainage of water and fertilizer.

When transplanting seedlings, it should be done in well-prepared land. The land should be as level as possible, as this is a factor that can determine success or failure. When the production land is uneven (falling to one side), the plants on the higher end of the land are usually weaker and grow slower, which often causes them to be attacked by trips. On the lower end of the land, the plants may become stunted and turn yellow due to water and nutrient logging. Onions fairly shallow root system may cause it to drown when heavy rain occurs over long periods. Good soil drainage is therefore very important.

When using precision planters, the seed should be planted in a well-prepared land that is as level and as smooth as possible to allow the seed to be planted at an even depth. Otherwise, water may dam in hollows in the production land, which is undesirable for this planting method. The aim is to have an even plant stand.

Graded, certified seed with a good seed germination count is important. The seeds are usually sorted by a gravity shaking machine which rejects dull and lighter seeds to improve germination.

29.5.2 CLIMATE REQUIREMENTS AND PLANTING TIMES
Also see Chapter 3 Climate.

Onion plants requires cool temperatures in its long growing season and enough soil moisture during the early stages of its development to allow the bulbs to begin growing. The ideal temperature for growth is between 18 and 22°C. Higher temperatures than 25 to 28°C may speed-up the bulb initiation process. Over long periods, low soil temperatures of 10 to 12°C may trigger bolting where bulbs develop a seed stalk and go-over to flowering. *See Chapter 3.10 for bolting of plants.*

29.5.2.1 DAY LENGTH
Day length is a primary factor for the bulb initiation process. It is mainly influenced by temperature. Onions are sensitive to periods that induce bulb development. All onion cultivars were originally cultivated for lengthy

days. When bulb initiation begins and the day length increases, the cultivars should adapt to this. If this does not occur and its bulb initiation is poor, it is usually because of the wrong cultivar choice.

In South Africa, short, long and intermediate day length cultivars are cultivated.

29.5.2.1.1 Onions are classified to specific day length periods as follows:
- Short day cultivars require 10 to 12 hours per day.
- Intermediate cultivars 12 to 14 hours,
- Long day cultivars require 14 or more hours per day.

29.5.2.1.2 The two main onion groups produced in South Africa are:
- Short-day cultivars which are produced in the northern areas (North from Welkom).
- Intermediate and long-day cultivars are produced in the southern areas (Welkom to the Cape Province).

When onion plants receive less day-length than a specific cultivar requires, then a high percentage will produce plants without a bulb formation and develop thick necks. When a short-day onion cultivar is exposed to a region that has long days, the result is usually small, narrow bulbs.

Certain cultivars are very day-length sensitive. It has been reported that a 2 to 3 week discrepancy in the planting time of certain F1 hybrid cultivars, can make the difference between success and failure. Always make sure that the F1 hybrid cultivars are established at the prescribed times for your area. Confirm this with your neighbours as well as local suppliers and seed companies that supply the F1 hybrid cultivars.

29.5.3 PH OF THE SOIL
Onion production is best in soil with a pH of 5.5 to 6.5. Crops may develop poorly in acid soil.

29.5.4 CULTIVARS
Many F1 hybrid cultivars are available from various Seed Companies. Some are adapted to certain areas and therefore the planting time is important. The open pollinated cultivars have been cultivated for many years and some of them are still popular. The F1 hybrid cultivars mentioned below represent only a sample of cultivars. Cultivars listed may already be outdated due to effective breeding programs as well as competition for this crop.

29.5.4.1 SHORT-DAY, OPEN-POLLINATED CULTIVARS
Radium, Hojem, Bon accord, Pyramid, Texas grano, Texas grano PRR (pink root resistant).

29.5.4.2 LONG-DAY CULTIVARS
Australian brown and Caledon globe.

29.5.4.3 F1 HYBRID CULTIVARS
Granex 33, Primavera, Dessex, Savannah, Gold rush etc.

Due to health benefits, half of the onion production in Australia is the red or purple-red cultivars. It is possible that the red cultivars could also become more popular in South Africa for this reason. The problem in South Africa regarding purple or red cultivars originates from the fact that the seed companies prefer to exclude the purple cultivars from their breeding programs. This is because it creates difficulties when onion seed production lands are cultivated. Purple cultivar seed production lands should be at least 300m from other lands to avoid cross pollination usually by bees. The purple onion cultivars are dormant and when cross pollination occurs, a bleached pinkish coloured onion bulb is the result.

29.5.5 SOWING SEED AND TRANSPLANTATION TIMES OF SEEDLINGS
Short-day onion cultivars could be established and harvested at different times in certain areas of South Africa:
- In the northern parts of SA, short-day cultivar seed is sown in February and the seedlings are transplanted

8 weeks later and harvested in September to October.
- In the middle veld and in certain parts of the Orange Free State, the short-day cultivar seed is sown in mid-March to April and transplanted 12 weeks later (due to the cold). The harvesting is normally in November to December.
- Indeterminate and long-day cultivar seed is sown in mid-April to the end of May and transplanted 6 weeks later. Harvesting is then in January.

29.5.5.1 SOWING TIMES FOR ONION PRODUCTION

REGION	SOWING TIMES
Northern parts with cold winters (short-day)	Feb-March
Northern parts with warmer winters (short-day)	March-April
Free State, Northern Cape and Western cape (intermediate)	April-May

29.5.5.2 OPTIMAL PLANTING TIMES FOR SHORT-DAY CULTIVARS

Areas with light to mild frost:	Mid Feb and May-June
Areas with heavy frost:	March-May
Frost-free areas:	Feb-August
Cooler areas with winter rainfall:	May-June

29.5.6 SEED P/HA
Foot paths take up a fifth of the land spacing, when the seedlings are spaced narrowly. When leaving out a row every 6 to 7 rows, the sowing rate and seed requirement could be 3kg p/Ha. This may be enough to produce 650 000 seedlings.

29.5.7 SEEDLING PRODUCTION
Seedling production in seed trays, seedlings in soil and sowing seed directly into the land.

29.5.7.1 SEEDLINGS PRODUCTION IN SEED TRAYS
It is not economical to produce onion seedlings in seed trays, due to the number of seedlings required. Nurseries use seed trays with smaller cavities, such as 200 cavities and then establish 4 to 6 seeds per cavity. This results in approximately 1 000 seedlings per seed tray. When 650 000 seedlings p/Ha are required, at least 650 seed trays will be required for the production of 1Ha.

A small number of small-scale farmers make use of seedlings from nurseries. The seedlings that are produced in seed trays do not develop very well and their sizes are smaller when compared with production in soil seedbeds. This is due to the dense population and small cavities. Many onion producers produce their own seedlings in well-prepared soil seedbeds, with great success.

29.5.7.2 SEEDLING PRODUCTION IN SOIL SEEDBEDS
Also see Chapter 8.4 to produce seedlings in soil.

- Onions have approximately 250 000 to 270 0000 seeds p/kg. Less than 10g seed per m² should be sown, provided that the germination is good.
- Apply fertilizer according to a soil analysis.
- Prepare a moist, smooth seedling bed 1 to 1.2m wide with paths 0.5 to 0.7m wide between the beds. The soil from the paths is used to create the raised beds which are 10 to 15cm high. If available, apply good compost at the rate of 80 ton p/Ha or 8 big spades per square meter.
- Create shallow furrows, 1 to 2cm deep across the beds and 10cm from one another.
- Sow the seeds in the furrows, approximately 2cm apart from one another and cover them with soil using a rake or spade.
- Irrigate 10mm water directly after sowing.
- If possible, during hot days the rows should be covered (mulched) to keep the top soil moist.

- Thin the plants to the required spacing, approximately 2 weeks after emergence.
- Some commercial farmers make seedbeds with a tractor and a bed maker.
- Tractor wheels are usually 1.2m wide and may also be used to make seedbeds.
- Monitor the plant colour and the way they are growing before applying nitrogen (N).
- Adequate trace elements should be incorporated into the seedbed before sowing. Foliar sprays are also an option.
- If the seed is sown too densely, spindly seedlings may result. The seedling plants will then compete for light in order to outgrow their neighbours. This may promote diseases due to the dense plant material, the over-occurrence of insects and the thin, soft skin of the plants.

29.5.7.3 SOWING SEED DIRECTLY IN THE PRODUCTION LAND

Almost all onion seeds are sown by means of some kind of mechanical planter. Brits planters are a good option. Tractor operated precision planters or hand push planters are available and very little thinning of plants is then necessary afterwards. Advanced commercial farmers use expensive precision planters to sow onion seed directly into the production land. However, this practice requires care as the slightest mistake may be disastrous.

When using planters, make sure that the land is well-prepared, smooth, moist and level. Irrigation is critical when large areas are established with planters. The irrigation water must be applied evenly across the entire land. When planters, especially the smaller ones are used, it is important that the germination of the seeds are good.

Good quality seed and a high germination percentage of more than 85%, is crucial. This represents the fine-line between failure and success. Seeds that germinate after 5 days are more likely to give rise to healthy plants, than seeds germinating after 2 weeks.

Pallet coated seed is available. It is expensive but a good option. It provides the best germination and even emergence of seedlings.

29.5.8 SPACING AND PLANT DENSITY IN THE LAND
29.5.8.1 SPACING
- In the production land, seedlings may be spaced 5 to 8cm in the rows and 25 to 45cm between rows.
- When a thick, flat, bulb cultivar is established and spaced at a close range, most of the bulbs will be flat and round.
- Consumers mostly prefer a medium-sized bulb, 50 to 70mm in diameter.
- General plant spacing is usually 7cm apart from one another in the row and 25cm between rows. The plant population is then approximately 57 plants per m^2 or 570 000 plants p/Ha.
- Do not establish less than 450 000 plants p/Ha.
- If every 7th row is used as a foot path, then 14 rows or 20 000 plants will not be planted. Therefore 570 000 minus 20 000 equals 550 000 plants p/Ha.
- When spacing is wider between the rows, then no foot paths are necessary. Because it facilitates easier weed control, many producers prefer 30 to 35cm between rows and 6 to 8cm in the row.
- Seedlings should be transplanted in smooth, well-prepared lands, in rows.
- Spacing in the rows should be wider for the giant Texas Grano families (sweet, juicy and ideal for salads) due to their larger bulb sizes.
- High plant populations affect onion bulb size which may affect the yields and also to a lesser degree, the maturing and harvesting time of the bulbs. It may also promote bolting of the plants.
- When 700 000 plants p/Ha are established, bulb sizes will not be as large. More than 40% will be small, pickle-sized bulbs.
- When 500 000 plants are established, then approximately 49% will be larger bulbs, 10% smaller and zero pickle bulbs may be harvested.

Trials have proven that certain plant densities result in certain bulb sizes. The table below shows the results of these trials. The trials are based on the common open-pollinated onion cultivar and the F1 hybrid cultivars because they are very uniform.

TABLE 26: PLANT DENSITIES AND BULB SIZES

PLANTS P/HA	BULB SIZE (%)				% BULBS	YIELDS TON/HA
	LARGE	MEDIUM	SMALL	PICKLES		
700 000	5.6	53.2	27.0	14.2	27 %	78.7
600 000	14.0	64.5	15.9	5.6	23.4%	92.5
500 000	17.2	66.7	12.9	3.2	10.8 %	90.5
400 000	39.7	67.1	8.2	0	10.8 %	87.5

29.5.9 TRANSPLANTING SEEDLINGS IN THE LAND
TRANSPLANTING SEEDLING FROM SEEDBEDS TO THE LAND

- Seedlings should be transplanted when they are approximately 10 to 12cm tall. Some producers prefer a seedling as thick as a pencil.
- Seedlings should be transplanted in well-prepared, moist lands.
- Make straight furrows, 3 to 4cm deep and 20 to 30cm from one another.
- The spacing of the seedlings in the rows can be 5 to 8cm from one another, but it depends on the cultivar's bulb size.
- Plant the seedlings in the furrows so that the seedlings stand upright against the sharp edge of the furrow.
- Usually 2 or 3 labourers place the seedlings in the furrows and another one covers the furrows with a rake and compacts the soil slightly with the back-side of the rake.
- As soon as possible, irrigate approximately 10mm water after transplanting.
- The soil should never dry-out when the seedlings are still young.

Some farmers still make use of a small, 2 scar, plough implement. The scars are 30cm from each other. The right hand-scar is used to cover the seedlings which have been placed in the furrow with soil. The left-hand scar creates a new furrow. When the furrows are covered, the plants lie almost flat on the ridge of the furrow. They will soon stand upright.

29.5.10 FERTILIZATION
Also see Chapter 15 Fertilizer.

Onion plants have a limited root system and respond well when a good fertilizer program is followed. Onion plants withdraw 2.85kg N, 0.63kg P and 3.88kg K per ton of plant material. Phosphates P do not move well in soil. The movement is especially slow in acid soil and therefore it is a good option to apply the P at an earlier stage, before establishing the crop.

If the soil is too acid, apply the correct type of lime at least one month before transplanting the seedlings. The crop is a heavy feeder of nitrogen N and potassium K, however, too much N late in the season may cause vigorous leaf growth and delay bulb development. Thick-necked plants may also result.

Two side/top dressings LAN/KAN and potassium chloride KCL can be applied 3 and 6 weeks after transplanting. The first N application should be supplied when the roots of the seedlings are well-developed. This is because in cooler conditions the roots develop slowly and when the roots are under developed, the N will drain-away out of the reach of the roots.

Apply 130 to 150kg LAN/KAN or KCL p/Ha. Based on the above, the following N, P and K should be applied to achieve yield targets of 40 to 50 ton p/Ha on loam soil: 170kg N, 90kg P and 230kg K. It is generally known that K plays an important role in the keeping quality of vegetables crops, especially for onion production. Experienced producers are aware that it is important to build P levels in the soil.

If small-scale farmers do not have access to a soil analysis, then the following fertilizer mixture is suggested: Broadcast 1 000kg 2:3:4 (30) Zn p/Ha and 2 top dressings of 130kg N, 3 and 6 weeks after transplanting the seedlings. If the crop is established in soil that leaches easily or if the leaves begin to turn a yellowish colour, then a side or top dressing of 130 to 150kg N p/Ha should be applied before bulb setting takes place.

Onions are sensitive to a molybdenum Mo, boron B, iron Fe and copper Cu shortages. It is also sensitive to a Zn shortage, which if it is a problem, should be included in the fertilizer program. According to some specialists, the crop is not a good candidate for foliar sprays because of its upright growth and waxy leaves. However, when spread open, the leaf surface of onion plants could cover a 10 Ha area and garlic leaves 12 Ha. Therefore, when using foliar sprays, make use of stickers in the spraying water together with the correct spraying apparatus (high pressure and fine droplets) and ensure that the water pH level is correct. The plants should be properly sprayed in order to cover the entire plant and at the correct stage of their development.

29.5.11 IRRIGATION
Also see Chapter 14 for Irrigation.

Onion roots penetrate the soil to a depth of more than 60cm, but mostly they feed in the top 30cm of the soil. Therefore, irrigate to at least 35cm deep into the soil.

Onions require approximately 500 to 600mm water during their long growing season. Irrigate 25mm the first 4 weeks after transplanting and 35mm per week thereafter, for the rest of the growing season. Keep records of rain water and adapt the irrigation accordingly. Most onion production occurs when the temperature is cool and little rain occurs, such as in the summer orientated areas.

The onion plants need soil moisture near the surface during early growth. New roots will not develop unless the soil is moist and therefore it should be kept moist for almost the entire growing season. As a result, the irrigation costs can become high. Stop irrigating at least 1 or 2 week before harvesting the bulbs, as they need to settle, dry-out and mature. If the soil is too dry before harvesting, irrigate it slightly to assist the harvesting process. In the long run, a center pivot system is generally one of the lowest cost irrigation systems, but it is expensive and can only be used in one land and in large commercial production lands.

When seed is established directly into the land with a precision planter, the irrigation must be spread across the entire land. The soil in the production land should be kept moist. When necessary, it is important to perform daily irrigations until the seedlings have emerged. This keeps the soil cool and soft. The soil should never be allowed to warm-up and damage the emerging seedlings.

29.5.12 CROP ROTATION
Also see Chapter 18 Crop Rotation and Soil Improvement.

There are a few related crops of onions. They include leeks, spring onions and garlic (which is classified as a herb). Onions should be rotated every 3 seasons with non-related crops.

29.5.13 WEED CONTROL
Also see Chapter 17 for weed control.

Onion plants have a fairly shallow root system and therefore shallow cultivation is necessary. This is when removal of small weeds and grasses between and in the rows, takes place. Deep and close cultivation, near the roots, may injure the roots.

Do not cultivate the soil deeper than 6cm. It is better to control weeds after irrigation when the plants are still young. Otherwise, the tender, young seedlings may easily be mistaken as weeds, especially when the weeds have overgrown the seedlings. The weeds should regularly be removed by hand in the rows. Because there is more space between the rows, the weeds may be cultivated mechanically until there is a risk the implements will begin damaging the plants and roots.

29.5.13.1 CHEMICAL HERBICIDE WEED CONTROL
Most Commercial farmers use herbicides to effectively control weeds. This is sometimes necessary. The seedlings are established in the production land usually from Feb to April in the summer-orientated areas, when the weeds may still germinate. Winter-orientated weeds are also waiting to emerge and the plants may have to compete against these.

There are various weed herbicides that are registered to control weeds for onions, but not for garlic. Weeds have to be properly identified before the correct herbicide may be applied. This is due to the long growing season. It is important to liaise with chemical companies about herbicides.

29.5.14 DURATION BETWEEN TRANSPLANTATION AND HARVESTING
Onion production has a long growing season of 180 to 220 days before maturity/harvesting. This is due to the cold conditions, day length and slow bulb formation during winter. Harvesting times also differ between cultivars.

29.5.15 HARVESTING AND DRYING
Also see Chapter 24 Harvesting, Handling and Packaging.

Onion bulbs comprise approximately 90% water. When onion bulbs are to be harvested, the mature bulbs must be dried before they are stored. The objective of curing onion bulbs after harvesting, is to dry the outer two to four scales and to seal any wounds.

The onion bulb is a series of concentric, thickened, juicy leaves attached to a short inner-base. The outer leaves are very thin and when they dry-out quickly, they change colour. When drying onion plants, they may be packed in wind rows or in pyramid heaps, outside. This drying process usually occurs near the production land or the pack store for approximately 7 days. The onions are stored for longer in pyramid heaps, as the top of the heaps are protected against rain.

When producing onions on a small-scale, the plants may be laid down in the store or in crates. They may also be hung from the roof or packed on mesh racks. When onion bulbs are left for too long in the soil before harvesting, the outer skin becomes dry and the outer layers may become soft and uniform in colour. The roots break-off easily when touched and the stem area shrinks in size, becoming soft and dry on the surface of the bulbs.

Air movement is important when curing onions. The air needs to move around the bulbs. Use a fan or power driven ventilation fan too eliminate losses from rotten bulbs. However, if the bulbs are to be stored for long periods, then too much wind movement in the store should be avoided, as the bulbs may lose too much moisture.

Onion bulbs may be cured in the field, when favourable weather conditions occur. After under-cutting and pulling the tops, the bulbs are left in the field for a few days. Care should be taken with this process, as the bulbs are sensitive to sunburn when left in the field for too long. Late onion maturing production can be prone to late season warm weather, which may allow bacterial diseases to become a problem.

Good labour supervision should be a priority during the harvesting, drying, loading and packaging operations. This prevents bruising and other damage from occurring. Producers normally harvest bulbs when 40% of the leaves have fallen over. This is usually associated with mild winter weather conditions, but the percentage of tops that fall over is not always a true indication of maturity. Cultivars differ and the tops may also fall over when heavy frost, strong winds, heavy rain or disease infections occur. Examine onions for softness in the neck as well as the size of the bulbs to indicate the correct time to harvest.

Late cultivars are susceptible to warm weather, bacterial diseases and may require harvesting before optimal maturity takes place. This is done to prevent widespread infections. Do not harvest onion bulbs too early, as the keeping quality will likely be poor. When the bulbs are harvested too early, they should be marketed immediately. Some producers harvest onions early to catch an early market, however the yields are lower as the bulbs are still in a developing stage.

Onions bulbs are lifted from the soil by loosening them with a fork or a tractor that is fitted with a under cutting soil blade. The cutting blade loosens the soil below the bulbs in order for the plants to be pulled out by hand. When lifted, producers normally dry the entire plant on the land or pack them in wind rows to dry. The bulbs and plants dry in approximately 5 to 7 days. Some producers harvest 8 to 10 rows and pack them in wind rows on the land to dry, provided that the soil is not too wet. After drying the plants on the land, the roots and leaves may also be cut on the land, provided that weather conditions permit it.

When the crop is established on a sandy or light loam soil, some producers establish a seedbed as wide as tractor wheels and sow 4 to 5 rows directly on these beds. They then use a cutting blade that undercuts the entire seedbed to loosen the bulbs at harvesting time. This makes it easy for the labourers to harvest the plants. Harsh topping or cutting of roots may lead to surface wounds and then to decay. Shock damage can occur when the packed bags are thrown onto trailers or on trucks. Pressure bruising occurs when labourers stand on lower bags to load higher bags. Harvest and handle the bulbs carefully to avoid damaging them.

It has been reported that large enterprise Australian farmers are more sophisticated when compared with their South African counterparts. They lift 50 tons of onion bulbs an hour, load them automatically on trailers, then transport and offload the bulbs in a shed, with only 4 men.

29.5.16 GRADING, PACKAGING AND STORAGE
Also see Chapter 24 Harvesting, Handling and Packaging.

After cutting the roots and leaf tops of the dried plants, the sorting, grading, classing and packing into different sizes according to market requirements, is carried out.

Early, short-day cultivars do not store well and should be supplied to the market within a few days after drying. To improve storability do the following:
- Harvest only mature bulbs.
- Do not over-fertilize the crop. Withhold nitrogen N at least 30 days before harvesting.
- Stop irrigation 1 to 3 weeks before harvesting.

29.5.16.1 GRADING AND PACKAGING
Onion bulbs are usually graded in class 1 and 2 and in other lower classes. They are packed in different sizes and classes. This is normally done in 10kg and more recently 7kg white bags. The dark-skinned onion cultivars look attractive in brightly coloured orange bags.

Specifications of the bulb sizes as well as the class packaging instructions are available on the Internet or at Prokon inspectors at the national markets.

29.5.16.2 STORAGE
Cured or dried onion bulbs may be stored for approximately 5 months or longer. It depends on the cultivar and its quality. Selected bulbs may be stored in *ventilated* stores on racks, in crates, hung from the roof and scattered on the floor. When storing onion bulbs, they continually lose moisture. However, the biggest losses occur from rotten and germinating bulbs. Some cultivars such as the Texas Grano families do not keep well and should be marketed as early as possible, as they can quickly spoil. Even when onion bulbs are stored under the best storage conditions, the bulbs can still lose approximately 15 to 20% moisture and as much as 30% when stored for very long periods. The exception is cold storage conditions.

29.5.17 YIELDS
A yield of 50 to 60 tons may be expected for a plant population of 600 000 plants p/Ha. However, some F1 hybrid cultivars have the potential to produce up to 80 tons p/Ha or more.

29.5.18 MARKETING
Also see Chapter 25 Management and Marketing.

When early harvesting of onion bulbs occurs in the Gauteng area, they should be marketed directly and as soon as possible. They should not even be stored for short periods. The reason for this, is that the keeping quality is poor and the evaporation of the bulbs moisture is rapid.

The largest suppliers of onions on the market are from the northern areas. They harvest in August and supply the markets from August to November. The Free State and Cape Province productions are marketed from October to December. At this time of the year, the markets are usually overstocked and prices may be low. Late cultivars from the Free State and Cape Province are usually supplied from February the following year. Producers may hold their produce back for long periods due to oversupply and overstocking of the market, in the hope of achieving better prices later.

29.5.19 INSECTS
Also see Chapter 20.5.

Onions have many narrow, thin leaves which are covered with a waxy layer. It is important to use chemicals effectively when spraying onion plants. It is better to space onion plants further apart from one another to receive more sunlight, as the growing season is from March to October in the Gauteng area, whilst it is still fairly cold. The plants therefore need maximum sunlight to produce good quality and high yields.

Controlling insects and diseases is also more effective when plants are spaced further apart from one another. To control trips, producers should spray chemicals early in the morning or late in the afternoon, as trips are active at night.

ROOT-KNOT NEMATODE: *See Chapter 20.5.2.1.*

TRIPS:
Trips are the main problem insect for onions and garlic and they are difficult to control. *Also see Chapter 20.5.2.4 for Trips.* Trips are difficult to control, but when controlled effectively other diseases will also be avoided.

Trips scrape the waxy layer from the leaves using their mouths parts. This leaves the surface unprotected and exposes the plant to infections such as:
- Downy mildew. The main disease of garlic and onion.
- Botrytis and Altenaria.

If trips are controlled effectively, then it may be difficult for diseases to penetrate through the waxy layer of the plants. Strong, healthy plant cell walls are thicker and tougher to penetrate by insects.

29.5.20 DISEASES
Also see Chapter 20.6 Diseases.

The most common diseases which affect onions are Downy mildew, Alternaria and Pink root rot.
POWDERY MILDEW: *See Chapter 20.6.3.5.*
PHYTOPHTHORA WILT: *See Chapter 20.6.3.16.*
ALTENARIA LEAF SPOT OR PURPLE BLOTCH: *See Chapter 20.6.3.2.*

PINK ROOT:
Pink root is a wide spread fungus. The roots are pink and later turn light pink. When harvesting, the roots shrink, turn brown and die. Heavy infections can affect yields. There are tolerant cultivars available. Avoid infected soil and rotate crops.

FUSARIUM WILT:
See Chapter 20.6.3.11.

FUSARIUM BASAL ROT:
This is a worldwide fungal disease and is a problem in the field and in the storage room. The fungus attacks the base of the bulb, causing browning and rotting of the basal end and roots. This results in the yellowing and death of the plants leaves. When onion plants reach maturity, the leaves die back from the top and the bulb begin to rot and in storage they rot completely. The disease is soil borne but could also be spread by water, implements or infected plant material.

DOWNY MILDEW:
Also see Chapter 20.6.3.6 Downy Mildew.

This disease, together with Altenaria leaf spot, is the most common onion disease in South Africa and also occurs throughout the world. Severe yield losses may occur as a result of the leaves being killed by these pathogens when optimal conditions occur such as cool, moist nights and mild overcast and rainy days. The disease favours warm, wet conditions.

Symptoms begin as small white lesions on leaves and stalks. They enlarge to cause white/purple sunken lesions that become brown and occasionally purplish. They also display brown concentric rings, especially in humid conditions and are more severe on the older leaves. Some producers use unregistered chemicals at their own risk such as Mancozeb, Ridimil, Gold and Bravo.

WHITE BULB ROT:
This disease is a serious problem and also occurs worldwide. The disease affects the bulbs and displays a characteristic white mould on the base of the onion bulbs. The roots become rotten and smell bad. The leaves turn yellow, become wilted and fall over. The disease may remain in the soil for more than 15 years and therefore these soil should be avoided when producing onions and garlic in the future.

BROWN RUST:
Also see Chapter 20.6.3.8.

This disease does not appear every year even if the disease has infected the plants previously. It favours high temperatures and high humidity in order for it to develop. If the plants are infected for long periods, it may become difficult to control due to the thousands of spores that are released. Therefore, it is necessary to control it in the early stage of development.

Symptoms of this fungal disease begin as small white spots on the leaves later developing into bright orange, open pustules which are slightly raised. The disease occurs when humidity is high and the conditions are dry. Some producers use unregistered systemic chemicals at their own risk, such as Folicur, Score, Bravo or Richter.

WHITE ROT:
White rot is a fungal disease and is widespread and destructive to onions and garlic. It may limit production and put lands completely out of production. A white fungal growth begins at the base of the bulb and stem. Small, round black balls occur, about the size of a match head which are usually in the fungal growth. Leaves of affected plants turn yellow and die. The bulb becomes rotten and smells bad.

CHAPTER 30

PRODUCTION GUIDELINES FOR SWEET POTATOES

272 Different sweet potato cultivars and breeding programs are currently the focus for ARC (Agricultural Research Council) to improve and breed new cultivars, especially those with orange flesh. Also, there is a training program to train disadvantaged and upcoming farmers to establish production within their own communities.

30.1 INTRODUCTION

The sweet potato is a tough vegetable crop and low risk to produce. It has a long growing season but requires warm soil and weather conditions to achieve the best yields. The crop is not affected by many insects and diseases. However, virus diseases are the main issue influencing sweet potato production and general propagation.

The crop is fairly easy to produce. Most producers are successful when producing this crop. If other vegetable crops fail due to hail storms, high winds and very hot conditions, then sweet potato is the answer. Many producers use sweet potatoes as a cover-up crop. This means that when other crops fail to produce, sweet potato will still thrive due to its edible roots under ground.

High yields can be expected and some farmers' retail the sweet potato directly from their farms which allows them to save labour, packaging and transportation costs. The crop does not require high input costs. If there is a market for this crop, then the low risk factor required to produce it, should be taken into account. Orange fleshed sweet potato cultivars are becoming popular in South Africa. They taste good and have high levels of beta carotene, energy and vitamins.

In most overseas countries, the orange sweet potato is already highly in demand. It should be noted that when sweet potato roots are harvested, they should go through a curing process for at least a week. This process facilitates the sweet potato's conversion to sugar which gives its characteristic taste. The roots may also be cured underneath the soil, after the vines die, usually due to frost.

273 Removing sweet potato vines from the land just before harvesting the roots. The vines are a good source of stock feed and ideal for making compost.

274 Sweet potato cuttings ready to be transplanted into the prepared land. The cuttings are usually 30cm in length.

275 Good quality sweet potato top growth. Labourers are in the process of removing the taller weeds by hand.

276 Sweet potato production land ready for harvesting after a 4.5 to 5 month growing period. Usually, the top growth is cut and removed from the land 3 days before harvesting the roots.

Sweet potatoes have a long growing season and are sensitive to low temperatures. When cloudy conditions occur over long periods, the plants grow slowly and very little energy is transferred to the roots. The crop is sensitive to virus diseases and when infected, the yields decline dramatically. If plants are cultivated optimally, with good insect control measures, even virus infected plants can achieve acceptable yields.

Some commercial producers establish disease-free plant material which is obtained from registered vine growers. When disease-free plant material is obtained from registered growers, the yields normally gradually decline after the second year due to virus infections. When insects are controlled correctly using an effective, scheduled chemical spraying program, the plant material may be kept for longer periods. This is because aphids and to a lesser degree trips, are the insect vectors that transfer viruses.

30.2 ESTIMATED PRODUCTION COSTS AND PROFITS (2014)

The production costs to produce vegetables may differ across the country due to the variances in cost of seed, seedlings, fertilizer, electricity/diesel, irrigation water, labour and so forth. Therefore, costs should be adapted accordingly.

The table below provides estimated costs, yields and profits for sweet potato production (2014). It will assist with designing a cash flow and business plan. Financial lenders require a business plan before they will consider lending money. Overhead costs are not included. Producers should do their own research on production costs when they produce vegetables.

The average yield of sweet potato as indicated in the table below is 35 tons p/Ha. Experienced producers claim to harvest up to 70 ton p/Ha or more.

The cost of tractors, implements, general equipment (wheel barrows, spades, forks, hoes, crates, twine, pegs etc.), irrigation equipment, water, water pumps, electricity, spraying chemicals, housing, toilets, telephone/computer costs, overalls, clothing for spraying chemicals and so on, are not included in the table below, nor the maintenance of tractors or implements. It is important to make sure that tractors are in good condition before production activity begins. Certain labour costs are also not included.

TABLE 27: ESTIMATED PRODUCTION COSTS AND PROFITS (2014)

ITEM	COSTS P/HA
LAND PREPARATION: Clean and level the land.	R950
SOIL PREPARATION: If necessary, rip, plough, disk, rotovate and make ridges. Apply lime according to a soil analysis test.	R2 500
SOIL ANALYSIS: At least 2 samples p/Ha (depends on the soil structure) = R290 per sample (2014) x 2 = R580.	R580
Interpretation of the soil analysis by a specialist approximately R200 per sample = R400.	R400
Fertilizer: If necessary, apply lime at least 5 to 6 weeks before establishing the crop. Apply fertilizer mechanically with a tractor and spreader. Broadcast fertilizer according to a soil analysis before making ridges. Suggested applications are: 1 200kg 2.3.2 (22) Zn @ R290 per 50kg bag. Require 24 bags x 50kg p/bag = 24 x R290= R6 960. 2 Top dressings LAN/KAN @ 150kg p/Ha = 6 bags x 50kg @ R240 per bag = R1 440. When fertilizer is band placed, then much less fertilizer is used.	R350 R6 960 R1 440
PLANT MATERIAL: Vines from vine growers (1 000 cuttings) = R300 per 1 000 (2013). Require 30 000 cuttings = R9 000.	R9 000
VINES: Establish 30 000 vines. Approximately 2 labourers plant 2 000 vines per day, therefore 30 labourers = 30 000 x R100 per labourer = 30 x R100 = R3 000.	R3 000
CHEMICALS: Insect and disease control, including labour.	R1 100
WEED CONTROL: Mechanical weed control between rows = R550.	R550
3 labourers control weeds in rows for approximately 4 days, 2 times per season = 3 x 4 x 2 = 24 x R100 = R2 400.	R2 400
Cut the top growth and remove it from the land before harvesting.	R1 000
IRRIGATION: Diesel, electricity and labour costs as well as shifting the irrigation system, cleaning and repairing it. Water is expensive when low rainfall occurs. Sweet potato is a fairly drought resistant crop. To irrigate 1mm water p/Ha, 10 000 litres water is required.	R2 400

HARVESTING AND CLEANING: 35 ton is to be transported to the store. Labour: 1 labourer harvests, loads and transports 2 tons of sweet potatoes per day, therefore 18 labourers are required for 35 tons = 18 x R100 per day = R1 800. Cleaning, washing and drying is an expensive process, approximately R2 000.	R1 800 R2 000
PACKAGING: Bags (10kg purple bags) required for 34 000kg yield = 3 400 of 10kg bags costing R0.60 per bag. When sold directly from the farm, transportation, packaging and transplanting costs are minimised.	R2 040
SORTING: Sort and pack produce in 10kg purple bags or containers according to market requirements.	R1 500
TRANSPORT: Transporting 10kg sweet potato bags to the market which is 100km away: Using a 10-ton truck which travels 4 times = 4 x 2 (there and back) x 100 = 800km @ R6 p/km = R4 800. Truck driver and assistant = R700.	R4 800 R700
Clean and plough residues into the soil. Remove and store irrigation system, etc. Establish a rotational crop.	R1 100
ESTIMATED PRODUCTION COSTS	**R46 570**

TABLE 28: ESTIMATE YIELDS AND PROFITS P/HA

YIELD: 40 ton p/ha, minus 4 000kg unmarketable roots = 36 000kg marketable roots @ R3 p/kg = R108 000. **MINUS PRODUCTION COSTS** **POSSIBLE PROFITS**	R108 000 R46 570 R61 430

30.3 PRODUCTION GUIDELINES FOR SWEET POTATOES

30.3.1 CLIMATE REQUIREMENTS AND PLANTING TIMES
Also see Chapter 3 Climate.

Sweet potatoes have a tropical origin and are classified as a warm weather crop. Warm days and nights as well as reasonable humidity are favoured by this crop. Growth and development of the plant is set back when cool or cloudy weather conditions occurs. Ideal temperatures are 20°C at night and 32°C during the day. Little growth takes place when temperatures are below 10°C and at 8°C. Growth normally stops completely. A light frost may damage the upper leaf tops and repeated light frost may kill all the tops as well as the vines. Heavy frost will kill all top growth.

30.3.1.1 PLANTING TIMES
Vines obtained from registered vine growers or other growers are usually not ready for establishing before October in the Gauteng area. However, vines that are produced in warmer winter areas such as the Lowveld, could be obtained much earlier. Most Commercial producers prefer to establish their planting vines in mid-September or earlier. This is not always possible, as vine growers in the summer orientated areas are not able to provide vines early enough for a September production, with the exception of producers that are situated in the warmer areas. *It is important to order vines at least 6 months before planting time to avoid disappointment and higher prices.*

30.3.1.2 PLANTING TIMES IN CERTAIN AREAS
Areas with light to mild frost: Oct-Jan
Areas with heavy frost: Nov-Dec
Frost-free areas: Aug-Mar
Winter rainfall areas: Nov-Dec

30.3.2 CULTIVARS

There are not many cultivars to choose from. The purple skinned, white fleshed cultivars are the most popular. Listed below are the main cultivars.

30.3.2.1 Blesbok

This is a purple-skinned, white fleshed cultivar that has thick brittle vines and is the most popular cultivar (2014). This is due to its higher yields, good taste and early maturity of approximately 4.5 months. Most other cultivars only mature after 5 months.

30.3.2.2 Bosbok

This is a purple skinned, white fleshed cultivar similar to Blesbok but matures 2 weeks later than Blesbok. This cultivar does well and is popular in the warmer winter areas. The vines are much thinner than the Blesbok cultivar.

30.3.2.3 Ribbok

This is a dull cream skinned cultivar which is most popular in the Western Cape due to its dry flesh root content. It has been replaced by the older, very dry, Borrie cultivar as many consumers in the Cape Province prefer a dry fleshed sweet potato.

30.3.2.4 Other cultivars

Koedoe is a tasty, yellow cultivar and is produced on a small-scale (2011).

Brondal is used for processing in the Eastern Transvaal and is preferred for its dry inner flesh colour, taste and shape. There are many old-fashioned cultivars from the 1940's, which are kept in specially designed glasshouses maintained by the ARC.

30.3.2.5 Orange fleshed cultivars

The Agricultural Research Council (ARC) is continuously improving orange fleshed cultivars. Most overseas countries prefer orange fleshed cultivars due to its high vitamin content and better taste when compared with local cultivars.

Producers should be aware of the yellow and orange fleshed cultivars. They has been specially bred to assist the poor, developing and upcoming farmers in KwaZulu-Natal and other similar areas. This is due to its high vitamin value and easy production in warm, humid climate areas which is ideal to improve the diet of poor rural communities.

30.3.3 PROPAGATION OF PLANT MATERIAL

Sweet potato production is propagated by means of vines. The biggest challenge when producing sweet potato, is the fact that all local sweet potato production is infected with viruses. Many producers obtain vines from other growers but they are also infected with viruses. An organization that is committed to supplying disease-free plant material was founded in 1991, called the National Vine Growers Association.

30.3.3.1 THE NATIONAL VINE GROWERS ASSOCIATION

Members of the National Vine Growers Association obtain disease-free plant material from the Agricultural Research Counsel (ARC) and propagate it under strict supervision on their premises. They then supply vines to commercial producers. Healthy, insect and disease-free plant material may be obtained from these registered vine growers.

Some commercial producers use these vines for 3 to 4 years before again re-ordering from the registered producers. If required, commercial producers are able to produce their own vines for the next two seasons. The vines are usually available from the registered producers in hessian bags or bound together in bundles. *See photo 251.*

30.3.3.2 DISEASE-FREE PLANT MATERIAL FROM ARC
Disease-free plant materials has been provided by the ARC (Agricultural Research Council) since 1972 and may be obtained in small quantities, directly from ARC Roodeplaat. They also supply disease-free vines of certain cultivars as well as older cultivars which have been maintained in protected Green Houses since 1940.

It is important that vines be ordered long before planting time. Some commercial producers obtain small quantities of the virus-free nucleus stock plants, directly from the ARC, which are supplied in plastic bags or seed trays. They then propagate their own plant material. *A vine grower's list for disease-free sweet potato plants is available from the Agricultural Research Counsel at Roodeplaat Pretoria.*

The main vectors that transmit viruses are aphids and to a lesser degree white fly. Plant breeding lands are usually preventatively treated to control these virus vectors. The chemical spraying programs are a priority in order to control aphids that carry the viruses which infect their lands. Research has shown that when sweet potato vines are infected with viruses and are established repeatedly, year-after-year, the yields may decline by approximately 30% after approximately 3 seasons. Yields will continue to decline every year thereafter, but this decline may stabilize. The reason for this, is that the viruses lie dormant in the plant until the plant experiences stressful conditions, at which time the virus takes advantage. When production is cultivated in optimal conditions, high yields may still be obtained, in spite of these virus infections.

30.3.3.3 AMOUNT OF VINES P/HA
Approximately 1 000 vines, cut directly from the land in order to fill one bag, is usually supplied by the members of the Association or other vine producers. This means, that approximately 30 to 32 bags are adequate to cover 1 Ha, when spacing the plants at 100 x 30cm which equals 30 000 plants p/Ha.

The vines are cut and stapled or pressed into hessian or other large bags. The vines are then cut again by the producers into pieces of approximately 25cm in length, in order to be established in the prepared production land. Vines are also supplied by neatly tying the cuttings into bundles of 1 000 which is more expensive. Receiving vines this way is more popular than in bags.

30.3.4 HANDLING OF VINES OR CUTTINGS
Sweet potato vines are cut into pieces of approximately 30cm in length, but some cultivar internodes are more closely spaced to one another, due to being produced under stressed conditions. The cuttings are then cut to 25cm and shorter. The distance between internodes can differ even for the same cultivar, which has been produced in a different region.

Make sure that at least three buds (internodes) are established below soil level and 3 above. It is important that at least 10cm of the bottom of the vine is under the ground. The cultivar Blesbok, has thick vines with internodes which are spaced close to one another.

When the vines are cut with the growing point still intact, it is called a top vine. Without a growing point it is known as a stock vine. There are no yield differences when establishing top or stock vines even if the stock vines are planted upside down. Top vine cuttings stabilize better and the survival percentage is also higher. The vines should be planted upright, with the internodes pointing upwards. The leaves of the stock vines must be removed before planting. A single leaf or two should be left at the top of the vine to indicate the top side. This makes it easier for the labourers to plant them upright.

The cuttings should never be subjected to stressful conditions just after planting. When vines are left-over which are not going to be planted on the same day, they should be placed in a container with about 10cm of water and kept in a cool place to be planted the next day. It is preferable that the cuttings do not develop roots in the containers, which usually occurs after 2 days during warm conditions. This is because the newly formed roots are very frail and may be damaged when transplanting them.

30.3.5 SOIL PH
The crop is sensitive to alkaline and brackish soil. Optimum pH is 5.6-6.5.

30.3.6 SOIL REQUIREMENTS
Also see Chapter 5.3 Soil Requirements.

Clay soil should be well-managed or preferably avoided, as it is difficult to clean the mud from the roots. When compost is introduced into the soil, it helps to loosen the structure of clay soil. Usually high applications of compost are recommended for other crops, but not for sweet potatoes.

Producers should avoid clay soil that has a high water holding capacity. These soil remain wet for too long, especially after long rainy periods. Sometimes clay soil may lead to coarse and miss-shaped roots which may be damaged during the harvesting process. It is better to establish vines on well-prepared ridges than on level beds.

Sweet potato roots tend to form trusses close together on the ridges underneath the plants. The harvesting process is much easier on sandy soil and less damage takes place. In winter, when the roots are stored in the soil for gradual marketing, the choice of soil type is very important. Sandy and sandy-loam soil is the best for high quality, clean and attractive roots. Finely structured soil tends to give the roots a course and a misformed appearance. If the top growth has died due to frost, the roots may rot quickly in clay type soil.

When producing sweet potatoes in a sandy soil (not coarse sandy soil), the colour of the roots is better. The yield in sandy soil tends to be lower and it might cause the roots to become long and sharp pointed. Irrigation should be adapted to minimize the long, sharply pointed roots. Roots may also become short when the top layer of the soil compacts. This may cause the shoulders of the roots to develop above soil level and become exposed to sunlight, which may result in the root tops turning green.

When planting on flat soil, the top layer of the soil should not form a hard crust. This can occur after heavy rains or irrigation. The crust should be broken as soon as possible.

30.3.7 SOIL PREPARATION
Also see Chapter 8.2 Soil Preparation.

Apply the correct types and quantities of fertilizer and lime according to a soil analysis. Apply lime at least 6 weeks prior to planting the vines. If a hard pan is present below the soil, it should be broken-up. Good drainage, aeration and root penetration is necessary to produce good yields.

It is preferable to make ridges 1m apart from one another, 20 to 25cm high and 30cm wide (or adapt to the wheel width of a tractor accordingly). Take note that when harvesting, the wheels of a tractor should remain between the ridges. The soil should be moist when planting the vines on ridges. The ridges should be made higher in clay type soil. It is not necessary to make ridges in sandy soil. Do not plough sandy loam soil too deep. Some producers claim the roots tend to become too long.

30.3.8 PLANTING METHOD AND SPACING
Also see Chapter 11 Spacing of Crops.

30.3.8.1 SPACING OF VINES
Recommended spacing of vines in the rows is 30 to 40cm from one another. A general spacing is 30cm within the rows and 1m between rows. This represents a plant population of 30 000 plants p/Ha for fresh market production.

The closer the spacing in and in-between the rows, the smaller the roots will be at harvesting time. The spacing between rows may vary from 0.9 to 1.5m. This also depends on whether a tractor is used to control weeds between the rows. The wheel width of most tractors is 1.2m apart.

When the roots are spaced 1.5m apart from one another, then the spacing in the rows should be closer to each other. If supermarkets require smaller sweet potatoes, the spacing could then be adapted to approximately 50 to 60 000 plants p/Ha.

30.3.8.2 PLANTING METHOD
The soil should be moist prior to planting the cuttings. It is recommended that the vines should be planted on broad, well-made ridges. Narrow ridges may dry out quickly and partly wash away. When using overhead irrigation, the ridges should be prepared 25cm high and at least 30cm wide.

30.3.8.3 PLANTING ON RIDGES (LAY DOWN METHOD)
Lay the cuttings at the correct spacing on the ridges. The bottom end of the vine cutting should be pointed at the exact position in the middle of the ridges where the cutting is going to be inserted. Create holes where the pointed ends of the vines are positioned. Use a round iron rod or a stick approximately 3 to 4 times thicker than the vines. Punch holes deep enough to cover at least 3 eyes of the vines, about 10cm deep. Be aware that cultivar internode distances differ from one another. Producers should use their own judgment to adapt the planting depth and the length of the cuttings during planting time.

After planting the cuttings in their holes, it is important to close the hole by firmly pressing down the soil around the stem of the vines with a round bar or stick. The round bar or stick is pressed into the soil a fewcm next to the vine and is moved sideways to close the hole.

It is recommended that most leaves are removed from the cuttings before planting. A few leaves should be left to indicate the upper side of the vine cutting.

30.3.8.4 PLANTING ON RIDGES (V-SHAPE METHOD)
Some producers plant vines that are cut 30 to 40cm in length and lay them down at the required spacing on the ridges. Simply use a v-shaped stick and press the vines halfway into the soft, moist soil. Next, compact the soil around the cuttings to remove any air pockets.

30.3.8.5 PLANTING ON FLAT SOIL
The yield is lower when planting the cuttings on flat soil. This method is only recommended on sandy types of soil.

30.3.8.6 FLOOD IRRIGATION
When flood irrigation is used, the cuttings should be planted on ridges spaced 80cm apart from one another. Shallow furrows are made on the ridges, 15cm wide. The cuttings should be planted half-way on the ridge, on the water-flow level mark.

30.3.9 PRODUCTION OF PLANT MATERIAL AND VINES
30.3.9.1 PLANT MATERIAL FROM REGISTERED MEMBERS OF THE NATIONAL VINE GROWERS ASSOCIATION (NPV)
Disease-free vines may be ordered directly from members of the Association. The Agricultural Research Council (ARC) also supplies small amounts of disease-free plant material from August to November. The producer then multiplies the vines and reproduces them in a nuclear plot. They are moved to a basic plot in November to December and then to the over-wintering land.

30.3.9.2 PLANT MATERIAL WHEN OVERWINTERING ROOTS
Do not make use of this method unless you are certain that all the requirements are in place. Contact experts from the Agricultural Research Council. Plant material in the Gauteng area and vines that are cut from other lands, are planted in a prepared land at the beginning of February and in the Lowveld, at the beginning of April.

Some producers produce their own vines in large lands after over-wintering. The land for vine reproduction should be chosen carefully and should not be too wet during the winter, otherwise the vines will develop to a certain stage, at which time the frost kills them. After the vines are killed by the frost, irrigation should be stopped. They should then be irrigated in August, to stimulate the roots to sprout. They should be ready to be cut and transplanted in October/November in the prepared land. Insects should be properly controlled.

30.3.9.3 PLANT MATERIAL FROM ROOTS THAT WERE STORED

Contact experts from the Agricultural Research Council for the best results, before using this method. The roots are stored at temperatures of about 15°C. The temperature should be low otherwise the roots may sprout. Be aware that temperatures which are too low, for example at 10°C, will negatively influence sprouting. Temperatures should be adjusted to about 30°C for 2 to 3 weeks prior to planting the roots in the production land.

Sweet potato roots may be stored for approximately 5 months. They are then normally planted laterally and slightly above the soil level, close to one another. Planting usually occurs in September in the summer-orientated areas. At this time, they begin to sprout quickly and after approximately 3 to 4 weeks, only the top vines are cut from the roots and planted into the production land.

30.3.10 FERTILIZER
Also see Chapter 15 Fertilizer.

Sweet potatoes are sensitive to alkaline and brackish soil and therefore these soil should be avoided. Use the correct type of lime according to a soil analysis. If lime is necessary, it should be worked into the soil at least 6 weeks before planting the vines. If a soil analysis is not available and if producing on fairly poor soil, producers may use mixed fertilizer as follows: apply approximately 1 000 to 1 300kg of the common, cheaper, balanced mix 2.3.2. (22) Zn. On sandy soil apply 1 600kg p/Ha. It is better to broadcast the fertilizer before making ridges.

Top dressing is important. Apply the correct type of N, for instance 130 LAN/KAN, at approximately 3 to 4 weeks. When the vine roots begin to develop after 5 to 6 weeks, a second dressing should be applied before the runners cover too much of the production land. On sandy soil, 3 top dressings may be necessary with less N used for every application.

30.3.11 IRRIGATION
Also see Chapter 14 for Irrigation.

Sweet potatoes are fairly resistant to drought conditions, but irrigation is essential in most areas. Over-irrigation may be harmful, especially once the roots become mature, as they may rot in clayish type of soil. Sweet potatoes require 450 to 550mm water during the entire, long growing season.

30.3.11.1 CRITICAL IRRIGATION TIMES
- After planting the vines in moist soil, the land should be irrigated to full capacity. If it is not possible to measure the water content of the soil, then irrigate approximately 5mm every second day. When very hot conditions occur, apply 3mm every day, especially on sandy soil, until the vines have properly rooted. It is not necessary to irrigate deeper than 35cm after the 3rd week.
- 45 to 55 days after planting, it is important that the roots are well spread. The application of water soluble fertilizer and top dressing should be planned carefully so that they are available to the plants in this time.
- 55-100 days after planting, the top vines begin to grow slower, but the roots are still expanding to maturity. A water shortage at this stage, is critical.
- When the roots are young and water stress occurs, this may encourage sand split or root cracking.
- Be careful to not apply too much water at the end of the growing season. The roots may rot or crack, especially in clayish type of soil. Instead, irrigate once in 10 to 14 days, however this also depends on the soil type and structure.

Some producers stop irrigation 2 to 3 weeks prior to harvesting, to mature the roots and to partly cure them underground. When the soil is hard at harvesting time, then it should be lightly irrigated to soften the soil in order to ease the harvesting process. The skin of sweet potato roots are very soft and thin at harvesting time and therefore they can easily become damaged if care is not taken.

If roots are damaged when they are still young, these marks enlarge as they grow and make the roots undesirable. If they are washed and harvested at an early stage when they are still young, they may lose most of their colour. Especially the purple skinned cultivars. This may be tested by scratching the skin using your nail.

It is quite well-known, that when producing sweet potatoes on wet soil or when too much soil moisture occurs over long periods, it may considerably improve top growth and at the same time reduce root development.

30.3.11.2 DRIP IRRIGATION
Drip irrigation is not recommended and rarely used for sweet potato production. Producers should seek advice from a specialist concerning this. Root crops, especially sweet potato, may become long and thin or become too short when less water is applied.

Be aware that when water scheduling is not balanced, then rooted crops will develop irregular shapes. The rain factor may also play an important role in this regard.

30.3.12 GROWING PERIOD
When sweet potatoes are cultivated in optimal temperature conditions, they should be harvested in a 4.5 to 5 month period while in cooler areas, a 5-6 month period. If the temperatures are low for lengthy periods during the growing season and if cloudy days persist, growth and quality will decline.

When roots are harvested early, be aware that the skin is very soft and bleach coloured. They should be marketed as soon as possible because they do not store well and are less nutritional.

30.3.13 WEED CONTROL
Also see Chapter 17 for weed control.

Weed control should be practiced during the early stages of production, as the labourers may otherwise damage the top vines, if the vines overgrow and cover the space between the rows. Some producers change the direction of the vines while they are still young or when they are overgrowing between the rows. They change the direction of the vines towards the ridges to control weeds. It also later facilitates application of N granular fertilizer without causing damage to the vines.

Herbicides may be applied before establishing the vines and should be worked into the soil. A disadvantage of using Eptam is that its working action is of short duration. Linuron is a long, active chemical that controls broad leafed weeds and also certain grasses. It should be applied just prior to planting the vines (but not on sandy soil). Remember that herbicides should be applied precisely according to the manufacturer's specifications. Make sure that herbicides are not beyond their expiry date.

30.3.14 CROP ROTATION
Also see Chapter 18 Crop Rotation and Soil Improvement.

Do not establish sweet potatoes on the same soil every year. The crop should be rotated at least 3 years before establishing the crop on the same soil again.

277 A good yield from different cultivars of sweet potato storage roots.

278 An excellent yield but some roots are too long, thin and pointed which is undesirable.

279 Different types of orange sweet potato cultivars.

280 Sand split disorder on sweet potato roots.

30.3.15 HARVESTING

Irrigation should be stopped approximately 14 to 20 days prior to harvesting in warmer areas. This is a method of curing the roots in the soil. The soil should be soft and irrigated lightly before harvesting the roots. This prevents breakage of the roots and the damaging of the soft root skin. Every attempt should be made not to damage or scratch the roots.

The top foliage should be cut and removed from the land 4 to 7 days prior to harvesting the mature roots. Be careful when temperatures are high and moisture is still in the soil. The roots may begin to sprout if left too long after the top vines have been removed. The underground roots should not be allowed to sprout after cutting the top growth.

Fungi and bacteria may cause the roots to quickly rot and lead to serious losses after harvesting. This occurs especially in storage, when the soft skin is damaged. When the underground roots show signs of sprouting after cutting the top growth, they should be harvested as soon as possible. If this is not done, the sprouts will change the quality of the roots, withdrawing moisture and nutrients.

Producers should wait until the top leaves of the vines have died due to frost, in order to avoid the roots sprouting. The storage roots may then be cured in the soil for several days, however, the soil should not be too wet and have good drainage abilities. The roots can then be harvested a month or two later.

In small production areas, the roots are harvested using forks by lifting them from both sides of the plants. Large commercial producers use modified potato lifters and undercutting blades to harvest the roots. Do not leave the roots on the land, in the sun for too long, as they may be damaged by sunburn. When night temperatures fall below 2ºC, storage roots which are left overnight on the land, may become chilled. In warmer conditions, the roots should be left out in the land in order to dry-out, only for a day or two.

Do not harvest the roots too early, as the skin may be too soft or bleached and they may be less nutritious and their storage ability may be poor. Even when the sweet potato roots are mature and ready to be harvested, the skins are still soft and care should be taken to avoid damage.

Vines are an excellent stock feed, which has a high nutritional value. It is also a valuable and excellent compost component.

30.3.16 STORAGE, PACKAGING, SORTING AND MARKETING
Also see Chapter 25 Managing and Marketing and Chapter 24 Harvesting, Handling and Packaging.

Store roots in a ventilated room. Remember that harvested roots should be stored for 2 to 3 weeks in order to cure them. This is done so that the starch may be allowed to change into sugar, which results in the characteristic sweet potato taste.

30.3.16.1 STORAGE OF SWEET POTATO ROOTS IN THE SOIL
Storage roots may be left in the soil for short periods and if necessary, for 1 to 2 months, provided that the following conditions are met:
- The right type of sandy loam soil, which drains well.
- Soil temperature should be below 8 to 9ºC. High temperatures and rain may cause root rot.

30.3.16.2 WHEN STORING ROOTS IN A STORE
- Select only medium and large roots, avoid the smaller ones.
- The roots should be dry and undamaged.
- Good air ventilation inside the store is required.
- Humidity can be provided in stores by placing wet sacks inside the store.

Sweet potatoes are normally packed in purple potato type bags which are 7 and 10kg and also in boxes. They are also sold separately and directly from the production land. Fresh, sorted, mature sweet potato roots may be stored in a cool, dry place for 5 to 6 weeks after harvesting.

30.3.17 YIELDS
Low yields: less than 30 ton p/Ha
Average yields: 35 ton p/Ha
High yields: 60 to 70 ton p/Ha

30.3.18 INSECTS
Sweet potatoes have relatively few insects and diseases. Application of chemicals is often unnecessary.

INSECTS: *See Chapter 20.5.*
ROOT-KNOT NEMATODE: *See Chapter 20.5.2.1.*
APHIDS: *See Chapter 20.5.2.3.*
CUTWORMS: *See Chapter 20.5.2.2.*
RED SPIDER MITE: *See Chapter 20.5.2.9.*

SWEET POTATO WEEVIL
The adult weevil is about 5 to 6mm long and is metallic-blue with a long snout. The weevil makes small, round holes in the leaves. The eggs are laid in small holes on the stems of the young roots. Within a week after hatching, the larvae begin to feed and then tunnel into the stems and tubers. The mature weevil can fly up to 3km and contaminate other lands. The main method by which the insect is spread is propagation of infected material.

SWEET POTATO MOTH
The moth lays eggs on the leaves. The larvae are brownish-green with a characteristic horn on the back side. The larvae grow fast and feed on the leaves. They can strip the entire plant of its leaves. The worms are easy to control and birds consume them if spotted by them.

BLOSYRUS BEETLE
The adult beetle is 8 to 9mm long and pale to dark grey in colour. The females lay their eggs near the plants. After hatching, the larvae fall to the ground and finds its way to the sweet potato roots. The larvae feed on the surface of the roots, causing paths on the skin of the roots.

30.3.19 DISEASES
DISEASES: *See Chapter 20.6.*
DAMPING OFF: *See Chapter 20.6.3.1.*
FUSARIUM WILT: *See Chapter 20.6.3.11.*
PHYTOPHTHORA WILT: *See Chapter 20.6.3.16.*
ALTENARIA LEAF SPOT OR PURPLE BLOTCH: *See Chapter 20.6.3.2.*
ANTHRACNOSE: *See Chapter 20.6.3.12.*

VIRAL DISEASES
Sweet potatoes are adaptable to several viruses, especially the mosaic virus. This may sometimes seriously influence the yield. However, when the plants are cultivated optimally, good yields can still be possible. Viruses are mainly transfer by aphids and when aphids are controlled properly the disease and virus infections will be minimal.

30.3.20 PHYSIOLOGICAL DISORDERS
SUN SCALD:
Usually the surface of the roots scald purple-brown when the shoulders are growing above the soil level. This is caused by exposing the roots to direct sun light and high temperatures.

WATER BLISTERS:
Small water blisters develop on the surface of the roots which are caused by wet conditions and a shortage of oxygen in the soil.

CHILLING DAMAGE:
Internal flesh becomes soft. The larger edible roots, when cooked, become hard and clot the flesh. It is caused by too much and then too little irrigation water being applied during the growing season.

SAND SPLIT:
This disorder occurs when too much water and then too little water is supplied. The roots develop a small crack which enlarges as the roots develop. The crack seals later, but it gives the roots a bad appearance which is not acceptable at the National markets. *See photo 280.*

CHAPTER 31

PRODUCTION GUIDELINES FOR PEPPERS, PAPRIKA AND CHILLIES

281 New and standard chilli cultivars displayed at a farmers day.

31.1 INTRODUCTION

In South Africa, green sweet pepper cultivars are produced and harvested while the fruit is still green. These green fruit will later change to a red colour after harvesting. Overseas countries generally prefer coloured pepper fruit. They are more flavourful and taste better because of the higher sugar content.

There are 7 different coloured sweet pepper cultivars available: green, red, yellow, orange, cream, purple and black. The coloured pepper cultivars have thick walls and some are block-shaped. They look attractive but the yields are lower, usually 20 to 30% less fruit per plant. Markets prefer 2 to 3 different coloured fruit combined in a package. Check the cultivar descriptions to obtain the best one for your particular market. Be careful as there are many types of cultivars including bullhorn and sharp-pointed types.

Peppers grow relatively slowly and one crop is usually established per year. It is important when producing peppers to manage them in such a way as to allow a natural balance to occur between roots, leaf area, flowering, fruit set and fruit load. The leaf canopy is determined by the size of the root system and the fruit load. Healthy growing plants usually exhibit larger leaves while smaller leaves are the result of adverse climate conditions and/or faulty production methods.

Fruit load is determined by the size and condition of the root system and the leaf canopy. Many types of chillies are available and become more pungent as they mature. Peppers are a slow growing crop and need protection against disease, especially virus infections. Production of peppers is similar to chillies, peppadews and paprika.

Paprika is actually a pepper that looks like a chilli. Originally from Hungary, it has a very thin fruit wall which dries out quickly after harvesting. Paprika is mostly produced under contract between a producer and buyer.

There is a small market for the finished product which is usually ground and refined by the buyer. The buyers usually prefer larger production lands and therefore not many small producers participate in the production of this crop. Paprika is harvested when the pods are fully red and they are usually dried on the farm.

Peppers, paprika, peppadew and chillies belong to the same family as tomatoes, eggfruit (brinjal), tobacco, gooseberries, ground nuts and potatoes. Hot peppers (chillies) are used for spices that include paprika. The most common Cayenne chilli is ground to produce the well-known red pepper.

Peppers are a relatively expensive crop to produce and have a long growing and harvesting period. Fertilizer, cultivation and labour requirements are high and there is little room for making mistakes when producing these capsicum crops.

282 Different chilli cultivars displayed at a farmers day. Some chilli cultivars are extremely hot. The Habanero small pointed type can burn through your skin.

283 Quality peppers harvested in the KwaZulu-Natal Pongola area, South Africa. Note the tomato production and sugar cane areas on the left. It may be a problem to produce peppers and tomatoes due to diseases.

284 Peppers are vulnerable to sunburn in warm areas. Sunburn can be a major problem when hot conditions occur even when under protection.

285 Different beautifully coloured F1 hybrid pepper cultivars. These cultivars mostly have thick inner walls and are blocky and produce lower yields.

286 A bullhorn pepper cultivar is kept upright by boxing them and using nylon twine to prevent the plants from falling over.

287 Hydroponics pepper production in an Open Bag System (OBS).

31.2 SUNBURN AND PEPPER PRODUCTION

Producers should be aware that sunburn of the pepper fruit is sometimes a huge problem. Peppers are susceptible to sunburn when they are exposed directly to sunlight. When the plants are not boxed correctly they may fall over which may make the fruit vulnerable to sunburn.

Many beginner producers lose a considerable portion of their production due to sunburn. This is also a problem in shade net houses and tunnels. Healthy, well irrigated and fertilized plants provide good leaf coverage against sunburn. Usually the plants start-off growing very well and setting good quality fruit, but then runs out of nutrients. This causes the leaves to underdevelop and the fruit to become exposed to sunlight. Producers should endeavour to ensure that the plants are always covered with leaves. The correct nitrogen N levels should be applied to promote healthy plant and leaf development. However, note that too much N may cause blossom drop.

Nematodes (eelworm) may also play a role in causing poor leaf development and subsequent sunburn of the fruit. In this case, the plants usually develop unevenly in some parts of the production land. The plants should never be water stressed as sunburn will take its toll. Don't be misled when it rains and wait too long before irrigating.

31.3 PRODUCTION OF PEPPERS UNDER PROTECTION
Also see Chapter 23.5 Pepper Production in Greenhouses.

Due to the demand for high-quality pepper fruit throughout the year, the trend is to produce peppers under protection, in partially heated tunnels. When producing peppers under protection (tunnels or shade net houses) the plants develop vigorously due to the improved humidity, cultivation practices and trellising of plants within favourable environmental conditions.

Humidity in shade net houses and tunnels is much higher than in the open field. When trellising and pruning plants in greenhouses, much higher and better quality yields are produced. Additionally, if they are managed well, pepper plants could grow for a period of 6 to 7 months. The harvesting period could begin after 3 months for a period of up to 4 months.

Plants are also established at a much higher plant population. This is due to the fact that they are pruned and trellised to more than one stem per plant. Approximately 25 000 plants p/Ha are established in a 10 x 30m tunnel and 26 000 plants p/Ha in a shade net house.

Production in tunnels and shade net houses is expensive, but shade net houses are the least expensive of the two. The risk factor in Greenhouses is almost zero as it offers protection against hail, wind, theft, large insects, birds, animals, sunburn (to a great degree) and to a lesser degree frost. Tunnels are isolated from one another which helps to prevent diseases from spreading. It is more comfortable for the labourers to work inside tunnels or in shade net houses due to the height of the trellised plants. Also, if bad weather conditions occur, labourers can still work inside tunnels but not in shade net houses.

Insects such as aphids are the main culprits in virus transmission. Trips and white flies may also transmit viruses but to a lesser degree. Producing peppers in tunnels and shade net houses as well as trellising and pruning them in the same way as for a hydroponic systems, allows the plants to be grown for up to 6 to 8 months. This is provided that an effective spraying program against diseases and insects is implemented. Controlling insects that transfer viruses should be a high priority. Virus tolerant pepper cultivars are also available. It may sometimes be necessary to use a preventative chemical spraying program, especially if problems related to early blight disease have occurred in the past.

When producing peppers in tunnels and in shade net houses directly in soil, in comparison to hydroponic systems, the yields are surprisingly high, provided that the correct cultivation practices are followed. Peppers plants are trellised in tunnels and in shade net houses and should be pruned to 2 to 3 stems. This is the same method as for hydroponic plants. This is easy because the structure of the tunnel and shade net houses facilitates the use of trellising.

Heat-controlled tunnels are suitable to produce high-value crops such as peppers. The crop can also be produced during the winter when heat control measures are implemented. This is however very expensive and producers should think carefully before producing crops with heat control.

Before being trellised, pepper plants are not particularly tall. The growing plants seldom reach heights of 2m after a 5 month growing period. Lower structures, such as half-moon tunnels may be considered, especially in cooler areas. However, sunburn of fruit may occur when hot conditions occur in low structured tunnels.

31.4 PRODUCTION COSTS FOR PEPPERS

The tables in this section will assist producers with designing a cashflow and business plan for pepper production.

Production costs may differ across the country. Therefore the costs indicated in the tables below should be adjusted accordingly. The tables provide only an estimated cost for pepper production (2014-15). Overhead costs are not included and producers should do their own research regarding production costs.

The plant population of peppers in shade net houses and in tunnels is approximately 26 000 plants p/Ha, while in the open field they are about 18 000 plants, but never more than 20 000. The average production yield for green peppers in the open field is approximately 8-9 fruit per plant over a 4 to 5 month growing period. Production in shade net houses, using the 2 to 3 stem trellising and pruning method, the yield may increase up to 13 peppers per plant over a 6 to 7 month growing period. The thick walled, large, blocky, coloured pepper cultivars yield less fruit per plant at approximately 7 to 8 fruit per plant and also grow lower. Perform your own evaluations to achieve the best plant populations for your area.

The following overhead costs are not included in the tables: tractors, implements, general equipment (wheelbarrows, spades, forks, hoes, crates, twine, pegs etc.), irrigation equipment, water, water pumps, electricity, spraying chemicals, housing, toilets, communication, etc. are not included. Also not included are maintenance costs of tractors, implements, spraying equipment and tools.

Labour costs are not included as labour costs differ from farm to farm and area to area. Unions are also constantly negotiating for better salaries and other advantages. In this example, wages of R100 per day would be the norm. Labour is expensive and should be used in a productive manner, especially temporary workers.

Financial lenders may require a budget and business plan before they consider lending money. Commercial and small-scale farmers should make use of a budget and cash flow plan in this instance. The tables below will serve to keep the producer informed and help him/her compile a financial cash flow plan. This table also approximates how much credit and capital is necessary to establish a 1Ha area.

TABLE 29: PRODUCTION COSTS FOR PEPPERS ON 1HA (2013)

ITEM	COSTS P/HA
Clean and level the 1 Ha production land.	R600
SOIL PREPARATION: Rip, plough, disk and rotovate.	R2 000
SOIL ANALYSIS: At least 2 samples p/Ha = R290 per sample x 2 = R580. Interpretation of the soil analysis by a specialist approximately R280.	R860
FERTILIZER: Apply lime if the pH is too low, as required by a soil sample analysis. Application of mechanical fertilization using a tractor and spreader = R350. Suggested fertilizer for broadcasting is as follows: 1 300kg p/Ha 2.3.4 (30) Zn @ R290 per 50kg bag, therefore 26 bags x 50kg p/bag = 26 x R290 = R7 540. 2 Top dressings LAN/KAN @ 150kg p/Ha = 6 bags x 50kg @ R240 per bag = R1 440. When placing fertilizer in a band formation, much less is necessary.	R9 330
SEEDLINGS: 20 000 seedlings (open pollinated cultivars from nurseries) @ R750 per 1 000 (2014) Hybrid seed and seedlings are more expensive.	R15 000
CHEMICALS: To control insects and diseases (including labour).	R2 000

TRANSPLANTING: For 20 000 seedlings and labour: If 2 labourers plant 2 000 seedlings p/day then 20 labourers plant 20 000 seedlings = 20 x R100 = R2 000. Seedling planters that transplant more than 3 000 seedling p/hour are available.	R2 000
WEED CONTROL: For mechanical weed control between rows (twice): R450 x 2 = R900. Control weeds in rows. 2 labourers for 4 days, twice = 16 days @ R100 per day = 16 x R100 = R1 600.	R1 600
IRRIGATION: Water can be costly when low rainfall occurs, especially with diesel, electricity and labour costs. To irrigate 1mm water p/Ha, 10 000 litre is required.	R3 900
TRELLISING MATERIAL: To box peppers and chillies, only for the first season. The poles are to be reused. Poles 1.5m long x 10mm diameter = 1 500 poles @ R5 p/pole = R7 500.	R7 500
Non-stretch nylon or baler twine = R900.	R900
If 2 labourers plant approximately 160 poles per day, then 20 labourers plant about 1 500 poles p/Ha = 20 x R100 = R2 000.	R2 000
HARVESTING: Harvest systematically using crates, 2 times per week and transporting the harvest to the pack store.	R2 000
SORTING: Grading and packing in different containers, according to market requirements.	R2 000
TRANSPORT: If 100km from the market and if travelling 3 times when using a 3 ton truck @ R4 p/km = R2 400. Labour, driver and assistant = R600. Total R3 000.	R3 000
Cleaning, removing poles, irrigation system and twine (to be reused).	R1 700
Plough in the debris of the crop as soon as possible after harvesting. Establish a rest cover crop to improve the soil fertility.	R1 800
PRODUCTION COST P/HA	**R58 190**

TABLE 30: POSSIBLE YIELDS AND PROFITS

YIELDS: 9 fruit per plant x 20 000 plants p/Ha = 180 000 fruit, less 5% unmarketable = 180 000 fruit minus 9 000 = 171 000 marketable peppers @ R0,80 per pepper	R136 800
PRODUCTION COST	R58 190
POSSIBLE PROFITS P/HA	R 78 610

31.5 PRODUCTION GUIDELINES AND REQUIREMENTS

31.5.1 CLIMATE REQUIREMENTS, MAIN AREAS OF PRODUCTION AND PLANTING TIMES
Also see Chapter 3 Climate.

The sweet pepper is a challenging crop to produce as they are more sensitive to climatic conditions than related crops such as tomatoes. The fruit is hollow and the plant's response to climatic conditions can be observed in the shape and quality of the fruit.

Climate requirements for chillies, peppers, peppadews and paprika are similar in that they all prefer warm weather conditions and have a long growing season. They are sensitive to cold or very high temperatures. Temperatures below 15°C may result in poor growth and when exposed to prolonged cloudy conditions, the development of the fruit may be retarded. The percentage germination of seed decreases when temperatures above 35° C occur. Temperatures above 33°C could cause flower drop and poor fruit set may be the result. When high temperatures occur over longer periods, combined with wind, some of the fruit may become malformed.

Temperatures between 8°C and 14°C at night result in poor plant growth. Cloudy weather influences fruit bearing and causes the plants to normally develop slower. Average temperatures should be a minimum of 15°C and optimally, 29°C. Small fruit pepper cultivars are more tolerant of high temperatures than the larger fruit. Therefore chillies and paprika tolerate higher temperatures better than peppers. Peppers have the same climatic requirements as tomatoes and eggplants (brinjals) but can withstand lower temperatures. Maturation times of chillies can vary throughout the season, due to the establishment of heat and cold tolerant cultivars. The habanera (very hot) chillies tend to be heat tolerant and have a longer growing season.

The chilli known as peppadew, originates from tropical areas. They also grow well in sub-tropical winters areas but set poor fruit in cool conditions. Higher yields are obtained when air temperatures are between 17 and 30°C during the fruit set. Seeds germinate slowly in cold weather or sometimes not at all. However, seeds emerge well when in a range of 21 to 29°C. Fruit that develops later is usually smaller and may have poor shape but could still be harvested for several months.

Don't make the mistake of transplanting peppers, chillies and paprika seedlings too early while it is still cold. They may stay alive but develop slowly.

Days required for peppers, chilli and paprika seed to emerge at various soil temperatures:

8°C	15°C	20°C	25°C	30°C
Too cold	26 days	15 days	10 days	too hot

Capsicum seed may not germinate when soil temperatures fall below 12°C.

31.5.1.1 OPTIMUM PLANTING TIMES FOR PEPPERS, CHILLIES AND PAPRIKA

Areas with light frost:	End Aug-Oct
Areas with heavy frost:	End Sept-Oct
Frost free areas:	Jan-Feb and July-Aug
Winter rainfall areas:	Aug-Oct

31.5.1.2 MAIN AREAS OF PRODUCTION

Peppers, chillies and paprika are established across most of the country, but irrigation is necessary. Most of the chillies for the fresh market are produced in KwaZulu-Natal due to the demand from the local Indian community. They are fond of freshly produced, hot chillies.

Chillies are also used for canning and making flavourful sauces. Chilli sauces can range from sweet to mild to hot. Some chillies are produced in the Brits area, Limpopo and Gauteng (all in South Africa). Dried chillies are generally produced under dry land conditions in Western Gauteng, Orange Free State, Groblersdal and Marble Hall areas (all in South Africa).

Winter production of peppers can be produced in the open field, tunnels or shade net houses in the warmer Lowveld areas of Mpumalanga, Limpopo (Messina Komatipoort), KwaZulu-Natal and in the Eastern Cape (all in South Africa).

31.5.1.3 TRANSPLANTING TIMES FOR SEEDLINGS

Chillies and peppers are usually transplanted from seedlings produced under warm and protected conditions, or from seedlings that were produced in warmer areas. Most producers establish pepper, chillies and paprika as early as possible to avoid high temperatures. This is because the plants are seriously affected by virus infections when temperatures rise.

Seedlings that are established in seed trays, in warm Greenhouses at the beginning of August may be transplanted during mid-September in the summer-orientated areas. Planting times for paprika production is not provided because it is produced almost exclusively for processing by arrangement with contractors.

31.5.2 CULTIVARS
31.5.2.1 CHILLI CULTIVARS
Chillies belongs to the Solanaceae family and are classified in 3 groups:
- Sweet peppers or green peppers
- Chillies
- Paprika

Many new sweet pepper cultivars are available. Peppers and chilli fruit vary in shape, thickness of the fruit wall, colour, flavour and pungency.

31.5.2.1.1 PEPPERS
Producers of sweet peppers are gradually changing to F1 hybrid cultivars due to the better quality, yields, grading sizes, uniformity, attractive colours, shapes and disease tolerance, especially virus tolerance.

Coloured sweet peppers are favoured by some consumers because they are blocky fruit, heavier and have thicker walls. They also have a better shelf life and are eye-catching. Blocky fruit are usually 11 x 11cm. Local markets sometimes prefer the smaller 8 x 8cm pepper fruit, as they are suitable for packaging into more manageable sizes.

Producers should determine the correct ratio of sweet pepper colours (red, yellow and orange) that customers prefer and then select the cultivars accordingly. Common open-field cultivars are: King Arthur, Moar and Paso real. New generation F1 hybrids are Rojo, Excursion, Aladin (orange coloured) and Glory.

31.5.2.1.2 CHILLIES
Producers may become confused because of all the different groups and families of chilli cultivars that are available. Make sure that you establish the correct cultivar.

The Jalapeno type, originally from America, are favoured and widely used for different processes. These plants produce large fruit and are less pungent than the Serrano's chilli types. The Thai chilli types are used fresh and for drying and are very pungent. Cayenne types are best known in South Africa and are ground into red powder known as red pepper.

The birds eye chilli type is normally dried and used for spices. It is rated as a highly pungency 80 000-140 000 scovilles. The Korean types are less pungent and ideal for sauces.

31.5.2.1.3 PAPRIKA
Paprika cultivars have been developed so that the fruit can dry fast. For this purpose, they should have a very thin skin. Paprika originates from Hungary. The fruit is sweet and 60 to 110cm long. Paprika is harvested when the fruit pods are bright red. It is usually delivered to Factories or Contractors for preparation of the red powder.

31.5.2.1.4 PEPPADEWS
Seed companies and local dealers do not supply peppadew seed to producers. The name peppadew belongs to a private company who owns the plant breeder's rights. Similar cultivars to peppadew are available from some leading seed companies.

31.5.2.2 CHILLIES BELONG TO ONE OF TWO MAJOR GROUPS
- Gossum group: mild - Sweet peppers and Spanish peppers.
- Longum group: hot - The Tabasco peppers are small fruit which are very hot and used to produce hot sauces.

31.5.2.3 CHILLI PUNGENCY IS RATED IN TERMS OF "SCOVILLE HEAT UNITS"
- Jalapeno and cayenne peppers range from 20 000 to 25 000 scoville units.
- Tabasco peppers between 60 000 to 80 000 scoville units.

- Habanera is almost the hottest, measuring nearly 400 000 scoville units.
- The sweet bell pepper and paprika are rated zero.

31.5.2.4 PRODUCE THE CORRECT CHILLIES OR PEPPERS FOR YOUR MARKET
- Know the market and the size, colour, fruit quality, packaging etc., the market prefers.
- Know the climate in your area.
- Know the most common diseases in your area.
- Gather as much information as possible. Different cultivars have their own requirements for ideal climate conditions, trellising methods, growth habits and disease resistance.
- Spread the risk and establish more than one cultivar.

31.5.3 SEED P/HA
Pepper and chillies (direct planting): 2kg p/Ha.
Transplanting seedlings: 0.5-0.7kg seed p/Ha.

31.5.4 YIELDS
Peppers: 20-25 ton p/Ha
Chillies (dry): 4-6 ton p/Ha
Paprika (dry): 4-5 ton p/Ha

31.5.5 SEEDLING PRODUCTION AND TRANSPLANTING OF SEEDLINGS
Transplanting old seedlings may reduce the yield. Plants mature quickly after transplantation. The faster the plant develops the more it is set back, as vigorous growth may affect yields to a large extent.

31.5.5.1 SEEDLING PRODUCTION
Seedlings are seldom produced in outside seedbeds in soil.

See Chapter 4.8 for how to produce your own seedlings. Most producers obtain their seedlings from Nurseries. **See Chapter 4.6 about seedlings obtained from nurseries.**

31.5.5.2 TRANSPLANTING SEEDLINGS TO THE PRODUCTION LAND
The first harvesting of peppers takes place 7 to 8 weeks from sowing the seed into the soil seed beds. When transplanting them under normal conditions, the first fruit could be picked at 9 weeks and the best fruit develops 4 to 5 weeks later.

To produce seedlings in soil see Chapter 8.2 Soil Preparation.

31.5.5.3 PLANT SEEDLINGS ON RAISED BEDS OR ON RIDGES
Producers mainly use seedlings in seed trays or from nurseries. The soil needs to be worked to a fine tilt, but should not be cloddy. Peppers can be grown on flat soil, raised beds or on ridges. Raised beds and ridges are sometimes unnecessary unless the soil does not drain properly.

Peppers that grow on ridges or raised beds are less likely to become contaminated with root rot (Phytophthora). This is a deadly disease that contaminates the root system. To avoid this, the crop requires good soil management, especially in soil without a loose structure. If there is a chance of Phytophthora contamination, implement raised beds or ridges.

Double or 3 rows could be established on each bed. When planting double rows, beds should be about 75 to 80cm wide and raised to a height of at least 15cm. Most producers establish 2 to 3 rows and box them to keep the plants upright to prevent them from falling over. **See photo 286.**

31.5.6 MULCHING
Also see Chapter 8.7 Mulching.

Peppers respond well to plastic mulching when installed with a dripper system and when using a weekly fertilization program designed by experts. Applications can be applied with a fertilizer spreader according to a soil analysis before mulching.

New types and colours of UV treated plastic covers are available. Producers should contact suppliers for plastic mulch coverings, due to the benefits of mulching.

31.5.7 SPACING, DENSITY AND TRELLISING OF PLANTS

Proper plant spacing is required to allow enough air movement and light around plants. This helps to reduce foliar diseases. Recommended plant spacing for peppers is 45cm in rows and 75cm between rows. This results in a plant population of 29 500 plants p/Ha. Do your own research to ensure the best plant spacing in your region and for your specific market requirements.

Some producers prefer that pepper rows are established in a north-south direction. Establishing the crop in other directions may lead to early infections of powdery mildew, especially when it is related to light conditions. An east-west row direction for winter production areas may also reduce infections.

It may be wise to reduce the plant density of peppers, chillies and paprika when temperatures are suitable for winter production. Some producers make use of dense plant populations, especially in the case of paprika, but this is not recommended. When wider spacing is implemented between rows, the spacing of plants can be closer within the rows, especially when producing in summertime when abundant bright light is present. The plant population for paprika can be spaced closer to improve the drying of the pods (they should be as red as possible) and high light intensity is favourable to achieve this.

Most producers have contracts with contractors to produce paprika and hot peppers, because of the drying and grinding process. Contractors normally advise the producer of the best cultivar that they require to be established as well as the required spacing.

31.5.7.1 SPACING CONSIDERATIONS FOR CAPSICUM CROPS

SPACING (cm)	IN ROWS	BETWEEN ROWS	PLANTS P/Ha
Sweet peppers	40-45	60-80	20 000-40 000
Chillies	40-45	45-75	30 000-55 000
Paprika double rows	3x45	100x150	20 000-30 000

31.5.7.2 TRELLISING CAPSICUMS

Different plant and stem populations are implemented by some producers. Producers should do their own research and evaluations on stem populations. Beginner producers should start-off with the best standards which suit their area.

Trellising peppers, chillies and paprika in the land is performed by planting poles 4m from one another. Use nylon non-stretching or baler twine and fasten it horizontally on both sides of the outer plants at intervals of about 15cm above the seedlings. Most producers establish peppers and chillies in double or 3 rows and then box them. The recommended spacing for double rows is 35 to 45cm between rows and 35 to 40cm within the rows. The beds are then 90 to 100cm from one another and the population will be 20 000 plants p/Ha.

When the plants grow taller they are boxed to prevent them from falling over. Alternatively, vertical strings for each plant are used in tunnel or in shade net houses. The main stem is tied to a stake on a vertical twine. During the season, several loose ties are placed around the whole plant and fastened as the plant grows taller. If a dominant leader shoot is produced it can be twisted around the vertical string. Twist shoots up individually on the vertical strings. One should limit the lateral leaders or shoots to 2 to 3 stems per plant.

The first 2 flowers that developed at the bottom of the plant should be removed. This will encourage the development of roots, lateral shoots and improve fruit size. Sunburn of pepper fruit is a problem, therefore foliage to cover the fruit is necessary. High leaf coverage can reduce sunburn to a great extent. Even when peppers are produced in shade net houses or tunnels, sunburn of the fruit may still occur. Production under cover could be as long as 8 months up until the last harvesting of fruit, provided that disease is properly controlled.

A suggested stem population of 70 000 to 75 000 p/Ha in the land is a good option for the Spanish trellising method. Prune 3 stems per plant and the population would then be 20 000 plants p/Ha. If implementing the Dutch trellising method and 3 stems per plant, the population could be up to 30 000 p/Ha. When using more stems per plant, fewer seedlings are required. The plant population is adapted according to the stem population. Producers should experiment with these different options.

31.5.8 FERTILIZATION
Also see Chapter 15 Fertilizer.

Less nitrogen N fertilizer will be needed if manure or legume cover crops are ploughed into the soil 3 to 4 weeks prior to establishing the crop. Make use of a soil analysis to verify soil requirements. Capsicums chillies require high N, P and K application levels. Peppers are sensitive to calcium Ca deficiency otherwise blossom end rot (BER) could become a major problem. It is therefore important to manage the uptake of calcium Ca carefully. Plants usually begin flowering when they are small.

The following fertilizer recommendation is a guideline: When applying N, as a rule, do not apply more than 100kg/Ha of N at a time during the growing season. The exception is when heavy rains occur, at which time the amount may be increased to 130kg p/Ha, due to leaching. Applying too much N may be wasteful and also delay fruit development. The table below provides a guideline for N application, when a soil analysis is not available. Make sure to use the correct type of N. **Also see Chapter 15.5 Using the correct type of lime.** Phosphorous and potassium 1:0:1 is generally recommended for side or top dressing applications.

In the case of basal treatments, the fertilizer should be worked into the soil just before transplanting the seedlings.

TABLE 31: EXAMPLE APLICATION RATES FOR N, P AND K (KG/HA)

BASAL KG/HA						
PRODUCING FOR MEDIUM YIELDS			PRODUCING FOR HIGH YIELDS			TIME TO APPLY FERTILIZER
N	P	K	N	P	K	
55	80	100	75	110	155	Just before transplanting
TOP DRESSING IN KG/HA			SIDE OR TOP DRESSINGS			
N	P	K	N	P	K	
20	0	20	20	0	20	4 Weeks after transplanting
20	0	20	20	0	20	8 Weeks after transplanting

Peppers and chillies require a well-balanced fertilizing program to achieve good plant development, quality and yields. If a soil analysis is not available, a fertilizer such as 2:3:4(30) could be used to improve a fairly poor quality soil, at a rate of 1 200kg p/Ha (two big handfuls, 100g per m^2 equals 1 000kg p/Ha).

If good compost is available, apply 80 ton p/Ha or 8kg per m² and band place it. *See Chapter 9.3 How to produce compost.*

Due to the long growing season of peppers and chillies, the crops may run out of nutrients towards the end of the growing period, especially if organic matter was not incorporated at planting time. Some crops can still be productive with lower rates of N, but not peppers. If the pepper plant is light in colour, yields are drastically reduced and the wall of the fruit will also be thin. This may be noticed after the first fruit is harvested. The blue green colour of a healthy plant should be maintained through the entire growing season.

For top dressing, on fairly poor soil, usually LAN/KAN should be applied 3 to 4 weeks and if necessary, again 6 to 7 weeks after transplanting, at a rate of 130-150kg LAN/KAN p/Ha. The roots of plants should be well developed before applying the first top dressing. Nitrogen N can be lightly worked into the soil (be careful not to damage the roots). Irrigate immediately after applications to dissolve the N into the soil. Remember to use the right type of N according to the pH level.

Organic producers often rely on the release of N from the breakdown of organic matter in the soil, but with peppers, additional N is usually necessary. This should be applied during the active growing period of the plants. Organic matter releases less N at later stages of plant development.

Make sure to get the N balance right. Pepper colour, fruit size, wall thickness and shelf life are all dependent on enough N being in the soil. The right balance of potash K and calcium Ca is also important. Ca levels should be correct at planting time. When Ca levels are low, developing fruit will be vulnerable to Blossom End Rot (BER). BER may also occur even with sufficient Ca levels. This occurs when the soil is too dry and the plants become water stressed. During hot periods, Ca is one of the elements that plants have a difficulty in taking-up, even in normal conditions.

Be careful not to apply to much N as it may cause flower drop and fruit abortion. The amount of pepper fruit that the plant provides depends on the amount of healthy leaves that convert sunlight into fruit development. Producers can make use of a weekly fertigation program (applying nutrients through the dripper system). However, it is recommended that the fertigation program is provided by fertilization experts.

288 Different pepper cultivars. Note the way that the plants are boxed to keep the plants upright.

289 New and standard cultivars inspected by farmers at a farmer's day. It is recommended that commercial farmers attend these days.

31.5.9 IRRIGATION
Also see Chapter 14 for Irrigation.

Peppers, chillies and paprika require about 400 to 550mm water during their long growing season in the summer-orientated areas. When viruses are controlled properly and when plants are boxed, certain pepper cultivars can be cultivated for as long as 8 months. These cultivars may also produce fruit for a second season when temperatures are favourable as it is a perennial crop.

Effective irrigation for peppers is essential to obtain the best yield and the ideal fruit quality. After transplanting, irrigate 10 to 15mm, 2 times a week for a period of 2 to 3 weeks. For the next 7-8 weeks, irrigate 35mm per week and thereafter 30mm per week.

Under controlled conditions, it has been found that peppers prefer a soil moisture content of 80% of the field's capacity. The seedling plants should be kept moist during the first 2 weeks and thereafter, water stress should be avoided at all times. These are critical times during the growing season. It is important that enough water be available to the plant for optimum yields.

31.5.9.1 CRITICAL IRRIGATION TIMES
FIRST: To develop strong and healthy seedlings, it is necessary to apply light irrigations after the pepper seedlings are transplanted.

SECOND: It is important that sufficient irrigation water be applied to fill the soil to a depth of about 40cm for the first half of the growing season. This would ensure favourable deep root development.

THIRD: During flowering and during fruit development near harvesting time, irrigation is necessary. Blossom end rot (BER) is a major problem when producing peppers. When the plants become water stressed, especially in the late flowering stage, the blossoms may drop. Estimate water stress by observing the leaves of the plants. This is a good indicator. When wilting of the leaves begin early in the day, it is an indication that the water content in the soil is unsatisfactory. Irrigation should then begin immediately. Irrigation should be decreased at the end of the growing season to promote the ripening of the fruit. A drip system is ideal. Apart from saving water, the plants will not be wet after irrigation, which may reduce the occurrence of diseases.

31.5.10 WEED CONTROL
Also see Chapter 17 for weed control.

Care should be taken not to damage the roots when weed control measures are implemented. This is because the upper roots are near the surface of the soil when the plants become mature.

Clean cultivation. The field should be kept clean from weeds for at least 6 weeks prior to establishing the crop. This practice controls cutworms and vectors that carry viruses.

31.5.11 CROP ROTATION
Also see Chapter 18 Crop Rotation and Soil Improvement.

Crop rotation is the best way to promote healthy plants and minimize nematodes and diseases. This applies especially to root rot diseases caused by soil borne pathogens. Tomatoes, potatoes, eggfruit, tobacco and gooseberries should not be included in a crop rotation program as they belong to the same family. Rotate and wait 3 years before establishing crops of the same family on the same soil again.

31.5.12 HARVESTING, COOLING AND STORAGE
Also see Chapter 24 Harvesting, Handling and Packaging.

31.5.12.1 HARVESTING PEPPERS

Cultivation problems such as drought, sunscald and diseases can affect the quality of the fruit. Similarly, issues may occur when the fruit is harvested too young and when they are overripe. Most of the pepper cultivars that are produced for the fresh market have thick fruit walls, which is preferable. The fruit of most pepper cultivars are harvested when they are still green and young. This is because the plant produces fruit over a long period of time and picking the young fruit stimulates the plant to flower and produce more fruit.

One should not work in the land when the leaves are wet, as bacterial pathogens or spores may spread rapidly. The first fruit is picked after approximately 75 to 85 days after transplanting the seedlings. After another 2 to 3 weeks, the best fruit becomes available and after another 2 to 3 weeks, the fruit begins to colour fully. In cooler areas maturity takes much longer.

The first harvesting period occurs after approximately 10 weeks, when ideal conditions occur. Picking the fruit in the early green stage stimulates the flowering of the plants and the production of more fruit. Harvesting could occur for several months, depending on the cultivar, training method and level of virus infections. Picking is continuous and is usually every 4-5 days. The plants are brittle and the branches break off very easily when picking the fruit. Do not use your hands to pick the fruit. Labourers should be taught to pick the fruit at the correct development stage. The stalk of the fruit should be cut with a sharp knife or scissors. Take care to cut the stalk in such a way that they do not show beyond the shoulder of the fruit. Otherwise they may damage other fruit when packing them.

Harvested fruit should immediately be moved to a cool, shady location. When pepper fruit is fully mature, the skin should be smooth, shiny and have a uniform dark green colour. The coloured fruit exhibit similar qualities. However, be aware that when the fruit is fully ripe, they are firm and brittle and can easily be damaged. Because pepper fruit may wilt, they should not be picked when it is warm or at any other time while the plant is in a wilted condition. Harvested fruit that is wilted, never regains its crispness and continues to slowly deteriorate.

Harvesting fruit at different stages of ripeness can result in uneven sizes when they are packed. Do not harvest the fruit when they are still too young, for the very first fruit. Although the size of the fruit may be adequate, its colour may still be too light and the walls too soft. Such fruit does not transport well and usually becomes softer. However, the sooner the fruit can be harvested, the higher the potential yield will be. Producers should pick as soon as the first fruit is marketable in order to keep the plant in a productive stage. The plants nature is to produce seed for the next generation and the sooner the fruit is picked, the more energy is available for the plant to propagate.

During harvesting, grading and packing the fruit should be handled with care. Fruit bruises easily and may rot. Harvested fruit should immediately be taken to the coolest part of the storage place or pack house. Coloured, thick-walled pepper fruit always fetches a better price due to their attractiveness when used in raw salads. Remember coloured type pepper yields are much lower and the market is limited. If fruit of the coloured types are left too long before harvesting (in order to wait for them to develop their colours fully), it will lower the yield and increase the risk of the fruit rotting and becoming prone to sunburn.

31.5.12.2 STORING PEPPERS

Storage is important to maintain quality for long periods for marketing. When storing peppers in cold rooms, remember not to store or transport green peppers below 6°C, otherwise they will be injured by the cold. Also, above 10°C they will soon ripen and turn red.

Most peppers may be stored for 2 to 3 weeks if kept cool at 7 to 8°C. Optimum storage temperature for peppers is 7 to 10°C and for humidity it is 85 to 90%. Cold injury may occur below 4°C and freeze damage occurs at or below 0°C. It is important that the fruit is kept cool otherwise when temperature are above 13°C and the water condensates on the fruit surface, it may cause fungal and bacterial rot, as well as soft fruit.

Keeping the pack house clean is important. Wash the packing surfaces and tools with sodium-hypochlorite (bleach) every day. This may prevent bacterial soft rot and other diseases.

Pepper fruit cannot be stored for too long without cooling. They will only last a week before wilting, especially in warm stores.

31.5.12.3 HARVESTING CHILLIES AND PAPRIKA

Paprika should be picked when the fruit is dark red and has a moisture content of 50-60%. At this stage the fruit is normally soft and should be carefully picked by using two hands to make sure that the stem is removed at the shoulder of the fruit. The plant should not be damaged in any way and labourers should be carefully trained.

Chillies used for drying are handpicked. After drying they are then sorted. Harvesting depends on the drying of chillies or paprika and should be handled according to the specifications required by the market. There are a number of paprika processing companies in South Africa.

31.5.13 MARKETING AND PACKAGING
Also see Chapter 24 Harvesting, Handling and Packaging.

There is always a good market for peppers however, it is important to pack them according to market requirements. The South African National markets generally prefer the California type of sweet pepper which is characterized as being blocky, weighing around 200 to 220 grams, is four-lobed and has thick flesh walls.

Some consumers prefer a firm, red pepper and are prepared to pay more, however, the red pepper has a limited shelf life compared to green peppers. The market for coloured peppers vs. green peppers has grown with red leading followed by yellow and then orange. Coloured peppers are tastier because of their higher sugar content.

Packaging the different coloured peppers in polyester containers should be carefully planned due to the limited markets. Produce the correct colours and take care not to produce too many of one colour as they may get dumped. Know what the best colours beside green are and what the market prefers.

31.5.14 INSECTS
INSECTS: *See Chapter 20.5.*
ROOT-KNOT NEMATODE: *See Chapter 20.5.2.1.*
CUTWORMS: *See Chapter 20.5.2.2.*
APHIDS: *See Chapter 20.5.2.3.*
LEAF MINER: *See Chapter 20.5.2.10.*
RED SPIDER MITE: *See Chapter 20.5.2.9.*
TRIPS: *See Trips Chapter 20.5.2.4.*
WHITE FLIES: Occurs especially in tunnels. *See Chapter 20.5.2.12.*
MEALIE BUG:
This bug is black with yellow spots. It is also called two-by-two's as these insects are often connected to one another. They prefer to consume the silk of the maize curds. *See photo 124.*

31.5.15 DISEASES
Also see Chapter 20.6 Diseases.

Disease resistant cultivars are available that are partially or fully resistant to viruses. Note that disease resistance is compromised as soon as the plant becomes stressed. The diseases that give the most problems are powdery mildew and at a later stage, Altenaria disease. Sunburn of the infected fruit then takes place.

DAMPING OFF: *See Chapter 20.6.3.1.*
EARLY BLIGHT: *See Chapter 20.6.3.3.*

LATE BLIGHT: *See Chapter 20.6.3.4.*
POWDERY MILDEW: *See Chapter 20.6.3.5.*
FUSARIUM WILT: *See Chapter 20.6.3.11.*
PHYTOPHTHORA WILT: *See Chapter 20.6.3.16.*
ALTENARIA LEAF SPOT OR PURPLE BLOTCH: *See Chapter 20.6.3.2.*
BACTERIAL WILT: *See Chapter 20.6.3.15.*
BACTERIAL SPOT: *See Chapter 20.6.3.10.*
Small brown spots on the leaves and fruit develop. Control measures should begin immediately after these symptoms are observed.

31.5.16 VIRAL DISEASES
Capsicums are susceptible to all the virus diseases that attack tomatoes. Viruses can be carried by humans and insects (aphids, trips or white flies). Disease/virus resistant cultivars are essential, if they are available.

TOBACCO MOSAIC VIRUS (TMV)
Also see Chapter 20.8.2.1 Tomato Mosaic Virus.

The most virulent virus threats originate from Tobacco Mosaic Virus (TMV) and Potato Virus Y (PVY). The latest development in this front is the resistance or partial resistance of cultivars to bacterial spot and tomato spotted wilt virus (TSWV). Also, be aware of root-knot nematode. This virus prefers lying latent in the tissues of the plant, especially in the capsicum families. This means that the virus affects the plant slowly, which is why pepper plant infections (besides nematodes (eelworm)) slowly cause a reduction in yield that can be severe at the end of the growing season.

Take note the TMV virus has similar outcomes as the PVY virus below. Evaluating cultivars that set fruit at low temperatures as well as cultivars with multiple disease resistance should be a high priority.

POTATO VIRUS Y (PVY)
See Chapter 20.8.2.2 Potato Virus.

VERTICILIUM WILT:
This viral disease is soil borne and infects the entire plant. The plant wilts as if it suffering from drought stress.

31.5.17 FRUIT DISORDER
Fruit disorder and other physiological problems may cause severe losses during some years.

BLOSSOM END ROT (BER)
See Chapter 20.9.1.3.

290 Blossom End Rot (BER) disorder on a pepper fruit blossom end.

CRACKED FRUIT
Also see Chapter 20.9.1.2.

Cracking of pepper fruit is also called sugar cracks. This is one of the most common disorders. Fine cracks are often found at markets. The cracks occur on the shoulder sides of the fruit as well as at the blossom end. The fruit of peppers are hollow and this makes it more susceptible to cracking. The main cause of cracking is the expanding of fruit during the day and shrinking at night. Developing fruit are less prone to cracking. When fruit becomes mature they may crack, especially the thick wall cultivars. Plant breeders' always select against fruit cracking, however, not all thick-walled cultivars are prone to cracking.

Causes: Low temperatures and high human traffic are the main causes of cracking. Also water stress during the day and sunlight directly on the fruit.

MALFORMED FRUIT
Malformed fruit develops only on one side of the fruit, however this problem seldom becomes severe. Poor pollination (mostly due to too low or high temperatures) and also bad fertilization practices cause the malformation. It could also be caused by water stress or a nutrient deficiency, especially nitrogen N and also when disease and insects infect the fruit.

EXCESSIVE ABORTION OF FLOWERS OR FRUIT AND POOR FRUIT SET
Mainly caused by too high or too low temperatures.

High quality seed & plugs.

Contact Ball Straathof for a unique collection of Vegetables, Herbs and Microgreens.

orders@ballstraathof.co.za
011 794 2316
www.ballstraathof.co.za

CHAPTER 32

PRODUCTION GUIDELINES FOR SPINACH

291 If spinach plants are left too long before harvesting the leaves become extremely large. Provided they are irrigated and fertilized properly these leaves can reach more than 1 meter in height. The leaves remain juicy but when warm conditions occur the plants may go over to bolting.

292 A healthy bunch of spinach leaves consisting of 14 leaves. However, be aware that the leaves may wilt quickly.

32.1 INTRODUCTION

Spinach belongs to the beetroot family, however it does not produce underground roots. It is incorrectly named in South Africa. Technically swiss chard is the correct name. True spinach leaves are smaller and rounder than those of swiss chard and the yield of swiss chard is also much higher. In this manual swiss chard is referred to as spinach.

Spinach is a popular crop amongst producers as it is easy to grow, matures fast and can be harvested 2 to 3 times. Production costs are lower and it has a faster turnover than many other crops. The crop is usually established using seed that is sown by hand or planted with a seed planter. Seedling plants are rarely used, except on a small-scale, because in order to establish 1 Ha seedlings approximately 220 000 seedlings p/Ha are required.

The broad leaves develop quickly. The first outer leaves can be harvested approximately 35 days after transplanting or 50 days after sowing the seed. Fresh leaves should be crispy with red, white, green, or multi-coloured stalks (petioles). Be aware that many producers produce this fast-growing cash crop and consequently markets may become overstocked. Spinach is a perishable crop and therefore it is difficult to supply distant markets due to the crops limited keeping ability. The crop should be marketed immediately after harvesting when cooling facilities are not available.

The leaves and stalks are a source of vitamins A, B, C and iron and therefore assist in strengthening the immune system. It is recommended to include spinach in a diet and for rural communities, it is an important vegetable crop. Spinach and beetroot seed are dried fruit and the larger seeds contain 2 to 4 seeds inside the seed cluster.

Uses: Spinach may be cooked with other vegetables such as potatoes, onions, tomatoes or used in combination with mince-meat. Spinach pies and cakes are also delicious.

32.2 PRODUCTION COSTS AND POSSIBLE YIELDS

The production costs to produce vegetables may differ across the country due to the variances in cost of seed, seedlings, fertilizer, electricity/diesel, irrigation water, labour and so forth. Therefore, costs should be adapted accordingly.

The table below provides an estimate for costs and profits for spinach production (2014-2015). Overhead costs are not included. The production cost of spinach is low in comparison with other vegetable crops. A cash flow plan should be considered before establishing the crop.

The costs indicated in the table below are based on a production of 20 tons p/Ha of spinach leaves including the stems. The cost of tractors, implements, general equipment (wheel barrows, spades, forks, hoes, crates, twine, pegs etc.), irrigation equipment, water, water pumps, electricity, spraying chemicals, housing, toilets, telephone/computer costs, overalls, clothing for spraying chemicals and so on, are not included in the table below, nor the maintenance of tractors or implements. It is important to make sure that tractors are in good condition before production activity begins.

Note that some labour costs are also not included in the tables below. Permanent labourers should be used and supervised, trained, especially tractor/lorry drivers. Producers should determine labour cost themselves, as costs differ from region to region. Labour is expensive and should therefore be well managed and made to be productive.

Producers may establish small pieces of land every 4 weeks in order to stay in production over an extended period.

TABLE 32: ESTIMATED PRODUCTION COSTS FOR SPINACH (2014)

ITEM	COSTS P/HA
Clean and level the production land.	R850
SOIL PREPARATION: Rip, plough, disk, rotovate till/harrow the soil, when necessary.	R2 800
SOIL ANALYSIS: At least 2 representative soil samples: 2 x R290 = R580. If the soil structure differs then more samples should be taken.	R580
Analyse the results using a reliable fertilizer expert who can guide the producer on the correct type and quantity of lime and fertilizer = R350.	R350
FERTILIZER: Apply lime according to the soil sample recommendation if the pH of the soil is too low. Apply the fertilizer mechanically: tractor and spreader R350. Broadcast fertilizer suggestion: 1 300kg p/Ha 2.3.2 (30) Zn @ R290 per 50kg bag. Require 26 bags x 50kg p/bag = 26 x R290 = R7 540. 2 Top dressings LAN/KAN (N) @ 150kg p/Ha = 6 bags x 50kg @ R240 per bag = R1440.	R8 980
DIRECT SOWING: Seed: 5kg for direct sowing in furrows @ R290 p/kg = R1 450 (2014). Labour: Making furrows, sowing seed, covering seed. 27 Labourers sow 1Ha @ R100 p/day = R100 x 27 = R2 700. **IF PLANTING SEED WITH A PLANTER:** Seed established with a mechanical planter should be graded. 4kg seed @ R290 p/kg (2014) to produce 1 Ha = R1 160 p/Ha. Planting seed with a mechanical hand punch planter: 2 labourers plant 6 days @ 12 x R100 p/day = R1 200. Seedling planters are available that can transplant 3 000 seedlings per hour and more. **IF USING SEEDLINGS FROM NURSERIES:** Seedlings from Nurseries = R140 per 1 000 seeds. Require 266 536 seedlings p/Ha = R140 x 266 = R37 240.	R1 450 R2 700

Labour: 1 Labourer plant 5 000 seedling p/day therefore 50 labourers plant 266 000 seedlings per day = 50 x 100 = R5 000. Total = R37 240 + R5 000 = R42 240.	
CHEMICALS: Spraying programs to control diseases and insects including labour.	R1 100
Weed control: Labour: Mechanically between the rows and by hand within the rows.	R1 900
THINNING: When sowing seed by hand, 3 labourers thin for 5 days = R100 x 15 = R1 500 Thinning of plants is not necessary if a mechanical planter was used. Thinning by hand in furrows can become an uncomfortable task. Herbicides to control weeds are a good option.	R1 500
IRRIGATION: Overhead sprinkler system. Diesel, electricity and labour costs as well as shifting the irrigation system, cleaning and repairing it. Water is expensive when low rainfall occurs. To irrigate 1mm water p/Ha, 10 000 litres water is required.	R2 800
HARVESTING: Labour: Cut the leaves and transport it to the pack house.	R1 000
SORTING AND TRIMMING LEAVES: Make bundles or pack in plastic bags.	R1 900
TRANSPORT: On the farm and to the markets.	R1 200
Clean the land, remove irrigation system and plough leftover plant debris into the soil. Establish a cover crop to improve the soil.	R1 200
TOTAL PRODUCTION COST	R30 310

TABLE 33: ESTIMATED YIELDS AND PROFITS FOR SPINACH

YIELDS: If established 333 200 plants p/Ha with spacing of 12 x 25cm minus every 5th row = 66 664. Total plants p/Ha = 333 200-66 664 = 266 536. 1 Plant produces about 14 medium/large size leaves which weight approximately 75g each. 1 Bundle contains 14 leaves and weigh approximately 700g. 266 536 plants p/Ha = 19 038 bundles @ R5 per bundle = R95 190	R95 190
MINUS PRODUCTION COSTS	R30 310
POSSIBLE PROFIT PER HA	R64 880

32.3 PRODUCTION GUIDELINES AND REQUIREMENTS FOR SPINACH

32.3.1 SOIL REQUIREMENTS
Also see Chapter 5.3 Soil Requirements.

A loam soil that is high in organic matter is preferable. The spinach crop also grows well in gravel, medium loam, sandy loam, sandy and even clayish soil. Sandy soil is sometimes desirable when establishing the crop in mild winter production areas as this type of soil is warmer and drains better. When clay soil is managed well, remarkable results may be achieved.

The crop responds well to manure and good compost. A heavy application of organic matter will suppress nematodes (eelworm} to some extent and provide a steady supply of nutrients over a long period. This is desirable as nematodes may otherwise multiply rapidly in sandy soil.

32.3.2 PH OF THE SOIL
A pH of 6.0 to 6.8 is acceptable. Spinach is sensitive to acid soil and does not thrive in soil more acidic than pH 5.5. If the soil pH is low then avoid these soil. As with beetroot, spinach copes with a fair amount of salt in the soil and prefers slightly brackish soil. It is however sensitive to a low boron soil content.

32.3.3 CLIMATE REQUIREMENTS AND OPTIMUM PLANTING TIMES
Also see Chapter 3 Climate.

Although spinach is a cool weather crop, cold damage may become a problem. For example, a light frost may cause damage if windy conditions occur. Therefore, when light frost begins to develop, irrigating water directly onto the plants may cause a slight rise in temperature and prevent frost damage. However, when heavy frost occurs, the irrigation water may freeze on the surface of the leaves, causing the plants to remain frozen for much longer resulting in frost damage. Therefore, the timing of irrigation is important to prevent plants from freezing.

When drought stress occurs, the plants are more prone to damage when it is cold. It is important to make sure that there is no water on the plants during cold spells and that the soil is not too wet. Winds hastens evaporation which may cause a temperature drop in the land.

32.3.3.1 OPTIMUM PLANTING TIMES
Areas with light to mild frost:	Aug-April
Areas with heavy frost:	Aug-Mar
Frost-free areas:	Mar-Aug
Winter rainfall areas:	Mar-April and Aug-Sep

32.3.4 CULTIVARS

293 Extremely large spinach leaves. Seedlings were transplanted in Feb 2010 when they were 10 weeks old. When grown in ideal conditions they are juicy and only 6 leaves are necessary to create a bundle.

294 Various plants established in a Gravel Film Technique (GFT) hydroponic system. Second from left are spinach plants that are only 3 weeks old and that were transplanted from healthy, strong seedlings.

32.3.4.1 FORD HOOK GIANT
Most commonly grown in South Africa and therefore the market leader. More resistant to bolting. Popular and commonly used because of the broad, dark green leaves, beautiful white stem (petiole) and dark green, shiny, bubbly leaves.

32.3.4.2 LUCULLUS
The leaves are light green. Mainly produced in cooler weather conditions. The plant is a good yielder but bolts easily when warmer conditions occur.

32.3.4.3 RHUBARB
Reddish broad leaves and light red stems (petiole). This cultivar yields well and is visually appealing when the leaves are still fresh, due to the bright red stems and broad bubbly leaves.

32.3.5 SOIL PREPARATION
Also see Chapter 8.2 Soil Preparation.

If the soil is too dry, or when necessary, irrigate the land. Do not cultivate dry soil. Determine if a hard pan is present below the soil and if so, it is important to break it up. Good land preparation is the first step to a successful, quality production. It is important to plough the soil long before establishing the crop as the soil needs time to settle.

Early soil preparation is a good option and breaks down the organic matter, improves moisture conservation and helps to control weeds. It also helps to control insects/diseases and to create smooth, aerated seedbeds. If using seedbeds, broadcast the fertilizer before making the beds.

Every so often, when heavy rain occurs after irrigation, the soil may form a hard crust on the top, approximately 2 to 3cm thick. This is undesirable and should be broken-up using a grub or similar implement to facilitate water and root aeration. Continue the process of soil improvement where necessary.

32.3.6 PLANTING METHODS AND SPACING
Also see Chapter 11 Spacing of Crops and Chapter 8.8 Transplanting seedlings from seed trays.

32.3.6.1 PLANTING SEED DIRECTLY IN THE LAND USING A MECHANICAL PLANTER
Also see Chapter 8.6 sowing seed directly in the land.

Most producers plant seed directly in the soil using a mechanical planter or the cheaper hand-push planter. Use certified and tested seed as the germination of certified seed is reliable. When using a seed planter or hand punch planter, the seed should be graded and should all be the same size. It is important when using planters that the soil is smooth and level.

32.3.6.2 SPACING AND SOWING SEED BY HAND IN THE PRODUCTION LAND
It is important to understand that when rows are established further apart from one another, then the plants are normally spaced closer to each other within the rows. Make sure that the germination of the seed is good and if possible use certified, tested seed. Irrigate the land thoroughly a day before sowing. Be careful of clay soil that may become muddy and slippery when too wet.

A good option between the rows, is to space plants 25 to 30cm apart from each other and approximately 10 to 12cm in the row. Make furrows 2-3cm deep when sowing the seed in the prepared land. Note that when cool conditions occur, it is not necessary to sow the seed too deeply. In hot weather conditions, sow the seed deeper into the soil and make use of frequent irrigation.

Train labourers to sow the seed approximately 4cm apart from each other, in the furrows. When using a spacing of 35 to 45cm between rows, a footpath between every 5 rows is not necessary. This is because the gaps are then wide enough to walk in and controlling disease/insects/weeds as well as harvesting is more comfortable. Do not sow the seed too densely as it is a time-consuming job to thin the seedlings. If sowing on flat ground, every 4-5 row could be left open for foot paths.

After sowing seeds in furrows, use a rake or spade to slightly cover the seeds with soil. Remove all clots that have landed on the covered furrows or else the clots may prevent the seed from emerging. Small-scale farmers can make use of dry or sandy soil from elsewhere to cover the seeds in the furrows.

Irrigate approximately 15mm directly after sowing. If producers make use of the full capacity water soil method before sowing, then no more water is required until the seed has emerged. Be aware when using the full capacity method, that the soil does not become muddy.

Apply cutworm bait as soon as the seedlings emerge.

32.3.6.3 SOWING SEED AND SPACING IN RAISED BEDS

Many producers sow seed in raised beds, using 4 to 5 rows or making beds with tractor wheels that are usually 1.2m wide. They establish 4 rows, starting the rows 10cm from the side of the bed, 25cm from one another and 12cm in the row itself. This produces 208 250 plants p/Ha. However, it is always better to do your own evaluations regarding spacing requirements in your area.

When sowing seed in raised seedbeds, the prepared land should be level and smooth without clots. Create furrows on 1 to 1.2m prepared raised beds, 2 to 3cm deep. Make sure that the furrows are all at the same even depth to avoid poor emergence. Sow the seed about 3 to 4cm from one another and cover them with soil from the furrows. Train one labourer to sow seed so that he/she may take full responsibility if the seeds are sown too densely, too loosely or when rows are skipped. It is important to irrigate the already moist land after planting, especially when hot conditions occur, to at least 10 to 15mm.

32.3.6.4 SPACING OF SEEDLINGS IN THE LAND

Spacing is not as critical as it is with other crops. Seedlings could be transplanted in the prepared land with 8 to 15cm spacing in the row and 25 to 45cm between the rows. Do your own research regarding spacing, however, be aware that closer spacing within and between the rows allows plants to develop upright but that smaller, narrower leaves can result.

Some producers prefer to space the plants 25cm between rows and 12cm in the row. This amounts to 333 200 plants p/Ha minus 66 664 as every 5th row should be left open for foot paths. The plant population would then be 266 536 plants p/Ha. On the other hand, a spacing of 35 to 45cm between rows and 8 to 10cm in the rows, provides a plant population of 285 700 plants p/Ha. Foot paths are not necessary due to wider spacing.

When weeds are expected, it is better for commercial producers to use herbicides prescribed by the manufacturers.

32.3.6.5 TRANSPLANT SEEDLINGS FROM SEED TRAYS

See Chapter 8.8 to transplant seedlings from seed trays and Chapter 4.6 about seedlings obtained from nurseries.

Spinach seedlings are normally not produced in seed trays due to the high amount of seeds necessary to produce a crop. If using seed trays, they should be irrigated before transplanting the seedlings. Knock the underside of the seed trays to loosen the seedlings. This eases the lifting of the seedlings from the trays even when the seedlings are well rooted. After removing the seedlings from the trays, transplant the seedlings as soon as possible.

32.3.6.6 TRANSPLANTING SEEDLINGS FROM NURSERIES

When seedlings are obtained from a nursery, they are usually supplied in large plastic bags and should be carefully handled. Do not transport them long distances in open vehicles or they may be damaged by the wind. The seedling medium may also dry out. The soil should be moist before transplanting the seedlings.

To transplant the seedlings, make a hole, large and deep enough with a hoe, garden spade, flat iron or other object. It is important to plant the seedling plug slightly deeper than the soil level. The seedling plug should be planted upright, directly into the hole, without bending the roots. Then press the soil slightly around the seedlings. Seedling planters are also available that can transplant 3 000 seedlings per hour. ***See photo 56 page***

Apply cutworm bait after transplanting the seedlings.

32.3.6.7 TRANSPLANTING SEEDLINGS FROM SOIL SEED BEDS
Also see Chapter 8.5 Transplanting seedlings from soil seedbeds.

This method is not practiced regularly due to the poor ability of the seedlings to survive the transplanting shock. The seedlings should not be too large or broad leafed. When transplanting seedlings from soil seed beds it is important that the soil of the production land should be well-prepared and irrigated to a depth of at least 20cm.

Carefully remove the seedlings from the seedbed with a spade, preferably with a small amount of soil on the roots. Place the seedlings in a wet hessian bag or bucket, plastic bags or other containers. Never allow the roots of the seedlings to become dry. Try not to harvest too many seedlings otherwise they may have to be carried over to the next day potentially drying them out. If this happens, add water to the buckets or containers to cover the roots and then store them in a cool place overnight.

When transplanting the seedlings, make proper holes in the production land at the required spacing. Make the holes deep enough so that the entire length of the plant roots can enter into the holes without bending them. Cover the holes with damp soil and compact it slightly around the transplants. It is important to compact the soil lightly to remove the air around the roots.

It is important that the soil be moist before transplanting the seedlings. The seedling plants should be irrigated as soon as possible after transplanting. It is advisable when producing large areas, to water each plant (about 0.3 litre per plant) individually after transplanting, by using a hose or watering can. This may not always be possible and is time consuming. When producing large lands in warm conditions, it is wise to irrigate small blocks of the transplanting lands behind the planting area.

Do not forget to apply cutworm bait after transplanting the seedlings.

32.3.7 THINNING OF PLANTS
Thinning the plants after sowing the seeds in rows is an uncomfortable and time consuming task, especially when weeds are already well established.

Thin the plants and remove all weeds about 2 to 3 weeks after emergence, to the required spacing. It is normally necessary to thin only a few plants when a seed planter was used.

32.3.8 SEED P/HA
Direct sowing for small seeds: 4kg p/Ha and for large seeds: 5kg.

32.3.9 FERTIGATION
Also see Chapter 15 Fertilizer.

Water and fertilization management is critical to producing high quality leaves. A soil analysis is the most accurate guide to fertilizer requirements due to the influences of soil type, climate conditions and other agricultural practices.

Spinach draws 5kg N, 0.65kg P and 5kg K per ton of plant material out of the soil during its growing period. On fairly poor soil, small-scale farmers may use the cheaper 2:3:2 (22) Zn fertilizer. Following this, 1 100kg p/Ha (100g or 2 hands full per m^2 which equals about 1 000kg p/Ha) is suggested. Apply the fertilizer with a calibrated fertilizer spreader. Small-scale farmers can spread fertilizer by hand as evenly as possible and work it into the soil to a depth of about 15cm using a fork or spade. ***See Chapter 15.10 for how to apply fertilizer.***

Nitrogen N is important for leaf growth. Apply N on fairly poor soil at a rate of at least two top dressings of 150kg p/Ha LAN/KAN. It should be applied 3 and 5 weeks after transplanting, but be aware LAN/KAN may lower the pH of the soil and make the soil more acid. This happens when using LAN/KAN repeatedly. Apply enough N as the leaves need to be stimulated in order to develop quickly to provide more and larger leaves.

Some producers harvest the outer leaves 2 to 4 times during the season and after each harvesting, apply smaller amounts of N. This boosts the younger leaves and gives them a chance to develop properly for the next harvesting session. Spinach belongs to the beetroot family and therefore boron should be present in the soil. Boron should not be applied generally but only when it is identified as a shortage. Foliar sprays can be effective, provided they are applied at the correct stage of the plants development.

32.3.10 IRRIGATION
Also see Chapter 14 for Irrigation and Chapter 13 Water.

Apply approximately 375mm water during the growing season. The first week 20mm and thereafter 35mm every week. The peak water usage of spinach in a summer-orientated area is from November to January. During these warmer months, maximum transpiration and plant growth is anticipated.

Keep the soil moist throughout the active growing period. It is better to apply 1 or 2 shorter irrigation periods per week rather than 1 heavy one. This is due to the crops shallow root system. Too much or too little irrigation water may reduce yields. When plants are water stressed at any stage it, will affect quality and vigorous growth. Drip irrigation is a good option. See Chapter 14.7 for drip irrigation.

32.3.11 WEED CONTROL
Also see Weed control. Chapter 15 page

Spinach has a shallow root system, therefore it is important to not damage the roots when practicing weeds control. Avoid lands that are known to have a high population of nutty weeds. Also, it is better to avoid land where no weed control has taken place or when there is an expectation of the shedding of thousands of seeds by weeds. However, using herbicides correctly could make a difference.

32.3.12 CROP ROTATION
Also see Chapter 18 Crop Rotation and Soil Improvement.

Do not establish spinach and related crops, such as beetroot, on the same soil every year. Wait 3 years before establishing the crop again.

32.3.13 YIELDS
2 to 15 ton p/Ha including the stems. 10 ton p/Ha average and 15 ton optimum.

32.3.14 HARVESTING AND MARKETING
Also see Chapter 24 Harvesting, Handling and Packaging.

It is better to harvest early in the morning, when the leaves are still cool. Harvesting is possible in sandy soil during rainy periods but not in clay types of soil. It may be impossible to harvest the leaves in clay soil when the conditions are too wet. Heavy clay soil takes a long time to dry and settle. It is important that leaves are kept free from leaf eating insects as the leaves are going to be sold. Leaves should not be full of holes caused by grasshoppers, leaf eating ladybirds and leaf miners.

The first harvesting of the outer leaves can be cut at approximately 50 days and in the case when seedlings have been transplanted, 30 to 35 days. This is when favourable conditions occur. When the leaves are harvested systematically and not too much top growth has been removed, the plants produce smaller leaves situated in the oxyles of the stems. They are stimulated to provide more young leaves and thereafter a longer harvesting time is possible.

The first outer leaves should then be harvested and cut with a sharp knife approximately 5cm above soil level, without damaging the buds of the smaller leaves. When the leaves reach a length of approximately 25cm then only cut the outer leaves and a few more. When the leaves are still young and small, much more leaves are necessary to produce a bundle. Harvest the outer leaves regularly to stimulate the regrowth of the smaller leaves. This is important and should be planned carefully.

Larger leaves look more attractive which is what the consumers prefer. Large leaves look more attractive, keep fresh for longer and less leaves are required to create a solid bundle. Producers should use common sense to provide the best quality leaves. A good yield is possible when good cultivation practices are followed. Harvesting too many leaves at one time causes the few remaining leaves to not be able to produce at maximum production. Be aware that the first harvest produces the highest quantity and best quality leaves. Thereafter, after each harvest, this declines systematically. Surprisingly, when the leaves are allowed to mature for longer and when ideal conditions occur, the leaves may grow to lengths of more than 1 meter. *See photo 266.*

The main spinach open cultivar produced in South Africa, is the well-known Ford Hook Giant, which to some extent is well adapted and does not go over to bolting easily. Some markets prefer the spinach stalk which are then tied in bunches. The leaves may also be placed in plastic bags without the stalks. For spinach, if sold in prepacks, consumers prefer the smaller, healthy fresh leaves.

Keep the harvested leaves as cool as possible and transfer them from the land to a ventilated store or cold room as soon as possible. A good option to keep the leaves fresh, is to cover them with crushed ice. However, beware of diseases when the leaves are treated in this way, as it could create fungal problems. Clean, healthy, non-wilted leaves are essential in order to obtain good prices as consumers are influenced by visual appearance. Therefore, make sure that your product is well presented.

32.3.15 STORAGE
Due to the high transpiration rate of the broad leaves, the keeping quality of spinach is very poor and they should be marketed immediately. If producers are able to store spinach in a cold room it could be stored for 7-8 days. Spinach may also be kept longer when sprinkled with cold or icy water.

32.3.16 INSECTS AND DISEASES
INSECTS: *See Chapter 20.5.*
ROOT-KNOT NEMATODE: *See Chapter 20.5.2.1.*
CUTWORMS: *See Chapter 20.5.2.2.*
APHIDS: *See Chapter 20.5.2.3.*
LEAF MINER: *See Chapter 20.5.2.10.*
RED SPIDER MITE: *See Chapter 20.5.2.9.*
TRIPS: *See Chapter 20.5.2.4.*
LAYDYBIRDS: *See Chapter 20.5.2.13.*
Ladybirds may be classified as good and bad. The good ones are shiny, with bright red and black patterns or spots. The bright colour species feed on aphids. The dull coloured ones consume plants and are unwanted. Spinach is very attractive for plant-eating ladybirds and they can do considerable damage when they consume the leaves.

DISEASES: *See Chapter 20.6 Diseases.*
DAMPING OFF: *See Chapter 20.6.3.1.*
POWDERY MILDEW: *See Chapter 20.6.3.5.*
FUSARIUM WILT: *See Chapter 20.6.3.11.*
PHYTOPHTHORA WILT: *See Chapter 20.6.3.16.*
ALTENARIA LEAF SPOT OR PURPLE BLOTCH: *See Chapter 20.6.3.2.*

CHAPTER 33

PRODUCTION GUIDELINES FOR BEETROOT

295 Different beetroot cultivars displayed at a farmers day using a drip system.

33.1 INTRODUCTION

Beetroot is a popular garden vegetable and is relatively easy to grow. Pests and diseases are not a major problem. Young beet tops are a good source of vitamins and are tasty when prepared in the same way as spinach. Beet is related to swiss chard (spinach), sugar beet and mangle beet. Mangle beets are grown for stock feed because of the high sugar 20% (energy) content. Two-fifths of the worlds sugar is produced from sugar beet. Beetroot is a good source of vitamin C.

F1 hybrid beetroot cultivar seed is expensive and is more adaptable to high temperatures. These cultivars have better yields and are more uniform. They also have more disease and temperature tolerance and a better internal colour in comparison with the local or open pollinated cultivars.

33.1.1 HEALTH AND FOOD VALUE

Beetroot is a good, healthy tonic containing high energy, mainly in the form of sugar. Raw beet juice is one of the healthiest products for health conscious people. The juice however tastes quite unpleasant and may be combined with other more palatable fruit juices. Beetroot juice has properties that cleanse the kidneys and gall bladder and is a rich source of natural sugar. It also contains sodium, potassium, phosphorus, calcium, magnesium, sulphur, chlorine, iodine, iron, copper and vitamin B1, B2, C and P.

New niche markets are developing in the fresh market beet arena. One such market is for baby round beets intended for both the local and export markets (2013). Baby beet cultivars are required to produce uniformly round beets that are harvested at diameters of 20 to 30mm. Research is taking place to use baby beet leaves of various colours in fresh salad packs. However, this is still being investigated.

296 Good quality beets. Note the thin leaf necks and thin tap roots that are desirable and a sign of good cultivation practices. This is especially applicable when establishing F1 hybrid cultivars out of season.

297 After frost, leaves will usually recover. Note the beetroot globe roots caused by seedling plugs. This is undesirable because when transplanting the seedlings, the globes develop hairy roots on their underside. It is better to wash them from the seedlings before transplanting them to avoid this problem.

298 Bad internal colour is mainly caused when temperatures are high. Some cultivars are more resistant to this bad colouring.

299 Various good quality summer F1 hybrid cultivars, displayed at a Farmers day.

300 The 2-beet globe on the left is ideally shaped. These beets have thin tap roots, thin necks and a dark internal colour.

33.2 ESTIMATED PRODUCTION COSTS, YIELDS AND INCOME

The production costs to produce vegetables may differ across the country due to the variances in cost of seed, seedlings, fertilizer, electricity/diesel, irrigation water, labour and so forth. Therefore, costs should be adapted accordingly.

The table below provides estimated costs and profits for beetroot production (2014). It will assist with designing a cash flow and business plan. Overhead costs are not included. Producers should do their own research on production costs when they produce vegetables.

The yield of beetroot as indicated in the table below is for 30 tons p/Ha. Some experienced producers harvest F1 hybrid cultivars up to 50 tons p/Ha or more.

The cost of tractors, implements, general equipment (wheel barrows, spades, forks, hoes, crates, twine, pegs etc.), irrigation equipment, water, water pumps, electricity, spraying chemicals, housing, toilets, telephone/

computer costs, overalls, clothing for spraying chemicals and so on, are not included in the table below, nor the maintenance of tractors or implements. It is important to make sure that tractors are in good condition before production activity begins.

Certain labour costs are not included in the tables below. Permanent labourers should be used as often as possible, so that they may supervise and train other labourers and also be tractor/truck drivers. Producers should determine labour cost themselves, as costs differ from region to region. Labour is expensive and should be well managed for increased productivity.

TABLE 34: ESTIMATED PRODUCTION COST FOR BEETROOT (2012)

ITEM	COSTS P/HA
Clean and level the production land.	R1 300
SOIL PREPARATION: Rip, plough, disk and create smooth seedbeds.	R2 800
SOIL ANALYSIS: At least 3 soil samples = 3 x R250 per sample = R750. Analyse the results by making use of a reliable fertilizer expert = R500.	R750 R500
FERTILIZER: According to the results of a soil sample, apply lime if the pH is too low. Broadcast fertilizer suggestion: 1 300kg p/Ha 2.3.4 (30) Zn @ R290 per 50kg bag. Require 26 bags x 50kg p/bag = 26 x R290 = R7 540 (2012). 2 Top dressings LAN/KAN @ 150kg p/Ha = 6 bags x 50kg @ R240 per bag= R1 440. Band placing of fertilizer uses less fertilizer.	R8 980
SEED: Open pollinated seed for 1Ha = 11kg p/Ha @ R300p/kg = R3 300	R3 300
PLANTING SEED: Planting seed with a mechanical planter: If planting seed with a mechanical planter: 2 Labourers plant 4 days to cover 1 Ha @ R100 per day = R800. **Direct sowing:** If sowing seed directly in furrows: 12kg seed p/Ha @ R300 p/kg = R3600. Labour to sow seed, make furrows and cover them: 18 labourers sow 1Ha @ R100 per labourer per day = R100 x 18 = R1 800 + R3 600 = R5 400. **Transplanting seedlings from nurseries:** Seedlings are rarely used by producers as they develop poorly with hairy roots. Also more than 450 000 seedlings p/Ha are required. Seedlings from Nurseries = R120 per 1 000. Require 450 000 seedlings p/Ha = R120 x 450 = R54 000. **Labour:** 1 Labourer plants 7 000 seedlings p/day therefore 64 labourers plant 450 000 to 500 000 seedlings p/day = 64 x 100 = R6 400. Total = R54 000 + R6 400 = R60 400. **Hybrid seed:** Hybrid seed and seedlings are much more expensive. Hybrid seedlings from Nurseries are R160 per 1000. Require 450 000 seedlings p/Ha = R72 000 + R6 400 = R78 400.	R800
CHEMICALS: Spraying program for disease and insect control. Costs and labour included.	R1 600
WEED CONTROL: Mechanical control between rows and hand control within rows. Herbicide control is much cheaper.	R2 900
IRRIGATION: Overhead sprinkler system. Diesel, electricity and labour costs as well as shifting the irrigation system, cleaning and repairing it. Water is expensive when low rainfall occurs. To irrigate 1mm water p/Ha, 10 000 litres water is required.	R3 900
HARVESTING: Harvest and transport to the pack store, including labour.	R1 800
PACKAGING MATERIAL: Red nylon bags (or other containers) 10kg = 2 800 x R0.60 per bag.	R1 680
Sorting: Binding bundles, cutting leaves and sorting in plastic bags.	R1 900
TRANSPORT: To the markets. If 100km from the market and transporting approximately 25 tons: Travel with 5 ton truck 5 times to the market and back =	R4 500

5 x 2 = 10 trips. 10 trips x 100km = 1000km @ R4 p/km = R4 000 plus lorry driver and assistant (R500) = R4 500.	
Clean the land, remove irrigation system and plough leftover plant debris into the soil.	R1 500
PRODUCTION COST P/HA	R38 770

TABLE 35: ESTIMATED YIELDS AND PROFITS FOR BEETROOT

YIELDS: 30 ton beetroot less 2 tons unmarketable = 28 tons. 10kg bags @ R35 per bag x 2 800 bags	R98 000
MINUS PRODUCTION COSTS	R38 770
POSSIBLE PROFIT PER HA	R59 230

33.3 PRODUCTION GUIDELINES AND REQUIREMENTS FOR BEETROOT

33.3.1 CLIMATE REQUIREMENTS AND PLANTING TIMES
Also see Chapter 3 Climate.

High quality beets are characterized as having a high sugar content, dark internal colour, thin tap roots, top leaf attachments and also dark, smooth, shiny globes. Beets are best produced in cool weather conditions. They may also be produced in fairly warm conditions and also in tropical areas. However, in warmer conditions, diseases may flourish and therefore disease control is critical. It is for this reason that little beetroot production takes place in warmer, hot areas or throughout the hot summer months. Similarly, little beet production takes place in dry and very wet conditions as the beet globes often have poor quality, or are malformed with small hairy globes.

F1 hybrid cultivars are available which are more adaptable and still provide a reasonable yield when warmer conditions occur. Producers should be aware that F1 hybrid cultivars do produce sweeter beet globes, better uniformity and higher yields. It is therefore important that producers evaluate different cultivars. When hot weather occurs, the quality of beets may be reduced as they may show white and coloured rings (evident when they are cut through). **See photo 298.**

The optimum planting time for beetroot is in spring and autumn. They also do well during summer in the Highveld and during winter in the warmer Lowveld and other warmer winter areas. High quality beets are best grown in cool climates where the air is not too dry. In cooler climate conditions the globes store a lot more sugar. The optimum temperature is 15 to 20°C. Beets are not particularly sensitive to moderate heat conditions provided that soil moisture is present. Although beetroot is resistant to moderately cold conditions, they grow slower in winter.

Beet is generally not adversely affected by winter cold and light frost, however, cultivars do differ. Frost just before harvesting may damage the leaves and retard growth. Maturity can be delayed when cold conditions are present and the top growth may also be reduced. When the climate is mild, beets can be grown throughout the year.

Sow seed directly in the land when temperatures are between 7 and 24°C. During hot sunny days, young plants may be injured if hot conditions persist over long periods. The globes develop white concentric rings which will become evident at a later stage. **See photo 298.** Most cultivar seed will germinate poorly when soil temperatures exceed 23°C. When establishing beet seed in warm conditions, irrigation should be applied in order to cool the soil. Germination will still occur at temperatures as low as 5°C. Under favourable growing conditions, the crop develops quickly. Some cultivars are ready to be harvested as early as 2.5 to 3 months. Over winter production, if warm weather begins again, crops may seed (bolt).

33.3.1.1 OPTIMAL PLANTING TIMES

The best time to establish the crop is in summer oriented areas from 15 August to 15 March and in winter from 15 March to 15 August. According to specialists, beetroot should not be established from April to the end of May in the summer orientated areas. This is because the quality and yield will be poor in these cold conditions and it may encourage the bolting of plants.

Some F1 hybrid cultivars are bred to be partially resistant to bolting.
Areas with light to mild frost: Feb-Mid-April and Aug-Oct
Areas with heavy frost: Aug-March (Highveld)
Frost-free areas: Feb-July
Winter rainfall areas: Feb-March and July-Nov

33.3.2 TYPES AND CULTIVARS

There are a number of different cultivars that vary in shape, colour and size. The round fresh market beet globes require 65 to 75 days to mature, when optimal conditions occur. Cylindrical beets (long shaped) required 75 to 85 days, but this also depends on planting date, soil type, desired size and sowing times. When producing beets in cold conditions, they will develop slower than in warmer conditions due to the shorter winter days.

33.3.2.1 FRESH MARKET BEETS

The fresh market concentrates mainly on bunching types of beets which are round, dark and red. It seems that the market is in the process of moving from open pollinated and local selections to F1 hybrid cultivars. Round, baby beets, both for the local and export market have become popular. They produce uniformly round beets that are harvested at diameters of 20 to 30cm.

Some producers make use of the smaller, thinner globes and market them as baby beets. Older, smaller beets may be fleshy and lose much of their sweetness. It is important to produce the sweet, juicy and crispy beetroot that the markets prefer.

33.3.2.2 PROCESSING BEETS

Processors are the largest users of beetroot production in South Africa. This is for pickling, canning, colour extracting and sugar content. Processors prefer beet globes with a small, thin tap root and small upper leaf attachments on the globe. Cylindrical beetroot cultivar's internal colour is very dark and is favoured for processing.

33.3.2.3 OPEN POLLINATED CULTIVARS

The most common cultivars are Detroit Strains, Crimson Globe and Early Wonder.

33.3.2.4 F1 HYBRID CULTIVARS (2011)

There are many F1 hybrid cultivars from various seed companies. Producers should contact them for advice. Hybrid cultivar seed is expensive but usually more adaptable to weather conditions. They are uniform, stable, sweeter, have darker inner colours and higher yields may be expected.

33.3.2.5 NOVELTY BEETS

There are also a number of novelty beets such as Formanova, Cylindra etc. and also others with yellow and white coloured roots and various shapes.

33.3.3 SOIL REQUIREMENTS AND PREPARATION
Also see Chapter 5.3 Soil Requirements and Chapter 8.2 Soil Preparation.

Good fertilizer practices, quality seed and the best care will not provide success unless the soil is of a good texture and well-prepared. A good stand and emergence can be achieved if the soil is of a crumbly structure and as level as possible.

Beetroot has a shallow root system and therefore the top 30cm of soil should not be allowed to dry-out. The soil should be kept moist to allow the seed to germinate properly. When establishing the crop in too dry or too wet conditions, poor quality and malformed hairy beets are the result. Good drainage is essential for this crop. Although average soil are acceptable to produce beetroot, they will grow better if cultivation practises are followed correctly. The crop thrives best in a deep, well-drained, sandy loam soil. The soil should be well-worked, free of stones and debris as the globes will then develop more quickly and uniformly.

Uniform soil moisture allows good quality beets to be produced across the entire land. Cloddy, stony or shallow soil are undesirable. Due to the beetroot's origin, it does not thrive well in acid soil. Avoid soil where compost or kraal manure has recently been incorporated as this will promote the formation of side roots. Instead, use the compost or kraal manure for other crops in the first season and then for the following season, beetroot may be produced.

If the soil is compacted or when the clay content is very high, the roots are likely to be deformed and may develop a tough texture that could reduce quality. It is difficult to clean the beets when harvested from muddy soil.

33.3.4 SEED P/HA
Small graded seed: 8kg p/Ha and larger seed: 12kg p/Ha. Table beet seed numbers are approximately 55 per gram when they are sized at 9 to 10 (4mm), then 60 000 to 70 000 seeds p/kg is the norm.

33.3.5 SEED SOWING METHODS AND SPACING
Also see Chapter 11 Spacing of Crops.

When sowing seed by hand or a mechanical planter, the producer should verify that the seed germination is good. If for example, the germination is 80%, the seed should be planted more densely to compensate for the 20% loss. When sowing seed by hand, try to sow them evenly apart from one another. Do not sow too densely, due to the uncomfortable task of thinning plants and removing weeds. Seed is expensive and it is best to use certified graded seed that has good germination thereby avoiding spot stands.

33.3.5.1 THE VARIOUS METHODS OF PLANTING BEETROOT SEED
- Sow the seed directly in large well-prepared lands.
- Sow seed in furrows.
- Sow seed in beds or raised seedbeds.
- Plant seed by using a calibrated seed planter.
- To control weeds more effectively the seed should be sown in straight well-spaced rows.

33.3.5.1.1 SOW SEED IN FURROWS
Do not sow seed too deep or too shallow. The depth should depend on climate conditions and soil temperature. Beets mature more quickly when the plant population is higher. Some producers make use of the full capacity irrigation method by irrigating their land thoroughly before planting the seed in the furrows. They then do not irrigate at all, until the seed has emerged. This method is used to prevent the top soil from forming a hard crust. However, care should be taken that the top soil never becomes dry when using this method. The seedbed should be irrigated properly a day before sowing the seed.

33.3.5.1.2 SPACING
Small-scale farmers make furrows 2 to 3cm deep and 30 to 35cm apart by drawing a rake or spade along a straight line in the soil. The wider the spacing between the rows, the easier it will be to control weeds and to thin the seedlings. If mechanical planters and implements are used to control weeds, some producers space the rows 45cm apart from one another to conform with the size of the mechanical implements.

33.3.5.1.3 SOWING SEED DIRECTLY IN THE PRODUCTION LAND
It is important that the land should be level, fine and smooth. The seed is sown by hand on prepared land and

worked into the soil with a small tractor and harrow. When it comes to irrigation, sprinklers are used to evenly irrigate the land almost every day using short irrigation cycles. This is important especially when hot, sunny conditions occur. Water is supplied in small quantities in order to keep the soil surface soft until the seedlings have emerged properly. This should be done accurately. When any aspect is neglected then failure will be the result. Many farmers fail when sowing directly in prepared land because they neglect to irrigate evenly over large areas or by not using herbicides or the wrong applications of herbicides. Weeds may then take over and the only way to control them is by pulling them out by hand.

33.3.5.1.4 FURROW FLOOD IRRIGATION
Producers use a grub that is set to about 7cm deep. They then sow the seed and cover the seed with 2 to 3cm soil. The irrigation water flows alongside these furrows.

33.3.5.1.5 PLANTING SEED WITH A SEED PLANTER
Accurate and precision planters are available that place the seed at fixed intervals in straight rows. When using these planters, attention should be given to soil preparation and seed size. It is important to establish a smooth and level seedbed when planting seed with a mechanical planter.

Producers should use certified graded seed when using seed planters as they are all the same uniform size. This makes the calibration of the planter quite simple in order to establish a proper plant stand. This should also ensure uniform emergence. Seed is usually graded to sizes 9 and 10 (4mm) which results in 65 000 to 70 000 seeds p/kg. The seedbed should be irrigated a day before planting the seed. Yields and grades are directly linked to the plant stand and harvesting date.

Beet seed numbers are approximately 50 per gram. Fresh market beets for small-scale farmers are normally spaced at 35 to 45cm apart from one another between the rows. When using a planter, the seed could be spaced 6 to 8cm in the rows depending on the size of the cultivar beet globes.

33.3.5.1.6 DOUBLE ROWS
Some producers prefer to establish double rows 7.5 to 10cm apart in the double rows and 35cm apart from one another between the rows. The spacing of the seed in the rows could then be 3 to 5cm depending on the globe size that the producer is looking to achieve.

33.3.5.1.7 RAISED SEEDBEDS
Raised beds increase the depth of the soil and have better drainage ability. They provide more top soil and oxygen around the roots, warm up slightly in spring and may reduce diseases such as phytium and damping off. It is beneficial to make raised beds in clayish and poorly draining soil types. Harvesting is also easier in raised beds especially when using the undercut blade method. The standard spacing for beetroot in seedbeds is usually 5 rows in the seedbed, spaced 25cm from one another. In winter, 4 rows are spaced 35cm from one another due to the limited sunlight in winter.

33.3.5.1.8 PLANT POPULATION AND SPACING FOR FRESH MARKET BEETS
Plants could be 40 to 45cm apart between rows and 6 to 8cm in the row. The planting season may be extended through the year however when trying to produce from March to April in summer orientated areas small beets, poor quality and bolting may be a problem. A spacing of 35cm between rows and 7cm in the rows provides a plant population of 410 000 plants p/Ha (450 000-500 000 plants is generally recommended).

33.3.5.1.9 BABY BEETS AND PLANT POPULATION
Baby beets that are produced for whole and pickled packs may be produced by reducing the spacing between the rows to about 25cm. The plant population p/Ha for baby beets is normally 600 000 to 750 000 plants p/Ha. A spacing of 25cm between rows and 6cm in the row provides a plant population of 666 400 plants p/Ha. If every 7th row is used for a foot path, the population will be approximately 600 000 plants p/Ha.

33.3.6 TRANSPLANTING SEEDLINGS
TRANSPLANTING SEEDLINGS FROM SEEDBEDS AND SEED TRAYS

More than 90% of all beetroot producers establish beetroot seed by using seed planters and planting seed directly in a well-prepared land. Seed may be sown in seedbeds in soil and be transplanted, but this practice is out of date due to the difficulty of transplanting seedlings. It is not practical or economical to produce beetroot seedlings in seed trays due to the high number of plants and the high production cost of producing 450 000 to 500 000 seedlings which are needed to cover 1Ha.

The quality of beetroot seedlings produced in seed trays is often not good and may influence the yield. The problem with beetroot seedlings produced in seed trays and transplanting them with the medium still on the seedling roots (plugs), is that the mature beetroot globes will have a mass of fine roots at the bottom of the globes, giving it an unattractive appearance. ***See photo 298.*** The plug of the seedling medium causes the growth of this mass of roots, especially as the seedlings become older. It is recommended that the seedling medium be washed from the seedlings plug before transplanting them.

301 Sowing graded carrot seed in a raised seedbed. The labourer is trained to sow seed and ensure that the spacing is not too dense or too sparse.

302 Modern bed maker. The bed maker may also be adjusted to make sweet potato ridges.

33.3.7 DURATION UNTIL MATURITY

Depending on the planting date, cultivar globe size and whether cool conditions occur, early cultivars may mature in 60 to 80 days. Cylindrical beets used for processing mature in 75 to 85 days. Beet produced in cool winter areas mature after about 120 days or more. Some F1 hybrid cultivars may mature earlier or later.

33.3.8 MULCHING
Also see Chapter 8.7 Mulching.

The purpose of mulching is to protect the soil from direct sunlight and wind thereby keeping the top soil moist and soft.

Small-scale farmers make use of mulching by using grass or other mulches after the seed has been sown, especially in raised beds. When warm or windy conditions occur, mulching should be seriously considered. The rows or seedbeds may be covered with mulch so that they may be kept moist until the seedlings have emerged. It is important to remove the mulch when the seedlings emerge.

When mulching is not practiced, the seedbed can be lightly irrigated every day, to keep the top layer cool, moist, soft and favourable for emergence. It is not necessary to cover the beds with thick layers of mulch. Just enough to get the job done.

33.3.9 FERTILIZER
Also see Chapter 15 Fertilizer.

A soil analysis is interpreted by a specialist, is essential to optimise fertilizer application. The ideal beet globes should be harvested with sweetness and good, dark, internal colour. Good production practices include the use of recommended cultivars, favourable temperatures, selection of good soil, favourable seedbed preparation, healthy, strong seedlings, harvesting at the correct stage (time) and good management of weeds/disease/insects.

High levels of nitrogen N when compared with phosphorus P and potassium K will result in vigorous leaf production and poor root development. Less nitrogen N fertilizer is needed when legume crops are ploughed into the soil, 4 to 5 weeks before establishing the crop. Soil analysis, field experience and knowledge of the specific crop requirements may assist the producer to determine the nutrient applications.

Optimum fertilization for beets is essential for high yields and good quality. Beets should grow quickly and continuously and should be high quality, sweet and juicy. This is especially important for consumers making beetroot juice for a healthy life style.

Some consumers prefer baby beets, however they should be aware that producers sometimes harvest these beets when they thin their production land and market them as baby beets. Sometimes they also supply small, unsweetened, stocky beet globes to the market. When beets are produced in favourable conditions, the large ones are just as good as the sweet baby beets. Beet bunches that are healthy and good tasting, can generally be identifief by their healthy top leaf appearance and by their globes which are large, clean and shiny.

Organic matter should not be incorporated into the soil when beetroot and carrots are to be established, as the beet globes will become hairy and have a bad internal colour. Beetroot production prefers well drained soil that has been well supplied with lime and potash K. Lime applications should be considered when the pH is below 5.8. Experienced producers apply 2 to 3 tons p/Ha of dolomite or limestone when their soil is acid or slightly acid. Slightly heavy or heavy clay soil do not normally run short of potash K. A lack of phosphorus P or nitrogen N will affect the growth negatively and produce a poor and slightly red/purple leaf colour. Beets can tolerate a high salt content in the soil but are sensitive to high acid soil and low boron B content.

One ton of beetroot plant material withdraws approximately 3.0kg nitrogen (N), 0.5kg phosphate (P) and 6.0kg potassium (K) per 1Ha, therefore if a 35 ton yield of beet production is the goal, the following elements are suggested:
N: 3.0kg x 35 t/Ha = 105kg/Ha
P: 0.5kg x 35 t/Ha = 17.5kg/Ha
K: 6.0kg x 35 t/Ha = 210kg/Ha

The producer should deduct the elements that are already in the soil according to the soil analysis. Top dressings of N are very important for these crops, however, make sure that the top growth develops well but not too vigorously. Top or side dressings of nitrogen N should be applied approximately 120 to 150kg p/Ha at the 3-leaf stage or 3 weeks after emergence of the seedlings. A second application of 100kg p/Ha may be applied 2 to 3 weeks later, but be aware that this amount depends on the fertility and soil status. Potassium K levels should be fairly high. The second top dressing could be 1:0:1 of potassium nitrate when the K level is low.

Beetroot is very sensitive to a boron B deficiency which may be identified when the young leaves fail to develop normally and begin to turn black. The internal colour will also show black spots. Foliar spray applications should help to prevent this condition.

When the leaves become reddish it is an indication of a phosphorus P shortage. This may be corrected with foliar sprays. Purple beet leaves are more visible in winter due to a shortage of phosphorus P and when soil temperatures are cool. The uptake of phosphorus P can also be controlled by using foliar sprays.

When a fertilizer analysis is not available, a small-scale farmer may use approximately 600 to 800kg p/Ha 3.1.5(26) Zn (or 70 to 90g or one and a half handfuls per m²) on poor soil. Most producers also use the cheaper 2.3.2 (22) Zn mixtures and apply it on poor soil at approximately 1 300 to 1 500kg p/Ha. Beetroot is sensitive to some trace elements deficiencies, therefore these applications are important.

33.3.10 IRRIGATION
Also see Chapter 14 for Irrigation.

Irrigation management is important during the early growth stages of plants and when root development begins. Beet seeds may not emerge when surface crusting occurs. In this case, light irrigations are essential to soften the soil surface and to assist emergence. Irrigation is necessary when beet roots develop in the last half of their growing period. During this time water fluctuation will have a negative influence on the yield. Irrigation should be applied early in the day to allow rapid top growth in order for the plants to dry completely before sunset to prevent diseases.

Poor quality and malformed hairy beets may result when the crop is established in too dry or too wet soil conditions. A good moisture balance should be maintained, however water should be minimized on the top growth canopy as this may promote fungal diseases, especially Altenaria. Irrigation should be managed carefully especially early in the growing season. Care should be taken not to over irrigate.

Slow water drainage may cause the beet leaves to turn red/purple and the plants may stop growing for a period. Beetroot has shallow roots and therefore it is very important that the top 30cm soil layer should never dry-out. In the growing season, 300 to 350mm water is the norm. Therefore, 20mm water for the first 2 weeks and 30mm every week thereafter for the rest of the growing season, is proposed. Beetroot uses an average of about 4mm water per day in moderate temperature conditions. When cold winter days occur, 2mm per day is more or less the norm but when long, hot summer's day occur, an average of 8mm would be normal.

Note that heavy irrigation or rain after an application of herbicide makes it less effective. When sowing beetroot seed, the soil should be damp and irrigated lightly to keep the surface of the soil cool. This will facilitate emergence of the seedlings especially when warm days occur. When the seed emerges, the growing points of the seedlings are very sensitive to hot top soil conditions. As a result, they may burn, if care is not taken. Soil that drain well and is irrigated regularly, ensures continuous growth of the plants. It is critical to irrigate the land in the last half of the growing season. Water shortage during this period may have a negative influence on the yield.

33.3.10.1 DRIP, SPRINKLER, OVERHEAD AND PIVOT IRRIGATION SYSTEMS
It is not recommended to use a drip irrigation system for the production of beetroot. This applies especially when seed has been sown by hand or with a planter as the seeds need to germinate. This system is not suitable to achieve good germination as the soil is irrigated in circles. Many dripper lines would also be necessary due of to the narrow spacing. This makes it difficult to shift the lines when controlling weeds. **See Chapter 17 for weed control.**

According to experts (Water Pleasure-Brits) the cost to establish a 1Ha dripper line system to produce beetroot is approximately R50 000 p/Ha in comparison with a tomatoes dripper system which costs approximately R25 000 p/Ha (2012). Small-scale farmers make effective use of cheaper sprinkler systems while commercial producers use overhead systems and pivot irrigation.

33.3.11 PH OF THE SOIL
Although an average soil may produce excellent beet, the best soil is a rich sandy loam soil with a pH of 6.5 to 7.5 (not lower than 6).

33.3.12 CROP ROTATION AND SOIL IMPROVEMENT
Also see Chapter 18 Crop Rotation and Soil Improvement.

Crop rotation is a method of planting different crops on the same soil, with the aim of retaining soil fertility and preventing diseases/insects/weeds proliferating. When repeatedly using fertilizer year after year, the soil may become toxic. Too much salt may be built up in the soil.

Do not establish beetroot on the same soil every year. Rotate crops with other or alternatively cover crops such as Legumes, Sun hemp, Black mustard, Cow peas, Green peas, Velvet beans etc. The best way to improve soil is to apply compost and to establish pod bearing green manure crops which can be incorporated into the soil.

33.3.13 THINNING OF SEEDLINGS
Also see Chapter 16.2 Thinning of Seedlings.

The seedling plants may be thinned to a spacing of 5 to 8cm in the rows. Hand thinning is time consuming, labour intensive and costs may run high. Seedling rates should be controlled as accurately as possible. It is important to purchase seed with good germination to create a good stand. Sometimes thinning can be combined with weeding but this is an entirely different job.

Hand thinning of seedlings when sowing by hand, is always necessary. Seedlings should be thinned when they are still young. Thin them no later than 2 to 3 weeks after emergence, to a distance of 5 to 8cm in the row. The distance will depend on the cultivar globe size, time of production and day length. If thinning is delayed until the leaves of the beetroot are 7 to 9cm high, the leaves which are removed can be cooked as greens, similar to spinach. The young beetroot leaves have an excellent taste.

Beetroot seed is actually a small fruit that contains more than one seed in a cluster of 1 to 5 seeds. When the seeds emerge, then seedlings may emerge in small groups. This is why the seeds are graded as producers prefer smaller seeds that contain no clusters. Smaller seeds however may be troublesome and emerge poorly. Beetroot cultivars that only have one seed per cluster are available and are useful when exact spacing is required using a seedling planter.

33.3.14 WEED CONTROL
Also see Chapter 17 for weed control.

When seeds are sown by hand, weed control is essential during the early stages (2 to 3 weeks after emergence). This gives the seedlings a chance to grow without competition. It is difficult to control weeds by hand, as it is an uncomfortable and time consuming task. When the seeds are sown by hand then time is going to be wasted to thin the plants to the correct spacing.

Herbicides are recommended to control weeds effectively. Hand weeding early, when the seedlings are still young, is the most effective way to control weeds. Young beetroot plants are sensitive when weeds are large and controlling large weeds may damage the seedlings roots.

33.3.14.1 CONTROLLING WEEDS WITH HERBICIDES
To avoid labour costs, chemical weed control is the best method to be used. Weed problems can be reduced by preparing the land a few weeks before sowing the seed. Use Roundup or Gramoxone in a seed bed system. Pre-emergence and early post-emergence Herbicides may control weeds for up to 8 weeks after application.

There are a few herbicides registered for beetroot production and producers should make use of them to control weeds effectively. *See Chapter 17 for herbicide-based weed control.*

33.3.15 YIELDS

Yields and grades are directly influenced by plant population and harvesting date. Yields differ from season to season or from year to year. A reasonable yield of beet bundles, when leaving the tops on, for open pollinated strains, is 25 to 30 ton p/Ha (40 ton p/Ha is a good yield).

F1 hybrid cultivars yield up to 55 tons p/Ha or more when spaced correctly and when using good cultivation practices. Yields of processing beets average more or less 40 ton p/Ha.

33.3.16 HARVESTING, HANDLING, MARKETING, STORAGE AND PACKAGING
Also see Chapter 24 Harvesting, Handling and Packaging and Chapter 25 Management and Marketing.

Some producers harvest beetroot by topping them with a self-built machine pulled by a tractor. The beets can also be undercut by using a tractor fitted with an undercut blade. Beets for processing are harvested into self-unloading trucks, bulk trucks or trailers and delivered to the factory to be immediately processed.

The crop can be lifted out of the soil by hand by pulling the tops of the plant when the soil is soft or sandy. After the dead and damaged leaves are removed the beetroots and leaves are tied in bunches. It is sometimes necessary to wash the globes, especially when producing them in clay types of soil. Beetroot globes should be smooth, firm and shiny. The medium-sized roots are the most tender and favourable.

Harvesting normally begins when the beets are 4 to 5cm in diameter. Most beets should be harvested when they reach the fully mature stage. This will obtain the maximum yield. The first harvesting is a basic thinning process while later, the beets are lifted when they are 5 to 7cm in diameter. This is after approximately 3 months in ideal summer conditions (4-5 months in winter). The internal colour of beetroot globes is better in winter or in cool areas than in summer or warm conditions.

Raised beds are most effective when harvesting with the undercut blade method. The beets are pulled out of the soil more smoothly and minimum damage will occur. The soil should be slightly moist before cutting or pulling the beets out of the soil. If the soil is too dry, the beet may be difficult to clean and to pull out using their tops.

33.3.16.1 MARKETING
When fresh markets prefer bunching types, it is important to use cultivars that have the ability to produce quality, dark, round beets. Improved F1 hybrid winter and summer cultivars are available.

33.3.16.2 REQUIREMENTS FOR PROCESSING BEETS
Producers should note that processing factories set high quality standards mainly relating to internal colour and fibre of the globes. The cultivar choice is normally determined by the processors. When beetroot globes are canned whole, then small beet globes (2 to 6cm across) are preferred. Producers should make sure that they meet the requirements for processing laid down by the factories contracting them.

33.3.16.3 HANDLING
Beets should be handled as carefully as possible after harvesting to avoid damage to the globes. Injuries reduce shelf life, increase decay and disease infections. Fresh market beets are especially susceptible to injury because they are harvested when fully mature in order to meet the market size requirements.

33.3.16.4 STORAGE
Beet globes are well adapted to storage. Fresh market beets may be stored for 10 to 14 days at 4°C and 98 to 100% relative humidity. Do not store in temperatures lower than 4°C as it may cause damage to the internal

globe flesh. The use of crushed ice is helpful to keep the bunched beet leaves fresh. Make sure that beet leaves are not covered with water as they will rot fast.

33.3.16.5 PACKAGING
Pack the beets according to market requirements, usually in plastic bags, purple potatoes bags etc.

33.3.17 INSECTS
INSECTS: *See Chapter 20.5.*
ROOT-KNOT NEMATODE: *See Chapter 20.5.2.1.*
CUTWORMS: *See Chapter 20.5.2.2.*
APHIDS: *See Chapter 20.5.2.3.*
RED SPIDER MITE: *See Chapter 20.5.2.9.*
LADY BIRDS: *See Chapter 20.5.2.13.*
Ladybirds may be classified as good when they feed on aphids and bad when they feed on plants. Beetroot and spinach is attractive to the plant eating ladybird.
LEAF EATING INSECTS:
Grass hoppers, plant eating lady birds and other insects feed on plants when they are produced in the summer. They attack the young leaves of plants.

33.3.18 DISEASES
DISEASES: *See Chapter 20.6 Diseases.*
DAMPING OFF: *See Chapter 20.6.3.1.*
LEAF SPOT: *See Chapter 20.6.3.13.*
POWDERY MILDEW: *See Chapter 20.6.3.5.*
PHYTOPHTHORA WILT: *See Chapter 20.6.3.16.*
The first symptom is usually small purple spots which when under moist conditions, become larger and produce leaf spots approximately 1cm in diameter. The margin of the leaves are purple surrounded with a yellow ring. The centre can be grey-brown-black and spores may be present in the centre. High humidity encourages this disease. This is also a seedborne disease.

303 These beetroot packs were bought at a large greengrocer. The logo indicates quality, farm fresh beetroot. Don't be fooled by the packaging, always check the quality of the actual produce where possible.

CHAPTER 34

PRODUCTION GUIDELINES FOR CARROTS

304 Open pollinated carrot and beetroot cultivars, selected to be transplanted in order to produce seed for seed production. To keep the cultivar stable, homogenous and strict true-to-type selection is necessary to ensure that future generations are true to the original cultivar description.

34.1 INTRODUCTION

Carrots are considered one of the most important vegetable crops. This rooted crop belongs to the celery and parsley family. The total market value is estimated at over R9 million (2008). The tap root anchors the plant and is also a storage place for starch and sugars. Side roots absorb nutrients from deep in the soil.

Heat, rain and leaf blights are the major causes of yield and quality reductions. The carrot root colour changes from pale yellow when they are young to dark yellow, orange and finally orange red as the carrot root matures. These changes are due to the changing carotene levels in the plants as they mature.

Carrots are a fairly expensive crop to produce, especially when smaller areas are established and when sowing seed by hand. This is due to the high plant population of more than 900 000 plants p/Ha and also the time consuming task of hand weeding, thinning and hoeing. Hand push or precision planters eliminate hand thinning of seedlings and also reduces the time and labour cost of thinning. Carrot tap roots take up very little space because of their downward development and provide high yields in relation to the area they occupy.

Carrots grow well in cool conditions and are resistant to cold and frost compared to other vegetables. Carrots are eaten raw, used in stir fry or cooked. They are rich in carotene, high in sugar and also contain vitamin A. Carotene is a pigment captured in the roots of the orange-fleshed cultivars and is a good source of vitamin B1, C, D, E, K, B1 and B6. Carrot juice provides anti-cancer protection, keeps the pancreas and liver clean and is an anti-oxidant. High quality, coloured roots are more nutritious and desirable than the lighter coloured ones. Success when producing a satisfactory carrot crop depends on good management, land preparation, planting, thinning, fertilization, irrigation and weed control (which should be a high priority).

CHAPTER 34: PRODUCTION GUIDELINES FOR CARROTS

305 Different carrot cultivars displayed at a Farmers day near Pretoria, South Africa. This is a summer production F1 hybrid cultivar of good quality.

306 Carrot seed production land which requires cross-pollination by insects, especially bees, in order to fertilize the crops.

307 Different coloured carrots. From left to right: yellow, purple and white. Different colours offer different nutrients which may be beneficial to consumers.

308 A new carrot F1 hybrid cultivar (2012). This is a good quality root which is crispy, sweet and eye catching.

309 This celery plant belongs to the carrot family. It shows a large, healthy root system, produced in a Gravel Flow Technique system. The gravel of this hydroponic system is only 4.5cm deep within the hydro lines (troughs).

310 Carrot roots with different root colours. Top left: orange. Bottom left: light yellow. Top right: cream and bottom right: white.

311 This carrot root was harvested with cabbage production after 4 months of being sowed. The cabbage was established from seedlings. The carrot was the only one in a small production land. This carrot was fairly juicy but had a dull colour. The reason for this may be due to cross-pollination. Be careful when reproducing seed from a large carrot root in order to improve yields, as usually, a mix of different shapes and sizes may occur due to the crop being cross-pollinated.

34.1.1 SUGAR (SWEETNESS) IN THE CARROT ROOT

The carrot is known as a storage root. More specifically, it is a tap root and stores healthy sugars, as the root mature. When carrot roots are ready to be harvested for the fresh market, they should be sweet and crispy. Producers should aim to produce high sugar content roots as this is desirable for the market and consumers. Low sugar results are due to poor cultivation practices.

Producers should be aware that it is the leaves that are responsible for producing high sugar content. Good top leaf growth should be maintained while leaf loss should be prevented. Consider establishing leaf disease resistant cultivars or make use of a good chemical spraying program. Disease resistant cultivars may be established for winter production. Carrots require specific fertilization to ensure high yields, good leaf growth and overall development.

34.2 ESTIMATED PRODUCTION COST FOR CARROTS

The production costs to produce vegetables may differ across the country due to the variances in cost of seed, seedlings, fertilizer, electricity/diesel, irrigation water, labour and so forth. Therefore, costs should be adapted accordingly.

The table below provides estimated costs, yields and profits for carrot production (2013). It will assist with designing a cash flow and business plan. Overhead costs are not included. Producers should do their own research on production costs when they produce vegetables.

The yield of carrots as indicated in the table below is for 40 tons p/Ha. Experienced producers harvest up to 55 tons p/Ha. However, in order to produce such large numbers, great care must be taken with managing soil improvement, disease/insect control, irrigation and fertilization.

The cost of tractors, implements, general equipment (wheel barrows, spades, forks, hoes, crates, twine, pegs etc.), irrigation equipment, water, water pumps, electricity, spraying chemicals, housing, toilets, telephone/computer costs, overalls, clothing for spraying chemicals and so on, are not included in the table below, nor the maintenance of tractors or implements. It is important to make sure that tractors are in good condition before production activity begins.

Certain labour costs are not included in the tables below. Permanent labourers should be used as often as possible, so that they may supervise and train other labourers and also be tractor/truck drivers. Producers should determine labour cost themselves, as costs differs from region to region. Labour is expensive and should therefore be well managed for increased productivity.

TABLE 36: ESTIMATED PRODUCTION COSTS AND PROFITS FOR CARROTS (2014)

ITEM	COSTS P/HA
Clean and level the land.	R900
Soil analysis: Take at least 2 samples @ R250 per sample = R500. Interpretation of sample by an expert = R200.	R500 R200
Soil preparation: Rip, plough, disk and rotovate/till to produce smooth fine seedbeds or raised seedbeds.	R3 500
Fertilizer: Apply lime according to the results of a soil sample recommendation (if the pH of the soil is too low). **Broadcast fertilizer suggestion:** 1 300kg p/Ha 2.3.4 (30) Zn @ R290 per 50kg bag. Require 26 bags x 50kg p/bag = 26 x R290 = R7 540 (2012). **2 Top dressings:** LAN/KAN @ 150kg p/Ha = 6 bags x 50kg @ R240 per bag =R1 440.	R8 980
Seed: Open pollinated cultivars @ R350 p/kg (2014). Require 3.5kg seed = R1 225. F1 hybrid carrot cultivar seed is much more expensive at R660 p/kg (2013) and special Hybrid cultivars = R3.80 per 1 000 seeds.	R1 225
Planting seed: In well-prepared land, use a mechanical planter, including labour.	R1 700
Chemicals: Insect and disease control, including labour.	R1 600
Weed control: Labour: 6 labourers to control weeds and thin seedlings in rows for	R3 000

approximately 5 days = 30 x 100 = R3 000. *Note:* When Herbicides are used correctly, less weed control is necessary. Thinning is more expensive when seed is sown by hand.	
Irrigation: Using an overhead sprinkler system. Diesel, electricity and labour costs as well as shifting the irrigation system, cleaning and repairing it. Water is expensive when low rainfall occurs. To irrigate 1mm water p/Ha, 10 000 litres water is required.	R3 700
Harvesting: Harvest and transport carrots to pack store.	R1 800
Sorting: Labour: Cut leaves, clean roots, pack in bags or make bundles or use other containers.	R2 900
Bags: 5kg orange potato type bags to be used to pack 35 ton carrots. Pack according to market requirements. Require 7 000 5kg bags @ R0.45 per bag = R3 150.	R3 150
Transport: Transport on the farm and to the markets including labour: If the market is 100km away and a 5-ton truck travel 5 times @ R5 p/km = 100km x R5 = R5 000 plus labour which is R1 000 for the truck driver and assistant and R500 for traveling on the farm. TOTAL = R5 000 + R1 000 + R500 = R6 500.	R6 500
Clean and plough residues into the soil. Remove and store irrigation system etc. Establish a rotational crop to improve your soil.	R1 200
ESTIMATED PRODUCTION COST FOR CARROTS P/HA	R40 855

TABLE 37: ESTIMATED YIELDS AND PROFITS FOR CARROTS

YIELD: 40 ton p/Ha less 5 tons unmarketable produce = 35 ton @ R15 per 5kg bag = 7 000 bags x R15 per bag = R105 000.	R105 000
MINUS PRODUCTION COSTS	R40 855
POSSIBLE PROFITS	R64 145

34.3 GUIDELINES FOR CARROT PRODUCTION

34.3.1 CLIMATE REQUIREMENTS AND OPTIMUM PLANTING TIMES
Also see Chapter 3 Climate.

Carrots can tolerate a certain degree of high or low temperatures and therefore may be cultivated throughout the year in most areas of South Africa. Carrots are a cool weather crop and not sensitive to frost. The crop develops slowly in winter and takes longer to mature. Optimal growth is 15°C at night and 23°C during the day. Temperatures higher than 32°C, over long periods, will influence quality negatively and cause poor internal colour and thicker roots. Some cultivars may bolt.

Temperature and soil moisture are the most important factors. Soil moisture also has an influence on the length of the growing period, colour and shape. A shortage of soil moisture causes longer tap roots to form. Moist or very wet soil causes lighter colouring and thicker tap roots. If carrots do not develop fully by the end of July in summer rainfall areas, it is possible that many plants will go over to bolting before they are harvested.

Carrot roots tend to become shorter during summer when the soil temperature is above 20°C. Carrot roots for the fresh markets normally mature in 3 to 3.5 months in favourable conditions. Day temperatures below 15°C may cause the roots to develop a poor colour. For the best colour and root development 15 to 20°C is ideal. The core of the carrot is thinner in winter than in summer. Forked and cracked carrot roots occur more in summer than in winter. Producers should establish recommended cultivars when warm conditions occur.

Carrot roots may become rough and unattractive when temperatures are high, especially when moisture conditions fluctuate.

Temperatures below 10°C tend to cause the roots to become long, slender and pale. Planting and harvesting carrots at extreme temperatures is not recommended. When very high temperatures occur during the growth period, most cultivars will produce an unfavourable colour and diseases such as Alternaria leaf blight may affect the plants. Lengthy, hot periods in the late development stage may retard growth, reduce yields, cause strong flavours and coarse roots. Some cultivars can withstand heat.

Most carrot cultivars are not affected by winter colds and light frost. However, heavy frost just before harvesting will damage leaves and retard growth. If temperatures drop below freezing point and the plants are irrigated late in the afternoon, the roots could be damaged as the soil may become frozen.

Carrot seed is influenced by temperature conditions. Most cultivars germinate poorly as soon as the soil temperature is above 27°C. Germination takes place as low as 8°C and may not take place at temperatures of 31°C and above. The exception may be if irrigation is used to cool the soil and any emerging plants. Experienced producers use this method effectively.

34.3.1.1 OPTIMUM SOWING TIMES

Select the best cultivar according to climate conditions for your area. Some cultivars are adapted to various temperature conditions.

Areas with light to mild frost:	Aug-April
Areas with heavy frost:	Sept-March
Frost-free areas:	March-July
Winter rainfall areas:	Jan-March and Aug-Nov

The critical planting periods for the summer rainfall areas is from May to the end of July. Otherwise when the carrot roots are not well-developed by late July, they may go over to bolting.

34.3.2 TYPES AND CULTIVARS

Carrot tap root lengths vary from very long (approximately 25cm), to very short (canning mini short). There are medium long conical shaped cultivars and cultivars with short, round roots. Early, medium to late and the well-known little finger cultivars are available as well as sweet thin medium-long cultivars.

Breeders' are mostly focused on better yields and quality. Tenderness, a high amount of sugar content (carotene, sugar), uniform skin colour, shape and size as well as tolerance to forking and bolting, smooth peeling and straight roots are all important characteristics of a good cultivar. Carrot types have been developed for specific markets such as Imperators for the mini packs that are popular in the USA. Larger Danver, Flakkee and Becilium types are used for processing while Kuroda and Chantenay are good for bunching and Nantes for pre-packaging.

The open-pollinated carrot cultivars such as Cape market, Chantenay, Karoo and Kuroda are not always suitable or visually appealing, as they may have poor internal colour, may not be sweet enough or may not be fleshy enough. This applies especially when establishing cultivars in the summer. There are however, F1 hybrid cultivars that may overcome these problems. A cultivar known as Sweet Candle has an excellent shape, good length, is very sweet and may be produced in most weather conditions. According to the experts, this cultivar has high yields (2010) but the seed is expensive.

When producing carrots in summer, an effective spraying program should be implemented as insects and diseases are active in warm summer conditions, especially in tropical areas.

34.3.2.1 THE FIVE MAIN CARROT TYPES OR GROUPS
34.3.2.1.1 NANTES TYPES
They are medium length (15 to 20cm) with a blunt point. They are used mostly for bunching, slicing, juicing, pre-packaging and mini carrots. They are good to eat, crispy and are well-suited for local markets.

34.3.2.1.2 DANVER TYPES
They have large roots, are medium length (18 to 22cm) and are used for processing and slicing. They require a long season to mature (120 days) and have high sugar content, however, markets are limited.

34.3.2.1.3 CHANTENAY TYPES
They have large shoulders, are quite short (12 to 15cm) and usually have a large, coloured, inner core. Chantenay types are old fashioned cultivars and are not of the quality required by processors. Small local producers and home gardeners are familiar with these types.

34.3.2.1.4 KURODA
They have large shoulders, are medium-short (15 to 18cm) usually with a conical shape, large core and blunt point. They are used for bunching, slicing, juicing and dicing. They normally mature before Imperator types. Kuroda is produced for fresh market production.

34.3.2.1.5 IMPERATOR TYPES
They are long (22 to 25cm) with small shoulders and have tapered and pointed roots used for fresh packs. A new trend is to use this type for export. Imperator is also used for cuts (mini fresh processed) for the fresh market and traditional fresh packs.

312 A bed maker connected behind a tractor. The blades can be adjusted for single rows.

313 Covering seedbeds with grass after seed has been sown.

34.3.3 SOIL REQUIREMENTS AND PREPARATION
Also see Chapter 5.3 Soil Requirements and Chapter 8.2 Soil Preparation.

Good fertilizer practices, good quality seed and the best care will not provide success unless the soil is of a good texture and well-prepared. A good stand and emergence can be achieved if the soil is of a crumbly structure and as level as possible.

The top soil should be kept moist at all times, to allow the seed to germinate properly. Although average soil may produce excellent carrot roots, the crop thrives best in a deep, well-drained, loose sandy loam soil. The root system (consisting of fine hairy roots) of carrots can penetrate the soil up to a depth of 60cm. The soil should be cultivated to be loose, crumbly and free from weeds, stones and debris. This is to allow the main tap root to develop unhindered.

Uniform soil moisture is essential for the best quality carrot roots. Cloddy, stony, trashy and especially, shallow soil are undesirable. Avoid soil with compost or kraal manure recently incorporated. This will promote the formation of side shoots and forked or malformed roots. Instead, use the compost or kraal manure for other crops in the first season and then for the following season, carrots may be produced.

Carrot roots are sensitive to soil compaction. The rows adjacent to where tractor wheels' travel (when making seedbeds with a tractor) normally produce more forked and stubbed roots. Movement of people and equipment across the land should be avoided for at least two weeks after establishing the crop. If soil are compacted or when the clay content is high, the tap root shoulders are likely to develop above soil level which will cause direct sun light on the shoulders to turn them green. It is also a difficult and costly task to clean the carrot roots when they are produced in clayish types of soil. The washing of the roots may also cause other problems.

34.3.4 PH OF THE SOIL
Carrots do well in slightly acid soil, with a pH of 6 to 6.5 and poorly in acid soil when the pH is lower than 5. Rather avoid brackish soil.

34.3.5 SEED SOWING METHODS, PLANT POPULATION AND SPACING
Also see Chapter 8.6 to plant seed directly in the land and Chapter 11 Spacing of Crops.

Rows should be at least 35 to 50cm apart. Closer spacing may make the harvesting process difficult and prevent air movement through the leaf canopies. When seed is sown in furrows, it is a good to wait for cooler conditions, when possible. Seed should be sown approximately 0.5 to 0.8cm deep in moist, well-prepared soil. If the soil is dry, irrigate the land immediately after seeding.

34.3.5.1 CARROT SEED
Carrot seeds are small compared to other vegetable seeds and emergence may be slower and more irregular. It is important to sow graded seed that is the same size, especially when using seed planters. Graded carrot seeds which are too small (smaller than 0.5mm), should be avoided when sowing or when using planters as small seeds often germinate poorly and negatively affect plant growth development. This may also be due to broken seeds passing through the sieve during the grading process.

Larger carrot seeds germinate better. Larger seeds have the advantage that they germinate more uniformly and produce stronger seedlings. The seed sizes of carrot are 1.8 to 2.2mm in diameter and the count is approximately 400 000 to 550 000 seeds p/kg. Young carrot seedlings are small, weak and fragile when they emerge. For this reason, sowing methods are important. Mistakes made during the seed sowing stage may lead to spotty stands or other defects.

Temperature and soil moisture is important. Most cultivars germinate poorly when the soil temperature exceeds 29ºC. They may germinate very poorly at 30ºC or not germinate at all, unless irrigation is applied to cool the soil. The seed may still germinate at temperatures as low as 7ºC.

When sowing seed by hand or with a vegetable seed planter producers should ensure that the seed germination is good. Considerable losses may be expected when seed is established on hot summer days especially if damping-off disease occurs. Carrot seedling plants cannot be transplanted.

34.3.5.2 PALLET SEEDS
Pallet coated seed, is seed coated to form a round shape approximately 2mm in diameter (the size of a small ball

bearing) they are coated with many bright colours. The bright colours of these coated seeds contain nutrients that give the germinating process a boost from the start.

Pallet seed is expensive. However, they areeasy to sow (because they are round), colourful and the germination is usually good. Only the best quality seed is used for this purpose. Pallet seed is ideal when using seed planters as the seeds are all the same size and fall smoothly through the calibrated planter plate.

34.3.5.3 SOWING SEED BY HAND

Irrigate the land before sowing the seed. Some producers use the full capacity irrigation method and only once the carrot seeds have emerged, will they be irrigated again. Make furrows 2.5 to 3cm deep and 35 to 40cm apart from one another. Sow the seeds 2 to 3cm from one another and for pallet seed, 3cm from one another. Use your common sense concerning spacing. After sowing the seeds, cover the furrows and make sure that the soil is lightly pressed down to obtain good soil-to-seed contact and air removal. It is important to irrigate the seedbed as soon as possible after sowing, especially during hot conditions.

Sowing seed requires experience and practice. When seeds are sown too densely, thinning will need to be done, which is an uncomfortable job, especially when weeds also need to be removed. It is advisable to use herbicides to control certain weeds effectively and to spare the producer the extra labour costs. Plants could be thinned to a spacing of 3 to 7cm from one another within the rows. However, spacing in the rows depends on the cultivar and shoulder size of the root. A plant population which is too high, will make harvesting and weed control difficult, preventing air movement which may facilitate blight leave disease infections.

One method of sowing seed by hand in furrows, is to hold some seeds between the thumb and fore finger and then to rub the these fingers together in order to drop them as evenly as possible in the furrow. Alternatively, seeds may be sown directly from a paper bag. To provide a clearer indication of seed density, one teaspoon of carrot seed could be mixed with 10 teaspoons of white mealie meal.

The quality, sweetness, flavour and beta carotene of the carrots depends on the plants receiving adequate amounts of sunlight. When the plant population is high it may influence the light intensity, especially in winter conditions. To produce larger carrot roots, smaller plant populations are suggested. Rather choose 1 100 000 plants p/Ha. Carrots established to a density of 100m² will grow more rapidly and be available earlier to the market. It is important to know that light levels differ in summer and in winter. Producers should therefore reduce plant populations in winter, cloudy and humid areas by at least 20%. In semi-cooler weather conditions, higher plant populations are suggested. When warmer conditions occur, the tops grow more vigorously.

If irrigation is not immediately available, plant seeds deeper into the soil, as the soil surface may become too hot during warm days. Remember plant populations determine the final carrot root size. Carrot plants compete with each other when the plant populations are high. The biggest mistake made is when carrot seeds are sown too densely.

It is important to irrigate the land a day before sowing, especially when hot, sunny days occur. After sowing, irrigate the land again. If fine, smooth seedbeds are created without clots, the moisture content especially in clay type soil, should be favourable. The correct amount of irrigation is important. The layout of the seedbed and foot path rows should be marked. Labourers should be trained not to step on the marked seedbeds under any circumstances, in order to avoid the soil being compacted.

34.3.5.4 PLANTING SEED WHEN USING SEED PLANTERS

Seed planters are commonly used. They vary from hand-pushed to highly sophisticated precision planters. Precision planters are very accurate, expensive and must be calibrated correctly for each crop. When using pellet coated seeds, a satisfactory stand can be obtained and the thinning of seedlings is limited. Accurate planters are generally used to plant seed at specific distances in rows. When using pellet coated seed of certain cultivars, a spacing of 25 to 35cm between the rows and 5 to 7cm in the rows is suggested.

Producers should experiment with the ideal spacing for cultivars in their region. They should adapt the spacing to meet the requirements of their regions. Carrots which are established at a density of 1 000 000 p/Ha will grow more rapidly and be harvested earlier. F1 hybrid cultivars have the advantage of developing more equally and uniformly due to their Hybrid vigour and uniformity of genes. Seed can also be established with a hand pushed seed planter that is calibrated correctly. Make sure to use graduated pellet coated seed.

34.3.5.5 SPACING WHEN USING PLANTERS
When using mechanical harvesting lifters, 3 rows are planted 35cm from each other. A planting density of 150 to 160 plants per m² (1 500 000 to 1 600 000 plants p/Ha) provides good results when using double rows.

Carrot seed that has been established at a lower density of 100 plants per m² (1 million plants p/Ha) will grow rapidly and be marketable at an earlier stage. When establishing 1 500 000 plants p/Ha, the top leaf growth is very dense and the plants may grow vigorously.

34.3.5.6 SPACING SINGLE ROWS
Single rows may be spaced 30 to 45cm from one another and 2.3cm apart in the row. The spacing also depends on the shoulder diameter of the specific cultivar.

34.3.5.7 SPACING DOUBLE ROWS
Some producers establish 3 to 4 double rows which is quite advantageous. The double rows are spaced 10cm from one another in the beds and 25 to 30cm between the rows. When using 3 rows in the seedbeds, the beds could be 100cm wide and when 4 rows are used, the beds could be 140cm wide. A planting density of 1 500 000 to 1 600 000 p/Ha provides good results when double rows are established.

34.3.5.8 PLANTING IN RAISED BEDS
Harvesting is easier when raised beds are used together with the under cutting blade method. Raised beds can be shaped with a bed maker. The bed maker provides an accurate and smooth seedbed and saves considerably on labour. **See photo 312.**

Seedbeds may be made 1 or 1.2m wide. It is ideal to make the beds 50 to 60cm from one another, when using a bed maker. Add fertilizer before making the beds according to a soil analysis recommendation. When making seedbeds by hand, the soil taken from between the beds and applied onto the 1 or 1.2m beds. When raised beds are created, it will increase the depth of loose soil. It will allow the soil to drain better and the concentrated topsoil around the root zone will provide more oxygen, aeration, nutrients and encourage healthy root development.

Sometimes producers make use of tractor wheels to create beds. The distance between the two wheels is usually 1.2m wide. Tractor wheels cut into the soil to make the foot paths. This method is most effective on sandy to sandy loam soil.

Research has shown that when establishing 4 rows in a seedbed, every 5th row is left open. Accordingly, the spacing between the rows in the seedbeds is 40cm in the summer and 35cm in the winter. This method is recommended due to the better sunlight and aeration it provides, thereby helping produce an even, lengthy carrot root.

34.3.6 MULCHING
Small-scale farmers should use mulch protection, such as grass, after the seed has been sown in the furrows of the seedbeds, especially during warm sunny conditions. Mulching keeps the soil moist, cool and soft and eases germination until the seedlings have emerged.

It is important to remove the mulch once the seedlings emerge. When mulching is not practiced, the seedbed should be lightly irrigated every day, or more than once a day, when hot conditions occur.

34.3.7 CROP ROTATION
Also see Chapter 18 Crop Rotation and Soil Improvement.

The main purpose of a crop rotational system is to control soil-borne diseases and insects. This process may include using a green-manure crop to increase the organic matter in the soil. Carrots and related crops should not be established in the same soil each year and should be rotated every 3 years.

Crops such as sun-hemp, black mustard, canola, eragrostis curvula, finger grass, lucerne (after harvesting several times), maize, babala, soybeans, lupines and so on, may be incorporated into the soil 4 to 6 weeks prior to planting the next crop. This reduces diseases and insects and keeps the soil fertile.

34.3.8 THINNING OF PLANTS
Also see Chapter 16.2 Thinning of Seedlings.

When carrot seed is sown by hand, it is necessary to thin the plants to the correct spacing. When using a seed planter, thinning is reduced to the minimum. Hand thinning of carrot seedlings can be disastrous when the seeds are sown too densely. The seedlings germinate at a rate of approximately 1.1 million plants p/H so hand thinning can be a problem. It is also a time consuming, difficult and expensive (in terms of labour) task. Labourers need to be trained to become familiar with sowing seed correctly.

The beginner should experiment with sowing densities in order to save seed and keep labour costs as low as possible. Un-thinned beetroot and carrots, when too dense, will be mostly unusable for marketing. In such conditions, the roots may become pale, have less sugar content and be quite hard, even when they are still young. It is therefore important to establish seed that is known to have good germination, in order to create a good stand and plant population.

When germination is low the stand will be uneven. Seedling rates should therefore be controlled as accurately as possible. Thinning may be combined with weeding, but it is an entirely different and uncomfortable job. The first thinning should be carried out when the seedlings are 3 to 4cm tall and the second when they are 7cm tall. Most weeds grow faster than seedlings and should therefore be removed before the plants are thinned. The seedling plants should be thinned 3 to 7cm from one another in the rows. This also depends on the cultivar shoulder size.

34.3.9 SEED PER HA
Approximately 2.8 to 3.5kg seed p/Ha when producing for the Fresh market and 1.4 to 2.5kg for the processing markets. Seed should be treated with Thiram just before sowing, according to specialist advice.

34.3.10 FERTILIZER
Also see Chapter 15 Fertilizer.

A soil analysis is important, as the fertilizer balance must be correct to produce healthy, sweet, juicy, colourful carrot roots. Optimum fertilization is therefore essential for the best quality and yields.

Nitrogen N is essential, especially for upper leaf growth. Be careful however, as too much N may increase leaf growth and reduce root development. This may affect the sugar content of the roots negatively. Too much N at the beginning of the season may promote splitting of the roots. Phosphorus P is responsible for the production of sugar and starch and because of this, P is very important. In the earlier stages of growth, the carrot plant reacts well to Nitrogen N. To increase sugar content after leaf development, phosphates P and potassium K levels should be increased. Calcium is also important to strengthen the cell walls of the carrot roots. Magnesium may also be applied in the later stages of growth to improve the colour.

Field experience and knowledge of the crop requirements will help to determine the nutrients required and the rate of application.

Phosphorus P is important when using virgin soil. Applying large amounts of potassium all at once, may decrease quality. Fresh manure or compost should not be used as it stimulates splitting as well as hairy roots. However, it would be ideal to establish the crop the following season, because by that stage, the microbe activity would have broken down the manure/compost.

It has been found that when applying N in the ammonia formation, when the plants are still young, it may increase forked roots, nematodes, un-decomposed organic matter and poor aeration. Fertilizer is very expensive so do not over fertilize your soil. Fertilizer is actually salt, therefore be careful that salts do not build up in your soil.

The estimated withdrawal of nutrients elements from the soil is as follows:
4.0kg Nitrogen N p/Ha
0.75kg Phosphate P p/Ha
5.0kg Potassium K p/Ha

TABLE 38: A FERTILIZER APPLICATION SUGGESTION FOR A YIELD OF 40 TON P/HA

If the goal is a 40 ton yield, the suggested fertilizer application could be: Direct seeding.	Macro Fertilizer kg/Ha		
	N	P	K
Basic p/Ha – poor soil	40kg	80kg	80kg
Basic p/Ha – fertile soil	15kg	24 g	30kg
3 weeks after emergence broadcast	50kg	-	-

The above serves only as an example. Confirm with fertilizer experts before applying anything. Top dressings should be applied at the 3 to 5 leaf stage of the seedlings, approximately 3 weeks after the seeds have emerged. Usually LAN/KAN is applied at a rate of 150kg p/Ha and if necessary, a second application 3 weeks after the first application at 120kg p/Ha, if on fairly fertile soil. Top dressings are very important for this crop, for a variety of reasons. The top growth needs to develop reasonably well but not too vigorously. Continue monitoring this and be sure that the correct applications of fertilizer and other cultivation practices are carried out.

The Potassium K level should be fairly high. A general fertilizer application recommendation for small-scale farmers on fairly fertile soil when a soil analysis is not available is as follows: Apply 1 200kg 2:3:4 (24) Zn just before establishing the seed. Work it into the soil to a depth of 15 to 20cm. Some producers use the cheaper and more common 2:3:2 (22) Zn at a rate of 1 500kg per/Ha, especially on clay soil. Clay types of soil usually have enough potassium and do not leach easily. Nematodes, un-decomposed organic matter, poor aeration and saturated soil may cause forked and stocky roots. Carrots are sensitive to a boron B deficiency.

34.3.11 IRRIGATION AND IRRIGATION SYSTEMS
Also see Chapter 13 Water and Chapter 14 for Irrigation.

Irrigation water should be well-distributed over the entire growing season. It is important that the land be irrigated regularly. Carrots need much more water when very hot or dry spells occur. Uneven irrigation results in rough and stocky carrot roots.

The norm for irrigation throughout the season is approximately 420mm. Irrigate 25 to 30mm the first week and 30mm every week thereafter. It is important to cut-back irrigation towards the end of the growing season when the roots begin to mature. Otherwise if too much water is supplied, the roots may crack. Temperature and soil moisture has a huge influence on the colour, shape and quality of the roots. The best product is obtained when all these factors are considered. The moisture content of the soil influences the length and colour of the roots. A shortage of moisture results in longer carrot roots. Wet conditions cause a shorter, thicker and lighter internal colour of root.

Over-irrigation leads to short carrot roots, especially in poorly drained soil. Therefore, it is better to establish the crop in lighter soil on raised beds in high rainfall areas. Experienced producers sometimes reduce the water during the early growing stage of the plant in order to establish a longer carrot root. The roots become longer when they search for water. However, be careful that the plants never wilt due to a lack of water. When heavy rains occur, the shoulders of the carrot roots may appear above the soil and should be covered again with soil. If this is not done, the sun may burn the shoulders changing it to a green colour.

34.3.11.1 DRIP, SPRINKLER, OVERHEAD AND PIVOT IRRIGATION SYSTEMS
It is not recommended to use a drip irrigation system as this system is not suitable to achieve good germination. Too much water through the dripper system may cause the roots to become too long. Too little water may cause the roots to be short. Too many dripper lines will also be required due to the narrowness and number of rows. It then becomes difficult to shift all the dripper lines when weeding needs to be done. Therefore, the cost of using so many dripper lines becomes high.

According to a Water Pleasure (experts in irrigation) claim, the cost to establish a 1Ha drip system for carrots is approximately R50 000 (2011) in comparison with tomatoes which is R25 000. Small-scale farmers make better and more effective use of cheaper sprinkler systems and some large commercial producers use overhead or pivot systems.

34.3.12 WEED CONTROL
Also see Chapter 17 for weed control.

Carrots are a slow growing crop that suffers severe yield losses when weed competition is a factor. The plant has thin leaves and does not overshadow weed competition. The seedlings grow slowly when they are young and are also thin and tender. Carrot production is limited by a broad spectrum of summer, winter, annual and perennial weeds.

Weed management early in the seedlings growth is critical in producing high yielding and quality carrot roots. Planning for weed control should be done ahead of planting time. Take care that lands are not infested with nutty weeds as they are difficult to control. Land such as this should rather be avoided. Hand weeding during the early stages of emergence must be frequent and shallow. This is essential for this crop. The seedlings grow slowly and therefore weed control during the early stages after the seeds emerge is very important.

Weeds should not at any stage compete with the crop. This is where most producers fail as weed control is an expensive, uncomfortable, time consuming and difficult job. During the early stages after emergence some producers control weeds with a light hoe between the rows. At the same time, they also cover the shoulders with soil, preventing the greening of the carrot root shoulders.

34.3.12.1 HERBICIDES
While the carrot seedlings are small, weeds are difficult to control by using machines. Hand weeding is expensive and therefore commercial farmers select soil with a minimum of weeds. They also make use of chemical herbicide control which eliminates the necessity for controlling weeds by hand.

Herbicide treatment reduces weed control when preparing the land. Applying them a few weeks before sowing the seed, keeps most weeds under control until the foliage of the carrot plants become large enough to cover and suppress the growth of the weeds. Many pre-emergence and early post-emergence herbicides are available. The common ones are **Roundup** and **Gramoxone** which control many weeds for up to 9 week after sowing the seed.

34.3.13 BOLTING
Bolting or seed formation occurs when the plant is stimulated to move from the vegetative stage to the seed reproductive stage.

Temperature, water deprivation, insufficient nitrogen N levels and the duration of daylight are some of the factors that may affect the bolting habits of some cultivars. *See photo 283.*

Bolting may become a problem with some cultivars that are vulnerable to bolting in certain locations. Bolting also occurs when the crop is established too late in the season or in warmer conditions and before the cold winter begins. Similarly, in warmer conditions after winter, some cultivars may go over to bolting. This also happens to other vegetable crops such as lettuce, beetroot, onion and cole crops.

Resistant or partly resistant F1 hybrid cultivars have been developed to control this problem to a large extent. Cold temperatures, insufficient nitrogen N levels and daylight may also trigger bolting of specific cultivars.

34.3.14 YIELDS
Yields can differ from season to season. The average yield of open pollinated cultivars is approximately 35 to 45 ton p/Ha. Hybrid cultivars in combination with precision planters, good fertilizer management and the correct plant population may yield up to 70 ton p/Ha or more.

34.3.15 DURATION BETWEEN PLANTING AND HARVESTING
Depending on cultivars, the climate and seasons, carrot roots are ready to be harvested after 3.5 to 5 months. There are early, medium and late cultivars as well as F1 hybrids available for improved winter and summer production.

Finger carrots may be harvested when the roots are approximately 12cm long within approximately 60 days. Other carrot cultivars should be allowed to grow until the roots have reached a minimum of 20cm. The Danvers types are ready to be harvested after approximately 115 to 135 days.

34.3.16 HARVESTING AND MARKETING
Also see Chapter 24 Harvesting, Handling and Packaging.

Harvested roots should be smooth, firm and shiny. Most roots should be harvested when they have reached the fully mature stage in order to obtain maximum yields. Fully mature roots have an increased shelf life. Mature roots can be harvested over a period of 3 to 4 weeks. The roots may be pulled out of the soil by hand in sandy and sandy loam soil.

Fresh market prices indicate that the highest prices are obtained in March to April (less in May) due to the difficulty of cultivating carrots during the summer months. Carrots for the fresh market are generally harvested before the roots are fully mature. They should have the following qualities:
- Mild, sweet flavour.
- High percentage of usable tap root.
- Vigorous plant growth.
- Uniform, deep orange or gold colour, including the inner core.
- Long (18cm or longer), slender, smooth roots with a small diameter leaf neck attachment.

Fresh market carrots should be tender, mild in flavour and brighter in colour than carrots harvested for processing. Care should be taken that the carrot roots never become soft or wilt. They will not recover if this happens. Also avoid large green shoulders.

When harvesting the roots, it is advisable to irrigate the soil lightly to make the harvesting (pulling) process more comfortable. If the soil is too dry when harvesting, it may be difficult to clean the roots and the tops may break off when pulling them out of the soil. Small-scale farmers use a fork to loosen the roots and then proceed to pull the top growth. Commercial farmers that produce carrots in heavier types of soil, undercut the roots with a blade and then pull the tops. Raised beds make the harvesting process easier, especially when an undercutting blade is used.

Do not harvest early in the morning when it is cold, as the roots may crack. After harvesting, protect the roots from the sun to retain their quality. Root injuries reduce the shelf life and increase the chances of decay.

34.3.16.1 MARKETING
Carrots should be packed or bundled in accordance with the market or supermarket requirements. Fresh markets prefer orange coloured roots. It is important to use cultivars that produce quality roots with long, bright, shiny roots, especially when establishing them in warmer conditions. Fresh market carrots are susceptible to injury, as they are often harvested before maturity. This is done as per the market requirements.

34.3.17 STORAGE
Most vegetable crops are perishable and can only be stored for a few days. It is best to harvest carrots and store them in a cool place or a refrigerator with a humidity of 95 to 99%. Carrots may be kept for 2 months or longer when stored in airtight plastic bags and then refrigerated. This maintains their moisture content. Processing carrots are harvested when they are fully mature and are less susceptible to injury. They can be stored for up to 3 months in cool storage rooms.

34.3.17.1 IN-GROUND STORAGE
Carrots may be stored in the land in certain soil and harvested as needed. This is possible because the carrot roots can withstand air temperatures as low as 5°C. For best results cover the roots shoulders with 2 to 5cm soil. The roots may be kept for up to 2 months in this way.

34.3.18 INSECTS
INSECTS: *See Chapter 20.5.*

ROOT-KNOT NEMATODE:
Damage and losses have been reported especially in summer. In spring and winter productions, the risk is much lower due to the cooler days/nights and the lower soil temperature. It is best to fumigate the soil before planting or sowing the crop.

APHIDS: *See Chapter 20.5.2.3.*

RED SPIDER MITES:
As a rule, this insect is not a serious problem for carrots, due to the numerous plants and the smothering top growth of the crop.

34.3.19 DISEASES
DISEASES: *See Chapter 20.6.*
DAMPING OFF: *See Chapter 20.6.3.1.*
ALTENARIA LEAF SPOT-PURPLE BLOTCH: *See Chapter 20.6.3.2.*
EARLY BLIGHT: *See Chapter 20.6.3.3.*
PHYTOPHTHORA WILT: *See Chapter 20.6.3.16.*
BACTERIAL BLIGHT:
Brown spots appear on the leaves and brown stripes on the leave stems. This disease makes horizontal lesions on the roots in a similar way as Altenaria blight. The disease prefers wet, warm weather and is difficult to distinguish from Altenaria.

CARROT ROOT ROT:
When lifted, the roots display brown rotten spots. This is caused by fungi and also bacterium Erwinia that causes the roots to become soft. The disease spreads rapidly after harvesting takes place and when the roots are supplied to the market.

CHAPTER 35

PRODUCTION GUIDELINES FOR BEANS

314 Runner beans trellised and produced in a shade net house. Note the vigorous plant development due to the favourable conditions.

315 A good, green bean cultivar which has an excellent yield. Note the rhizobia knots on the roots, not to be confused with nematodes. When squeezing the knots, a typical purple colour would become visible. This is a newly bred cultivar (2010). The problem with this particular cultivar is that the pods are too bleached.

35.1 INTRODUCTION

Green beans are part of the plant family known as legumes. All legume plants produce pods. Examples are peas, green beans, dry beans, sun hemp, black mustard, lucerne, lupines, soybeans, velvet beans etc. Pod bearing plants develop numerous small knots on their roots, especially when favourable conditions exist. These knots are sometimes mistaken as nematodes (eelworm). These knots are called *rhizobium* and when the plant is incorporated/ploughed into the soil it becomes beneficial for the next crop. The reason for this is that rhizobium supplies natural N, bound from the atmosphere.

Legume vegetables, shrubs and trees that bear pods are almost all rhizobium binding. Some producers inoculate seed of pod bearing crops to stimulate the rhizobia and help the plant to produce the root binding rhizobium. The most challenging aspect of harvesting fresh bean pods, is that labourers are required to pick the pods by hand. This is an onerous task as they are bending down, sometimes all day long. When the pods are mature, producers only have 3 weeks to pick them. After 3 weeks, the pods become stringy, hard and jelly inside the pods result.

Green beans are a warm weather crop and are sensitive to low temperatures. Runner, pole or climbing beans need to be supported by trellising them. In this way, it becomes effortless to harvest the pods due to the upright positioning of the plants. Much more work is necessary to grow runner beans using the trellising system. The growing period of this bean is also much longer. Yields for runner beans are more than 3 times better than garden beans per plant. However, due to the trellising method and the spacing requirements, the yields per square meter of the two are similar.

Most bean cultivars are cultivated for the fresh markets, freezing (dehydration) and producing dry beans for canning (baked beans). The white seeded cultivars are more popular as they can be cooked when the pods are dry. Similarly, for dry or baked beans. Many bean cultivars are available, but the markets and consumers prefer the narrow, straight, dark green, potted cultivars.

316 Green beans which are slightly pale green in colour. This occurs when a shortage of N is present. However, too much N may cause flower drop.

317 A runner bean cultivar. Salamanca is popular and produces high yields. The internodes are much closer to each other and are the reason for the high yields. The pods are also narrow and dark green which is advantageous.

318 Various bean cultivars. Consumers prefer dark green pods. These pods were displayed at a farmer's day. Top left is a yellow potted cultivar.

35.2 ESTIMATED PRODUCTION COSTS AND PROFITS

The production costs to produce vegetables may differ across the country due to the variances in cost of seed, seedlings, fertilizer, electricity/diesel, irrigation water, labour and so forth. Therefore, costs should be adapted accordingly.

The table below provides estimated costs, yields and profits for bean production (2013). It will assist with designing a cash flow and business plan. Overhead costs are not included. Producers should do their own research on production costs when they produce vegetables.

The yield of green beans as indicated in the table below is for 14 tons p/Ha. Experienced producers harvest up to 20 tons p/Ha. However, in order to produce such large numbers, great care must be taken with managing soil improvement, disease/insect control, irrigation and fertilization.

The cost of tractors, implements, general equipment (wheel barrows, spades, forks, hoes, crates, twine, pegs etc.), irrigation equipment, water, water pumps, electricity, spraying chemicals, housing, toilets, telephone/computer costs, overalls, clothing for spraying chemicals and so on, are not included in the table below, nor the maintenance of tractors or implements. It is important to make sure that tractors are in good condition before production activity begins.

Certain labour costs are not included in the tables below. Permanent labourers should be used as often as possible, so that they may supervise and train other labourers and also be tractor/truck drivers. Producers should determine labour cost themselves, as costs differs from region to region. Labour is expensive and should therefore be well managed.

TABLE 39: ESTIMATED PRODUCTION COSTS AND PROFITS FOR BEANS (2013)

ITEM	COSTS P/HA
Clean and level the production land.	R800
Soil analysis: 2 samples @ R290 per sample = R580 and interpretation by an expert = R250.	R830
Soil preparation: Rip, plough, disk and rotovate/till to produce a smooth, fine seedbed. Apply lime if the pH of the soil is too acidic. Labour included.	R1 910
Fertilizer: Apply lime according to the results of a soil sample recommendation (if the pH of the soil is too low). Apply fertilizer mechanically using a tractor and spreader = R350. Broadcast fertilizer suggestion: 1 100kg p/Ha 2.3.4(30) Zn @ R290 per 50kg bag. Require 22 bags x 50kg p/bag = 22 x R290 = R6 380 (2012). 2 Top dressings LAN/KAN @ 100kg p/Ha = 5 bags x 50kg @ R240 per bag = R1 200.	R7 930
Plant material: Seed: 130kg p/Ha @ R25 p/kg. F1 hybrid bean seed cultivars are more expensive.	R3 250
Plant seed: Using a mechanical seed planter and labour.	R680
Chemicals: Insect and disease control and labour.	R900
Weed control: Mechanical: Between rows and labour = R450. 3 labourers control weeds in rows for approximately 4 days @ R100 per labourer = 12 x R100 = R1 200 = R1 550.	R1 550
Irrigation: Overhead sprinkler system is used. Diesel, electricity and labour costs as well as shifting the irrigation system, cleaning and repairing it. Water is expensive when low rainfall occurs. To irrigate 1mm water p/Ha, 10 000 litres water is required.	R3 500
Harvesting: Hand picking 14 tons of bean pods. Packing in crates, loading and transporting to the packing store. One labourer picks approximately 300kg bean pods per day; 48 labourers pick and transport 14 400kg per day = 48 day @ R100 per day = R4 800.	R4 800
Sort and pack in 5kg green potato bags or boxes.	R1 400
Packaging material: Packaging according to market requirements: 5kg green potato bags or boxes = 1 400 x 5kg boxes or green potato bags; R0.60 per bags/box = 0.60 x 1 400 = R840. Other containers and bags may be used.	R840
Transport: If 100km from the market. Travelling with a 3-ton truck 3 times = 600km @ R5 p/km = R3 000 plus R600 for truck driver and assistant.	R3 600
Plough the remaining debris into the harvested land and establish a soil improvement rotational crop.	R1 450
ESTIMATE PRODUCTION COST FOR BEANS P/HA	R33 440

TABLE 40: ESTIMATED YIELDS AND PROFITS FOR BEANS

YIELD: 14 Ton green bean pods, less 1 000kg unmarketable = 13 000kg @ R6.50 p/kg = MINUS	R84 500
PRODUCTION COSTS	R33 440
POSSIBLE PROFITS	R51 060

35.3 PRODUCTION GUIDELINES FOR GREEN BEANS

35.3.1 SOIL REQUIREMENTS
Also see Chapter 5 Soil Requirements.

Green beans are successfully grown in soil that ranges from sandy to clay soil (even on some heavy, clay types of soil). The crop prefers deep, well-drained soil that hasgood water holding ability and drainage. Bean roots penetrate the soil to 60cm deep in soft loose soil but most of the roots are found in the top 30cm layer of soil. Beans do not develop well when soil are too wet and this should be avoided.

Soil with a good, uniform texture, promotes uniform crop development, maturity and harvesting of the pods. Overall, less than 2 pickings stages are required. When picking the young, tender, juicy bean pods at an early stage, it will stimulate the plant to produce more flowers and pods which will encourage higher yields. Producers should decide for themselves when to pick the pods so that it may suit the market. When germination, fertilization, irrigation and spacing is managed well, the producer will be rewarded with attractive, juicy, green bean pods.

35.3.2 PH OF THE SOIL
Beans are best produced in soil that is neutral to slightly acid. Lime should be applied if the pH level is below 5.8. The best results are achieved in medium loam soil with a pH of 5.8 to 6.6. Good yields are seldom possible when producing beans in heavy, clay types of soil.

Alkaline soil with a pH above 6.8 may cause a manganese Mg deficiency. Acid soil with a pH of 5.5 or below should be avoided unless it is treated by using lime. Beans are sensitive to high boron B and brackish conditions, hence these soil should be avoided.

35.3.3 SOIL PREPARATION
Also see Chapter 8.2 Soil Preparation.

A range of seed planters are available for mechanical planting. Seed can be planted in moist or fairly dry soil. This should be followed by frequent overhead irrigation until the seed has emerged.

During the planting process, the following should be kept in mind:
- All seed should be planted at the same depth (not deeper than 4cm) and care should be taken to ensure that the seed skin is not cracked or broken.
- When mechanical seed planters are not calibrated correctly, the correct seed plate is not installed or the rotation speed of the plate is not correctly controlled, the outer skin of the seeds may be damaged.
- When bean seeds begin to germinate, it forces the seed lobs above the soil, including the first leaves. It is therefore critical to make smooth seedbeds without any clots. The soil should also not form a hard crust after rain or irrigation. It is better to use the full irrigation capacity method for this crop.
- Too much irrigation is also a problem, as the thin seed skin may crack when the soil is too wet.

To ensure a good stand the following is important:
- The soil should be properly cultivated to prevent large clots.
- The soil surface should be kept soft but not wet. It is better to use light irrigations until the seedlings emerge.
- If the soil is raked slightly after planting, the water will collect in these shallow furrows and be more evenly distributed.

Poor soil preparation will hamper production and also limit root penetration.

35.3.4 CLIMATE REQUIREMENTS AND PLANTING TIMES
Also see Chapter 3 Climate.

Beans are a warm weather crop. The ideal temperatures are 18°C at night and 28°C during the day. When temperatures fall below 10°C, plant development is retarded. When temperatures are above 35°C and dry winds occur, flowers drop and tender young pods abort.

Beans are sensitive to cold weather and the seed may germinate poorly when the soil is too cold. Make sure that the danger of frost has passed when producing beans. This time is usually 2 to 3 weeks after the last frost is expected. Green beans may be produced as a winter crop in frost-free areas such as the Lowveld. The summer months in the Lowveld are too hot. Night temperatures of 7°C and below, will result in pods that develop well, but may contain a small number of seeds in their pods. Short and malformed pods may also be the result. Good yields cannot be expected in areas where day temperatures are above 30°C.

High humidity and heavy rain normally encourages diseases. In clay types of soil, rain would delay harvesting. Beans germinate quickly, from planting to emergence it takes only 6 to 10 days. However, this depends on soil temperatures and when low temperatures occur, emergence is only after 16 days. At the optimum soil temperature of 24 to 30°C the seed will germinate quickly and seedlings may emerge in approximately 8 days. For temperatures above 30°C, the seed may emerge in 6 days. At a low soil temperature of approximately 15°C the seed will germinate, but slowly and may only emerge after approximately 16 days. Seeds will not germinate below 8 to 10°C.

35.3.4.1 OPTIMUM PLANTING TIMES

Areas with light to mild frost:	Sept-Feb
Areas with heavy frost:	Sep-Jan
Frost-free areas:	Feb-April and July-Aug
Winter rainfall areas:	Sept-Feb

35.3.5 CULTIVARS

Older local bean cultivars still exist, such as the cultivar called Wintergreen. These are still popular cultivars and may be a good option to produce. Improved F1 hybrid cultivars are available from various seed companies.

35.3.5.1 GREEN BEANS (GARDEN OR BUSH BEAN CULTIVARS):

Wintergreen (Black seeded), Champ Strike, Contender, Provider, Harvester, Tongati, Sodwana etc.

35.3.5.2 FINE POTTED CULTIVARS

Often called 'Haricot beans'. The pods are deep-green and very tender. Assegai, Bonzai, Primero, Twiggy, Tokai etc.

35.3.5.3 RUNNER BEANS:
- **Witsa:** A good yielder. The pods are light green, bent slightly, are round and long.
- **Lazy Housewife:** Long, broad and flat. Most of the pods are bent and it is not a good yielder.
- **Salamamca:** F1 hybrid cultivar which has excellent yields due to its shorter internodes. It is popular due to its dark, narrow, green pods. The seed is expensive because it is a F1 hybrid cultivar.

Green bean pods are classified by their pod length (cm) and diameter (mm):

	Length	Diameter
Extra fine pods	10-11cm	5mm
Very fine pods	10-14cm	6-8mm
Fine pods	11-14cm	7-9mm
Medium fine pods	11-13cm	8-10mm
Medium large pods	12-14cm	9-11mm
Large pods	13-15cm	10-12mm

35.3.6 PLANTING METHOD, SPACING AND DENSITY
Also see Chapter 8.4 to produce seedlings in soil.

35.3.6.1 PLANTING METHODS

A number of excellent hand-push or mechanical planters are available. Special care should be taken that these planters do not damage the seed by calibrating them correctly. Vacuum planters accurately plant individual seeds with minimum damage to the seed.

Seedlings should emerge in approximately 10 to 14 days. Any delay in emergence of the seed may cause them to rot in the soil. Uniform emergence of beans is important, especially when huge plant populations are established. A delay of only 2 to 3 days for emergence (or uneven emergence) may be a problem, as the plants that emerge later may suffer from poor plant development. Always handle bean seeds with care. Handling seeds in a rough manner, lowers the germination percentage and increases the number of weak seedlings. If a bean seed falls only 1m, the tensioned seed coat may crack which will influence germination. When the seed coat is cracked, the seed will rot in the soil.

Temperatures above 30°C may create plant water stress even with adequate soil-water levels. This is because the root system may not be able to handle the high transpiration rates of the top growth and blossom drop may occur. Several light irrigations may be necessary to avoid the top soil from crusting. During the rest of the season, retain moisture in the top 30cm of the soil or above 50% of the soil-water content. The critical periods for water stress during the development of the plants, are at the bloom and pod-set stages. It is advisable to irrigate early in the morning to allow the plants to dry out before nightfall.

35.3.6.2 DEPTH OF PLANTING

Depth of planting is important to ensure uniform germination and emergence. Depth of planting may be from 3cm to 4cm depending on soil type. Also at the time of planting, temperature and moisture conditions are important. Generally, beans are planted deeper in sandy soil, especially when dry and hot conditions occur. Shallower planting depths are suggested for heavier soil when the soil is cool and the soil moisture is high. Uniform planting depth leads to uniform emergence. Irregular planting depths may slow emergence and encourage root rot.

35.3.6.3 SPACING

Spacing is determined by factors such as the viability of the seed, the availability of implements and harvesting methods, especially when a mechanical harvester is used. Spacing also depends on soil type and irrigation. When harvesting mechanically, closer row spacing is not practical, as harvesters need wider spacing. Too close row spacing reduces air movement and causes high humidity during rainy periods. It also promotes diseases.

On sandy soil, spacing may be 25 to 30 plants per m² or 35 to 50cm between rows and 6-10cm in the rows. This results in a plant population of 250 000 to 350 000 plants p/Ha. The seeding rate (kg/Ha of seed) used to achieve a certain plant population depends on the germination percentage and seed size. Seed size varies considerably between different cultivars. Generally, when bean seeds are spaced wider between the rows, then narrower spacing could be used within the rows. A spacing of 5cm in the row and 60cm between rows, provides good results. This results in a plant population of 332 000 plants p/Ha. Do your own spacing research and evaluation. The spacing for runner beans is about 20 x 100cm. Establish 2 seeds per stake.

35.3.6.4 PLANTING SEED

When the bean seeds emerge, the entire seed lob is pushed above the soil level. Therefore, a fine, soft, smooth seedbed is required. Plant the seed shallower when early planting times are required. Do not establish bean seeds in wet, clayish types of soil. The seed's cover may crack and poor germination may result. A simple test is to place a bean seed in your mouth. The seed cover will crack after approximately 6 minutes.

Seedlings should emerge in approximately 8 to 12 days when warm conditions occur. Uniform emergence of the seed is important, especially with high plant populations. A delay of emergence of only 2-3 days may result in uneven emergence.

When seeds are planted by hand, a fine, smooth seedbed should be prepared. The seed may be planted in furrows or by making holes with a stick along the planting line, approximately 2cm deep. For small areas, press the seed 2 to 3cm deep into the soil with your finger.

Producers may make furrows at the required spacing with a spade or a modified plough grub pulled behind a tractor. The soil should be moist before making the furrows. Place the seed at the required spacing in the furrows and cover them with a rake. Irrigate 10ml after planting. Plants spaced too far away from one another will have more pods, however this may cause top-heavy plant growth, which may then fall over, especially when windy conditions occur. Plants may even break clean off at the soil level.

Some producers irrigate the soil thoroughly and do not irrigate until the seed emerges. This prevents the top soil from forming a crust. However, the top soil should never be allowed to dry out.

35.3.6.5 PLANTING AND TRELLISING RUNNER BEANS
There are many ways to plant and trellis runner beans. Any kind of trellising material can be used as long as the stakes are 2m or higher. The main factor is to bind the stakes together. They should be driven deep enough into the soil to strengthen the overall structure. It should be kept in mind that when the plants cover the entire length of the trellising poles, the pressure caused by the wind is high, which may result in damage to the system.

35.3.7 AMOUNT OF SEED P/HA
Seeds are one of the most important factors when producing green beans. The seed must be of a high quality and be treated with insecticide and fungal chemicals. Direct planting for the fresh market results in 50 to 100kg p/Ha. Seed size varies considerably due to the variety of cultivars that are available in the trade. Fine potted bean seeds require 40 to 55kg p/Ha. Remember to select the best cultivar for a specific area, planting time and season.

35.3.8 DAYS TO HARVEST
Green beans are an early maturing crop and the first harvest may be in 60 to 75 days after establishment.

35.3.9 FERTILIZATION
Also see Chapter 15 Fertilizer.

A soil analysis is important and is the most accurate guide for correct fertilization application. One ton of bean plant material, may withdraw approximately the following from the soil:
9.0kg Nitrogen N p/Ha
0.9kg Phosphate P p/Ha and 6kg Potassium K p/Ha

If a yield of 15 ton p/Ha is the goal, apply the following suggested fertilizer applications:
N: 9.0kg x 15 t/Ha expected yield = 135kg per Ha
P: 0.9kg x 15 t/Ha expected yield = 13.5kg per Ha
K: 6.0kg x 15 t/Ha expected yield = 90kg per Ha

This is an example of how much fertilizer may be used. Confirm with fertilizer experts before apply anything. A suggested fertilizer mixture applied on a fairly fertile soil could be as follows: 800 to 1 000kg 2:3:2: (22) Zn broadcasted and worked it into the soil just before establishing the seed. When the soil is already rich with phosphate P or potassium K then the commonly used 2.3.2 (22) Zn will not provide much benefit. For small-scale farmers, approximately 600-800kg/Ha or 60-80gr per m^2 of 2.3.4(30) Zn is suggested.

Due to the short growing period of beans, it is important to apply an early N top dressing 2 to 3 week after emergence. Be careful not to apply too much N as this may stimulate vigorous plant growth and flower drop may result. Generally, crops that are lush and healthy are a requirement for most crops, but not for beans. Top or side dressings using Nitrogen N should be approximately applied as follows: 120kg p/Ha at the 3 leaf stage, approximately 3 weeks after emergence, then 80kg p/Ha 2 to 3 weeks later. Experienced producers look at the plant colour. It should not be too light or dark. Beans grow well when they follow heavy crop feeders such as cabbage.

35.3.10 IRRIGATION
Also see Chapter 14 for Irrigation.

Irrigation and water consumption differs in warmer areas. Irrigation will be applied more in warmer areas than in the cooler Highveld. Apply 350mm water - 20mm the first week and 30mm every week thereafter, until harvesting. Beans are sensitive to moisture stress in their 2-month growing period. Therefore good soil and irrigation management is very important to ensure optimum yields and good pod quality. Overhead sprinkler irrigation is commonly used, especially on sandy types of soil. Soil moisture management when planting beans is necessary to ensure uniform germination.

Effective moisture management, controlled through the irrigation system will also increase flower and pod development. Planting bean seeds in soil which is too wet, causes the bean seed coat to crack and produce abnormal seedlings. Alternatively, the seeds may rot. This applies especially in clayish types of soil. Adequate water produces larger and more vigorous plants, a greater number of pods, of which fewer are young pods. However, over-watering may result in excessive growth, which will make it difficult to harvest mechanically and it will also encourage weed growth and diseases. The relatively short growing period and shallow root system of beans, makes it sensitive to moisture stress. The soil should not dry out at any stage of production. It is essential to have strong, growing plants before they bloom and at the pod stage.

Sometimes, irrigation can be used to change temperatures. High temperatures are more of a problem than low temperatures. At high temperatures, the plants use water faster than the roots can handle. If this type of stress occurs while the plants are blooming, then an early pod-set will result. Constant rain during the flowering and harvesting period can lead to problems, such as the rotting of the young pods and hampering the harvesting process. Frequent irrigation and low application rates of 2.5mm water during periods of high temperatures, can reduce the temperature in the land and prevent or reduce flower and pod drop.

35.3.11 CROP ROTATION
Also see Chapter 18 Crop Rotation and Soil Improvement.

The main purpose of a crop rotation, is to control soil-borne diseases and insects. Crop rotation may include a green-manure crop to increase the organic matter in the soil. Beans and related crops should not be established on the same soil each year, but instead should be rotated every 3 years.

Crops such as sun-hemp, black mustard, canola, eragrostis curvula, finger grass, lucerne (after harvesting several times), maize, babala, soybeans, lupines etc. may be incorporated into the soil, 4 to 6 weeks before planting the next crop. This reduces diseases and insects by allowing the bacteria to do their work.

35.3.12 WEED CONTROL
Also see Chapter 17 for weed control.

Bean seed emerges with two lobs that are pushed above soil level. Be aware that damage to the brittle stems of the seedlings whilst controlling weeds, will result in a loss of plants.

35.3.13 YIELDS
Yields 12 to 20 ton/Ha. Experienced producers yield up to 20 tons of pods p/Ha. The yield of runner beans is similar to garden beans, due to the wider spacing.

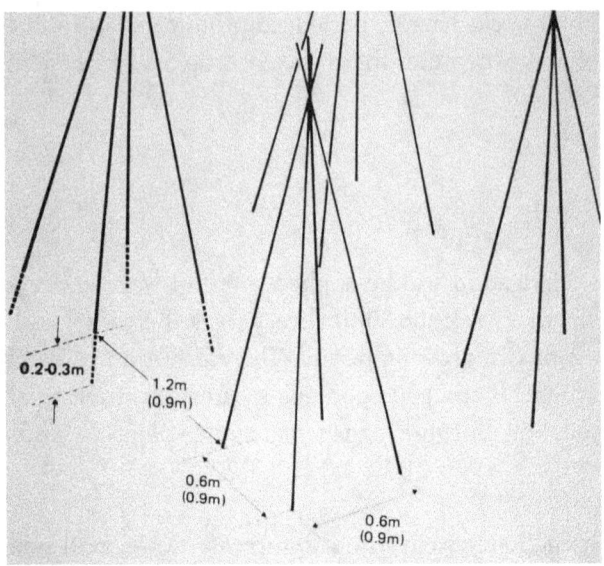
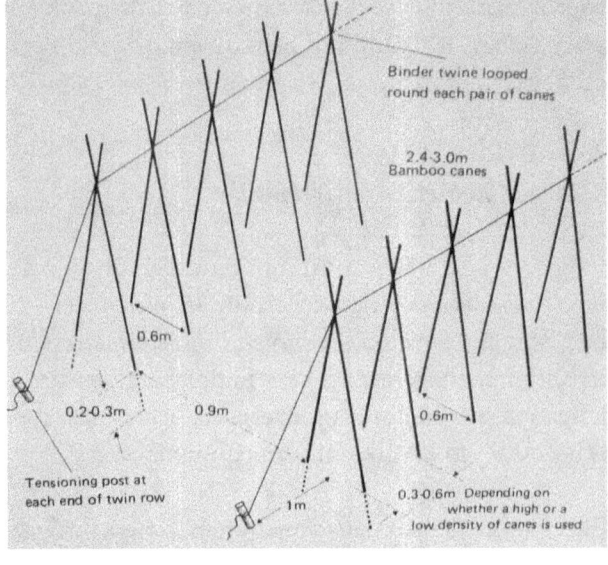

319 A common way to trellis runner beans.   *320 An alternative way to trellis runner beans.*

35.3.14 HARVESTING, STORAGE, COOLING AND PACKAGING
Also see Chapter 24 Harvesting, Handling and Packaging.

35.3.14.1 HARVESTING
Green beans are harvested for short periods, for only 3 to 4 weeks, from the first harvesting to the last. Harvesting is performed before the jelly disappears inside the pods. Afterwards the pods become hard with little/no jelly. Some producers establish beans at intervals of 3 to 5 weeks to extend the harvesting period, but this method needs to be planned carefully.

The various production lands should not be established too close to one another, as disease/insects may spread from older lands to the newer lands. Do not harvest beans when the plants are wet, as this may spread bacterial blight and other diseases. Be careful not to damage bean plants when working in the land. Some bean cultivars cannot be harvested mechanically. The quality of beans depends on the pods ripening and maturing evenly. If beans are to be harvested mechanically, make sure to produce the correct cultivar. This is because the bean plants should carry the maximum pods possible, in order to make the process economical.

When transporting beans from cold rooms to supermarkets, refrigerated trucks should be used to maintain the freshness and extend the keeping quality of the beans. Bean rows, may for instance be 28m long. When the pods are picked, the length of the rows should allow a crate to be filled for the first harvesting. The second harvest could also fill one crate. Do your own evaluations and planning in order to achieve this.

35.3.14.2 PICKING STAGE
The picking stages of bean pods is one of the most important decisions that a producer needs to make. Picking bean pods is an expensive and uncomfortable task. Producers should time the harvesting carefully so that the pods are firm and crispy. This must happen before the seed in the pods begin to harden and the jelly begins to disappear, otherwise it is too late to be harvested.

Harvesting the young pods during the early stages of the plants development, will stimulate them to produce more blooms and pods which ensures higher yields. These young pods are popular. Labourers should be trained to understand this fine balance, as harvesting only takes place during a short period and therefore there is little time to waste. The appearance of the young pods after picking is important.

Producers should pick a sample of pods, cut them and look at the size of the developing seeds inside the pods. They should be a fair size, but not too large. When breaking the new pods, they should break cleanly, with a good appearance of jelly. The crop should be picked regularly, sometimes 3 times per week. Make sure that the pods do not over-develop and lose the jelly inside the pods. Otherwise, they will be of little use.

35.3.14.3 HANDLING

It should be taken into consideration that after harvesting, green bean pods are still in a stage of transpiration. If they are left in the crates in the warm sun, even for an hour or two, the pods may wilt and the pressure on the bottom of the crates will increase. Heat and moisture may build-up quickly and lead to early spoiling of the pods.

35.3.14.4 STORING AND COOLING

Bean pods intended for storage, should be free of breaks, scars or cuts, which may lead to fungal growth. Such conditions hasten wilting and rotting of the pods. Be aware that the little stalks should be left on the pods when harvesting beans. Green beans are best kept between 4ºC and 7ºC, in about 95% humidity to prevent them from wilting. However, cold temperatures may cause chilling, injuries and loss of moisture.

Bean pods have a shelf life of 7 to 14 days but bruises and other damage will show after three days. Stack beans in such a way, so as to allow maximum air circulation in a cold room.

35.3.14.5 PACKAGING

Always confirm with market agents or supermarkets what kind of packaging is required. The most common way to pack green beans is in 8 and 10kg carton boxes.

35.3.15 INSECTS
INSECTS: *See Chapter 20.5.*
ROOT-KNOT NEMATODE: *See Chapter 20.5.2.1.*
CUTWORMS: *See Chapter 20.5.2.2.*
APHIDS: *See Chapter 20.5.2.3.*
RED SPIDER MITE: *See Chapter 20.5.2.9.*
BROWN BEATLE: Active at night and causes leaf damage.
BOLWORM: *See American bollworm. Chapter 20.5.2.7.*
CMR BEETLE: Small to large beetles with bright yellow and black bands across the wings. They fly slowly making a buzzing noise. They mainly prefer green bean flowers. *See photo 116.*

35.3.16 DISEASES
DISEASES: *See Chapter 20.6.*
DAMPING OFF: *See Chapter 20.6.3.1.*
POWDERY MILDEW: *See Chapter 20.6.3.5.*
PHYTOPHTHORA WILT: *See Chapter 20.6.3.16.*
ANTRACNOSE, HALO BLIGHT AND COMMON BLIGHT: *See Chapter 20.6.3.12.*
Some of these diseases are also seedborne and cannot be treated with chemicals. Certified seed is always treated with seed surface chemicals.

RUST:

Bean rust attacks the plant above the ground. This is most commonly seen on the underside of the leaves. The disease occurs worldwide but is most common in humid tropical or subtropical areas. The common sign of rust is the reddish brown, circular pustules on the leaves or pods. As much as 2 000 spores may be seen on a single leaf. Heavy infections of leaves become yellow and fall off. The disease begins on the underside of the leaves and is difficult to control. Remember there are tolerant cultivars available.

BACTERIAL BLIGHT:

Bacterial blight is a serious disease that damages plants, especially when the disease infected plants are still young. Small water-soaked areas on the leaves change to larger brown lesions with a lemon-yellow margin. Most diseases favour high humidity and hot wet weather.

ANGULAR LEAF SPOT:
Angular leaf spot is a major problem in tropical and subtropical areas. The disease affects foliage and pods of beans throughout the growing season. It is particularly destructive in areas where warm, moist conditions occur. Infested plant residues and contaminated seed carry the disease. Leaf lesions show grey/brown irregular spots that may be surrounded by sclerotic circles. On primary leaves, the lesions are circular rather than angular. Pod lesions are large, oval to circular, reddish brown spots, usually surrounded by a darker, coloured circle. Stem lesions occur after about 48 hours of the infection, especially in humid or moist conditions.

CHAPTER 36

PRODUCTION GUIDELINES FOR LETTUCE

321 Lettuce production using a drip system. When the dripper lines are long, they stretch or shrink and change their original position. The lines should be fastened using wire hooks to keep them in position.

36.1 INTRODUCTION

Lettuce is a popular crop amongst health conscious people. Lettuce is ideal in a diet, due to its low kilojoules, nutritional qualities and low energy level. It is important for a diet due to the calcium, iron and vitamin A it contains when eaten raw.

Lettuce is fairly easy to produce and matures quickly. Any setback however will affect the crispness of the heads as well as the yields. When producing leaf lettuce from seed trays and transplanting them in ideal conditions, they can be harvested after 4.5 to 6 weeks. Head or crispy lettuce may be harvested after 6.5 to 7 weeks.

There is not a huge demand for head lettuce, but in overseas countries it is different. For example, in England the production of crisp or head lettuce is 10 times that in South Africa. It is worthwhile for small-scale producers to include lettuce in their production program. Leaf lettuce should be marketed as soon as possible after harvesting, as it may wilt quickly due to its poor keeping quality. This applies especially when hot conditions occur. Cooling facilities are an ideal option.

Do not over produce lettuce. The markets are limited and competition is high, as it is an easy and fast cash-crop to produce. Sometimes when the holiday season begins (especially in December) the market demands more head lettuce. Choose the best summer orientated crisp head lettuce cultivar that is tolerant to high temperatures.

322 Different lettuce types and cultivars. Some are bolting early, especially the leafy types on the far right.

323 Crisp or head lettuce production using hydroponic Gravel Film Technique (NFT). This is an ideal system to produce head lettuce and leaf lettuce. Harvesting may take place 4 weeks after transplanting the healthy seedlings.

36.2 ESTIMATED PRODUCTION COSTS AND PROFITS

The production costs to produce vegetables may differ across the country due to the variances in cost of seed, seedlings, fertilizer, electricity/diesel, irrigation water, labour and so forth. Therefore, costs should be adapted accordingly.

The table below provides estimated costs, yields and profits for lettuce production (2014-2015). It will assist with designing a cash flow and business plan. Financial lenders require a business plan before they consider lending money. Overhead costs are not included. Producers should do their own research on production costs when they produce vegetables.

The estimated cost to produce 1 Ha head lettuce as indicated in the tables below, is based on an average yield of 25 ton p/Ha or 85 000 marketable lettuce heads. Experienced producers achieve yields of up to 30 ton p/Ha or more. However, be careful of overproduction. 85 000 heads p/Ha is a considerable amount and therefore planning in order to market them effectively, should be a high priority. Due to the effort involved in handling such large quantities of lettuce at the same time, producers usually establish production in stages. Seedlings or planting seed may for example, be established every 3 to 5weeks.

The cost of tractors, implements, general equipment (wheel barrows, spades, forks, hoes, crates, twine, pegs etc.), irrigation equipment, water, water pumps, electricity, spraying chemicals, housing, toilets, telephone/computer costs, overalls, clothing for spraying chemicals and so on, are not included in the table below, nor the maintenance of tractors or implements. It is important to make sure that tractors are in good condition before production activity begins.

Certain labour costs are not included in the tables below. Permanent labourers should be used as often as possible, so that they may supervise and train other labourers and also be tractor/truck drivers. Usually 2 permanent and 2 temporary workers can manage 1Ha for a period of 4 to 5 months. During peak times, more temporary labourers are necessary. Producers should determine labour cost themselves, as costs differs from region to region. Labour is expensive and should therefore be well managed.

TABLE 41: ESTIMATED PRODUCTION COSTS AND PROFITS FOR LETTUCE (2014)

ITEM	COSTS P/HA
Clean and level the land	R800
SOIL PREPARATION: Rip, plough, disk or rotovate. Apply lime if the pH is too low. Labour included.	R1 600
SOIL ANALYSIS: A farmer's package consists of at least 2 soil samples @ R290 per sample = R580. If the land is uneven, take more samples. Interpretation of the soil analysis = R280.	R860
FERTILIZER: Apply lime according to the results of a soil sample recommendation (if the pH of the soil is too low). Apply fertilizer mechanically using a tractor and spreader = R350. Broadcast fertilizer suggestion: 1 100kg p/Ha 2.3.4 (30) Zn @ R290 per 50kg bag. Require 22 bags x 50kg p/bag = 22 x R290 = R6 380. 2 Top dressings LAN/KAN @ 100kg p/Ha = 5 bags x 50kg @ R240 per bag = R1200.	R7 930
USING SEED: Establish lettuce using graded or paletted seed and seed planters. 1.5kg seed for 1 Ha = R290 p/kg = R435.	R435
Planting seed: Using a seed planter and labour: 2 labourers for 4 days @ R100 per labourer per day = R800. **Seedlings from nurseries:** R120 per 1 000; Require 90 000 seedlings p/Ha: R120 x 90 = R10 800 (2014). **Transplanting seedlings:** To transplant 90 000 lettuce seedlings: 1 labourer transplants 3 000 seedlings per day therefore 30 labourers transplant 90 000 seedlings per day x R100 per labourer per day = 30 x R100 = R3 000. Mechanical seedling planters are also available. 2 labourers transplant approximately 3 000 seedlings per hour and some planters more.	R800
CHEMICALS: Insect and disease control including labour.	R900
WEED CONTROL: Mechanical weed control between rows = R450. Labour: Include 5 labourers to control weeds in the rows for approximately 2 days = 5 x 2 x R100 = R1 000. Consider weed control using herbicides. It is cheaper.	R1 450
IRRIGATION: Diesel, electricity and labour costs as well as shifting the irrigation system, cleaning and repairing it. Water is expensive when low rainfall occurs. To irrigate 1mm water p/Ha, 10 000 litres water is required.	R3 500
HARVESTING: Harvest and transport lettuce heads to the store. Labourer to harvest and load: For 3 000 heads p/day. 30 labourers harvest 90 000 @ R100 p/labourer p/day = 30 x R100 = R3 000.	R3 000
SORTING: Cut stems and remove outer leaves and pack in containers.	R2 000
CONTAINERS: Polystyrene wrapped in plastic film.	R1 500
TRANSPORT: Transport 20 tons of head lettuce to the market which is 100km away: Using a 3-ton truck which travels 4 times = 4 x 2 (there and back) x 100 = 800km @ R5 p/km = R4 000. Truck driver and assistant = R400. Total = R4 000 + R400 = R4 400.	R4 400
Clean the lands and plough residues into the soil. Remove and store irrigation system etc. Establish a cover crop after harvesting.	R1 800
PRODUCTION COST WHEN USING SEED	R30 975

TABLE 42: ESTIMATE YIELDS AND PROFITS FOR LETTUCE

YIELD: 90 000 lettuce heads, less 5 000 unmarketable heads = 85 000 heads per Ha @ R1.50 per head	R127 500
MINUS PRODUCTION COST	R30 975
POSSIBLE PROFITS	R96 525

36.3 PRODUCTION GUIDELINES FOR LETTUCE

36.3.1 SOIL REQUIREMENTS
Also see Chapter 5.3 Soil Requirements.

Lettuce should grow rapidly if they are to be sweet, juicy and crispy. This requires fertile soil which is continuously improved and moist soil conditions. Lettuce may be produced in all types of soil, from heavy clay, turf to light sandy soil. However, the best results are obtained from fertile, well-drained sandy loam soil which has a good water holding capacity, especially for the crisp or head lettuce types.

Lettuce has a shallow root system and may be produced in shallow soil, however the top surface of the soil should be well broken and should not form a hard crust. Water should also penetrate comfortably into the soil. Soil that compact easily will affect plant growth. Lettuce is sensitive to salt/acid concentrations which may affect plant growth. When the pH is lower than 5.5, then lime should be applied.

Producers should make sure that the plants have a fast start. Even growth uniformity is not possible when soil are compacted. Plants should never be stressed during their short growing season. Well-prepared and aerated soil allows fast root growth.

36.3.2 SOIL PREPARATION
Also see Chapter 8.2 Soil Preparation.

Lettuce seeds are small and seedlings are not normally transplanted. Producers normally sow seed directly into a well-prepared land. To do so, it is important to prepare a smooth, level land with minimum clots.

36.3.3 SOIL PH
Lettuce is sensitive to acid soil. The pH is best at 5.8-6.

36.3.4 CLIMATE REQUIREMENTS AND PLANTING TIMES
Also see Chapter 3 Climate.

Lettuce prefer a cool climate, but can be produced throughout the year. Summer and winter cultivars are available. Ideal temperatures are 7 to 13°C at night and 19 to 28°C during the day. The crop grows quickly in warm weather conditions, however, when very hot weather conditions occur over long periods, the heads of crisp/head lettuce may lose their firmness and become loose and soft.

Lettuce is sensitive to different climate conditions and may sometimes bolt during its recommended seasonal planting times. If abnormal growing conditions occur, it is recommended that more than one cultivar should be produced, as some cultivars are more resistant against bolting. It is therefore sensible to perform your own evaluation with different cultivars in your area.

Cold damage and light frost may cause damage when the frost becomes harsh and windy conditions occur. Optimal soil germination temperatures in order to establish seed are between 15 to 20°C. The seed will germinate in seed trays at 10°C and emergence will occur after 14 days, while at 25°C, emergence occurs after 6 days. At low temperatures germination is slow and may increase the chance of fungal disease.

When light frost occurs, irrigation water may be applied to the plants. This should cause a slight rise in temperature and prevent frost damage. When heavy frost occurs, water on the plants may freeze which could do even more damage. It is important to avoid water on the plants or wet soil, as the wind chill and moisture may cause the temperatures in the land to fall. Therefore, the timing of irrigation is important.

Plants under drought stress will be more prone to cold damage. Frost covers are also an option for various situations, especially for crops such as lettuce and spinach, which are moderately frost tolerant. Frost covers protect the crop against sudden intense drops in temperatures, while promoting vigorous plant growth and providing protection for frost susceptible crops when light frost occurs. Plastic tunnels only provide a few degrees of frost protection, but covers used inside tunnels provide much more protection.

36.3.4.1 OPTIMUM PLANTING TIMES

Areas with mild frost: Feb-Sept
Areas with heavy frost: Jan-Feb and Aug-Sept
Frost-free areas: March-May
Winter rainfall areas: Aug-May

36.3.5 GROWTH PERIOD

Head lettuce: 55-65 days from when seedlings are transplanted.
Leaf lettuce: 35-50 days from when seedlings are transplanted.
Butter lettuce: 45-55 days from when seedlings are transplanted.

36.3.6 IRRIGATION
Also see Chapter 14 for Irrigation.

Lettuce is a water loving plant which is why it is an excellent crop to produce in hydroponic systems (such as in a Nutrient Film Technique (NFT) water and nutrient recycling system). In this system, the leaf lettuce type may be harvested in only 25 days from the time the healthy seedlings were transplanted.

Irrigation is the most important factor when producing lettuce. Producers should remember that to produce a quality crop, soil moisture should be kept optimal for the entire growing season. When young and nearly mature plants are exposed to uneven soil moisture conditions, it has a negative influence on head or crisp lettuce. Lettuce has a swallow root system. The majority of roots are not deeper than 30cm in the top soil. The irrigation water should match this depth.

In dry areas, it is important to sprinkler irrigate the tender soft seedlings 2 times a day. After the seedlings have settled, irrigate after 7 to 14 days. This also depends on the weather conditions. Irrigate 340mm during the growing season. That is 15 to 20mm the first week after sowing and 30mm every week thereafter, for the rest of the growing season. Irrigation is critical, the first week after transplanting the seedlings. Any water stress may cause the plants to develop unevenly. The soil should always be moist around the roots. Light irrigation is required until the roots begin to develop.

Almost all locally produced lettuce is sprinkler irrigated. It is best to irrigate using sprinklers early in morning. Over-watering at the end of the growing season may cause rotting of the lower leaves. Due to the fact that lettuce is a fast-growing crop and has a shallow root system, it requires frequent irrigations.

Fluctuations in soil moisture, especially during the later stages of development, are detrimental to optimal growth and head formation. Too much (in clayish soil) or too little water during this period, along with high temperatures, may result in loose heads and induce premature bolting.

Lighter types of soil require more water applications but less volume per application. Once the heads and loose leaves become fully developed, irrigation should be reduced, in order to avoid the bottom loose leaves from being infected and rotting. In hot conditions the plants may be covered with a 40% shade net cloth which will assist and improve yield and quality.

36.3.6.1 DRIP IRRIGATION
Lettuce is a good candidate for drip irrigation. Many producers produce head lettuce using drip systems. Many dripper lines are necessary and therefore planning should be done properly long before planting time. Some producers establish 7 rows, 20 to 25cm from each other and 3 dripper lines between every second row. The eighth row is left open for a foot path. Producers should experiment with this. It also depends on soil structure. Some producers use a dripper line with distances of 30cm between each opening. The openings usually deliver 2 to 4 litres of water per hour. **Also see Chapter 14.7 for drip irrigation systems.** Contact irrigation experts when installing dripper systems.

36.3.7 FERTILIZER
Also see Chapter 15 Fertilizer.

A soil analysis should be a high priority. Lettuce has a poor root system and therefore care should be taken to provide the soil with enough nutrients. Fertilizer should be applied shallowly in comparison to other crops. Lettuce picks up very little of the nutrients that are applied.

Lettuce responds well to N fertilizer. Slightly less nitrogen N and phosphorus K is required for late December and early January establishments. A suggestion for small-scale farmers when a soil analysis is not available is as follows: On fairly poor soil apply 800-1 000kg 2.3.4 (22) Zn. That is 80 to 90g per m² or 2 handfuls per m² which equals 1 000kg p/Ha. Apply 2 top dressings of N, 2.5 and 3.5 weeks after the seed has emerged, or 2 to 2.5 weeks after transplanting the seedlings. Repeat this 3.5 to 4.5 weeks later. Potassium is important and if necessary apply 100 to 200kg potassium chloride p/Ha.

The top dressing may be worked into the soil and irrigation should take place immediately afterwards. Do not apply to much N top dressing when the plants are older, as the outer, older leaves of crisp lettuce are near the soil surface, which creates a canopy. Then, when applying irrigation water, the fertilizer underneath may damage the outer, bottom leaves.

If excessive amounts of manure are applied, especially chicken manure, the crop may grow too lush and the outer leaves may develop too quickly. In hot weather conditions the heads may also become loose. Lettuce is sensitive to shortages of manganese Mg, boron B, copper Cu and molybdenum Mo. Copper Cu is often found in acid soil and manganese Ma and molybdenum Mo in alkaline soil.

36.3.8 TYPES OF LETTUCES AND CULTIVARS
36.3.8.1 THERE ARE FOUR MAIN LETTUCE GROUPS THAT ARE OF IMPORTANCE
1. Crisp head
2. Leaf lettuce
3. Butter head lettuce
4. Cos lettuce

36.3.8.1.1 CRISP (HEAD) OR ICEBERG TYPES
Crisp head is the most common lettuce grown in South Africa. It has a large, firm head with outer leaves. The leaves are usually broad and the outside leaves are dark green. The heads are firm with a crispy texture.

36.3.8.1.2 LEAF TYPES
Leaf lettuce is a large group and varies in size, shape, colour and taste. Leaf lettuce forms a rosette of leaves that are frilly to smooth. They are dark or light green with or without a red or purple colour. Because of its open growth habit, it has fewer bleached inside leaves, as sunlight causes the coloured pigment to occur when the leaves fall open. Leaf lettuce usually has a stronger taste and also a higher vitamin and mineral content.

36.3.8.1.3 BUTTER HEAD TYPES
This type is known by its smooth, oily texture. It develops a soft head. The outer leaves can be light to dark green and the inner leaves creamy. Some cultivars are purple/reddish and creamy inside.

36.3.8.1.4 COS (ROMAINE) TYPES
Cos (Romaine) lettuce heads are long and loaf-shaped with closed or relatively open soft heads. The leaves are longer with a coarse and crisp texture. The outer leaves are dark green and the inner leaves yellowish. The taste is usually sweeter than the crisp head lettuce type.

36.3.8.1.5 OTHER TYPES
The Latin lettuce which is similar to the Romaine and Butter types. The Batavia type is softer and smaller than the crispy types. The least known lettuce is the oilseed type, which goes over to flowering or bolting quickly. The seeds are used to produce oil.

36.3.8.2 CULTIVARS
A wide range of winter and summer crisp head lettuce cultivars are available as well as promising F1 hybrid cultivars. Producers should evaluate head lettuce cultivars to determine the best ones for their region, also taking into account the best times for establishing these cultivars. In summer conditions, sometimes head lettuce become loose. This is due to the cultivar choice and the time of establishing the crop.

36.3.8.3 HEAD, CRISP, BUTTER AND LEAF LETTUCE
Choose the correct head lettuce cultivar and planting time (winter or summer orientated) for your region. Leaf and butter lettuce are adapted to all regions. When establishing the different coloured leaf lettuce cultivars, note that they are darker and more colourful in summer due to the bright sunlight which causes the pigment colouring of the specific cultivar.

Leave lettuce and to a lesser degree butter lettuce, is used by restaurants. This is to decorate the customer's food plates in order to make them attractive and colourful. The leaves are also consumed as they are juicy and crispy. There are also colourful butter lettuces and mini butter lettuce cultivars available (mainly purple). Leaf lettuce is not as juicy and crispy as head lettuce. It also has a stronger lettuce taste. There are numerous seed companies that can supply the above mentioned types.

36.3.9 SEED P/HA
For direct planting: 1.3kg seed p/Ha.
Seed to produce seedlings in seed trays: 0.4-0.6kg p/Ha.

1kg seed contains approximately 600 000 to 800 000 seeds. Seed size between cultivars may differ. Lettuce seed is normally graded small, medium and large.

36.3.10 SOWING METHODS, SPACING AND PLANTING
36.3.10.1 SOWING SEED DIRECTLY IN THE SOIL
It is recommended to produce lettuce seedlings directly in the land. Seeds are sown in rows by hand or with seed planters. The rows are usually 35 to 45cm from one another and the plants are 25 to 30cm from one another in the row. Usually producers sow 4 to 5 graded seeds per 30cm row and for paletted seed, only 2 seeds per 30cm.

When using a seed planter in a well-prepared, smooth land, much less seed is necessary, especially when using paletted seed. Planting depth is important and therefore the land should be level and smooth without debris or clots. The seeds should not be established deeper than 1.5cm.

36.3.10.2 SEEDLINGS FROM SEED TRAYS
High temperatures may cause seedlings to produce too much top growth in seed trays. When the cavity sized

in seed trays are large, then mostly softer seedlings will the result. Seedlings produced in cold conditions are generally better than seedlings produced in warmer conditions.

In summer, 4-week-old seedlings are able to be transplanted. In colder conditions this may take as long as 7 weeks. The result of transplanting older seedlings which have been produced in smaller seed tray cavities, especially when they are left for too long in the seed trays, is that they mature faster. The faster the maturity of a crisp head lettuce is, the more it is set-back when establishing old root bound lettuce seedlings.

Always use fresh, certified, tested lettuce seed which has good germination characteristics, as otherwise, lettuce seed may lose its germination efficiency quickly. Most lettuce seedling producers or nurseries use standard 200 or sometimes 300 cavity seed trays thereby saving money and space. When larger 128 cavity seed trays are used, then larger, healthier and better quality seedlings result, which means that the seedlings can remain longer in the seed tray cavities and therefore require less time in the land. More often than not, this is more economical. However, it is cheaper to use seed trays with 200 or 300 cavities as less seed trays and less medium is used. Producers should do their own research with regards to seed tray cavity size.

Approximately 90 000 head lettuce seedlings are necessary to produce 1Ha lettuce plants in the production land. Therefore 450 seed trays with 200 cavities are necessary. When using 300 cavities, 300 seed trays are required and for 128 cavities, 703 seed trays are required. The size of the seedlings when transplanting is important. A smaller seedling is able to withstand high and low climactic conditions better. *See Chapter 4 Seed Nurseries and Chapter 8.8 Transplanting seedlings from seed trays.*

36.3.10.3 SPACING LETTUCE
Head lettuce seedlings may be spaced 20 to 30cm in the rows and 30 to 45cm between the rows. A spacing of 35cm between the rows and 25cm in the rows provides a plant population of 114 000 p/Ha, less every 7th row which is used for foot paths. 57 rows = 22 000 plants and the total plants p/Ha would be 92 000.

Some head lettuce cultivars develop large head sizes. Baby types produce small heads and therefore producers should adapt the spacing according to the size of the heads. Raised beds, especially for small-scale farmers, are a good option.

36.3.11 CROP ROTATION
Also see Chapter 18 Crop Rotation and Soil Improvement.

The main purpose of a crop rotation system is to control soil-borne diseases and insects. Crop rotation may include a green-manure crop to increase the organic matter in the soil. Lettuce and related crops should not be established on the same soil each year and instead should be rotated every 3 years. Crops such as sun-hemp, black mustard, canola, eragrostis curvula, finger grass, lucerne (after harvesting several times), maize, babala, soybeans, lupines etc. could be incorporated in the soil 4 to 6 weeks before planting the next crop.

36.3.12 WEED CONTROL AND THINNING OF PLANTS
Also see Chapter 17 for weed control.

Due to the shallow root system of lettuce, it is necessary to practice shallow weed control. Weeds should be controlled between the rows using a hoe or spade. It may also be helpful to break the soil crust, especially after heavy rain. On large lands, the weeds are controlled between the rows by using a small tractor and grub. Weeds that are in the rows are controlled by hand hoeing. Herbicides may be used to control certain weeds effectively.

36.3.12.1 THINNING OF PLANTS IN THE LAND
Young plants should be thinned at about 14 days after the plants have emerged, when they have been sown by hand in the production land. Thin the head lettuce seedlings so that they are 20 to 30cm from one another in the rows depending on the head size. The thinned plants may be transplanted to rows that have uneven emergence.

Do not wait too long before thinning the plants as weeds may become a serious problem. Keep the rows clean at all times and avoid deep cultivation.

36.3.13 YIELDS
Crisp or head lettuce yield 20 to 30 ton p/Ha.

36.3.14 BOLTING OR PREMATURE SEEDING
Lettuce plants sometimes begin to develop seed stalks and go-over to flowering due to the plants desire to reproduce themselves. Some types are able to go-over to flowering very early, especially when high temperatures occur. *See photo 322.*

36.3.15 HARVESTING, PACKAGING AND MARKETING
Chapter 24 Harvesting, Handling and Packaging and Chapter 25 Management and Marketing.

Head lettuce should be harvested when the heads become mature and firm. Not all heads of some cultivars become firm at the same time and therefore they should be harvested continuously. Certain F1 hybrid cultivars mature uniformly and can be harvested at the same time.

As soon as the crisp heads of some cultivars begin to turn lighter in colour, it is an indication to begin harvesting. The mature heads should be cut with a sharp knife, just above soil level. This also depends on the market requirements. After harvesting the lettuce heads, they should be taken to the pack store immediately. The outside leaves of crisp and butter lettuce should be removed and packed in crates or other containers.

When harvesting head lettuce, the soil moisture should be favourable as the lettuce plants should be crispy and juicy. Do not expose leaf and butter lettuce to broad sunlight for too long. Leaf and butter lettuce should be immediately transported to the pack store and should be kept in a cool place before marketing. Some producers use ice water for short periods to keep head leaf lettuce cool and crispy.

Most commercial producers use cooling facilities and refrigerated trucks to supply lettuce to the market. Leaf and butter lettuce is usually delivered to restaurants and supermarkets. Head lettuce is delivered in crates, packed and wrapped in plastic film or sold separately. Always adhere to market requirements.

36.3.16 INSECTS
INSECTS: *Also see Chapter 20.5.*
Due to the short growing period of lettuce, it is not usually affected by many insects/diseases. When insects and diseases begin to attack the plants, it is almost time to harvest.
Downy mildew is the biggest problem. It may develop quickly and prefers cooler conditions. *See Chapter 20.6.3.6.*
ROOT-KNOT NEMATODE: *See Chapter 20.5.2.1.*
CUTWORMS: *See Chapter 20.5.2.2.*
APHIDS: *See Chapter 20.5.2.3.*

36.3.17 DISEASES
DISEASES: *Also see Chapter 20.6 Diseases.*
DAMPING OFF: *See Chapter 20.6.3.1.*
PHYTOPHTHORA WILT: *See Chapter 20.6.3.16.*

DOWNY MILDEW/POWDERY MILDEW: *See Chapter 20.6.3.5 and 20.6.3.6.*
Every lettuce and Cucurbit producer knows that they will have problems with mildew sooner or later. It is difficult to identify the difference between the two mildews. Downy mildew prefers cool and moist weather conditions, while powdery mildew prefers hot and dry conditions. It is more difficult to control powdery

mildew as it lies latent in the plant tissues. Therefore, contact spraying of chemicals is not an option. Instead systematic chemical control measures should be a priority.

SPOTTED WILT VIRUS
The western trip is a virus vector for this disease. Spotted wilt attacks a wide range of crops and weeds. Weeds are the reservoir for the virus while the trips carry the virus into the land. It is a good option to spray a broad leaf herbicide on weeds that are growing near the land. This is a good practice and reduces many diseases and insects.

LETTUCE MOSAIC VIRUS (LMV)
Unlike spotted wild lettuce virus, this is a seedborne disease. Therefore, good quality seed from reliable seed companies should be established. Infected plants develop a yellowish colour while the edges of the leaves bend downwards. When infested leaves are held up to the light you will see a mosaic pattern. Aphids are the main vector. Spotted wild aphids carry the virus through the winter and then infect the plants in spring.

36.3.18 PHYSICAL DISORDERS
Also see Chapter 20.9 Physical Disorders.

36.3.18.1 TIP-BURN
Tip-burn occurs when the plant cells collapse on the edge of the leaf margins. This disorder may also occur outside or inside the lettuce head. Tip-burn is a calcium Ca related disorder and is associated with low levels of Ca in the leaf tissues. Warm temperatures, excessive fertilization, increasing light intensity and also rapid growth may encourage its development. Other factors that reduce the uptake of calcium Ca, such as high salt concentrations in the soil and high humidity, may enhance this problem. Good cultural and management practices which reduce rapid plant growth, may limit this problem.

36.3.18.2 BOLTING
Also see Chapter 3.10 Bolting.

Bolting is the elongation of the seed stem. Because of the reproductive (flowering) process, head formation is prevented and the plant develops a flowering stem. Cultivars that are more resistant to climate and seasonal variations are available. However, if temperatures are higher than usual in a region, then bolting is a factor to be reckoned with.

CHAPTER 37

PRODUCTION GUIDELINES FOR SWEET CORN AND GREEN MEALIES

324 New and old sweet corn cultivars displayed at a farmers' day.

325 New and standard F1 hybrid sweet corn cultivars displayed at a farmers' day.

326 Green mealies (Border King). Left and right: A sweet corn F1 hybrid cultivar intercropping with beans. Some home owners establish a maize plant. When the maize plant is nearly fully grown, they plant a runner bean which is then trellised up the maize plant.

327 A Hybrid green mealie cultivar. The plants with the tassels are the male plants which pollinate the female plants (detasselled). This is cross-pollination which introduces the next generation of Hybrid cultivar.

328 A good combination of crops. That which limits the risk factor, is the intercropping of sweet corn/green mealies combined with sweet potatoes.

329 Sweet corn cultivars which average 1.7 cobs per plant. Hybrid cultivars sometimes produce more than 2 cobs per plant.

37.1 INTRODUCTION

Sweet corn and green mealies are relatively easy and cheap to produce. It is also less prone to disease and fairly drought resistant. The crop belongs to the grass family and is bred to produce sweetness in the grain.

Sweet corn and green mealies are warm weather crops popular in home gardens which have sufficient space. The maize cobs which are produced taste good when harvested fresh. It is important for small-scale farmers to produce sweet corn and green mealies in series. Establish pieces of land every 3 to 4 weeks to ensure fresh marketable cobs are produced over longer periods. Sweet corn cultivars with white and yellow kernels are available, but the white cultivars are sweeter.

37.1.1 CLASSIFICATION OF SWEET CORN SWEETNESS
Sweet corn is classified in four degree of sweetness:
1. Standard sweet
2. Sweet
3. Extra sweet
4. Super sweet

Cultivation methods, harvesting and handling also has an influence on the sweetness of the grain. Producers should be aware that there is a difference in sweetness in the grain of green mealies when compared to the standard sweet and super sweet corn cultivars.

Different cultivar groups should not be established together due to cross pollination, as this may influence the starch factor that in turn may influence the taste. Different cultivars which provide pollen and which are established at the same time, should be located at least 100m from each other. Alternatively, use time-isolation by establishing sweet corn every 3 to 4 weeks. Early and late cultivars may also be planted together. Some sweet corn cultivars are widely used for processing, especially for frozen and whole kernel canning.

37.1.2 GREEN MEALIES
Green mealies are white seeded, some cultivars such as Border King have large flat seed and other cultivars such as SR 52-Hickory King, PP-Potch Paarl and white Perde Tant has smaller medium-large kernels. Green mealies are not as sweet as sweet corn but is sweeter and tastes better than the white seeded maize which is ground to produce mealie meal. Commercial white and sometimes yellow maize kernels are mixed with white maize and ground for mealie meal.

Green mealies is popular among small-scale farmers, who produce the crop and sell them in the townships, alongside roads or in stalls. Some retailers alongside the roads and townships, cook the cobs with the leaves still on and sell them when they are still warm. They taste good when harvested at the correct time of kernel development, when they are still soft and not too hard.

Production planning should be a priority when producing green mealies on a continuous basis. Proper planning allows the crop to be produced over long periods (especially for the street market), from November to April in the summer orientated regions.

37.2 ESTIMATED PRODUCTION COSTS AND PROFITS (2014)

The production costs to produce vegetables may differ across the country due to the variances in cost of seed, seedlings, fertilizer, electricity/diesel, irrigation water, labour and so forth. Therefore, costs should be adapted accordingly.

The table below provides estimated costs, yields and profits for maize production (2014-2015). It will assist with designing a cash flow and business plan. Financial lenders require a business plan before they will consider lending money. Overhead costs are not included. Producers should do their own research on production costs when they produce vegetables.

The yields indicated in the tables below are determined is for 25 tons of sweet corn cobs p/Ha. This is the average and includes the outer leaves of the cobs.

The cost of tractors, implements, general equipment (wheel barrows, spades, forks, hoes, crates, twine, pegs etc.), irrigation equipment, water, water pumps, electricity, spraying chemicals, housing, toilets, telephone/

computer costs, overalls, clothing for spraying chemicals and so on, are not included in the table below, nor the maintenance of tractors or implements. It is important to make sure that tractors are in good condition before production activity begins.

Certain labour costs are not included in the tables below. Permanent labourers should be used as often as possible, so that they may supervise and train other labourers and also be tractor/truck drivers. Usually 2 permanent and 2 temporary workers can manage 1Ha for a period of 4 to 5 months. During peak times, or when selling on the streets or in communities, more temporary labourers are necessary. Producers should determine labour costs themselves, as costs differs from region to region. Labour is expensive and should therefore be well managed.

TABLE 43: ESTIMATED PRODUCTION COSTS AND PROFITS FOR SWEET CORN AND GREEN MEALIES (2013)

ITEM	COSTS P/HA
Clean and level the land.	R650
SOIL PREPARATION: Rip, plough, disk or rotovate. Apply lime if the pH is too low. Labour included.	R2 500
SOIL ANALYSIS: A farmer's package consists of at least 2 soil samples @ R290 per sample = R580. Interpretation of the soil analysis = R280.	R860
FERTILIZER: 1 000kg p/Ha 2.3.2 (22) @ R290 per 50kg bag. Require 20 bags x 50kg p/bag = 20 x R290 = R5 800. 2 Top dressings LAN/KAN @ 110kg p/Ha = 2 bags x 50kg @ R240 p/bag = R480. Mechanical application of fertilizer, tractor and spreader = R300.	R6 580
PLANT MATERIAL: Sweet corn seed cultivars: 12kg per Ha @ R70 p/kg = R840. Hybrid seed is more expensive.	R840
PLANT SEED: Using a maize planter including labour.	R600
CHEMICALS: Insect and disease control including labour.	R800
WEED CONTROL: Mechanical (between rows) = R450 including labour. If 3 labourers control weeds in rows for approximately 3 days = 9 x R100 per day = R900.	R1 350
IRRIGATION: Diesel, electricity and labour costs as well as shifting the irrigation system, cleaning and repairing it. Water is expensive when low rainfall occurs. To irrigate 1mm water p/Ha, 10 000 litres water is required.	R2 800
HARVESTING: Transport 60 000 x 1.7 cobs = 102 000 cobs to store. 1 labourer harvests, loads and transports 6 000 cobs per day, therefore 17 labourers required @ R100 = R1 700.	R1 700
POTATO BAGS: Sorting, cutting leaves and stems and packing in bags or other containers, including labour: 100 000 cobs and 30 cobs per bag = 3 300 bags @ R0.60 per bag = R1 980.	R 1 980
TRANSPORT: Transporting 20 tons of maize to the market which is 100km away: Using a 3-ton truck which travels 4 times = 4 x 2 (there and back) x 100 = 800km @ R4 p/km = R3 200. Truck driver and assistant = R400. Total = R3 200 + R400 = R3 600.	R3 600
Clean and plough residues into the soil, remove and store irrigation system etc. Establish a rotation crop to improve soil fertility, if necessary.	R1 400
ESTIMATED PRODUCTION COST FOR SWEET CORN AND GREAN MEALIE - 1HA	R25 660

TABLE 44: YIELDS AND POSSIBLE INCOME FOR SWEET CORN AND GREEN MEALIES

YIELDS: 60 000 sweet corn plants p/Ha @ 1.7 curds/cobs per plant = 102 000 curds (cobs)-Minus 2 000 unmarketable cobs = 100 000 Cobs @ R0.90 per cob.	R90 000
MINUS PRODUCTION COSTS	R25 660
POSSIBLE PROFITS	R64 340

37.3 PRODUCTION GUIDELINES FOR SWEET CORN AND GREEN MEALIES

37.3.1 SOIL REQUIREMENTS
Also see Chapter 5.3 Soil Requirements.

Sweet corn and green mealies grow well in a wide range of soil. The best results are most likely to be obtained when preparing a deep, well-drained, medium loam to loam type of soil. Early sweet corn planting can be established on a light, well-drained, sandy loam soil, especially in a warm location. Early harvesting may be expected when sandy soil warms-up quickly.

Irrigation is sometimes necessary when dry weather occurs, particularly on light sandy soil. It is a critical time for irrigation and keeping the soil moist, when the silk of the cobs begin to develop. Sweet corn and green mealies are susceptible to soil compaction and therefore soil that is known to form a hard crust, should be broken-up as soon as possible using a grub. Ploughing the remaining corn residues into the soil is recommended, however this is also valuable food for cattle, pigs, sheep and goats.

37.3.2 SOIL PH
Low pH levels are the most frequent cause of poor corn yields. Acid soil may result in a reduction of root development and nutrient availability. For good plant growth, maintain a soil pH of 5.8 to 6.5.

It is recommended to base any lime applications on a soil analysis. Lime will be most efficient if it is applied at least 6 weeks prior to planting. This will raise the Ph. Use the correct type of lime, when it is necessary. **See Chapter 15.5 Using the correct type of lime.**

37.3.3 CLIMATE REQUIREMENTS AND PLANTING TIMES
Also see Chapter 3 Climate.

Sweet corn and green mealies are warm weather crops. Optimum temperatures are 12 to 29°C. Soil temperature should be about 15°C. When the temperature falls below 10°C, this may negatively affect plant growth. When sweet corn seed is established in a soil temperature of 10°C, the seed will emerge after approximately 20 days, at 15°C after approximately 12 days and at 20°C after only 7 days.

Temperature has an effect on the development of this crop. When temperatures above 32°C occur and when periods of high rainfall or cloudy conditions occur, especially over longer periods, this will affect pollination and cause poor kernel set. In general, high temperatures hastens the growing tempo of the plant. The harvesting period is then also earlier. However, lower temperatures cause slower plant growth.

In the summer orientated areas, sweet corn and green mealies are usually established in October to December. Planting later, may provide poor yields. However, when very early plantings are established, the crop will take much longer to mature. Poor pollination may occur when temperatures rise above 35°C and when it is windy. Pollen may burn when hot conditions occur over long periods, which may result in a poor kernel set.

In order to continuously supply sweet corn and green mealies throughout the summer, producers should plan their production properly. They could for example, establish early, medium or late developing cultivars at the same time, in order to extend the harvesting period. Alternatively, every 3 to 4 weeks establish small pieces of land when temperatures are favourable.

37.3.4 SOIL PREPARATION AND PLANTING
Also see Chapter 8.2 Soil Preparation.

Seed may be planted by hand, by hand operated planters, power-drawn planters or with a mealie planter. The seed surface of sweet corn is coarse. They shrink due to the sugar content inside the seed. Therefore, the sweeter the cultivar is, the more the seed shrinks. The seed planter should be calibrated correctly, the tractor speed should be correct and everything must be carefully monitored when planting the seed.

Sweet corn seed should not be planted deeper than 3.5cm, even in moist soil. Hand planting requires the making of furrows. Alternatively, holes may be created in the soil using an iron rod attached with a stopper. This allows the planting depth of the seed to be even, especially when establishing smaller areas.

Maize has an effective root depth of 50 to 60cm. Take into consideration that the root system of sweet corn and green mealies is sensitive to hard layers below the surface of the soil. Such hard layers should be broken-up with a ripper. This will promote good drainage, root penetration and aeration. Although the root system of maize is strong, some cultivars produce even stronger roots in the upper soil surface, which also helps to anchor the plant. The main roots of maize are found in the top 30cm of the soil surface, but the roots may penetrate the soil more deeply.

Enough moisture should be present when preparing the soil. Never prepare soil when it is dry, as this has a damaging effect. Soil should be cultivated when the correct soil moisture is present. A fairly fine, smooth seedbed should be in place to obtain a uniform emergence. It is important that the soil should be as level and fine (smooth) as possible, especially when a seed planter is used, as the seeds should be planted at an even depth.

37.3.5 CULTIVARS

Most of the sweet corn cultivars produced are F1 hybrids which are a cross between two or more inbred lines. Recent F1 hybrid green mealie cultivars which are in production, should be better than the older open pollinated cultivars, such as Border King, SR 52, Perde tand, Potch Parl PP and Hicory.

The main aim for a green mealie is sweetness in the grain, more cobs per plant and therefore higher yields. The open pollinated cultivars such as Border King which have huge, flat kernels are popular. The cultivar SR 52 is also reasonably popular. F1 hybrid cultivars which taste good and produce higher yields with more uniformity, are also available. The seed however cannot be reproduced like the open pollinated cultivars due to their hybridity (crossing of different parents).

Hybrid green mealie cultivars yield more than the open cultivars and usually produce 2 cobs per plant or more. The problem with F1 hybrid cultivars is that the producer will not be able to reproduce the seed for the next season and must obtain seed from a supplier again. Open pollinated green mealie cultivars such as Border King and SR 52 are not always easy to obtain. Producers should order seed long before planting time. Choose the correct sweet corn cultivar for your purpose. Be aware that some sweet corn cultivars are very sweet and not recommended for people who have diabetes problems.

37.3.6 IRRIGATION
Also see Chapter 14 for Irrigation.

Sweet corn and green mealies are a fairly deep rooted crops. The roots develop mainly laterally, approximately 30 to 40cm from the stalk and downward to a depth of more than 80cm in ideal soil conditions. About 90 percent of the roots are found in the top 30 to 40cm of the soil. In the course of the growing season, approximately 40% of the water used by sweet corn and green mealies are from the topsoil.

30% of the roots penetrate 40cm deep into the soil, 20% penetrate 50cm deep and less than 10% penetrate 80cm deep. The depth and water holding capacity of soil has a great influence on how often irrigation is required. Soil texture determines the amount of water that the soil can hold. The greater the water holding capacity of the soil in the root zones, the less frequent irrigation should take place. It is important to avoid water stress, especially when the silk becomes visible. Sweet corn has relative high water efficiency and is not tolerant to drought.

37.3.6.1 WATER STRESS MEANS YIELD LOSSES
Periods of drought, especially during the silk development and pollen shedding stages, will affect the germination of pollen. It restricts silk development, sometimes seriously. This will lead to poor kernel-set which will result in many curds being unmarketable.

Sweet corn requires approximately 400mm water through its growing season and green mealies 500mm. Irrigate 25mm the first week and thereafter, 30mm per week. Irrigation may only be necessary, when dry weather conditions occur in high rainfall areas. This is particularly applicable on light soil and when silking of the kernels begin. Uniform distribution of rainfall or irrigation is necessary to produce maize in dry land conditions. Only at certain times may producers gamble when producing maize on dry lands, for example when they know the weather conditions well.

Light irrigation should be applied at regular intervals until the seed has emerged. If the soil moisture is low when hot weather conditions occur, then the plants may not receive sufficient moisture for normal growth and the yields will decline. Critical irrigation times are two weeks before and two weeks after silk development.

37.3.7 FERTILIZER
Also see Chapter 15 Fertilizer.

Green mealies and sweet corn production require high levels of nitrogen N, potassium K and phosphate P. This is given as a base dressing according to a soil analysis. Experienced producers know how to minimize leaching of nitrates especially on sandy type soil.

Nitrogen N applications should be divided between pre-plant and top dressings which are applied at a later stage, before the plant reaches a height of 30cm. Nitrogen N stress begins as a yellowing on the tip of the leaves and then moves along the middle part of the leaves. N deficiency is aggravated by cold, floods or sandy soil that is low in organic matter.

If a soil analysis is not available, a suggestion for small-scale producers, in soil that is fairly fertile is as follows: 1 000kg p/Ha of the common, balanced, cheaper fertilizer mix 2.3.2(22) Zn which equals two handfuls or 100gr per m^2. Fertilizer can be applied in wider furrows and is much cheaper. Be careful that the seed in the furrows does not make contact with the fertilizer. Sweet corn and green mealies respond well to top dressings and therefore farmers usually apply 2 top dressings N of 150kg LAN/KAN p/Ha. Make sure to use the correct type of N. *See Chapter 15.6 for top and side dressings.* Trials have shown that the ideal time to apply the top dressings is 2 to 3 and 5 to 6 weeks after emergence. The Nitrogen N application should be completed after eight weeks from emergence in order to achieve good quality and sweetness in the grain.

37.3.8 GROWING PERIOD
The growing period for sweet corn until the first curds are ready for harvesting is 80 to 100 days and for green mealies 95 to 120 days. When the silk of sweet corn begins, the first harvest is only 20 to 26 days away and for green mealies 30 to 49 days away.

Sweet corn cultivars are classified as early, medium and late cultivars. Therefore, these different cultivars may be established at the same time in order to achieve different harvesting times. However, this requires careful planning.

37.3.9 PLANT POPULATION AND SPACING
37.3.9.1 PLANT POPULATION
A suggested spacing for sweet corn is 18cm in and 100cm between rows. This provides a plant population of 60 000 plants p/Ha. A high plant population will reduce cob size and maturity will be delayed. It is crucial to obtain an optimum stand.

37.3.9.2 SPACING
Spacing in the row can be adapted according to producer's requirements. Planting densities lead to a reduction in yield as well as large variations when the plants begin to mature. When the plant population is reduced, plants compensate by producing multiple cobs.

Sweet corn could be spaced 15 to 20cm in the row and 70 to 100cm between rows. This provides a plant population of 50 000 to 60 000 plants p/Ha. Green mealies may be spaced 17 to 24cm in the row and 90 to 120cm between rows. This provides a plant population of approximately 50 000 plants p/Ha.

When planting early in the season when the soil is still cool, more seed should be established to achieve a good stand.

37.3.10 POLLINATION
The male and female parts of maize are on the same plant. The female part (cob) of the plants produce sticky silks. A single silk requires only one pollen grain to fertilize a kernel. The pollen is received from the male tassel.

The spacing for sweet corn and green mealies is important and must be understood to achieve 100% pollination. Maize is cross-pollinated, mainly by the wind. The pollen drizzles down to fertilize the silk of the cobs. When maize is established in single rows, the chance that proper pollination will take place, is poor. It is therefore better to plant maize in square blocks or 5 to 6 rows next to one another in order to ensure good pollination opportunities.

A moderately dry atmosphere is preferable during the pollination period. Pollen drizzles down to the silk and fertilizes the curds. The plant can also pollinate itself or it may receive pollen from other plants via cross pollination. The pollen is shed by each plant for a period of one week. The silk of the cob appears first and then the tassels release the fertile pollen which needs to land on the silk of the cobs. Only one pollen grain is necessary to fertilize one silk. The pollen germinates and grows alongside the silk to fertilize the undeveloped kernel in the curd.

If cold, wet periods and heavy winds occur during flowering time, it may result in poor pollination and the cobs not fully developing. Efficient pollination and fertilization is essential for good quality cobs. Beware of the mealie bug and corn borer. The moth lays eggs in the sheets of the plant and at night, the maggots can do considerable damage within a short period of time. They begin to consume the kernels from the silk side of the curds and may lay thousands of eggs.

37.3.11 PLANTING DEPTH
The planting depth should be 2.5 to 3.5cm in moist soil. Shallow planting may result in problems if the topsoil dries out during hot conditions.

37.3.12 SEED P/HA
Direct seeding for sweet corn ranges from 10 to 13kg p/Ha = 8 000 seeds p/kg. 7 to 8kg is required for a plant stand of 60 000 to 70 000 plants p/Ha.
Super sweet cultivars: 8-10kg p/Ha.
Green mealies: 15-20kg p/Ha.

Seed sizes are graded into different classes. The flat seeds are usually in the middle of the curd and the round ones are at the silk end of the cob.

37.3.13 WEED CONTROL
Also see Chapter 17 for weed control.

Various registered weed control herbicides are available.

After planting, the primary purpose of soil preparation is to control weeds. The secondary purpose is to break the hard layer that forms on the surface, especially on heavier types of soil. The wide spacing makes it easier to control weeds between the rows mechanically using a small tractor and grub. In the rows, labourers can make use of hand tools such as hoes. When the plants are still young, the weeds should be controlled shallowly near the main stem, as the roots of maize are extremely shallow.

37.3.14 YIELDS
Sweet corn: 10 to 25 ton p/Ha with an average of approximately 20 ton p/Ha or 1.3 ears per plant.
F1 hybrid **sweet corn cultivars:** Up to 35 ton p/Ha or 1.7 ears per plant.
Green mealies: 25 to 30 ton p/Ha is a reasonable yield.

37.3.15 HARVESTING AND GRADING
Experience is necessary to determine the correct time to harvest the cobs of sweet corn and to a certain extent green mealies as well. The young grain kernels have a watery endosperm. This becomes thick and creamy as the plant develops, which is known as the milk stage. It is followed by the dough stage at which time the grain is no longer tender.

The time to pick sweet corn and green mealies is when the cobs are in the milky stage. Experience is required to judge this correctly. A trial should be made by pressing one or two grains with the thumb nail and the contents of the kernels should be spurt out. Note that the second curd on the plant may be younger and therefore too young to be harvested. Maize growers have discovered that when the silks of the ears become brown and the cobs feel plump, then they are ready to be picked. When the kernels of the crop are still young and small, the juice is clear and watery and the corn is said to be in the pre-milk stage.

Do not pick the cobs too young as they are not acceptable at the markets as the tender kernels will shrink quickly. After the ears have been harvested, the sugar inside the juicy kernels normally converts to starch, due to temperature changes. The higher the temperature, the faster the change from sugar to starch will occur, resulting in the quality of the juicy kernels declining.

When sweet corn is produced for processing and to a certain degree for the fresh markets, it is desirable to harvest the entire crop in one picking. This requires uniform maturity. Most F1 hybrid sweet corn cultivars meet this requirement. Small-scale farmers often spread harvesting of the crop over a period of several days. Therefore, it is an advantage to use an open pollinated cultivar that can be harvested over a longer period due to the uneven ripening of the curds. Care should be taken when the cobs are harvested, to achieve optimum kernel maturity. Picking early in the morning is a good option.

Packaging sweet corn in plastic containers with most of the leaves removed, exposes the kernels quality. This is preferable for green grocers and supermarkets. This packaging should be stored in a cold room. Later the produce should be transported in a refrigerated truck to a refrigerated supermarket store. Never leave the cobs in the sun, as producers will be surprised how fast the kernels shrink and become unmarketable.

The growing period of green mealies is much longer than that of sweet corn. Green mealies should be harvested long before the plant reaches maturity, which is approximately 3 weeks after flowering. Experienced farmers feel the firmness and thickness of the cobs by hand. They also look at the development of the brown silk. To make sure they are ready for harvesting, open the outer leaves of a few ears and inspect the kernels. Well-filled and uniform ears of optimum maturity should be selected for the best results. Poorly filled, caterpillar damaged, overripe and undersized ears should be discarded.

37.3.16 STORAGE

When harvesting maize cobs for the fresh market, the cobs should be as juicy as possible. Do not allow the plants to become water stressed at harvesting time, as the kernels need a maximum supply of moisture to maintain their milky kernels. Do not store the ears too long after harvesting, as they will only last for 10 days under refrigerated conditions of 4°C.

When supplying sweet corn and green mealies to the super markets, producers should take care that they follow the stipulated regulations. If the kernels shrink and become too starchy then the supplier will have to bear the loss. Green mealies can be stored longer, especially when they are cooked day to day and sold retail.

37.3.17 MARKETING
Also see Chapter 25 Management and Marketing.

It is important to pick the sweet corn at the correct stage and to market it as soon as possible after harvesting, as the sugar in the grain changes to starch if left too long. Also, do not leave harvested maize cobs in the heat of the sun as the kernels will form indentations as moisture is lost.

37.3.18 INSECTS
BIRDS:
Birds can be a problem, especially when planting late. Once the birds discover the succulent young kernels on the silk side of the curds they may cause quite a lot of damage.
INSECTS: *See Chapter 20.5.*
CUTWORMS: *See Chapter 20.5.2.2.*
ROOT-KNOT NEMATODE: *See Chapter 20.5.2.1.*
RED SPIDER MITE: *See Chapter 20.5.2.9.*
AMERICAN BOLWORM: *See Chapter 20.5.2.7.*

STALK BORER:
Moths lay their eggs between the leaf sheets of the maize plant. The young hatching larvae then climb up and feed for a short period on the leaves. Once they become adults, they bore into the growing points or on any place on the maize stems. The pink caterpillar can grow as large as 3.5cm.

CORN BORER:
At night, moths lay their eggs on the leaf sheets of the maize plant. The young hatching larvae then bore into the cobs and damages the silk. Next, they begin to consume the kernels from the silk side of the curds and can lay thousands of eggs. *See photo 125.*

MEALIE BUG BEETLE
Black with yellow spots (called two by twos) as they are often connected to each other. They consume the pollen of the tassels and the silks of the cobs. *See photo 124.*

37.3.19 DISEASES
DISEASES: *See Chapter 20.6 Diseases.*
DAMPING OFF: *See Chapter 20.6.3.1.*

RUST:
This fungal disease causes small white spots which begin on the leaves and later develop into bright orange open pustules, which are slightly raised. The disease occurs when humidity is high.

NORTHERN LEAF BLIGHT:
Long grey, green or tan coloured lesions, 1 to 2cm in length. Concentric grey/black rings of the spores develop into lesions, normally in damp conditions. Lesions begin on the lower leaves and move upwards.

37.3.20 VIRUSES
MAIZE DWARF MOSAIC VIRUS
Affected plants have fine, dark green streaks and light coloured leaves while older plants become stunted. Aphids transmit the virus from grass or other plants to the maize plants.

MAIZE STREAK VIRUS
Yellow arrow streaks are distributed all over the leaf surface. Ear sizes become reduced and cobs may only be partially filled.

CHAPTER 38

PRODUCTION GUIDELINES FOR GARLIC

330 Garlic bulbs packed in white onion type bags and sorted in extra-large, large, medium and small bulbs. Always sort, class and pack garlic bulbs according to the market requirements.

331 Controlling trips in a garlic land. Note the protective clothing when working with chemicals to control insects and diseases.

332 Cleaning and cutting garlic roots and stems. This is followed by sorting the bulbs according to the market requirements. Also, stock plant material may be kept for the next planting season.

38.1 INTRODUCTION

Garlic is not an easy crop to produce commercially due to the low yields, harvesting, drying, cleaning, sorting and packaging processes. The market requirements are also strict. Almost all garlic produce is delivered to the national markets with a small amount sold to the Indian community in KwaZulu-Natal.

Garlic has been grown and used for thousands of years and is classified as a herb. It is an important flavouring ingredient and also a traditional herb which is used for its medicinal and health properties. Consumers, markets and the industry prefer larger garlic bulbs and cloves, as the cloves are mostly cleaned by hand. Currently only 1 cultivar (the Egyptian white) is produced commercially in SA (2014). Small bulbs produce smaller cloves which lowers the price at the National markets. The aim of the producer should be to produce large, good quality bulbs.

Garlic is a management, labour and a capital-intensive crop. The crop has a long growing season (approximately 6 months) and is established in autumn, developing through the winter, up until the beginning of spring. The growing period of the elephant giant is much longer (up to 9 months). Asia is the world's largest producer (2008) with 8 186 million tons per year (81%) while Europe is the 2nd largest producer with 0.767 million ton (8%). Africa produces 0.375 million ton (4%) and America 0.727 million ton (7%).

Production of garlic in South Africa is small, at approximately 3 900 tons per year (2010). Beginner farmers are advised to become members of the South Africa Garlic Growers Association (S.A.G.G.A). S.A.G.G.A normally advises beginners and other producers about the best marketing, production areas, cultivation practices and other successful producers in the garlic industry. Together, SAGGA and the Author have trained up-and-coming farmers in the Limpopo area (2008) to produce garlic. Good results have been achieved and currently these farmers' are successfully producing on large commercial lands.

333 A garlic land infected with a disease called red rust.

334 Garlic cultivated in isolation in order to produce disease-free plant material.

335 Garlic hanging from a roof as well as lying on a mesh table, in the process of drying. Once dried, the leaves and roots are cut so that they may be packed for marketing.

38.1.1 MEDICAL USES

Garlic is used for medicinal purposes e.g. high blood pressure and anti-inflammatory properties. There has been an increase in research into its medical and health properties. It has been found that the crop reduces cholesterol, improves the immune response and prevents stomach cancer. HIV infected people are advised to consume one or two raw cloves of garlic every day, in order to improve their immune systems. However, consuming too much garlic may adversely affect a person's health.

336 On the left, a garlic leaf is infected with a disease called red rust. It has germinated and has grown through to the bulb stage. This usually happens late in the growing season, when warmer conditions combined with too much irrigation water or rain, occurs.

337 A garlic land production in Polokwane district, South Africa, ready to be harvested.

338 Breaking garlic bulbs into cloves and sorting the cloves into large, medium and small cloves before planting time.

38.2 PLANNING FOR GARLIC PRODUCTION

The table below may be used as a planning guide for garlic production.

TABLE 45: PLANNING FOR GARLIC PRODUCTION

	ACTIONS FOR ESTABLISHING GARLIC	ACTION PERIOD
1	Do financial planning according to the two tables that follow this table. Producers should design a cash flow plan so that they know approximately how much money is needed during the growing season. Design a financial plan if a loan is needed. Decide how big the production area is going to be and determine how much it will cost. See the tables below for estimated costs and yields. Determine if you will have enough cash available, at all times, during the growing period (6 months) as well as the harvesting, drying and marketing process. Determine the estimated income of the crop by referring to the tables further down. With all these factors in mind, the producer can now make decisions to achieve a successful, economical production, which is within budget.	4 months prior to planting
2	**SEED** If you do not have your own seed then this must be ordered well ahead of time, especially for more expensive seed cultivars. Do not wait too long as seed prices are usually higher just before planting time due to the high demand. Some producers get hold of planting seed at the National market. This should be from reliable producers, mostly in the Polokwane (Pietersburg) area. Only one cultivar called Egyptian White is produced in South Africa.	6 months prior to planting
3	**SOIL AND SOIL ANALYSIS** Select medium texture sandy loam soil, if possible. This soil will allow good root penetration, drainage (aeration) and will have good moisture-holding capacity. Take soil samples for a soil analysis test and hand it in for testing.	Weeks prior to planting
4	**SOIL PREPARATION** Perform soil preparation. Clean, level the land, remove weeds, stones, bushes etc. If necessary, irrigate the land if it is too dry. Do not cultivate dry soil. Early soil preparation is important to facilitate the breakdown of organic matter, improve moisture, control weeds and insects. Most importantly, an early soil preparation allows the creation of a good smooth seedbed. Determine if a hard pan is present below the soil surface. It is important to break the hard pan using a ripper to improve the drainage. Plough the land and apply lime if necessary (use the correct type). **See Chapter 15.5 to use the correct type of lime.** When it is found that the pH is too low, plough lime into the soil at least 5 to 6 weeks before establishing the crop.	6 weeks prior to planting
5	**FERTILIZER** The land has already been ploughed 5 to 6 weeks previously. The soil should be irrigated, if it is too dry. Apply N, P, K and other elements according to the soil analysis. This will avoid excessive applications of fertilizer and reduce production costs. It is important to ask a reliable fertilizer expert to interpret the soil analysis and advise the producer as to which fertilizer/lime should be used and how much of each. Just before planting, apply the fertilizer according to the soil analysis. Use a calibrated fertilizer spreader. Small-scale farmers usually apply the fertilizer by hand. **See Chapter 15.10 for how to apply fertilizer.** It is now necessary to disk, till or rotovate the land. The fertilizer should be worked into the soil with a disk, tiller, harrow or rotovator to a depth of 15 to 20cm before planting. When loam or clay type of soil is present, it is best to use a rotovator to prepare a fine seedbed. If a soil analysis is not available, a suggested fertilizer application on fairly fertile soil is as follows: 1 300kg p/Ha 2.3.2(22) and 2 Top dressings LAN/KAN @ 130kg p/Ha, applied 3 and 5 weeks after the cloves have emerged.	1 week prior to planting
6	**SORTING THE CLOVES** Break the selected seed bulbs (that have been stored from the previous season),	2 weeks prior to planting

	into cloves. Sort the cloves into large, medium and small sizes. Eliminate the small cloves as only large and medium cloves should be established in the land.	
7	**PREPARE THE LAND** Make furrows in the prepared land and place the cloves at the required spacing in the furrows. Furrows can be made using a modified 3-tooth grub fitted on the back of a small tractor. Some producers create seedbeds with tractor wheels. The wheels are usually 1.2m wide. Therefore, producers establish 5 rows in the seedbed and space the cloves 25cm between the rows, starting 10cm from the sides and 7 to 8cm from one another in the row. Plant the large and medium cloves separately in furrows, in the prepared land.	2 days prior to planting
8	**IRRIGATION:** Irrigate 25mm the first week and thereafter 30mm. This depends on weather conditions. Less water is necessary during cold winter months. Measure the soil-water content. The most effective measurements are made by tension meters or data loggers. Most producers over-irrigate garlic lands.	Weekly applications
9	**SCANNING** Scan for insects and diseases one or two times per week. Control insects and diseases using chemicals that are used on onions. Look for insects/diseases and control them as soon as they appear, especially trips. They hide between the bottoms of the garlic leaves.	Scan weekly after planting
10	**CONTROL WEEDS** Control weeds when necessary. Do not allow weeds to overgrow the young plants. ***See Chapter 17 for weed control.***	When necessary
11	**TOP OR SIDE DRESSINGS** Use 2 top or side dressings of the correct type (100 to 150kg p/Ha, 3 and 5 weeks after emergence), if necessary. This depends on the soil fertility. The roots should be well developed before the first top dressing is applied. It is of no use applying top dressings when the roots are under developed, as the Nitrogen N fertilizer in the soil dissolves and drains away quickly. The N fertilizer should be applied before the first heavy cold arrives. The plant grows slowly during cold conditions and therefore takes up the N fertilizer slowly. Always look at the plant colour. Do not apply N late in the growing season.	3 and 5 weeks after emergence
12	**HARVESTING** Harvest at the right stage. Use a cutting blade. After approximately 6 months, small-scale farmers use forks to lift the bulbs.	After harvest
13	**DRYING OF PLANTS** Dry the plants on the land. See ***Chapter 38.4.16 for how to dry plants.***	
14	**STORAGE** After drying, store the plants in a store. When the plant is completely dry, cut the stems and roots neatly.	After harvest
15	**SORTING AND PACKAGING** Sort and pack the garlic bulbs according to market requirements. Use the correct size and colour bags or carton boxes according to market requirements. Sorting, packaging and classification requirements are available from SAGGA or PROKON inspectors at the markets.	In ventilated stores
16	**RETAIN PLANT MATERIAL (SEED)** Plan the production area for the next season. Also, retain some plant material (seed) for the following season. Sort the bulbs according to their quality (only use good quality bulbs for your seed for the next season). Store the garlic seed bulbs	Store in a ventilated room

apart from the garlic which is intended for market. Approximately 1 ton garlic bulbs are required to establish 1 Ha for the next season.	
17 MARKETING Bulbs may be stored for 6 months in cool, ventilated places, before they are marketed. Producers should be constantly monitoring the market to evaluate seasonal price trends. Market your garlic systematically, according to the market trends.	In ventilated stores
18 Clean the production land and work-in the debris. If necessary, establish a rotation crop to improve the soil fertility.	After harvesting

38.3 ESTIMATED PRODUCTION COSTS AND PROFITS

The production costs to produce vegetables may differ across the country due to the variances in cost of seed, seedlings, fertilizer, electricity/diesel, irrigation water, labour and so forth. Therefore, costs should be adapted accordingly.

The table below provides estimated costs, yields and profits for garlic production (2014). It will assist with designing a cash flow and business plan. Financial lenders require a business plan before they will consider lending money. Overhead costs are not included. Producers should do their own research on production costs when they produce vegetables.

Production costs to produce garlic are high. Planting material is expensive and cultivation is labour intensive. The average yield of garlic as indicated in the table below is 6 tons p/Ha. 0.5 ton will be discarded due to various factors such as mechanical damage, being too small or dull, disease infection, open cloves and burst bulbs. Also, 1 ton bulbs are retained for the following seasons planting material. Therefore, if a producer harvests 6 ton p/Ha, then only 4.5 ton will be available to be marketed. Experienced producers claim to harvest up to 10 ton p/Ha.

The cost of tractors, implements, general equipment (wheel barrows, spades, forks, hoes, crates, twine, pegs etc.), irrigation equipment, water, water pumps, electricity, spraying chemicals, housing, toilets, telephone/computer costs, overalls, clothing for spraying chemicals and so on, are not included in the table below, nor the maintenance of tractors or implements. It is important to make sure that tractors are in good condition before production activity begins.

Certain labour costs are not included in the tables below. Permanent labourers should be used as often as possible, so that they may supervise and train other labourers and also be tractor/truck drivers. Usually 2 permanent and 2 temporary workers can manage 1Ha for a period of 7 months. During peak times (when breaking the bulbs, sorting, planting cloves, weed control, harvesting, cutting the roots/stems, packaging, sorting etc.), more temporary labourers are necessary. Producers should determine labour cost themselves, as costs differ from region to region. Labour is expensive and should therefore be well managed.

TABLE 46: ESTIMATED GARLIC PRODUCTION COSTS PER HA (2006 VS 2014) (SUPPLIED BY THE GARLIC ASSOCIATION)

	2006	2014
SOIL PREPARATION: Level, rip, plough or disk the production land, including labour.	R565	R3 000
SOIL ANALYSIS: 2 soil samples @ R290 per sample. Use a fertilizer expert to interpret the results = R250 per sample.	R100	R1 080
FERTILIZER: Prices rose drastically from 2009. 1 200kg p/Ha 2.3.4 (30) @ R290 per 50kg bag. Require 24 bags x 50kg p/bag = 24 x R290 = R6 960. 2 Top dressings LAN/KAN @ 150kg p/Ha = 6 bags x 50kg @ R240 per bag = R1 440. Apply fertilizer mechanically-tractor and spreader = R350.	R2 200	R8 750

PLANT MATERIAL (Seed): Using your own plant material (Egyptian White cultivar from the previous season). The producer requires 900 to 1 000kg bulbs = R14 000. When obtaining plant material from other producers, the prices in 2013 were approximately R35 p/kg or more. For 1 000kg planting material = R35 x 1 000kg = R35 000. Therefore, it is much cheaper to use your own plant material. Retail prices in 2013 were approximately R70 p/kg.	R8 600	R14 000
CHEMICALS: To control insects and diseases.	R3 000	R3 500
SPRAYING COSTS: Labour and equipment.	R520	R1 000
PLANT CLOVES: Establish cloves in the land by hand. Labour included.	R700	R3 000
ENERGY: Ventilation in stores for approximately 1 to 4 months and to retain seed for 6 months for the following season. Includes labour and maintenance.	R820	R2 500
IRRIGATION: Diesel, electricity and labour costs as well as shifting the irrigation system, cleaning and repairing it. Water is expensive when low rainfall occurs. To irrigate 1mm water p/Ha, 10 000 litres water is required. The period from planting to harvesting is about 6 months.	R1 650	R4 900
WEED CONTROL: Control weeds in and between the rows by hand. Labour included.	R280	R1 600
HARVESTING: Harvest using an undercut blade and pulling the tops by hand.	R2 100	R3 900
PACK IN HEAPS: Pack in wind rows, pyramid heaps or in flat crates to dry the bulbs. Labour included.	R1 020	R2 000
CUT ROOTS AND LEAVES: Cut the roots and the stems to market requirements.	R1 800	R2 500
CLEANING PROCESS: Clean bulbs, trim roots, cut stems, remove outer skins and sort into different classes.	R5 700	R6 100
PACK IN CONTAINERS: Carton boxes, white onion bags etc.	R1 200	R2 200
TRANSPORT: Transportation on the farm and to the Markets. Depends on the distance to the markets. The distance from Polokwane (Pietersburg the main production area) to Pretoria requires a 3 ton truck to travel 2 times = 350km x 4 (there and back) = 1 400km @ approximately R6 p/km including driver and assistant costs of R800. Total = R8 400 + R800 = R9 200.	R4 200	R9 200
CARTON BOXES: Usually 10kg white onion bags or 5kg specially designed boxes. Contact Procon inspectors at the market or the Garlic Association for more information.	R3 000	R3 900
COST PER HA	R37 455	R73 130

TABLE 47: ESTIMATED YIELDS AND INCOME FOR GARLIC (2014)

YIELD: 6 ton p/Ha minus 0.5 ton waist material minus 1 ton seed to be retained for the next season = 4.5ton @ R45 p/kg = R202 500. Garlic prices for the 2011 season doubled. The import of garlic from China was cancelled because of production problems in the June/July season. Retail selling prices at certain markets was R90 p/kg and more (2014).	R202 500
MINUS PRODUCTION COST	R73 130
POSSIBLE INCOME P/HA	R129 370

38.4 PRODUCTION GUIDELINES FOR GARLIC

38.4.1 SOIL REQUIREMENTS
Also see Chapter 5.3 Soil Requirements.

Garlic may adapt to a wide range of soil. However, production is best in a soil that has a good texture such as a sandy loam to loam soil that drains properly. Heavy loam clay type soil should be avoided. In such soil, when the bulbs are harvested, it is difficult to clean them. When the thick outer skin of the bulbs are dry, they peel-off easily to leave the bulbs clean and white.

Top crust soil formation is a serious problem and may hamper the emergence of the cloves. It is better to avoid soil that form a hard crust. If a hard crust is developed after emergence the top soil should be broken in and between the rows.

38.4.2 SOIL PREPARATION
Also see Chapter 8.2 Soil Preparation.

Soil preparation should begin well before establishing the crop. Good soil preparation is important, especially if pre-emergent herbicides are going to be used. A soil analysis is essential and soil samples should be handed in for analysis at least 8 weeks prior to establishing the cloves. If lime is necessary, the correct type should be incorporated 6 weeks prior to planting the cloves. **See Chapter 15.5 to use the correct type of lime.** The soil analysis must be interpreted by a reliable fertilizer expert so that the correct fertilizer may be applied.

Any hard pans below the soil surface should be broken using a ripper. This will improve aeration, drainage, root penetration and root development. Good drainage is essential. A smooth, level, fine grained seedbed should be prepared free of clots, stones and plant material. This is important especially when chemical control (herbicides) is going to be used.

When applying the first top dressing 3 weeks after emergence, it is good practice to apply N by grubbing it into the soil. This may control some weeds as well.

38.4.3 PH OF THE SOIL
The ideal pH of the water-soil content should be between 6.5 and 7.5. Do not consider soil with a pH lower than 6.0 or higher than 7.5.

38.4.4 CLIMATE REQUIREMENTS
Also see Chapter 3 Climate.

Garlic is a cool weather crop and is sensitive to day length. It can tolerate fairly heavy frost conditions. Garlic prefers fairly cold winters because a rest period for bulb formation is preferable. The optimum monthly temperature is 14 to 25°C. The maximum 30°C and minimum 7 to 8°C.

The growing tempo of garlic bulbs is improved when longer day length and higher temperatures occurs. In cooler regions, the cloves could be established earlier in the season, than in the warmer regions. It is better to produce strong plants with a strong root system before the first frost occurs. The size of the plants will determine the bulb size. Therefore, smaller plants produce smaller bulbs.

When plants grow vigorously and thick stems are evident during harvesting, the plants should be dried slowly, preferably in the shade to prevent the bulbs from bursting. If temperatures are above 27 to 30°C for long periods in the land and/or in storage, the bulb formation may be seriously affected. These bulbs should not be used for planting material, but rather sent to the market as soon as possible.

38.4.5 AMOUNT OF SEED PER HA

Select approximately 900 to 1 000kg good quality and large bulbs to be retained for the next season. It is important to break the bulbs into cloves just before planting time, to avoid moisture loss by the cloves.

38.4.6 CULTIVARS

All garlic sources in South Africa are infected with viruses. Most of the viruses are latent and become active when the plants are stressed, especially during drought stress or fluctuations in water applications. The viruses may then negatively influence the yield and quality.

The cultivar, Egyptian White, is the main cultivar produced in South Africa. The cultivar Douglas Pink or 'small pink' is produced on a small-scale in the Western and Eastern Cape areas. This cultivar is not suitable for other areas due to shorter day lengths in those regions. When the bulbs of the small pink cultivar are stored in cool conditions (below 10ºC) for 1 month before planting, this stimulates bulb formation and allows for establishment in short day areas. No research however has been done to prove this.

Many other garlic cultivars are produced on a small-scale. There is a reasonable market in KwaZulu-Natal. Indians consumers prefer purple cultivars due to the stronger flavour. The purple cultivar is a bit knobbier and irregular in its shape. It does not have a good visual appearance for the National markets and therefore is sometimes not accepted. Some garlic cultivars have large cloves with a lesser number harvested. This is not economically viable to produce, as double the number of bulbs should then be retained for seed. For example, the large planting clove cultivars have only 20 cloves, while the Egyptian White has 40 cloves.

The Egyptian White cultivar has approximately 40 cloves per bulb and of these cloves, 26 are large to medium-large and 14 are small (under 0.7g) and therefore not suitable to plant. Germination would be very poor if these smaller cloves were used. It is not proven that the small-to-medium cloves develop smaller bulbs. However, the author has noticed that there is a slight difference in plant development. It is therefore not necessary for producers to select only large or medium cloves, as they can be mixed and planted together. The small cloves should not be established due to their poor germination. Some producers plant the small cloves in separate lands and then reject the very small ones, however germination rates and plant development remain poor. Do not plant the smaller cloves too deeply. After planting, they should be irrigated immediately and on a frequent basis until they emerge.

Some farmers and home owners produce various white and pink/purple cultivars on a small-scale, especially the pink/purple types. The elephant giant cultivar is also produced on a limited scale because of its long growing season and its bitter taste. The cultivar is grown in some home gardens due to its huge bulb size. There is an export market for the elephant giant in certain European countries, were the cloves are ground into a pulp. This cultivar is produced under contract with the dealers.

Elephant garlic is not classified as true garlic, but rather a type of leek that produces large cloves, often only 3 or 4 cloves per bulb. Several small bulb-lets (called beads) may also develop outside the large bulbs. The plant also produces a large seed stalk that may be cut and sold to florists. Elephant garlic is not generally used for dehydration. It has become popular for 'medicinal' purposes and the flavour is mild and tastes slightly bitter.

38.4.7 PLANTING TIMES

February to March seems to be the best planting time. The author prefers to establish the Egyptian White cultivar in the middle of March in the Gauteng area. Good yields are achieved when planting mid-February in the Polokwane (Pietersburg) area.

It is important that the plant and root system should be developed before the first cold or frost conditions occur. Planting in cold areas should be done earlier, so that the plants develops a strong root system before the first frost occurs. The Elephant Giant is established later in the season, March to April.

339 High yields are expected from this Egyptian White cultivar. These plants are 4.5 months old. Established in mid-March and harvested mid-September

340 Planting cloves in furrows in a well-prepared land. Spacing is 45 x 6cm thus producing 370 000 plants p/ Ha.

341 Good quality garlic. Note the size of the bulbs. Very few small bulbs are present. The plants are going to be packed in wind rows.

342 Left: The bulbs are rotten; 3rd from left: the cloves begin to emerge inside the bulb; 4th from left: black colouring due to a disease that has infected the outer skin of the bulbs. This occurs especially during wet conditions, just before harvesting.

343 A cross-cut of an Egyptian White garlic bulb. Note the different sized cloves, many of which are too small to use. This bulb has 16 large, 10 medium and 14 small cloves. The large and medium cloves may be used for plant material while the small ones should be discarded.

344 Garlic plants packed in windrows to dry them before transporting them to the pack house or store room. The plants can also be stored in pyramid heaps.

38.4.8 PREPARATION OF BULBS AND PLANTING METHODS
38.4.8.1 PREPARING THE BULBS FOR PLANTING MATERIAL

Garlic is produced by means of cloves and do not produce seed as in the case of onions. Garlic bulbs are broken into separate cloves just before planting time. They are sorted and classified in different sizes which are to be established during the most optimum time in the season. The cloves are known by producers as seed.

Do not break the bulbs into separate cloves too early before planting time, as they may lose most of their moisture. Sort the cloves into small, medium and large sizes. The small cloves should be discarded, as they do not always germinate well. Only the medium and large sized cloves should be used for planting material.

Plant the medium and large cloves separately or mixed. Some producers prefer to plant them separately but this is not important. Small cloves may also be planted but they should be planted separately as they do not germinate well and develop slowly. Otherwise discard these very small cloves.

38.4.8.2 PLANTING METHOD

In South Africa, garlic is usually planted by hand. The bulbs of the planting stock must be broken up shortly before planting. Cloves of the Egyptian White cultivar may be established 3 to 4cm deep in furrows and should be covered with a 3 to 4cm layer of soil.

Make straight furrows in the moist soil and insert the cloves at the required spacing. Some commercial producers use a modified, 3-tooth grub attached to a small tractor to make the furrows. Close the furrows directly after planting. Compact them slightly and irrigate as soon as possible. Some producers make seedbeds in light loam soil using the tractor's wheels. The wheels are usually 1.2m wide and therefore they establish 5 rows. Next, space the rows 25cm apart between the rows, starting the first row 10cm from the sides and 8 to 9cm from one another in the rows. Keep the soil moist for the first week or two, before planting the to ensure emergence of the cloves.

38.4.9 SPACING

It is recommended to produce large bulbs as the markets and the consumers prefer this. Therefore, the spacing of garlic is important. When the plant population is high then the bulbs will be smaller. When spacing the plants closer to one another, the light intensity needs to be good. Generally, close spacing results in high yields of smaller bulbs while wide spacing results in slightly higher yields of larger bulbs.

A spacing of 45cm between rows and 6cm in the rows for the Egyptian White cultivar is a good choice. This should produce a considerable amount of large and extra-large bulbs. Some experienced producers use a high plant population, such as 450 000 plants p/Ha and still achieve higher yields than normal. This is due to the high plant population as well as the presence of medium sized bulbs which are of the best quality. To achieve such high plant populations however, ideal conditions are necessary, especially with regards to a balanced irrigation/fertilization program that includes recommended top dressings and foliar sprays.

Producers should perform their own research regarding spacing. This also depends on the climate, because for very cold areas, the plant population should be minimized.

38.4.9.1 CLOVES REQUIRED FOR DIFFERENT SPACING

Note: space for walking has not been considered.

Between		in row	Cloves p/Ha
cm		cm	
30	x	6	= 555 556
30	x	7	= 476 190
35	x	6	= 476 190
40	x	7	= 357 143
40	x	6	= 416 667
35	x	7	= 408 126
45	x	6	= 370 370
45	x	5	= 444 445

Walking space for weeding and insect control can be achieved by leaving out every seventh row or adapting the spacing to accommodate tractor wheels (usually 1.2m wide).

A guideline for the spacing of Egyptian White is 35 x 7cm which is 408 000 plants p/Ha. When walking space is included (by leaving out every 7th row), the plant population will be 368 000 plants p/Ha. Some producers do not leave the 7th row open since the overhead irrigation system does not need to be shifted and few weeds are expected, especially if herbicides are going to be used. If done this way, movement in the production land will not be necessary and therefore foot paths are also not necessary.

Due to the size of the Elephant Giant garlic plant, its planting density is much lower.

38.4.10 CROP ROTATION
Also see Chapter 18 Crop Rotation and Soil Improvement.

A crop rotation of at least 3 years should be followed with related crops such as onions and leaks.

38.4.11 WEED CONTROL
Also see Chapter 17 for weed control.

Weed control during plant establishment is critical, as otherwise the weeds could overgrow the young garlic plants after they have emerged. Seed emergence occurs quickly (from 7 days) followed by rapidly growing shoots. The growing period of the Egyptian White garlic is about 6 months. During this period, proper weed control should be implemented, especially when temperatures rise. In winter, weeds are usually not a problem to control.

Avoid lands that contain nutty weeds. They give producers problems and sometimes grow straight through the bulbs. When controlling weeds with a hoe in the rows, it helps to loosen the soil which promotes aeration and water penetration.

38.4.11.1 HERBICIDE WEED CONTROL
Some producers use herbicides such as Goal, directly after planting.

38.4.12 YIELDS
Experienced producers yield up to 10 ton p/Ha from the Egyptian White cultivar. 6 ton garlic bulbs p/Ha is a reasonably good yield. Yields less than 5 ton p/Ha are poor.

Remember that approximately 1 ton of the garlic bulbs will be retained and stored to be used as seed for the following season. At least 0.5 ton will be discarded as waste material due to factors such as dull, undersized and broken bulbs, mechanical damage, splitting, discolouration, disease infection, etc.).

38.4.13 IRRIGATION
Also see Chapter 13 Water and Chapter 14 for Irrigation.

- Well managed irrigation and irrigation scheduling is one of the most important factors for garlic production..
- The land should be irrigated before planting the cloves. This will keep the soil moist and cool when warm conditions occur.
- Irrigate as soon as possible after planting the cloves, to keep the top soil moist, especially when the soil forms a crust. The cloves should be able to emerge with ease in order to maintain a good stand.
- Do not over or under-irrigate garlic. Make use of tension meters or data loggers to determine the water content in the soil.
- Producers often over-irrigate their lands. Note that the plants do not transpire a large amount of water during winter or in cold conditions.
- At all times, ensure that the soil is moist, but never wet, up to a 30cm depth.
- Approximately 400 to 500mm water is required for the entire growing season.
- Elephant Giant needs more water due to its 8 to 9 month growing period.
- During the warmer months, up to 30mm of water per week may be necessary.
- Depending on climate conditions, the land should be irrigated at least once a week. However, note that two light irrigations are better than one.
- Do not over-irrigate when the plants are fully developed. Stop irrigation about 2 to 3 weeks before harvesting, especially in loam and clay type soil, otherwise the outer leaves of the bulbs may become black. This occurs due to fungi in the soil which turn the bulbs black when too much moisture is present late in the season.

Also, this may become a problem the following season when early rains occur.
- When the soil is too wet and when temperature rise after the winter, single cloves of the bulbs which are below the soil, may begin to individually germinate and grow through the bulb tissues. This occurs mostly at the end of the growing season. It is in not necessary to irrigate at the late stage as the plants are already mature. Rain may exacerbate this problem.
- A light irrigation just before harvesting, could be applied to ease the lifting of the garlic plants.

Large commercial producers make use of pivot systems and also top dressing fertigation systems. However, overhead sprinklers are used mostly.

38.4.14 FERTILIZATION
Also see Chapter 15 Fertilizer.

Garlic producers strive to achieve high yields and good quality. A reasonable yield is 6 ton p/Ha. The bulbs should be healthy and have good storage abilities. Storage of the bulbs is affected by fertilization. Therefore, nitrogen N, potassium K and calcium Ca levels require special attention.

Garlic withdraws 8 to 12kg N, 2kg P and 7.5kg K per ton of plant material (including the bulbs), from the soil. As for all crops, no single fertilizer program can be recommended. The result of a soil analysis will aid the producer in applying the required amount of fertilizer. A soil analysis can result in a reduction of fertilizer, an increase in yields and a saving in money. A soil analysis should be interpreted by a fertilizer expert to indicate the correct amounts of different fertilizer which are required. If a soil analysis is not available, a suggested fertilizer application, on fairly poor soil or new soil may be as follows: 1 300 to 1 500kg/Ha 2:3:2: (22) Zn.

38.4.14.1 TOP DRESSINGS
Two top dressings should be applied 3 to 5 weeks after emergence of the cloves at a rate of 100 to 150kg p/Ha. It is better to apply the top dressings before cold conditions begin, as the plants do not transpire well when cold conditions occur. The plants also develop more slowly, especially when frost occurs. Therefore, it is recommended to apply top dressings when it is still reasonably warm.

Care should be taken that the top-dressing N, is not applied too late in the growing season. Too many thick necks and bushy plants may result.

On sandy soil, 3 top dressings may be necessary due to the leaching of these soil. Too much N may influence the uptake of Copper Cu and Calcium Ca. The elements phosphorous P end zinc Zn do not move much in the soil and it is sometimes necessary to apply these elements when preparing the soil.

38.4.14.2 FOLIAR-SPRAYS
Foliar-sprays are a good option when the plants need to be sprayed with certain elements at the correct stage of their development, usually before the 4th leaf stage.

Some producers establish garlic by cultivating a high plant population such as 450 000 plants p/Ha in order to achieve a high percentage of medium-sized bulbs. It is their claim, that due to the high population of medium-sized bulbs, a higher yield is produced, with a good market value. However, this is only achievable if the crop is subjected to a fertilizer program.

Foliar sprays could make significant difference to most crops. Be careful of high plant density in areas that are too cold, or which have lengthy cold perieods.

38.4.15 HARVESTING THE PLANTS
Also see Chapter 24 Harvesting, Handling and Packaging.

Small-scale producers may harvest garlic by loosening the bulbs with a fork on both sides of the rows and then pulling the plants from the soil, by hand. Commercial producers use a cutting blade pulled behind a tractor which loosens the soil below the bulbs and allows the plants to be lifted out of the soil by hand. Some commercial producers use modified potato lifters, which lift the plants out and on top of the soil, thereby making it unnecessary to lift the plants out of the soil by hand.

Garlic should be handled with care as the bulbs may bruise and become infected. A general guideline for the duration between planting and harvesting garlic cloves, is approximately 5.5 to 6 months and for the Elephant Giant, 8 to 9 months. Irrigation should be stopped approximately 2 weeks before harvesting, to allow the bulbs to mature and dry-out. If necessary, lightly irrigate the soil to soften it, before harvesting. This eases the undercutting and lifting process.

The timing of when to harvest garlic is an important decision. When garlic bulbs are harvested too early and the cloves are still immature, they may lose moisture, weight and dry-out quickly. When harvesting bulbs early, they should be marketed immediately. Producers should also be careful not to harvest garlic too late, as a fungus may infect the bulbs and turn the outer skins black. It can be quite challenging to determine when to stop irrigating and when the correct time for harvesting is.

38.4.15.1 EXPERIENCED PRODUCERS CONSIDER THE FOLLOWING ASPECTS TO DECIDE WHEN TO HARVEST

- Plants become a dull colour at the end of the growing season.
- The tips of the leaves dry-out.
- The neck of the plants, just above the bulbs, begins to soften.
- When plants begin to fall over. Usually when 50% of the plants fall over, it is an indication that they are ready to be harvested. This may differ according to planting dates and seasonal temperatures. Sometimes only a small number of plants fall over, as the plants may grow vigorously when N top dressings are applied.
- Break the bulbs and inspect the cloves. They should be well formed and firm.

38.4.16 DRYING OF PLANTS

When garlic is harvested, it should be dried before cutting the roots and leaves. If a store room is not available for drying, the garlic should then be dried outside.

Garlic plants can be packed in wind rows next to the production land. After pulling the tops they should be packed in such a way that the leaves of the front bundle covers the bulbs in the back part of the row. This protects the bulbs from sun burn. **See photo 344.** Bulbs should have dried-out after 5 to 7 days. If it should rain during the drying process, care should be taken to immediately turn the garlic, otherwise the whole bunch may rot.

Garlic may also be stacked in pyramid heaps and left to dry for longer. The top of the pyramid heaps may be covered with plastic sheets to prevent the rain from penetrating into the heap. Plants may also be dried directly on the floor or on mesh racks inside the store. Alternatively, they may be hung from the roof when producing on a small-scale.

38.4.17 CLEANING, SORTING, GRADING AND PACKAGING

When the garlic plants are dry enough (usually 3 to 4 weeks after harvesting), then the roots and stems may be neatly cut to market requirements. Cut the roots and necks with sharp pruning shears. It is important to leave a short 2cm piece of the neck attached, otherwise the bulbs may burst if the cut is too close to the bulb. This is also the market requirement.

345 Good garlic bulbs which are dry enough to have their roots and stems cut.

346 Garlic sorted into 3 classes and packed in white 7kg onion bags. From left to right large, medium and small bulbs. When the bulbs are dry, the bags weigh just over 6kg. When the bags are sent to market, they should be packed so that they weight just a little more than 6kg, in order to compensate for any moisture loss which might occur. Underweight bags will be rejected or degraded by the market.

Remove the thin, outer sheets of the bulbs. These outer leaves are usually a white bleached colour and quite loose. When these bleached outer leaves are removed, the white bulbs are exposed. Never wash the bulbs as the layers of the bulbs may become infected. When the bulbs are completely dry, rub the thin outer layer of the sheet off the bulbs.

Garlic should be sorted and graded according to different sizes or classes and then packed in the correct containers (boxes or bags). White onion bags and suitable boxes of different sizes are used when marketing at the National markets. It is important that the market requirements are followed precisely and carefully, as the batches will be inspected by Prokon inspectors and if they find something unusual, they may degrade the total batch.

Garlic bulbs lose moisture after packaging and therefore the containers should be packed a little heavier than what is required to compensate for any weight loss. When the containers are underweight, the batch is going to be degraded at the producers cost.

38.4.18 STORAGE

Garlic should be dry before it is stored. The bulbs may be stored for 6 to 7 months while the Elephant Giant may be stored for up to 10 months, before being marketed.

After the plants stems and roots have been cut, they may be stored in a well-ventilated store in various ways- in crates, onion bags, loose in heaps, in chicken wire or mesh racks. They may also be hung from the roof or stacked in heaps. Ventilation may be improved by using extractor fans inside the store. The wind speed should not be too high, as the bulbs may lose moisture and dry out. Optimum temperatures inside the store should be 19 to 21ºC and when the bulbs are to be sent to the market just after harvesting, the store can be 24-26 ºC.

The humidity inside the store should not be too high, ideally between 60 and 70%. Too high humidity may promote diseases. Humidity may be increased by wetting hessian bags and placing them on the floor inside the store.

Do not stack the bulbs layers too thickly on racks or in crates, as air circulation between the bulbs is important. Garlic planting seed material should never be stored in refrigerators, as this may result in the next generation of garlic having rough bulbs, sprouting side-shoots and maturiting too early. When the seed is stored at temperatures that are too high, it may result in delayed sprouting and late maturity.

Care should be taken to use disease resistant planting material. The bulbs and cloves used for planting may carry and transmit diseases such as white and basal rot. The garlic should be inspected regularly in the store for diseases and rotting of the bulbs. Eliminate infected or rotten bulbs immediately.

38.4.19 MARKETING
Also see Chapter 25 Management and Marketing.

Garlic is mainly produced for the National fresh market and it may be kept for more than 6 months before marketing. Batches can be sent to the market at any time when the producer chooses or when best prices are expected. Producers should always investigate and follow the market trends. The biggest bulbs usually receive the best prices.

Market requirements are strict and the garlic producer is advised adhere to them. The South African market is small and the export of garlic to other countries is not possible. This is due to the high production costs in South Africa, the shipping costs and strict export requirements. Some overseas countries subsidize their local garlic producers which often results in much cheaper garlic from other counties being sold on the South African markets. The Garlic Association does their best to prevent illegal and cheap garlic from entering the country in order to protect the local producer's price index.

It is law that garlic which enters the country must be x-rayed. This damages future germination in order to prevent local producers from propagating the imported cultivars. It is found that the keeping quality of garlic is poor after they have been x-rayed.

Producers in South Africa cannot supply the markets with quality garlic bulbs from June to July, as the bulbs are harvested in September/October and cannot be stored for longer than 6 to 7 months. No effective way has yet been found to store garlic for longer than 7 months. Therefore, fresh garlic is imported from June to August from other countries to fill this gap.

38.4.20 QUALITY STANDARDS
Prokon inspectors are responsible for inspecting all the garlic at the National markets, from all producers across the country.

Standards, rules and regulations are then laid down. Producers should follow the standards and rules carefully, otherwise their garlic batches may be rejected or degraded to a lower class, at the cost of the producer. A colour photo brochure explaining the standards is available from Prokon inspectors located at the National markets or at S.A.G.G.A. The standards advise the producer as to the correct sorting, grading, class (sizes) and packaging according to the market requirements. The author recommends this brochure., especially for beginner producers. Good quality garlic which is packed correctly, should meet the standards on the markets. Second and third grade garlic does not achieve good prices.

38.4.21 PLANT MATERIAL (SEED) SORTING AND STORAGE
After harvesting and drying, the garlic bulbs have to be selected, sorted and classified into different sizes. Only suitable bulbs for propagation (planting material) should be retained to be established for the next season. These propagating bulbs should be stored for approximately six months before planting. The bulbs must be broken into cloves, just before planting time.

When selecting bulbs for planting material, bulbs with the qualities mentioned below should not be included in the stock planting material:
- Bulbs with little or thin outer leaves.
- Bulbs with open or loose cloves.
- Badly coloured and crimped bulbs.
- Mechanically damaged bulbs.
- Bulbs with cloves that have germinated.
- Dull bulbs.
- Bulbs that are disease infected.

38.4.22 INSECTS
Also see Chapter 20.5.

Garlic producers sometimes use unregistered chemicals to control insects and diseases. This is due to the fact that no registered chemicals are registered for garlic (2014). Instead, chemicals registered for onion production are used.

Garlic has thin flat leaves covered with a waxy layer on the outer surface. It is important to use chemicals which are effective despite this waxy layer. The flat, narrow and numerous leaves on the garlic stem do not have a huge leaf surface. In fact, if the leaves of 1Ha garlic were spread-open, the leaves would cover at least an 11 Ha area.

It is better to make use of lower plant populations when very cold conditions occur in your area. Some producers employ lower plant spacing of about 450 000 plants p/Ha. More than 80% medium bulbs may be harvested in the warmer areas when managed well, especially when proper irrigation and fertilizing (from a soil analysis) practices are followed. Producers may also apply fertilizer through the irrigation system and make use of foliar sprays.

The crop goes through a rest period during the colder months and therefore requires maximum sunlight when the temperatures begin to rise during the cooler months. Wide spacing makes disease and insect control much more effective. It is important to spray chemicals early in the morning and late in the afternoon as some insects (such as trips) are active at night.

ROOT-KNOT NEMATODE: *See Chapter 20.5.2.1.*
RED SPIDER MITE: *See Chapter 20.5.2.9.*

TRIPS: *See Chapter 20.5.2.4 Trips.*
The biggest problem when producing onion and garlic is caused by these small insects. They are difficult to control. If trips are controlled effectively, you will be able to overcome 60% of the other disease related problems. Trips are the main insect that transmits viruses. They feed and scrape the waxy layer from the leaves with their mouths parts. This leaves the surface of the plant unprotected and opens the door for other infections such as:
- Downy mildew (the main disease on garlic and onions)
- Botrytis
- Alternaria Salami

If trips are controlled effectively, the above-mentioned diseases find it difficult to penetrate through the waxy layer of healthy plants. Remember, strong healthy cell walls are thicker and tougher, preventing diseases from penetrating the plant.

38.4.23 DISEASES
Also see Chapter 20.6 Diseases.

POWDERY MILDEW: *See Chapter 20.6.3.6.*
PHYTOPHTHORA WILT: *See Chapter 20.6.3.16.*
ALTENARIA LEAF SPOT OR PURPLE BLOTCH: *See Chapter 20.6.3.2.*

FUSARIUM WILT OR BASAL ROT: *See Chapter 20.6.3.11.*
This is a worldwide fungal disease that is a problem in the field and during storage. The fungus attacks the base of the bulb, causing browning and rotting of the bulb as well as yellowing and eventually death of the leaves. The disease is soil borne and spreads through the soil. It infects the roots and bulbs. Plant material, water and implements that carry infected soil should be disinfected.

BROWN RUST: *See Chapter 20.6.3.8.*
This fungal disease causes white spots, beginning on the leaves. They develop into bright orange pustules, which are slightly raised. The disease occurs when conditions are dry and humidity is high. The disease may not appear every year. It depends on temperature/high humidity. If this disease become active, it is difficult to control, as thousands of spores are set free.

It is important to control this disease early. The disease mostly appears late in the season when temperatures and humidity are high. Sometimes it does not do much harm, especially later in the season just before harvesting.

Some producers use unregistered systematic chemicals at their own risk such as Folicur, Score, Bravo or Richter.

PINK ROOT:
This is a fairly wide spread fungus. The roots turn light pink and later shrink and turn brown after harvesting. Pink root affects the yield and quality of the bulbs. When the infectionrate is high, the plants show signs of poor development.

DOWNY MILDEW:
Also see Chapter 20.6.3.6 Downy Mildew.

This fungal disease occurs throughout the world. Severe yield losses may occur as a result of leaves being killed by this pathogen. The results are destructive, causing the leaves to die, especially when optimal conditions occur for the fungus (cool, moist nights and mild, overcast, rainy days). The disease favours warm wet conditions.

Symptoms begin as small white lesions/spots on the leaves and stalks. These then enlarge to cause white/purple sunken lesions. These lesions later become light/dark brown (occasionally purplish) showing brown concentric rings in humid conditions. This condition is more severe on the older leaves. Some producers use unregistered chemicals at their own risk such as Mancozeb, Bravo, Ridamil Gold, Bravo etc.

WHITE BULB ROT:
The disease can become a serious problem when producing onions and garlic. This is also a worldwide disease. The disease affects the bulbs and shows a characteristic white mould on the base of the bulbs. Roots begin rotting and the leaves turn yellow, wilt and fall over. The white fungus growth at the bottom of the bulb becomes rotten and smells bad. The disease may remain in the soil for 15 years or more. Therefore, onion or garlic cannot be produced again when the soil is affected.

VIRUSES:
All garlic sources in South Africa are infected with one or more viruses. Most viruses are latent (dormant) and do not make a difference on yields when the crop is cultivated properly. The main virus (not yet classified) may set back yields and influence about 18% of the crop and then stabilize. When using your own plant material, year-after-year, the viruses will remain latent, but the yields will still decline to a certain degree. However, when the plants are stressed during the cultivation process, the virus has a highly negative influence on yields.

Improved garlic material supplied by the ARC (Agricultural Research Council), improves the yields and quality by more than 16% in 2006.

Garlic viruses are not visible to the naked eye except when observed closely for certain symptoms. The most common is the mosaic virus infection which causes discolouration of the leaves which includes mosaics, flecking, streaking and moulting. The ARC is busy with a garlic virus elimination program and researchers are aiming to supply virus and disease free material to the Garlic Association for the purpose of propagation. The planting material is to be supplied to commercial as well as small-scale and up-and-coming farmers.

Together with researchers from ARC, the author was involved for more than 20 years in research programs attempting to eliminate virus infections of the Egyptian White cultivar. They began with Elisa tests and later made use of electron microscope tests. This was partly successful, but a microscope researcher later discovered that the electron microscope did not show some viruses as they combine together with the plant sap.

Although the electron microscope can magnify virus strains that are as small as 0.00001mm (i.e. magnifying the virus strains to the size of a match stick), the electron microscope could still not provide a reliable indication of virus infections. The electron microscope did however assist in eliminating most viruses which resulted in better yields. When the improved plant material was supplied to the producers, it improved quality and increased yields by more than 16%.

REFERENCES

Adams P. 1980. Nutrient uptake by cucumbers from recirculation solutions.

Albert A. Allnet-Shade net suppliers. Info of shade net and construction of shade net houses.

Elbie. Kameelfontein. Organic nursery. Info for organic seedling production.

Hoffman D. Bouquet Garni. Herbs nursery. Info herbs and herbs fore the NFT hydroponics system.

Dalzell H W, Gray K. R. and Biddlestone A. J. 1979. Composting in tropical agriculture. England.

De Bertoldi M. 1991. The control of the composting process London SW8 5DR 1991.

De Briun C. Gundle plastics. Info tunnel plastics, hytex dams, hydro lines.

De Klerk J. Plant Protection Institute. Info chemical equipment and how to use it.

Du Plessis H. Knitex-Shade net suppliers. Info of shade net and construction of Shade Net Houses.

Du Plessis H. Ag-Chem Africa. Info of irrigation open field, especially potatoes and Fertigation of Hydroponic systems.

Erasmus B. Rhino Tunnels and plastics. Installation and plastics info.

Groenewalt C. General Manager Tshwane market. Marketing info.

Halley R.J. 1982. The agricultural notebook (17th edition). Primrose Mc Connell's.

Harris D. 1992. Hydroponics. The complete guide to gardening without soil.

Henning G. Ag-Chem Africa. Spraying equipment and insect control.

Hever C. Constructions. Tunnel construction heating/cooling systems info.

Howard M.R. 1981. Hydroponics food production.

Jackson D. C. 1976. Besproening gids vir groente verbouing.

Joubert T la G. 1975. The production of carrots, beetroot and cole crops- Capsicums in SA.

Kerr B. Info of breeding new cultivars and production of vegetable crops. Manager of Alpha Seed-Meyerton.

Knitex. Guide for the use and installing shade Net. M Rotcher 1996.

Korkie J. Milner chemicals. Info chemicals, foliar sprays, general info of hydroponic production.

Kotze F. Info of vegetable production. Klein Karoo Koop Brits.

Mac Gilivray. 1952 California Vegetable production.

Nel A and Krause M. A guide for the control of Plant Pests. National Dep of Agricultural. 2004.

Nelson P.V. 1991. Greenhouse operations and management.

Nieuwhof M. 1969. Cole crops. Leonard Hill. London

Wagner. All about tomatoes 1973.

Rose C. Crop Protection 1963.

Ryall A. Handling transportation and storage of fruit and vegetables 1772.

Stephan Wallner. Post Handling and Storage of Vegetables for the Fresh Market 1972.

Schwarz M. 1968. Guide to commercial hydroponics. Jerusalem.

Njoroge J. W. 1996. Training manual on organic farming in medium and high potential areas. Kenya.

Valenzuela H. Krakty B. Lettuce Production Guidelines for Hawaii Crop Production Guidelines. 1993.

Vicky. Info of soil sampling and analysis. Institute for Soil Water and Climate.

Victor R.S. 1973. Growing tomatoes using gravel with high saline water.

Van Niekerk A. C. 1883 The cultivation of crops. Vegetable and Ornamental research institute. Pretoria.

Walker J. Diseases of vegetables Crops, 1952 Hill Brook Co.

Watters C. Tomato diseases. A practical guide for seed men 1985 California.

INDEX

A

acidity 1, 48, 49, 51, 108, 142, 205, 253
agents 9, 12, 13, 14, 192, 193, 206, 355
air temperature 16, 38, 66, 151, 177, 230
analysis 1, 7, 13, 43, 45, 48, 49, 51, 52, 53, 62, 66, 70, 97, 98, 101, 102, 103, 104, 106, 108, 109, 111, 112, 114, 115, 171, 172, 181, 183, 191, 195, 196, 200, 203, 204, 207, 208, 209, 215, 216, 219, 228, 238, 250, 254, 261, 270, 273, 276, 283, 287, 289, 297, 302, 303, 311, 316, 321, 327, 328, 334, 340, 341, 342, 348, 352, 359, 362, 369, 370, 372, 379, 381, 383, 388, 392, 395
aphids 67, 84, 121, 130, 131, 135, 137, 138, 140, 142, 149, 150, 162, 185, 222, 244, 246, 264, 265, 282, 286, 292, 293, 296, 307, 308, 318, 331, 345, 355, 365, 366, 376

B

bacteria 35, 37, 43, 45, 59, 69, 72, 73, 76, 87, 113, 121, 126, 127, 129, 130, 133, 135, 142, 145, 148, 155, 188, 190, 266, 291, 353. See disease
bacterial cancer 127, 146, 148, 223
bacterial spot 147, 148, 149, 223, 308
bacterial wilt 127, 148, 223, 308
beans 5, 12, 14, 15, 17, 18, 19, 26, 27, 29, 39, 45, 49, 65, 70, 79, 80, 81, 83, 87, 93, 102, 105, 106, 109, 110, 113, 114, 119, 125, 126, 127, 129, 139, 142, 153, 165, 177, 186, 188, 204, 205, 262, 329, 346, 347, 348, 349, 350, 351, 352, 353, 354, 355, 356, 367
beetroot xxiv, 14, 15, 17, 18, 21, 26, 27, 28, 29, 44, 49, 53, 62, 65, 66, 79, 80, 84, 93, 105, 119, 125, 126, 127, 147, 149, 192, 204, 310, 312, 317, 319, 320, 321, 322, 323, 324, 325, 326, 327, 328, 329, 330, 331, 332, 341, 344, 395
biofumigation 1
black rot 145, 146, 255, 265
blossom drop 17, 87, 153, 156, 296, 351
blossom end rot 87, 88, 151, 152, 166, 171, 182, 185, 210, 216, 224, 240, 303, 305, 308
bollworm 139, 140, 355
bolting 1, 20, 21, 87, 251, 254, 255, 266, 267, 268, 271, 274, 310, 313, 318, 323, 325, 335, 336, 343, 344, 358, 360, 361, 363, 365, 366
botrytis 147, 148, 223, 240, 279, 392
brassicas 28, 248, 262. See cole
breeding programs 5, 16, 33, 149, 199, 206, 272, 281
broccoli 15, 20, 21, 26, 53, 173, 248, 249, 251, 253, 254, 255, 257, 258, 259, 260, 261, 262, 263, 264, 265, 266. See cole
brown rust 145, 148, 280, 393
brussels sprouts 26, 248, 251, 253, 256, 257, 260, 261, 262, 264, 266. See cole
business plan viii, 12, 13, 195, 200, 203, 227, 249, 269, 283, 297, 320, 334, 347, 358, 368, 381

C

cabbage xxv, 2, 3, 9, 12, 13, 18, 20, 21, 26, 29, 48, 49, 53, 59, 64, 79, 80, 93, 94, 96, 126, 127, 139, 145, 146, 151, 159, 165, 173, 226, 248, 249, 250, 251, 252, 253, 254, 255, 256, 257, 258, 259, 260, 261, 262, 263, 264, 265, 266, 333, 353. See cole
capsicums 26, 28, 34, 39, 106, 109, 110, 125, 142, 144, 147, 149, 302, 303, 308, 395
carrots xxiv, xxv, 8, 14, 15, 21, 24, 26, 27, 28, 29, 49, 53, 62, 65, 66, 79, 80, 84, 88, 93, 103, 105, 116, 119, 125, 126, 127, 149, 151, 192, 193, 204, 327, 332, 333, 334, 335, 336, 337, 338, 339, 340, 341, 342, 343, 344, 345, 395
cash flow viii, 4, 11, 12, 13, 59, 195, 200, 203, 227, 249, 269, 283, 297, 311, 320, 334, 347, 358, 368, 379, 381
cauliflower 15, 18, 21, 23, 26, 47, 49, 53, 79, 81, 139, 145, 165, 173, 248, 249, 251, 252, 253, 254, 255, 256, 258, 259, 260, 261, 262, 263, 264, 265, 266. See cole
centre pivot 94, 261
certification v, vi, 24, 76
chemicals 3, 7, 11, 13, 24, 35, 57, 65, 76, 77, 115, 123, 131, 133, 134, 136, 137, 138, 139, 140, 142, 143, 144, 147, 154, 155, 157, 158, 159, 160, 161, 162, 163, 185, 203, 227, 228, 245, 249, 250, 269, 270, 279, 280, 283, 292, 297, 311, 312, 320, 321, 334, 347, 348, 352, 355, 358, 359, 366, 368, 369, 377, 380, 381, 382, 392, 393, 395
chillies 18, 26, 27, 39, 49, 65, 77, 84, 137, 149, 177, 188, 198, 218, 226, 294, 295, 298, 299, 300, 301, 302, 303, 304, 305, 307. See capsicums
chinese cabbage 21, 165, 173, 248, 251, 253, 256, 257, 262. See cole
chisels 55
climate vii, 4, 5, 8, 12, 15, 16, 21, 34, 71, 72, 113, 128, 144, 177, 193, 205, 229, 230, 237, 254, 263, 285, 294, 301, 316, 322, 324, 336, 344, 360, 366, 271, 284, 386, 387, 298, 313, 322, 335, 349, 360, 370, 383, 395
cole 14, 17, 27, 30, 34, 41, 48, 52, 53, 63, 65, 68, 79, 105, 106, 116, 125, 129, 139, 143, 145, 156, 165, 248, 249, 251, 253, 254, 255, 256, 257, 259, 260, 261, 262, 263, 264, 265, 266, 344, 395
commercial producers v, 3, 29, 36, 40, 44, 53, 60, 62, 63, 65, 71, 96, 112, 144, 157, 165, 176, 192, 200, 206, 214, 225, 282, 284, 285, 286, 292, 315, 328, 343, 365, 386, 388, 389
communication 6, 14, 203, 297
compacted soil 47, 58
compost vii, 1, 5, 7, 33, 43, 44, 57, 62, 69, 70, 71, 72, 73, 74, 75, 76, 112, 122, 127, 171, 178, 183, 195, 196, 207, 209, 211, 215, 216, 235, 239, 240, 252, 253, 273, 282, 287, 292, 304, 312, 324, 329, 338, 342
condition xxv, 31, 43, 44, 45, 71, 109, 112, 113, 129, 142, 148, 153, 187, 191, 197, 219, 225, 240, 242, 262, 269, 283, 294, 306, 311, 321, 327, 334, 347, 358, 369, 381, 393. See disease
control v, vii, 1, 2, 7, 8, 9, 11, 12, 13, 17, 20, 23, 24, 34, 38, 42, 47, 55, 56, 57, 59, 62, 71, 75, 76, 77, 80, 81, 83, 84, 102, 115, 119, 120, 121, 122, 127, 128, 130, 131, 133, 134, 135, 136, 137, 138, 140, 141, 142, 143, 144, 145, 146, 147, 148, 149, 151, 154, 155, 157, 158, 159, 160, 161, 162, 165, 166, 169, 172, 173, 175, 176, 179, 181, 185, 186, 187, 189, 192, 195, 198, 201, 204, 205, 208, 210, 212, 214, 216, 217, 227, 228, 233, 234, 237, 241, 245, 246, 249, 250, 254, 259, 262, 266, 267, 270, 274, 276, 277, 279, 280, 282, 283, 286, 287, 290, 293, 296, 297, 298, 305, 308, 312, 314, 317, 321, 322, 324, 325, 328, 329, 330, 332, 334, 335, 339, 341, 343, 344, 347, 348, 353, 355, 359, 364, 365, 366, 369, 374, 377, 379, 380, 381, 382, 383, 386, 387, 392, 393, 395
cooling vii, 1, 67, 166, 170, 172, 176, 189, 221, 305, 307, 310, 354, 355, 357, 365, 395
cool weather crops 39
corn viii, 12, 15, 17, 27, 45, 59, 66, 93, 113, 116, 125, 141, 186, 198, 261, 367, 368, 369, 370, 371, 372, 373, 374, 375. See maize
cracking 87, 88, 151, 152, 173, 181, 185, 202, 204, 210, 223, 240, 244, 247, 261, 263, 289, 308, 309
crop rotation 1, 3, 5, 44, 75, 122, 125, 126, 127, 128, 143, 145, 171, 195, 218, 241, 262, 276, 290, 305, 317, 329, 341, 353, 364, 387
cucumbers 8, 33, 34, 39, 40, 65, 79, 81, 83, 84, 99, 103, 105, 106, 114, 115, 149, 153, 165, 170, 172, 173, 174, 175, 176, 177, 179, 182, 183, 184, 187, 190, 199, 216, 223, 225, 229, 230, 232, 233, 236, 237, 238, 239, 242, 243, 244, 395
cucurbits 5, 15, 26, 29, 34, 39, 69, 87, 123,

125, 139, 142, 144, 149, 150, 156, 225, 227, 229, 230, 231, 233, 234, 235, 237, 238, 240, 241, 242, 243, 245, 246
cultivar vi, xxiv, 3, 5, 8, 9, 12, 13, 18, 19, 20, 21, 23, 24, 25, 28, 30, 31, 32, 33, 79, 83, 102, 119, 125, 136, 151, 152, 169, 170, 173, 174, 175, 176, 178, 187, 191, 192, 195, 196, 197, 199, 200, 201, 202, 203, 204, 206, 207, 210, 211, 212, 215, 218, 219, 220, 222, 225, 226, 228, 231, 232, 234, 235, 237, 242, 248, 251, 252, 253, 255, 256, 257, 258, 262, 263, 267, 269, 270, 272, 273, 274, 275, 278, 285, 286, 288, 294, 295, 300, 301, 302, 306, 314, 318, 319, 322, 323, 325, 326, 329, 330, 332, 333, 334, 336, 339, 340, 341, 346, 347, 350, 352, 354, 357, 360, 363, 367, 368, 371, 374, 377, 379, 382, 384, 385, 386, 387, 394
cutworm 65, 138, 141, 218, 233, 235, 315, 316

D

damping off 36, 143, 222, 245, 265, 293, 307, 318, 325, 331, 345, 355, 365, 375
day length 18, 20, 267, 268, 271, 272, 277, 329, 383
daylight 230, 344
determinate 17, 30, 79, 83, 91, 122, 165, 170, 197, 199, 200, 201, 202, 205, 206, 209, 210, 211, 212, 213, 214, 215, 218, 219, 220, 221
diamondback moth 139, 265
discs 55
disease vi, vii, 1, 2, 7, 8, 11, 17, 30, 32, 33, 36, 41, 42, 75, 76, 77, 81, 83, 84, 109, 113, 119, 123, 127, 130, 133, 134, 135, 136, 139, 142, 143, 144, 145, 146, 147, 148, 149, 150, 154, 155, 157, 158, 159, 170, 174, 175, 178, 179, 185, 186, 187, 195, 198, 199, 200, 204, 206, 210, 212, 214, 216, 217, 218, 223, 229, 233, 235, 242, 245, 246, 254, 264, 265, 266, 267, 270, 277, 279, 280, 282, 283, 285, 286, 288, 293, 294, 296, 300, 301, 303, 307, 308, 309, 314, 319, 321, 322, 327, 330, 331, 334, 338, 339, 345, 347, 348, 354, 355, 356, 359, 360, 366, 367, 369, 375, 378, 381, 385, 387, 391, 392, 393, 394
disorder 151, 152, 153, 185, 223, 246, 291, 293, 308, 366. See disease
downy mildew 145, 146, 227, 245, 265, 279, 280, 365, 392, 393
dripper system xxv, 94, 96, 97, 98, 103, 110, 111, 112, 114, 169, 171, 172, 197, 209, 210, 211, 216, 219, 235, 239, 302, 304, 328, 343
droplet size 154, 160, 161

E

early blight 126, 142, 143, 144, 145, 155, 218, 222, 296, 307, 345
equipment vii, viii, 1, 2, 4, 6, 7, 8, 13, 14, 39, 46, 47, 54, 55, 56, 57, 130, 133, 134, 158, 159, 163, 189, 191, 203, 227, 249, 269, 283, 297, 311, 320, 334, 338, 347, 358, 368, 381, 382, 395

F

F1 hybrid 1, 3, 12, 18, 21, 23, 25, 30, 32, 33, 37, 38, 47, 62, 83, 104, 134, 148, 156, 181, 193, 196, 197, 199, 201, 203, 207, 211, 212, 213, 215, 219, 221, 223, 228, 230, 231, 234, 235, 236, 238, 242, 248, 251, 252, 253, 254, 255, 257, 259, 263, 264, 267, 268, 270, 272, 274, 279, 295, 300, 319, 320, 322, 323, 326, 330, 333, 334, 336, 340, 344, 348, 350, 363, 365, 367, 371, 374
F1 hybrid cultivar 18, 25, 83, 197, 211, 242, 248, 267, 333, 350, 367
F1 hybrid seed 1, 62, 203, 207, 235, 236, 270
fertilizer vii, 1, 3, 5, 6, 7, 8, 11, 15, 19, 33, 34, 36, 37, 38, 39, 41, 42, 43, 44, 45, 48, 49, 51, 52, 53, 55, 56, 57, 60, 61, 62, 66, 70, 71, 72, 73, 75, 85, 86, 96, 97, 98, 101, 102, 103, 104, 106, 107, 108, 109, 110, 111, 112, 113, 114, 115, 121, 126, 127, 143, 162, 171, 172, 173, 178, 179, 181, 183, 191, 195, 196, 197, 198, 200, 201, 203, 204, 207, 208, 209, 215, 216, 219, 227, 228, 233, 234, 235, 236, 238, 239, 240, 241, 247, 249, 250, 261, 269, 270, 271, 273, 275, 276, 283, 287, 289, 290, 295, 297, 302, 303, 311, 314, 316, 320, 321, 323, 327, 328, 329, 334, 337, 340, 341, 342, 344, 347, 348, 352, 358, 359, 362, 368, 369, 372, 379, 380, 381, 383, 388, 392
fertilizing 46, 49, 98, 104, 204, 208, 238, 239, 265, 303, 392. See fertilizer
financial planning 203, 379
flower drop 110, 139, 153, 205, 238, 298, 304, 347, 352
foliar sprays 1, 7, 103, 105, 113, 114, 124, 161, 209, 216, 239, 262, 274, 276, 317, 328, 386, 388, 392, 395
frost 15, 16, 18, 19, 67, 95, 176, 177, 186, 206, 225, 229, 230, 234, 243, 254, 255, 273, 277, 281, 284, 287, 289, 292, 296, 299, 313, 320, 322, 323, 332, 335, 336, 350, 360, 361, 383, 384, 388
fruit drop 102, 153, 156, 209, 242
fungi 23, 35, 36, 42, 45, 62, 77, 84, 86, 121, 130, 133, 135, 143, 148, 149, 189, 190, 241, 246, 345, 291, 387
fungus 17, 135, 142, 143, 144, 147, 148, 280, 389, 393. See disease
furrow 63, 95, 119, 236, 240, 258, 275, 325, 339

G

garlic vi, viii, 7, 12, 15, 17, 18, 58, 65, 77, 86, 93, 105, 106, 109, 110, 114, 125, 136, 137, 138, 139, 145, 146, 148, 149, 150, 186, 187, 188, 204, 269, 276, 277, 279, 280, 377, 378, 379, 380, 381, 382, 383, 384, 385, 386, 387, 388, 389, 390, 391, 392, 393, 394
germination vi, 16, 23, 24, 25, 27, 28, 29, 34, 35, 36, 37, 39, 47, 60, 64, 66, 86, 94, 95, 113, 116, 119, 120, 122, 230, 235, 271, 273, 274, 298, 314, 322, 324, 328, 329, 336, 338, 339, 340, 341, 343, 349, 351, 353, 360, 364, 372, 384, 391
gmo crops 135
grading vi, vii, 1, 3, 187, 190, 192, 195, 198, 278, 298, 300, 306, 338, 374, 389, 391
grafting 32, 199, 200
greenhouse vii, 83, 84, 173, 175, 176, 187, 190, 395
growing chamber 37
guidelines v, viii, 1, 93, 154, 157, 163, 195, 196, 205, 225, 229, 248, 253, 267, 271, 281, 284, 294, 298, 310, 312, 319, 322, 332, 335, 346, 349, 357, 360, 367, 370, 377, 383, 395

H

hardening 1, 34, 40, 41, 55
harrows 56
harvesting vii, 1, 3, 5, 6, 7, 8, 9, 11, 12, 13, 14, 16, 20, 21, 24, 32, 57, 59, 71, 73, 80, 83, 84, 86, 87, 91, 93, 97, 126, 128, 129, 130, 133, 134, 146, 152, 159, 174, 175, 178, 181, 186, 192, 195, 198, 199, 202, 205, 210, 212, 215, 217, 219, 220, 221, 226, 228, 229, 232, 237, 238, 239, 243, 244, 246, 249, 250, 252, 253, 255, 256, 257, 258, 259, 262, 263, 264, 265, 266, 267, 268, 269, 270, 271, 273, 274, 276, 277, 278, 279, 280, 282, 283, 284, 287, 289, 290, 291, 292, 294, 295, 296, 298, 301, 303, 305, 306, 307, 310, 312, 314, 317, 318, 321, 322, 325, 327, 330, 335, 336, 338, 339, 340, 341, 344, 345, 346, 348, 349, 350, 351, 353, 354, 355, 357, 358, 359, 364, 365, 368, 369, 370, 371, 372, 374, 375, 377, 379, 380, 381, 382, 383, 385, 387, 388, 389, 390, 391, 393
heat control 172, 296
herbicide 120, 121, 122, 123, 124, 150, 153, 267, 277, 321, 328, 330, 343, 366, 387
housing 6, 203, 227, 249, 269, 283, 297, 311, 320, 334, 347, 358, 368, 381
humidity 1, 16, 17, 27, 28, 35, 39, 102, 109, 110, 114, 142, 144, 145, 147, 148, 151, 153, 155, 165, 166, 170, 173, 176, 177, 185, 187, 189, 205, 206, 210, 229, 230, 233, 235, 242, 254, 261, 265, 280, 284, 292, 296, 306, 330, 331, 345, 350, 351, 355, 366, 375, 390, 393
hydroponic v, vi, xxiv, xxv, 1, 28, 32, 43, 73, 75, 83, 104, 127, 128, 146, 151, 167, 169, 170, 175, 176, 177, 178, 185, 195, 199, 216, 217, 296, 313, 333, 358, 361, 395

INDEX

I

implements viii, 1, 2, 3, 6, 7, 8, 11, 12, 13, 14, 54, 55, 56, 59, 63, 80, 121, 122, 130, 133, 135, 160, 187, 191, 199, 203, 214, 227, 241, 249, 269, 277, 280, 283, 297, 311, 320, 321, 324, 334, 347, 351, 358, 368, 369, 381, 393

indeterminate vii, xxiv, 30, 79, 81, 83, 84, 91, 99, 111, 127, 128, 151, 165, 166, 167, 169, 170, 172, 173, 174, 176, 178, 181, 182, 186, 196, 197, 199, 200, 201, 202, 206, 207, 209, 210, 211, 212, 213, 215, 216, 217, 218, 219, 220, 273

insect 1, 2, 4, 7 8, 11, 13, 30, 32, 33, 75, 76, 77, 83, 115, 119, 127, 130, 134, 136, 137, 138, 141, 150, 154, 155, 157, 158, 159, 162, 175, 179, 186, 187, 195, 199, 200, 206, 210, 212, 214, 216, 217, 218, 225, 244, 265, 267, 270, 279, 282, 283, 285, 293, 321, 334, 345, 347, 348, 359, 369, 386, 392, 395

irrigation v, vii, viii, 1, 2, 3, 6, 7, 8, 11, 12, 13, 14, 18, 19, 28, 29, 33, 36, 37, 38, 39, 40, 41, 42, 44, 47, 51, 55, 59, 60, 62, 63, 65, 66, 67, 73, 84, 85, 86, 87, 88, 89, 91, 92, 93, 94, 95, 96, 97, 98, 99, 102, 103, 109, 114, 115, 119, 121, 130, 134, 135, 137, 138, 145, 146, 151, 152, 158, 161, 162, 166, 171, 172, 173, 177, 178, 181, 185, 191, 195, 201, 203, 204, 205, 208, 209, 210, 214, 216, 217, 218, 219, 226, 227, 228, 234, 235, 236, 237, 238, 239, 240, 241, 243, 249, 250, 252, 253, 259, 260, 261, 266, 269, 270, 274, 276, 278, 283, 284, 287, 288, 289, 290, 291, 293, 297, 298, 299, 305, 311, 312, 313, 314, 317, 320, 321, 322, 324, 325, 328, 332, 334, 335, 336, 338, 339, 342, 343, 347, 348, 349, 351, 353, 358, 359, 361, 362, 368, 369, 370, 371, 372, 378, 381, 382, 386, 387, 388, 389, 392, 395

L

labour v, vii, 1, 2, 3, 4, 6, 8, 9, 11, 12, 13, 14, 31, 34, 54, 56, 59, 66, 67, 71, 72, 74, 88, 91, 94, 95, 115, 119, 121, 123, 133, 135, 138, 170, 179, 181, 191, 193, 198, 199, 201, 202, 203, 210, 213, 216, 218, 220, 227, 228, 249, 250, 262, 269, 270, 271, 277, 281, 283, 284, 295, 297, 298, 311, 312, 320, 321, 329, 332, 334, 335, 339, 340, 341, 347, 348, 358, 359, 368, 369, 377, 381, 382

ladybird 265 ,331

late blight 144, 222, 308

leaf miner 140, 222, 244, 265, 307, 318

leaf spot 144, 147, 223, 279, 280, 293, 308, 318, 331, 345, 356, 393

legume 69, 70, 75, 129, 238, 303, 327, 346

lettuce xxiv, 17, 18, 19, 20, 21, 24, 26, 27, 28, 29, 34, 39, 40, 45, 49, 62, 65, 66, 68, 79, 84, 93, 97, 104, 105, 106, 119, 125, 149, 165, 177, 186, 204, 225, 253, 344, 357, 358, 359, 360, 361, 362, 363, 364, 365, 366, 395

light 16, 18, 19, 28, 29, 34, 46, 60, 64, 67, 77, 79, 81, 93, 95, 102, 105, 109, 112, 116, 123, 126, 139, 140, 141, 142, 149, 150, 151, 153, 159, 163, 166, 172, 177, 178, 185, 206, 209, 214, 217, 221, 222, 232, 236, 240, 243, 253, 255, 260, 267, 273, 274, 278, 280, 284, 293, 299, 302, 304, 305, 306, 313, 314, 322, 323, 328, 333, 336, 338, 339, 343, 349, 350, 351, 353, 360, 361, 362, 363, 366, 370, 372, 376, 386, 387, 388, 393

lime 1, 33, 43, 48, 49, 51, 52, 53, 60, 62, 71, 74, 108, 113, 153, 171, 181, 183, 201, 203, 204, 207, 209, 228, 250, 254, 261, 270, 275, 283, 287, 289, 297, 303, 311, 321, 327, 334, 348, 349, 359, 360, 369, 370, 379, 383

list 2, 6, 54, 107, 286

long shelf-life 243

looper 140

M

maize vi, 18, 26, 29, 39, 49, 65, 73, 96, 101, 104, 105, 106, 113, 119, 129, 135, 136, 138, 141, 142, 162, 186, 198, 204, 234, 307, 341, 353, 364, 367, 368, 369, 371, 372, 373, 374, 375, 376

manure 1, 5, 62, 69, 70, 71, 73, 75, 81, 106, 112, 122, 126, 127, 128, 129, 202, 218, 235, 239, 240, 253, 303, 312, 324, 329, 338, 341, 342, 353, 362, 364

market 1, 2, 3, 5, 8, 9, 11, 12, 13, 14, 26, 39, 40, 83, 103, 104, 107, 122, 134, 154, 159, 170, 174, 175, 176, 177, 181, 186, 187, 188, 189, 190, 191, 192, 193, 197, 198, 199, 202, 203, 205, 206, 212, 213, 220, 221, 222, 228, 230, 231, 237, 241, 243, 248, 249, 250, 255, 260, 263, 264, 265, 267, 269, 270, 278, 279, 281, 284, 287, 294, 295, 298, 299, 301, 302, 306, 307, 313, 319, 321, 323, 325, 327, 330, 331, 332, 333, 335, 336, 337, 339, 341, 344, 345, 348, 349, 352, 355, 357, 358, 359, 365, 368, 369, 375, 377, 379, 380, 381, 382, 383, 384, 388, 389, 390, 391, 395

marketing v, vii, 1, 3, 8, 9, 13, 76, 83, 188, 189, 191, 192, 203, 205, 222, 226, 249, 256, 262, 264, 267, 268, 279, 287, 292, 306, 307, 317, 330, 341, 365, 375, 377, 378, 379, 390, 391, 292, 395, 307, 317, 330, 344, 345, 365, 375, 381, 391

microelements 1, 33, 43, 52, 103, 107, 113

monocropping 8

mulching 1, 62, 64, 66, 67, 83, 177, 178, 195, 207, 208, 216, 218, 219, 235, 236, 302, 326, 327, 340

N

nematode 115, 122, 137, 138, 139, 149, 200, 222, 223, 244, 259, 265, 279, 292, 307, 308, 318, 331, 345, 355, 365, 375, 392

NFT hydroponics 97, 216, 395

nurseries vi, 1, 7, 12, 13, 23, 29, 30, 31, 34, 35, 37, 39, 40, 95, 99, 181, 201, 204, 207, 250, 258, 270, 273, 297, 301, 311, 315, 321, 359, 364

nutrients 8, 30, 32, 35, 36, 37, 38, 39, 41, 43, 44, 45, 46, 48, 49, 51, 52, 59, 60, 62, 66, 69, 71, 72, 73, 75, 76, 79, 81, 85, 86, 92, 102, 103, 104, 105, 108, 109, 112, 113, 114, 115, 121, 125, 126, 128, 137, 142, 151, 171, 178, 214, 239, 240, 251, 258, 260, 261, 291, 296, 304, 312, 332, 333, 339, 340, 341, 342, 362

O

onions 7, 12, 13, 14, 18, 26, 27, 28, 29, 34, 63, 64, 65, 79, 80, 84, 86, 93, 105, 109, 110, 125, 136, 138, 146, 165, 173, 186, 189, 204, 267, 269, 270, 271, 272, 273, 276, 277, 278, 279, 280, 310, 380, 385, 387, 392, 393

organic farming 1, 5, 75, 76, 134, 395

P

packaging v, 1, 3, 8, 9, 13, 76, 131, 176, 181, 186, 187, 188, 189, 190, 192, 193, 195, 198, 202, 205, 206, 210, 219, 243, 249, 262, 264, 277, 278, 281, 284, 292, 300, 301, 305, 307, 317, 321, 330, 331, 336, 337, 344, 348, 354, 355, 365, 374, 377, 380, 381, 388, 389, 390, 391

paletted seed 24, 25, 62, 64, 359, 363

paprika 27, 39, 65, 84, 149, 218, 294, 295, 298, 299, 300, 301, 302, 305, 307. *See* capsicums

peppadews 294, 298, 300

peppers 5, 7, 8, 15, 26, 27, 31, 32, 39, 40, 63, 65, 79, 80, 81, 83, 84, 87, 99, 114, 115, 131, 139, 142, 147, 149, 150, 151, 152, 153, 156, 165, 170, 172, 173, 174, 175, 176, 177, 178, 179, 181, 182, 183, 184, 186, 188, 198, 199, 205, 208, 211, 216, 217, 218, 223, 226, 294, 295, 296, 297, 298, 299, 300, 301, 302, 303, 304, 305, 306, 307, 308. *See* capsicums

pH vii, 1, 6, 33, 36, 42, 46, 48, 49, 51, 53, 74, 75, 76, 98, 102, 105, 108, 109, 113, 114, 116, 123, 142, 150, 158, 161, 163, 171, 173, 181, 195, 201, 204, 205, 209, 229, 250, 253, 254, 261, 262, 270, 272, 276, 287, 297, 304, 311, 312, 317, 321, 327, 329, 334, 338, 348, 349, 359, 360, 369, 370, 379, 383

phytophthora 23, 84, 146, 148, 216, 222, 245, 265, 279, 293, 301, 308, 318, 331, 345, 355, 365, 393

planning v, 1, 3, 9, 11, 12, 13, 97, 177, 189, 195, 203, 227, 230, 241, 249, 255, 269, 343, 354, 358, 362, 368, 372, 378, 379

planter 6, 23, 42, 56, 58, 62, 65, 119, 234,

236, 270, 274, 276, 310, 311, 312, 314, 316, 321, 324, 325, 328, 329, 334, 338, 339, 340, 341, 348, 359, 363, 369, 371
pollination vii, 16, 17, 19, 26, 93, 150, 153, 156, 170, 185, 225, 230, 241, 242, 247, 252, 272, 309, 333, 367, 368, 370, 373
potato virus 150, 223, 308
powdery mildew 126, 135, 144, 146, 222, 232, 245, 279, 302, 307, 308, 318, 331, 355, 365, 393
puffiness 153
purple blotch 144, 279, 293, 308, 318, 345, 393

Q

quality v, vi, vii, 4, 5, 7, 8, 9, 11, 12, 13, 14, 15, 16, 18, 19, 20, 23, 24, 25, 30, 32, 33, 34, 41, 42, 43, 44, 45, 46, 47, 48, 51, 58, 62, 64, 65, 70, 72, 73, 75, 76, 81, 83, 84, 85, 87, 88, 91, 92, 101, 102, 104, 109, 110, 113, 119, 129, 133, 134, 142, 155, 165, 173, 174, 175, 176, 182, 185, 186, 187, 188, 189, 191, 192, 193, 196, 199, 206, 208, 209, 210, 212, 213, 215, 216, 217, 218, 219, 226, 239, 248, 251, 252, 254, 255, 256, 258, 260, 261, 263, 264, 265, 267, 269, 274, 275, 278, 279, 282, 287, 290, 291, 295, 296, 298, 300, 301, 303, 305, 306, 314, 316, 317, 318, 320, 322, 323, 324, 325, 326, 327, 328, 330, 331, 332, 333, 335, 336, 337, 338, 339, 341, 342, 343, 345, 352, 353, 354, 357, 361, 362, 364, 366, 372, 373, 374, 377, 380, 384, 385, 386, 388, 391, 393, 394

R

rainfall 1, 4, 13, 15, 16, 19, 44, 48, 85, 87, 92, 93, 109, 112, 156, 166, 169, 176, 201, 204, 206, 210, 216, 228, 229, 233, 234, 236, 238, 240, 250, 255, 265, 270, 273, 283, 284, 298, 299, 312, 313, 321, 323, 335, 336, 343, 348, 350, 359, 361, 369, 370, 372, 382
record keeping 1, 3, 11
red spider mite 222, 244, 265, 292, 307, 318, 331, 355, 375, 392
regulations 5, 7, 14, 24, 45, 75, 76, 134, 136, 143, 154, 158, 375, 391
resistance 1, 20, 25, 32, 33, 134, 135, 136, 137, 139, 149, 151, 155, 161, 163, 175, 200, 206, 267, 301, 307, 308
ridges 7, 40, 56, 63, 65, 66, 70, 81, 83, 84, 94, 114, 125, 127, 128, 149, 166, 169, 170, 171, 172, 173, 178, 179, 183, 195, 196, 197, 199, 204, 208, 209, 210, 211, 213, 214, 215, 216, 217, 219, 226, 229, 283, 287, 288, 289, 290, 301, 326
ridging 1, 83, 84, 236
rippers 55
risk factor 281, 296, 367
root-knot nematode 137, 139, 149, 222, 223, 244, 259, 265, 279, 292, 307, 308, 318, 331, 345, 355, 365, 375, 392
roots 5, 16, 19, 24, 25, 28, 30, 31, 34, 35, 38, 40, 41, 43, 45, 46, 48, 53, 56, 59, 64, 68, 69, 70, 71, 74, 79, 80, 81, 86, 87, 88, 92, 93, 97, 102, 105, 108, 109, 113, 115, 116, 119, 121, 122, 123, 124, 126, 129, 130, 131, 135, 137, 139, 142, 147, 149, 160, 178, 189, 198, 200, 208, 210, 229, 230, 233, 235, 240, 241, 257, 259, 262, 269, 270, 271, 275, 276, 277, 278, 280, 281, 282, 284, 286, 287, 288, 289, 290, 291, 292, 293, 294, 303, 304, 305, 310, 315, 316, 317, 320, 321, 322, 323, 324, 325, 326, 328, 329, 330, 332, 333, 335, 336, 337, 338, 339, 341, 342, 343, 344, 345, 346, 349, 353, 361, 371, 372, 374, 377, 378, 380, 381, 382, 389, 390, 393
rotation 1, 3, 5, 44, 47, 75, 122, 125, 126, 127, 128, 138, 143, 145, 171, 195, 198, 205, 216, 218, 262, 305, 329, 241, 349, 262, 276, 290, 305, 317, 329, 341, 353, 364, 369, 381, 387

S

scanning 1, 130, 131, 136, 195, 380
scouting 1, 248
seed v, vi, xxv, 1, 2, 3, 6, 7, 9, 10, 12, 13, 15, 16, 17, 20, 21, 23, 24, 25, 26, 27, 28, 29, 30, 31, 32, 33, 34, 35, 36, 37, 38, 39, 40, 41, 42, 44, 47, 56, 58, 59, 60, 62, 63, 64, 65, 66, 67, 68, 74, 84, 86, 87, 95, 101, 110, 113, 119, 120, 122, 129, 134, 136, 138, 143, 144, 145, 147, 149, 150, 153, 155, 175, 191, 195, 196, 199, 200, 203, 204, 205, 206, 207, 208, 210, 212, 213, 218, 221, 225, 227, 228, 229, 230, 231, 233, 234, 235, 236, 238, 241, 242, 243, 245, 249, 250, 254, 255, 257, 259, 261, 262, 266, 268, 269, 270, 271, 272, 273, 274, 276, 283, 286, 297, 298, 299, 300, 301, 306, 310, 311, 312, 314, 315, 316, 319, 320, 321, 322, 323, 324, 325, 326, 328, 329, 332, 333, 334, 335, 336, 337, 338, 339, 340, 341, 342, 343, 346, 347, 348, 349, 350, 351, 352, 353, 354, 355, 356, 357, 358, 359, 360, 362, 363, 364, 365, 366, 368, 369, 370, 371, 372, 373, 374, 379, 380, 381, 382, 384, 385, 387, 391, 395
seedbeds xxv, 55, 56, 58, 59, 61, 62, 63, 64, 65, 66, 67, 68, 120, 204, 207, 208, 214, 250, 258, 259, 270, 273, 274, 275, 301, 314, 315, 316, 321, 324, 325, 326, 334, 337, 338, 339, 340, 349, 380, 386
seedling 1, 6, 24, 25, 28, 29, 30, 31, 32, 33, 34, 35, 36, 37, 38, 39, 40, 41, 42, 56, 58, 62, 63, 64, 67, 68, 75, 99, 102, 119, 121, 143, 148, 149, 166, 172, 195, 201, 204, 207, 218, 230, 235, 240, 250, 254, 258, 259, 261, 271, 273, 274, 275, 298, 301, 305, 310, 311, 312, 315, 316, 320, 326, 329, 338, 341, 359, 364, 395

318, 331, 345, 355, 365, 375, 392

shade net house xxiv, 7, 24, 32, 38, 42, 80, 127, 128, 165, 166, 167, 168, 169, 170, 171, 172, 174, 176, 177, 179, 181, 182, 196, 197, 211, 212, 216, 217, 296, 346
sidedressing 109
small-scale farmers 3, 6, 11, 25, 32, 33, 37, 39, 56, 61, 65, 67, 69, 71, 79, 96, 99, 112, 125, 162, 176, 186, 209, 218, 227, 255, 259, 273, 276, 297, 314, 316, 324, 325, 326, 328, 340, 342, 343, 344, 352, 362, 364, 367, 368, 374, 379, 380
soil vi, viii, xxiv, 1, 3, 4, 5, 6, 7, 8, 11, 12, 13, 15, 16, 17, 19, 20, 23, 24, 25, 29, 31, 34, 35, 38, 39, 42, 43, 44, 45, 46, 47, 48, 49, 51, 52, 53, 54, 55, 56, 57, 58, 59, 60, 61, 62, 63, 64, 65, 66, 67, 68, 69, 70, 71, 72, 73, 74, 75, 76, 79, 80, 84, 85, 86, 87, 88, 89, 91, 92, 93, 94, 95, 97, 98, 99, 101, 102, 103, 104, 105, 106, 107, 108, 109, 110, 111, 112, 113, 114, 115, 116, 119, 121, 122, 123, 124, 125, 126, 127, 128, 129, 130, 131, 133, 135, 137, 138, 141, 142, 143, 145, 146, 147, 148, 149, 150, 151, 153, 155, 158, 161, 162, 163, 165, 166, 167, 168, 169, 170, 171, 172, 173, 175, 178, 179, 180, 181, 182, 183, 184, 191, 195, 196, 198, 199, 200, 201, 202, 203, 204, 205, 207, 208, 209, 210, 211, 212, 214, 215, 216, 217, 218, 219, 223, 227, 228, 229, 230, 233, 234, 235, 236, 237, 238, 239, 240, 241, 244, 246, 247, 249, 250, 251, 253, 254, 257, 258, 259, 260, 261, 262, 264, 266, 268, 270, 271, 272, 273, 275, 276, 277, 278, 280, 281, 283, 284, 286, 287, 288, 289, 290, 291, 292, 293, 296, 297, 298, 299, 301, 302, 303, 304, 305, 308, 311, 312, 313, 314, 315, 316, 317, 318, 321, 322, 323, 324, 325, 326, 327, 328, 329, 330, 332, 334, 335, 336, 337, 338, 339, 340, 341, 342, 343, 344, 345, 346, 347, 348, 349, 350, 351, 352, 353, 359, 360, 361, 362, 363, 364, 365, 366, 369, 370, 371, 372, 373, 374, 379, 380, 381, 383, 386, 387, 388, 389, 392, 393, 395
sowing 1, 3, 12, 13, 34, 37, 56, 58, 62, 63, 64, 65, 67, 68, 74, 119, 121, 122, 195, 204, 205, 207, 255, 262, 265, 272, 273, 274, 301, 310, 311, 312, 314, 315, 316, 321, 323, 324, 325, 326, 328, 329, 332, 336, 338, 339, 341, 343, 345, 361, 363
spacing vii, 1, 3, 12, 13, 25, 34, 42, 59, 63, 65, 67, 68, 79, 80, 81, 84, 96, 114, 119, 121, 131, 165, 171, 172, 179, 180, 183, 189, 195, 197, 199, 204, 207, 210, 211, 212, 213, 214, 215, 216, 217, 228, 233, 234, 235, 236, 237, 238, 240, 241, 255, 256, 259, 260, 273, 274, 275, 286, 287, 288, 302, 312, 314, 315, 316, 324, 325, 326, 328, 329, 338, 339, 340, 341, 346, 349, 350, 351, 352, 353, 363, 364, 373, 374, 380, 385, 386, 392

spider track 152, 153

spinach xxiv, xxv, 3, 5, 14, 17, 18, 19, 20, 21, 27, 28, 29, 34, 49, 62, 65, 66, 84, 93, 103, 105, 106, 119, 125, 141, 146, 147, 149, 165, 169, 186, 204, 310, 311, 312, 313, 315, 316, 317, 318, 319, 329, 331, 361

spotted maize beetle 142

spraying 3, 14, 17, 54, 77, 84, 113, 114, 115, 123, 134, 137, 144, 147, 154, 157, 158, 159, 160, 161, 163, 203, 218, 227, 229, 239, 245, 248, 249, 269, 276, 279, 282, 283, 286, 296, 297, 311, 320, 321, 334, 336, 347, 358, 366, 368, 369, 381, 395

spraying equipment 14, 54, 158, 159, 163, 249, 297, 395

sprinklers 28, 34, 36, 38, 39, 62, 73, 87, 91, 92, 94, 95, 96, 99, 158, 210, 240, 325, 361, 388. See *irrigation*

storage vi, 7, 27, 43, 46, 85, 86, 163, 186, 187, 189, 195, 243, 264, 267, 278, 280, 291, 292, 305, 306, 318, 330, 332, 333, 345, 354, 355, 375, 380, 383, 388, 390, 391, 393, 395

sun scald 166, 186, 224, 293

suppliers 33, 36, 49, 88, 130, 166, 184, 193, 208, 272, 279, 302, 395

sweet potatoes 8, 15, 39, 56, 79, 80, 84, 86, 93, 125, 126, 149, 151, 186, 189, 192, 204, 281, 282, 284, 287, 288, 289, 290, 292, 293, 367

T

temperature 1, 13, 15, 16, 18, 19, 21, 23, 24, 25, 26, 27, 28, 29, 31, 35, 36, 37, 38, 39, 44, 66, 67, 70, 71, 72, 73, 74, 91, 102, 110, 139, 142, 149, 151, 153, 158, 165, 167, 170, 172, 173, 176, 177, 178, 185, 187, 189, 195, 197, 203, 205, 220, 229, 230, 237, 238, 242, 243, 251, 254, 260, 261, 265, 266, 267, 271, 276, 289, 290, 292, 306, 313, 319, 322, 324, 328, 335, 336, 338, 342, 344, 345, 350, 351, 353, 361, 370, 374, 383, 388, 393

tension meters 7, 63, 87, 88, 89, 115, 172, 210, 380, 387

thinning 1, 56, 81, 119, 120, 274, 312, 316, 324, 329, 330, 332, 335, 339, 341, 364, 365

tip burn 116, 266

tomatoes xxv, 2, 5, 7, 8, 12, 13, 15, 26, 28, 31, 32, 34, 39, 45, 49, 56, 63, 65, 68, 79, 80, 81, 83, 84, 87, 88, 98, 99, 102, 103, 105, 106, 109, 110, 114, 123, 125, 126, 131, 139, 141, 142, 143, 144, 145, 146, 147, 149, 150, 152, 153, 155, 156, 165, 166, 167, 169, 170, 172, 173, 174, 175, 176, 177, 178, 179, 181, 182, 183, 184, 185, 186, 188, 193, 196, 197, 198, 199, 200, 202, 203, 204, 205, 206, 207, 208, 210, 211, 212, 213, 214, 215, 216, 217, 218, 220, 221, 222, 223, 242, 267, 269, 295, 298, 299, 305, 308, 310, 328, 343, 395

tomato mosaic virus 149, 150

topdressing 49, 102, 108, 109, 111, 115

tractors vii, 6, 12, 13, 47, 54, 55, 203, 214, 227, 228, 249, 250, 269, 283, 287, 297, 311, 320, 321, 334, 347, 358, 368, 369, 381

training v, vi, vii, xxv, 1, 3, 10, 14, 54, 83, 84, 112, 174, 198, 199, 213, 215, 218, 281, 306, 395

transplanting 1, 3, 8, 12, 13, 29, 30, 31, 33, 34, 35, 36, 38, 39, 40, 41, 42, 56, 58, 62, 63, 64, 67, 68, 74, 103, 110, 121, 138, 143, 155, 173, 174, 185, 195, 201, 203, 204, 205, 207, 208, 209, 210, 214, 217, 220, 232, 234, 235, 243, 246, 249, 258, 259, 261, 262, 265, 271, 275, 276, 284, 286, 298, 299, 301, 303, 304, 305, 306, 310, 314, 315, 316, 320, 321, 326, 357, 358, 359, 361, 362, 364

trays 1, 23, 24, 25, 28, 29, 30, 31, 33, 34, 35, 36, 37, 38, 39, 40, 41, 56, 58, 63, 67, 68, 95, 143, 149, 195, 204, 207, 208, 229, 230, 235, 254, 259, 261, 262, 273, 286, 299, 301, 314, 315, 326, 357, 360, 363, 364

trellising vii, 1, 3, 7, 8, 17, 79, 83, 84, 130, 150, 165, 170, 174, 176, 178, 181, 182, 184, 195, 198, 199, 200, 201, 202, 205, 206, 208, 210, 211, 212, 213, 214, 215, 216, 218, 219, 233, 296, 297, 298 301, 302, 303, 346, 352

trips 121, 130, 131, 135, 137, 138, 139, 145, 148, 149, 151, 153, 162, 222, 244, 245, 265, 271, 279, 282, 296, 307, 308, 318, 322, 366, 377, 380, 392

tunnel xxiv, xxv, 24, 31, 37, 39, 42, 81, 131, 165, 167, 169, 170, 172, 173, 175, 176, 177, 182, 183, 184, 185, 197, 206, 213, 217, 233, 293, 296, 302, 395

turnip 257. See *cole*

V

vines 58, 80, 121, 147, 149, 150, 155, 175, 236, 237, 239, 240, 241, 247, 281, 282, 283, 284, 285, 286, 287, 288, 289, 290, 291, 292

virus vi, 4, 23, 33, 123, 127, 130, 135, 142, 143, 149, 150, 155, 158, 182, 199, 223, 233, 238, 246, 281, 282, 286, 293, 294, 296, 299, 300, 306, 308, 366, 376, 393, 394

W

warm weather crops 39, 367, 370

water 1, 2, 3, 4, 6, 7, 8, 12, 13, 15, 18, 19, 20, 23, 26, 28, 30, 32, 33, 34, 35, 36, 37, 38, 39, 40, 41, 42, 44, 45, 46, 47, 48, 49, 51, 52, 54, 56, 57, 58, 60, 61, 62, 63, 64, 66, 67, 68, 69, 71, 72, 73, 74, 75, 79, 84, 85, 86, 87, 88, 89, 91, 92, 93, 94, 95, 96, 97, 98, 99, 102, 103, 104, 105, 107, 108, 109, 112, 114, 115, 116, 119, 121, 122, 123, 127, 128, 129, 130, 135, 137, 142, 143, 145, 147, 148, 151, 153, 155, 158, 159, 160, 161, 162, 163, 166, 167, 169, 170, 171, 172, 173, 176, 177, 178, 180, 181, 185, 187, 189, 197, 198, 200, 201, 203, 204, 206, 210, 212, 214, 216, 217, 219, 225, 227, 228, 229, 233, 235, 236, 237, 239, 240, 241, 243, 247, 249, 250, 251, 253, 254, 259, 260, 261, 264, 265, 266, 269, 270, 271, 273, 274, 275, 276, 277, 280, 283, 286, 287, 288, 289, 290, 293, 296, 297, 298, 304, 305, 306, 309, 311, 312, 313, 314, 315, 316, 317, 318, 320, 321, 325, 328, 331, 334, 335, 342, 343, 344, 347, 348, 349, 351, 353, 355, 358, 359, 360, 361, 362, 365, 368, 369, 371, 372, 375, 378, 380, 381, 382, 383, 384, 387, 393, 395

weeds v, 2, 3, 4, 5, 6, 7, 14, 47, 55, 56, 57, 58, 59, 62, 63, 66, 72, 73, 80, 102, 119, 120, 121, 122, 123, 124, 125, 126, 128, 133, 139, 150, 178, 181, 192, 201, 203, 205, 208, 214, 219, 228, 233, 237, 241, 250, 262, 267, 270, 276, 277, 282, 283, 287, 290, 298, 305, 312, 314, 315, 316, 317, 324, 325, 327, 328, 329, 330, 334, 338, 339, 341, 343, 348, 353, 359, 364, 365, 366, 369, 374, 379, 380, 382, 383, 386, 387

white bulb rot 149, 280, 393

white fly 286

wind damage 19, 179, 241

Y

yields v, vii, viii, 1, 2, 3, 4, 5, 8, 9, 11, 12, 15, 16, 18, 19, 20, 32, 33, 34, 43, 44, 45, 46, 47, 48, 53, 66, 76, 79, 80, 81, 83, 84, 87, 88, 91, 101, 102, 104, 108, 109, 110, 111, 113, 114, 127, 128, 134, 135, 136, 150, 165, 174, 175, 176, 177, 182, 183, 184, 195, 196, 197, 198, 199, 200, 202, 203, 205, 206, 208, 210, 211, 212, 213, 216, 217, 218, 219, 227, 230, 237, 238, 241, 242, 243, 249, 251, 253, 255, 259, 260, 261, 262, 269, 271, 274, 275, 278, 279, 280, 281, 282, 283, 284, 285, 286, 287, 292, 293, 294, 295, 296, 298, 299, 300, 301, 303, 304, 305, 306, 311, 312, 314, 317, 319, 320, 322, 323, 325, 327, 330, 332, 333, 334, 335, 336, 341, 344, 346, 347, 348, 349, 350, 353, 354, 357, 358, 360, 365, 368, 370, 371, 372, 374, 377, 379, 381, 382, 384, 385, 386, 387, 388, 393, 394

Z

zipping 152, 153

Bayer Crop Science Division Southern Africa

Facebook: Bayer Crop Science Division Southern Africa
Twitter: @bayer4cropssa
www.cropscience.bayer.co.za
www.bayer.co.za

Science For A Better Life

FUNGICIDES

Product	Reg. No.	ASPARAGUS	Cabbage	Broccoli	Cauliflower	Brussel Sprouts	Kale	Other (Brassica)	Onion	Spring Onion	Leek	Chives	Garlic	Other (Bulb)	Cucumber	Squash	Melon	Pumpkin	Gherkin	Watermelon	Butternut	Hubbard Squash	Other (Cucurbit)	Tomatoes	Pepper	Eggplant	Hot Pepper	Capsicums/Pepper Group	Other (Fruiting)
Antracol contains Propineb (CAUTION)	Reg. No. L2065 Act / Wet No. 36 of / van 1947																							X					
Folicur contains Tebuconazole (CAUTION)	Reg. No. L3857 Act / Wet No. 36 of / van 1947								X															X					
Infinito contains Flupicolide and Propamocarb (CAUTION)	Reg. No. L8470 Act / Wet No. 36 of / van 1947																							X					
Melody Duo contains Iprovalicarb and Propineb	Reg. No. L6714 Act / Wet No. 36 of / van 1947																							X					
Nativo contains Tebuconazole and Trifloxystrobin (CAUTION)	Reg. No. L8942 Act / Wet No. 36 of / van 1947		X	X	X				X	X			X	X															
No-Blite contains Fenamidone and Mancozeb (CAUTION)	Reg. No. L6681 Act / Wet No. 36 of / van 1947																							X					

INSECTICIDES

Product	Reg. No.	ASPARAGUS	Cabbage	Broccoli	Cauliflower	Brussel Sprouts	Kale	Other (Brassica)	Onion	Spring Onion	Leek	Chives	Garlic	Other (Bulb)	Cucumber	Squash	Melon	Pumpkin	Gherkin	Watermelon	Butternut	Hubbard Squash	Other (Cucurbit)	Tomatoes	Pepper	Eggplant	Hot Pepper	Capsicums/Pepper Group	Other (Fruiting)
Belt contains Flubendiamide (CAUTION)	Reg. No. L8860 Act / Wet No. 36 of / van 1947		X	X	X	X	X	X	X						X	X	X	X	X	X	X	X	X	X	X	X	X	X	X
Biscaya contains Thiacloprid (HARMFUL)	Reg. No. L9350 Act / Wet No. 36 of / van 1947		X	X	X			X	X						X	X	X	X	X	X	X	X	X	X	X	X	X	X	X
Bulldock contains Beta-cyfluthrin (HARMFUL)	Reg. No. L7612 Act / Wet No. 36 of / van 1947	X	X	X	X	X	X	X	X	X	X	X	X	X	X	X	X	X	X	X	X	X	X	X	X	X	X	X	X
Confidor contains Imidacloprid (HARMFUL)	Reg. No. L8685 Act / Wet No. 36 of / van 1947		X	X	X	X	X	X	X	X	X	X	X	X	X	X	X	X	X	X	X	X	X	X	X	X	X	X	X
Curaterr contains Carbofuran (HARMFUL)	Reg. No. L871 Act / Wet No. 36 of / van 1947		X																										
Decis contains Deltamethrin (HARMFUL)	Reg. No. L6563 Act / Wet No. 36 of / van 1947	X	X	X	X	X	X	X	X	X	X	X	X	X	X	X	X	X	X	X	X	X	X	X	X	X	X	X	X
Movento contains Spirotetramat (CAUTION)	Reg. No. L8559 Act / Wet No. 36 of / van 1947	X	X	X	X	X	X	X	X	X	X	X	X	X	X	X	X	X	X	X	X	X	X	X	X	X	X	X	X
Velum Prime contains Fluopyram (CAUTION)	Reg. No. L9565 Act / Wet No. 36 of / van 1947																							X					

HERBICIDES

Product	Reg. No.	ASPARAGUS	Cabbage	Broccoli	Cauliflower	Brussel Sprouts	Kale	Other (Brassica)	Onion	Spring Onion	Leek	Chives	Garlic	Other (Bulb)	Cucumber	Squash	Melon	Pumpkin	Gherkin	Watermelon	Butternut	Hubbard Squash	Other (Cucurbit)	Tomatoes	Pepper	Eggplant	Hot Pepper	Capsicums/Pepper Group	Other (Fruiting)
Ronstar contains Oxadiazon (CAUTION)	Reg. No. L417 Act / Wet No. 36 of / van 1947								X	X																			

SEEDLINGS

Product	Reg. No.	ASPARAGUS	Cabbage	Broccoli	Cauliflower	Brussel Sprouts	Kale	Other (Brassica)	Onion	Spring Onion	Leek	Chives	Garlic	Other (Bulb)	Cucumber	Squash	Melon	Pumpkin	Gherkin	Watermelon	Butternut	Hubbard Squash	Other (Cucurbit)	Tomatoes	Pepper	Eggplant	Hot Pepper	Capsicums/Pepper Group	Other (Fruiting)
Previcur N contains Propamocarb	Reg. No. L3394 Act / Wet No. 36 of / van 1947		X					X	X	X	X	X	X										X	X	X		X	X	X

«Change your green into gold»

with GreenGold® 30

GreenGold® 30 is a dry fertilizer applied as top dressing after planting and post emergence with the three-in-one winning combination of nitrogen, calcium and boron.

GreenGold® 30:

- Comprises a combination of nitrogen, water-soluble calcium and water-soluble boron for rapid uptake and quick plant reaction – all in one application.
- Nitrogen is water soluble and partially quickly available and partially gives you a longer reaction.
- Enhanced efficiency can result in higher yield and profitability.
- Is highly suitable for vegetables and fruit but also suitable for grain crops, sunflower and pastures.

GreenGold® 30 is a must for vegetable growers.

Kynoch – enhanced efficiency through innovation.

011 317 2000 | info@kynoch.co.za | www.kynoch.co.za
Not trading in the Western Cape.
GreenGold® 30 K9750, Act 36 of 1947.

Farmisco (Pty) Ltd t/a Kynoch Fertilizer
Reg no. 2009/0092541/07

NOTES

www.ingramcontent.com/pod-product-compliance
Lightning Source LLC
Chambersburg PA
CBHW081424300426
44108CB00016BA/2300